5-11-66

SEMICONDUCTOR SURFACES

SEMICONDUCTOR SURFACES

BY

A. MANY, Y. GOLDSTEIN, AND N. B. GROVER

The Hebrew University, Jerusalem, Israel

1965

NORTH-HOLLAND PUBLISHING COMPANY - AMSTERDAM

*No part of this book may be reproduced in any form
by print, photoprint, microfilm or any other means
without written permission from the publisher*

PUBLISHERS:
NORTH-HOLLAND PUBLISHING COMPANY - AMSTERDAM
SOLE DISTRIBUTORS FOR U.S.A.:
INTERSCIENCE PUBLISHERS, a division of
JOHN WILEY & SONS, INC. - NEW YORK

PRINTED IN THE NETHERLANDS

PREFACE

Surface research in semiconductors is motivated both by the basic scientific importance of the field and by the prominent role played by surface effects in device performance. The intensive effort devoted to the study of surfaces is reflected in the vast literature accumulated on the subject during the past fifteen years. As is usually the case with a rapidly growing field, most of this literature is in the form of original papers published in scientific and technical journals. Although there are a number of excellent review articles, these treat only specific aspects of the field and there is no text covering a major part of the subject in an integrated fashion. This book is an attempt to provide the reader with a source of organized, basic information on the physics of semiconductor surfaces. We hope that it will prove useful both to the physicist and chemist active in surface research, and to the device engineer whose interest in the surface is primarily of a practical nature. It is also our intention that the book serve as a text for graduate students with a general background of solid-state physics who are about to make a special study of surface or interface phenomena.

The theoretical basis required for a proper understanding of surface phenomena is discussed in some detail. Wherever applicable physical considerations accompany the more rigorous derivations. On the whole, however, the book is experimental rather than theoretical in tone, and two entire chapters are devoted to an analysis of the various measuring techniques. Considerable space is given over to a review of the available data on surface structure, accompanied by extensive references to the literature covering the period up to the end of 1964. Although the accent is placed on the electrical properties of the surface, a fairly detailed summary of the lattice structure and chemical reactivity of the surface is provided. In order to avoid the necessity of having to refer constantly to general textbooks on semiconductors, a résumé of bulk properties is also included.

The MKS system of units has been employed throughout, but the numerical values of physical parameters are cited in the accepted units. An effort has been made to develop a consistent terminology and set of symbols that coincide, whenever possible, with those commonly used in the various branches of the literature.

We are grateful to the authors and publishers of many papers for permission to reproduce figures; in each case the original source appears in the caption. We also wish to thank the following, each of whom has read one or more chapters of the manuscript and offered helpful comments: A. Amith, P. Feuer, L. Friedman, R. F. Greene, Y. Margoninski, M. Simhony, H. S. Sommers, Jr., P. K. Weimer, R. Williams, and J. N. Zemel. A substantial portion of the book was written while two of us (A.M. and Y.G.) were at the RCA Laboratories, Princeton, N. J. The generous help of the Laboratories, both technical and otherwise, has been a significant factor in the early completion of this work. Finally, we would like to express our sincere appreciation to the staff of the North-Holland Publishing Company and in particular to Mr. W. H. Wimmers, managing editor, for their cooperation and infinite patience in the preparation of the manuscript for publication.

A. Many
Y. Goldstein
N. B. Grover

Jerusalem, March 1965

CONTENTS

Preface v
List of Principal Symbols xii
Values of Physical Constants xvi

CHAPTER 1

INTRODUCTION   1

1. Historical notes 2
2. Recent developments in surface research 8
References 13

CHAPTER 2

THE SEMICONDUCTOR BULK — A RÉSUMÉ   15

1. Wave-mechanical approach to the problem of electrons in solids 15
    1.1. Free and bound electrons 16
    1.2. One-electron models 18
2. The band model for a one-dimensional periodic lattice 19
    2.1. The Kronig–Penney model 19
    2.2. The motion of electrons 25
    2.3. Positive holes 30
3. Energy bands in a three-dimensional lattice 32
    3.1. Motion of electrons and holes 33
    3.2. Shape of the energy bands 34
    3.3. Band structure of germanium and silicon 37
    3.4. Optical excitation 41
4. Lattice vibrations, impurities, and defects 43
    4.1. Lattice vibrations 43
    4.2. Impurities 46
    4.3. Lattice defects 50
5. Occupation statistics for semiconductors 52
    5.1. The Fermi–Dirac distribution law 52
    5.2. Intrinsic semiconductors 55
    5.3. Semiconductors with localized levels 57

6. Carrier transport. 61
   6.1. Scattering processes 61
   6.2. The Boltzmann transport equation 67
   6.3. Galvanomagnetic effects 71

7. Non-equilibrium phenomena 75
   7.1. Lifetime of excess carriers 75
   7.2. Recombination processes 76
   7.3. Drift and diffusion of excess carriers 79
   7.4. P–n junctions 86
References 89

CHAPTER 3

LATTICE STRUCTURE AND CHEMICAL REACTIVITY OF THE SURFACE   90

1. Preparation of clean surfaces 92
   1.1. Cleavage 92
   1.2. Crushing 92
   1.3. Hydrogen reduction 93
   1.4. Heating in high vacuum 94
   1.5. Ion bombardment and annealing 95

2. Measurements on clean surfaces. 96
   2.1. Low-energy electron diffraction 96
   2.2. Photoelectric emission and contact potential 99
   2.3. Field emission and field ion microscopy 100
   2.4. Auger ejection of electrons by noble gas ions 101
   2.5. Adsorption measurements. 103
   2.6. Heat of adsorption 105

3. Clean surface structure of diamond-type semiconductors 105
   3.1. Lattice structure of clean surfaces 105
   3.2. Chemisorption on clean surfaces 113

4. Real surfaces 117
   4.1. Sample preparation 117
   4.2. Chemical etching 119
   4.3. Electrolytic etching. 121
   4.4. Preferential etching and etch pits 122
References 124

CHAPTER 4

THE SURFACE SPACE-CHARGE REGION   128

1. The origin of the space-charge region. 129
   1.1. External field 129
   1.2. Contact potential and the metal–semiconductor contact. . . 131
   1.3. Surface states 134

CONTENTS ix

2. The space-charge density and the shape of the potential barrier 136
   2.1. Concepts and definitions 136
   2.2. Poisson's equation 138
   2.3. Approximate solutions for extrinsic semiconductors 141
   2.4. Numerical solutions 143
   2.5. Extension for degenerate surface conditions 145
3. The excess surface-carrier densities ΔN and ΔP 149
   3.1. Definitions 149
   3.2. Graphical representation and approximate expressions 150
4. The space-charge region under non-equilibrium conditions 156
5. The space-charge region in the presence of deep traps 158
   5.1. Discrete set of localized levels 159
   5.2. Continuous distribution of traps 162
References 163

CHAPTER 5

SURFACE STATES 165

1. Introduction 165
2. Tamm and Shockley states 166
   2.1. One-dimensional periodic potential 166
   2.2. Extension to three-dimensional crystals 171
3. The tight-binding approximation 174
   3.1. Tamm states 174
   3.2. Shockley states 178
4. Occupation statistics of surface states 182
   4.1. Single-charge surface states 182
   4.2. The variation of surface potential with external field 184
5. Occupation statistics for complex surface states 185
   5.1. General case 185
   5.2. Single- and double-charge centres without excited levels 187
   5.3. Single-charge centre with excited levels 188
6. The interaction of surface states with a single band 189
7. Surface recombination 194
8. More sophisticated models of surface recombination 201
   8.1. Double-charge centres 201
   8.2. Single-charge centre with excited levels 204
References 208

CHAPTER 6

EXPERIMENTAL METHODS I — THE FIELD EFFECT 209

1. Introduction 209
2. Surface conductance 211

x   CONTENTS

3. Surface capacitance 220
4. Dc field effect. 225
5. Low-frequency field effect. 226
6. Pulsed field effect 229
   6.1. Charge relaxation time 229
   6.2. Surface-state time constants. 234
   6.3. Measurement of majority-carrier surface mobility 238
   6.4. Relaxation processes involving both carrier types 240
7. High-frequency field effect 244
8. Channel conductance. 249
References. 254

CHAPTER 7

EXPERIMENTAL METHODS II — OTHER MEASUREMENTS 257

1. Surface recombination velocity 257
   1.1. Sample effective lifetime 258
   1.2. Measurement of effective lifetime 263
2. Contact potential and surface photovoltage 273
3. Photoelectric emission 279
4. High-field effects. 286
5. Optical measurements 292
6. Magnetic measurements 295
   6.1. Galvanomagnetic effects 295
   6.2. Paramagnetic resonance 297
7. Noise. 299
References. 301

CHAPTER 8

TRANSPORT PROCESSES 304

1. Diffuse and specular scattering 305
2. Simple considerations 307
   2.1. Thin slabs 307
   2.2. Thick samples. 308
   2.3. Partially specular scattering. 311
3. Calculations of average and surface mobilities 312
   3.1. The average mobility. 312
   3.2. The surface mobility. 314
   3.3. Partially diffuse scattering 322
4. Galvanomagnetic surface effects 323
   4.1. Solution of transport equation 323
   4.2. Effective mobility formalism 326

CONTENTS xi

5. Experimental results 329
   5.1. Conductivity mobility 329
   5.2. Hall effect and magnetoresistance 337
6. Concluding remarks 342
References 344

CHAPTER 9

THE ELECTRONIC STRUCTURE OF THE SURFACE   346

1. The surface of a semiconductor 347
   1.1. The space-charge layer 349
   1.2. Fast and slow surface states 355
   1.3. Surface recombination 363
   1.4. Noise 366
   1.5. Adsorption and electrical surface properties 371
2. Real germanium and silicon surfaces 377
   2.1. Characteristics of fast states on etched surfaces 377
   2.2. Effect of surface treatment on fast states 394
   2.3. Slow states 408
   2.4. Semiconductor–electrolyte interface 415
3. Clean surfaces 425
   3.1. Experimental procedures 425
   3.2. Bombardment–annealed germanium 429
   3.3. Cleaved germanium 437
   3.4. Bombardment–annealed silicon 440
   3.5. Cleaved silicon 443
   3.6. Effect of adsorption on electrical properties 449
4. Summary and conclusions 454
References 459

AUTHOR INDEX 471
SUBJECT INDEX 482

LIST OF PRINCIPAL SYMBOLS

Boldfaced type represents vector quantities. The entries in parentheses following some of the definitions refer to the pages or equations where the symbols concerned appear for the first time.

A_n, A_p	capture cross section for electrons, holes (p. 76)
A'_n, A'_p	effective capture cross sections (p. 386)
$\mathbf{a}, a$	acceleration
$\mathbf{B}, B$	magnetic induction
b	ratio of electron to hole mobility
C_g	geometric capacitance (p. 221)
C_0, c_0	effective integral, differential surface capacitance (eqs. (6.25), (6.26))
C_s, c_s	integral, differential surface capacitance (pp. 220–222)
C_{sc}, c_{sc}	integral, differential space-charge capacitance
C_{ss}, c_{ss}	integral, differential surface-state capacitance
$\mathbf{c}, c$	electron or hole velocity
$\langle c \rangle$	thermal velocity (eq. (2.124))
$\bar{c}_z$	unilateral mean velocity in z direction
D	ambipolar diffusion constant (eq. (2.195))
D_n, D_p	electron, hole diffusion constant (p. 80)
E	energy
E_A, E_D	acceptor, donor energy level
E_c, E_v	energy of conduction-, valence-band edge
E_F	Fermi level
E_F^*	steady-state mean Fermi level (eq. (4.61))
$E_F^{(i)}$	intrinsic Fermi level (eq. (2.88))
E_g	forbidden gap, $E_g \equiv E_c - E_v$
E_1	energy line that runs parallel to band edges and coincides in the bulk (assumed homogeneous) with $E_F^{(i)}$ (p. 81)
E_n	neutral level of surface states (p. 428)
E_T	threshold energy for photoelectric emission (pp. 281–283)
E_t	energy of localized levels
E_t^t	effective energy of localized levels (eq. (2.74))
e_g	$\equiv E_g/kT$
$\boldsymbol{\mathscr{E}}, \mathscr{E}$	electric field
$\mathscr{E}_s$	electric field just below semiconductor surface
F	space-charge function (eq. (4.16))
$\mathbf{F}$	force
F^+, F^-	approximations for F for extrinsic semiconductors in accumulation, depletion layers (eqs. (4.24), (4.26))
F_n, F_p	quasi Fermi levels for electrons, holes (eq. (2.189))

LIST OF PRINCIPAL SYMBOLS

F_s	value of F at surface
f	frequency
f	Fermi–Dirac distribution function (p. 52)
f_n, f_p	distribution functions for electrons, holes (eqs. (2.66), (2.67))
f_0	thermal-equilibrium distribution function
G	conductance
G^+, G^-, g^+, g^-	functions expressing excess surface-carrier densities (eqs. (4.40)–(4.51))
g_0, g_1, g_2	numbers of degenerate quantum states of localized levels (pp. 54, 187)
h	Planck's constant
$\hbar$	$\equiv h/2\pi$
I	current
$\langle i^2 \rangle$	mean square noise current (p. 299)
J	current density
J_n, J_p	electron, hole component of current density
$\mathbf{K}, K$	phonon wave vector
K_n, K_p	capture probabilities for electrons, holes (pp. 77, 78)
k	Boltzmann's constant
$\mathbf{k}, k$	electron wave vector
L	effective Debye length (eq. (4.12))
L^*	steady-state effective Debye length (p. 157)
L_c	effective charge distance (eq. (4.20))
L_D	Debye length (p. 139)
l	mean free path (eq. (2.107))
$\mathscr{L}, \mathscr{L}_s$	bulk, surface generation rate of carriers by external excitation
m	electron mass
m^*	effective mass
m_l, m_t	effective mass component in longitudinal, transverse direction
m_n, m_p	electron, hole effective mass
m_x, m_y, m_z	effective mass components along principal axes
N_A, N_D	acceptor, donor concentration
N_c, N_v	effective density of states in conduction, valence band (p. 56)
N_t	concentration of localized states
$\overline{N}_t$	concentration of localized states per unit energy (pp. 162, 427)
n, n^*	electron density under equilibrium, non-equilibrium conditions
n_b, n_s	bulk, surface electron density under equilibrium conditions
n_b^*, n_s^*	bulk, surface electron density under non-equilibrium conditions
n_i	intrinsic carrier density
n_i^*	steady-state intrinsic carrier density, $n_i^* = (n_b^* p_b^*)^{\frac{1}{2}}$ (eq. (4.63))
n_t, n_t^*	density of occupied centres under equilibrium, non-equilibrium conditions
n_1	emission constant for electrons from localized centres (eq. (2.173))
p, p^*	hole density under equilibrium, non-equilibrium conditions
p_b, p_s	bulk, surface hole density under equilibrium conditions
p_b^*, p_s^*	bulk, surface hole density under non-equilibrium conditions
p_t, p_t^*	density of unoccupied centres under equilibrium, non-equilibrium conditions

LIST OF PRINCIPAL SYMBOLS

p_1	emission constant for holes from localized centres (eq. (2.172))
Q_s	surface charge density
Q_{sc}	surface space-charge density
Q_{ss}	surface-state charge density
q	absolute magnitude of electronic charge
R	resistance
R_f	filament resistance
R_H	Hall coefficient (eq. (2.159))
r	$= \lambda/L = \mu_b(m^*/2\pi\kappa\varepsilon_0)^{\frac{1}{2}}(n_b+p_b)^{\frac{1}{2}}$ (eq. (8.13))
$\mathbf{r}$	position vector
$\mathbf{r}_{lmn}$	equilibrium position of atoms in crystal lattice
S	sticking coefficient (p. 99)
s	surface recombination velocity (pp. 196, 197, 259)
s_{max}	maximum value of s
T	absolute temperature
t	time
U	potential energy
U	steady-state carrier recombination rate
U_n, U_p	electron, hole recombination rate
u	dimensionless potential, $u = q\phi/kT$
u^*	steady-state potential, $u^* \equiv (E_F^* - E_1)/kT$ (p. 156)
u_b, u_s	bulk, surface potential under equilibrium conditions (value of u in bulk, at surface)
u_b^*, u_s^*	bulk, surface potential under non-equilibrium conditions (value of u^* in bulk, at surface)
u_k	periodic part of wave function
u_0	$\equiv \frac{1}{2}\ln(K_p/K_n)$ (p. 196)
V	potential barrier, $V = \phi - \phi_b$
V_a	bias across junction
V_{cp}	contact potential
V_D	Dember potential (p. 276)
V_E	electrode potential (p. 417)
V_H	potential drop across Helmholtz layer (p. 416)
V_s	barrier height (value of V at surface)
V_{sm}	value of V_s for $\Delta\sigma = \Delta\sigma_{min}$
v	dimensionless potential barrier, $v \equiv qV/kT$
v^*	steady-state potential barrier, $v^* \equiv u^* - u_b^*$
v_s, v_s^*	dimensionless barrier height under equilibrium, steady-state conditions (value of v, v^* at surface)
v_{sm}	$\equiv qV_{sm}/kT$
W_b	energy difference between Fermi level and majority-carrier band edge
W_{bn}, W_{bp}	energy difference between Fermi level and conduction-, valence-band edge
W_ϕ	work function
w_b	$\equiv W_b/kT$
x, y, z	coordinates

LIST OF PRINCIPAL SYMBOLS

Y	photoelectric yield (pp. 280–283)
Z	atomic number
$\Delta N, \Delta P$	excess surface electron, hole density (eq. (4.35))
ΔQ_s	surface charge density relative to its value at $V_s = 0$
ΔQ_{ss}	surface-state charge density relative to its value at $V_s = 0$
$\Delta \rho / \rho B^2$	transverse magnetoresistance (eq. (2.161))
$\Delta \sigma$	surface conductance (eqs. (6.1), (6.2))
$\Delta \sigma_{min}$	minimum value of $\Delta \sigma$
δf	$\equiv f - f_0$
δf_b	δf in bulk
δf_b^H	δf_b in presence of magnetic field
δG	excess conductance under non-equilibrium conditions
$\delta n_b, \delta p_b$	electron, hole density in excess of thermal equilibrium density
δP	change in ΔP (p. 242)
δP_{tot}	total number of excess holes in sample (eq. (7.22))
$\overline{\delta p_b}$	average value of δp_b (eq. (7.5))
$\delta Q_{sc}, \delta Q_{ss}$	change in Q_{sc}, Q_{ss}
δV_{cp}	change in contact potential with illumination (p. 275)
δV_s	surface photovoltage (pp. 157, 275)
δv_s	$= (q/kT) \delta V_s$
$\delta \sigma$	change in surface conductance $\Delta \sigma$
$\delta \sigma$	excess conductivity under non-equilibrium conditions
$\overline{\delta \sigma}$	average value of excess conductivity (p. 270)
ε_0	permittivity of free space
θ_n, θ_p	Hall angle for electrons, holes
κ	dielectric constant
λ	wavelength
λ	unilateral mean free path (eq. (8.7))
μ	mobility
μ	ambipolar mobility (eq. (2.196))
$\overline{\mu}$	average mobility (pp. 306, 308)
μ_b, μ_s	bulk, surface mobility (pp. 306, 308)
μ_{fe}	field-effect mobility (p. 244)
μ'_{fe}	real part of field-effect mobility (p. 244)
μ_H	Hall mobility (p. 74)
μ_n, μ_p	electron, hole mobility
$\overline{\mu}_n, \overline{\mu}_p$	average electron, hole mobility (eq. (8.1))
μ_{nb}, μ_{pb}	electron, hole bulk mobility
μ_{nH}, μ_{pH}	electron, hole Hall mobility
μ_{ns}, μ_{ps}	electron, hole surface mobility
ν	frequency
ρ	volume charge density
ρ	resistivity
ρ_i	intrinsic resistivity, $\rho_i = [q(\mu_n + \mu_p) n_i]^{-1}$
σ	conductivity
$\overline{\sigma}$	average conductivity (p. 296)

$\bar{\sigma}_{\min}$	minimum of $\bar{\sigma}$
σ_0	equilibrium conductivity
τ	lifetime of excess carriers (p. 76)
τ	relaxation time (pp. 62, 67)
τ_b	bulk relaxation time (p. 307)
τ_c	charge relaxation time (p. 232)
τ_{eff}	sample effective lifetime (p. 258)
τ_n, τ_p	lifetime of excess electrons, holes
τ_1, τ_{ss}, τ	surface-state time constants (pp. 193, 245–246, 359)
ϕ	potential, $\phi = (E_F - E_1)/q$ (p. 136)
ϕ_b, ϕ_s	bulk, surface potential (value of ϕ in bulk, at surface)
χ	electron affinity
Ψ, ψ, ψ_k	wave functions
ω	probability that carrier colliding with surface be reflected specularly (p. 311)
ω	angular frequency, $\omega \equiv 2\pi f$
ω_0	cyclotron frequency (p. 40)

Values of physical constants

Electronic charge q	1.60×10^{-19} coul
Mass of electron m	9.11×10^{-31} kg
Mass of proton	1.67×10^{-27} kg
Planck's constant h	6.62×10^{-34} joule · sec
$\hbar \equiv h/2\pi$	1.054×10^{-34} joule · sec
Bohr radius	0.529×10^{-10} m
Boltzmann's constant k	1.38×10^{-23} joule/deg
Avogadro's number	6.02×10^{23} mole^{-1}
Permitivity of free space ε_0	8.85×10^{-12} farad/m

1 eV = 23.05 kcal/mole
1 eV corresponds to a temperature of 1.16×10^4 °K
(0.026 eV corresponds to a temperature of 300 °K)
1 eV corresponds to a wavelength of 1.24 μ

CHAPTER 1

INTRODUCTION

Surfaces play an important role in a large range of phenomena, extending from everyday experience to catalysis of chemical reactions and the most complex life functions. A considerable amount of effort has been devoted to the study of these phenomena, embracing the various fields of science and technology. Progress in our understanding of many of the fundamental processes has been slow, however, and our knowledge of the interface is far less extensive than that of the bulk. This is due in large measure to the fact that the physical and chemical processes at the surface are inherently more difficult to analyse. Consider, for example, a single crystal medium. The bulk, which consists of those regions of the crystal sufficiently removed from the surface so as not to be affected by it, may be looked upon as an infinite, uniform periodic structure; as such it is readily amenable to theoretical treatment. At the surface, on the other hand, the forces acting on the atoms are no longer symmetrical, so that even if the bulk is perfect in every respect the surface atoms are usually displaced from their ideal lattice positions, giving rise to a rather complex two-dimensional structure. Moreover, merely the fact that the surface constitutes an abrupt termination of the crystal lattice results in a deformation of the crystal potential — its periodic nature is lost at the surface. This has far-reaching consequences on the electronic processes in the underlying region of the crystal close to the surface. At the same time, the unsaturated bonds of the surface atoms make them highly reactive towards the various species outside the crystal. Thus, except when produced and maintained in ultra-high vacuum, the surface is covered by one or more layers of foreign matter. All of these characteristics make the surface a totally different entity from the bulk and require a high degree of sophistication, both experimental and theoretical, for its study.

The term surface or the equivalent term interface, in its broad sense, describes any boundary region between different media. Usually, however, one is primarily interested in those boundaries in which the change in one or more of the properties characterizing each medium takes place over a region narrow compared to the spatial extent of the system considered. It is these abrupt boundaries that are generally identified with the surface or the interface. The most common interfaces are those separating any two of the solid, liquid, or gaseous phases in contact. Here the change in the most obvious characteristic property, the state of aggregation, occurs over atomic dimensions. If the phases are reasonably uniform, this abrupt discontinuity is far more significant than any other inhomogeneity in the system. Additional interfaces often occur, however, within each phase. In the solid these may be associated with the boundaries between crystallites of different crystallographic orientation, or between sections of different composition. Metal–semiconductor contacts and p–n junctions are important examples of such solid/solid interfaces. In the former case the two sections consist of entirely different substances, while in the latter the host crystal is the same and it is only the impurity distribution that varies across the interface.

One of the most important manifestations of the interface is the change in electrostatic potential associated with the transition from one medium to the other. In a semiconductor the potential variation may extend considerably into the underlying bulk, even if all other properties change abruptly. This accounts for the controlling effect of the interface on many of the electronic processes in a semiconductor. From the experimental standpoint, such a penetration is an invaluable asset in any quantitative investigation of surface phenomena.

The present book is concerned with the interfaces between a semiconductor and a vacuum, gas, or liquid. Such interfaces will be referred to as *free* surfaces or simply as surfaces, while the term interface itself will be reserved for solid/solid systems. Historically interfaces, in the form of metal–semiconductor contacts, were the first to be investigated, and we begin by reviewing this early work. We then proceed to outline the recent developments in the study of free surfaces, emphasizing those topics which will be treated in subsequent chapters.

1. Historical notes

The first reported phenomena directly related to the field of semiconductor surfaces date back to the end of the last century when rectification effects were discovered. In 1874 Braun [1] observed that the current-voltage characteristics of a system composed of a metal in contact with a sulphide such as galena were asymmetric with respect to the voltage polarity. Similar effects were observed at about the same time by Schuster [2] with copper–copper oxide contacts. Shortly afterwards Adams and Day [3], working on selenium, discovered the photovoltaic effect — the appearance of a voltage across a rectifier upon illumination. The role of the interface in these phenomena was not appreciated, however, until much later. By the 1920's, copper oxide and selenium rectifiers and photocells were manufactured on a commercial basis. The work at this stage was largely empirical, involving much art and little understanding of the underlying physical principles. Such an understanding had to await the advent of quantum mechanics and the general theories of Sommerfeld [4], Bloch [5], and Wilson [6] on electron behaviour in metals and semiconductors.

These theories form the framework of modern solid state physics, and even today most of the basic ideas and concepts date back to that era. Many characteristic properties of semiconductors can be correlated by the Bloch band model, in which electrons in a solid are distributed among allowed energy bands usually separated by forbidden zones. The concept of a free hole was introduced by Wilson, who also showed how impurities were able to control the densities of electrons and holes. Certain impurities (donors) give rise to n-type conduction, in which the electrons are the majority carriers and the holes the minority carriers, while other impurities (acceptors) result in p-type conduction, the roles of electrons and holes being interchanged. Once the possibility of the coexistence of electrons and holes in a semiconductor was recognized, it required just one more step to understand photoconductivity in terms of hole–electron generation and recombination processes. Due to the finite recombination lifetime of the excess hole–electron pairs generated by the light, the overall carrier density in the semiconductor and hence its conductance are increased. A deeper insight into such non-equilibrium processes was provided by Frenkel's analysis [7] of the problem of the diffusion of excess carriers in a concentration gradient and their drift under electric and magnetic fields. Much the same reasoning was used later in analysing non-equilibrium phenomena at the surface.

In the early 1930's it became increasingly apparent that rectification and photovoltaic effects were intimately associated with the interface between a metal and a semiconductor (or between two different semiconductors). It was realized that these effects could arise from the difference in work functions of the two solids. Upon contact, electrons flow from the solid with the lower work function into the solid with the higher one. As a result, two space-charge regions (of equal and opposite charge) are formed, one on each side of the interface. This process continues until thermal equilibrium is attained, when the potential barrier set up by the accumulated space charge is sufficient to prevent further charge transfer. The magnitude of the potential barrier is equal to the difference in work functions and is referred to as the contact potential. In a metal–semiconductor contact the space-charge region extends much deeper into the semiconductor side of the contact (because of the much smaller carrier concentration), and it is there that most of the contact potential drops.

In 1939 Schottky[8], Mott[9], and Davidov[10] independently formulated theories of rectification based on such considerations. The Schottky barrier model assumed that at the semiconductor side of the contact there existed a region from which majority carriers had been repelled, leaving behind the uncompensated donor (or acceptor) ions. In an n-type semiconductor such a depletion layer would arise when the work function of the semiconductor is smaller than that of the metal. An applied voltage across the rectifier would then modulate the built-in contact potential barrier, lowering it in the forward polarity (metal positive with respect to the semiconductor) and enhancing it in the reverse polarity. The former is the direction of low resistance to the flow of electrons, the latter the direction of high resistance. A similar situation obtains when the semiconductor is p-type if the work function of the semiconductor is larger than that of the metal, the forward current corresponding to the metal negative with respect to the semiconductor.

The theories of Schottky, Mott, and Davidov were very successful in accounting quantitatively for the basic features of rectification, and they undoubtedly constituted a major advance in our understanding of interface phenomena. This work was also of great value in the ensuing technological development of rectifiers for radar detection and other high frequency applications during the Second World War. As it turned out later, however, two very important links were missing in the theory, one pertaining to the origin of the space-charge region, the other concerning the role of the minority carriers in the current flow.

The Schottky barrier was supposed to arise from the difference in work functions of the metal and semiconductor. The degree of rectification was expected, then, to depend on the polarity and magnitude of the contact potential. For example, higher work-function metals should have given rise to higher potential barriers at an n-type semiconductor and therefore to better rectification. Although such a trend was discernible in some cases, the variation in rectification characteristics was far less than predicted, and in many cases no correlation whatsoever was found. In particular, for metal point contacts on silicon and germanium, the work function of the metal made little or no difference. A similar lack of correlation between rectification characteristics and contact potential was found for semiconductor–semiconductor contacts. Benzer [11] observed that a contact between two similar n-type or two similar p-type germanium samples showed high resistance for both polarities, even though the contact potential was very small or zero. This behaviour indicated the existence of two potential barriers of opposite polarity, one on each side of the interface. Measurements of contact potential by Meyerhof [12] also failed to substantiate the theoretical predictions: the observed contact potential between n-type and p-type silicon samples was found to be much smaller than expected.

This group of puzzling phenomena was clarified in one stroke by the far-reaching hypothesis of Bardeen [13] in 1947 that the potential barrier at the semiconductor surface was produced by surface states rather than by the contact potential between the semiconductor and the metal in contact with it. These states were assumed to be localized at the surface and to have energies in the otherwise forbidden energy gap. The possibility that localized states can exist at the surface had previously been pointed out by Tamm[14] and by Shockley [15] on purely theoretical grounds, but the important role played by such states in surface phenomena was not recognized at the time. Bardeen's idea was that electrons from the semiconductor interior can be trapped in the surface states, leaving behind an equal and opposite charge to maintain overall neutrality. Thus a space-charge region and associated with it a potential barrier form near the surface. If the density of surface states is sufficiently large, the potential barrier remains essentially unaltered when contact is made with a metal (or with another semiconductor). The charge exchange due to the difference in work functions between the two solids can be effected almost entirely by a relatively small change in the electron occupation of the states. It was deduced that a density of states which is much lower than that of the surface atoms is sufficient to produce such a screening. Only if the density of states is negligibly small, or if the metal is able in some manner to nullify the effect of the surface states, can the contact potential modify the space-charge region and thus control to any appreciable extent the rectification characteristics.

A considerable surge of activity, mostly at Bell Telephone Laboratories, followed Bardeen's work on surface states, and huge strides in the understanding of surface and interface phenomena were made in a very short time. The main work at that period was directed towards the investigation of surface states on the one hand and of the detailed characteristics of germanium point-contact rectifiers on the other. It was in the course of this work that transitor action was discovered by Bardeen and Brattain [16].

The existence of surface states on a *free* surface was demonstrated by Shockley and Pearson [17], who were also able to estimate their density, by a very direct experiment. This, the field-effect experiment, has since become one of the most important tools in surface studies. In it a space-charge region is produced at the surface by an externally applied electrostatic field in an analogous way to the space charge set up by the proximity of a metal of a different work function. A semiconductor slab was used as one plate of a parallel plate capacitor, the other plate being a metal electrode. By measuring the change in conductance of the slab in a direction parallel to the surface as a function of the voltage applied across the capacitor, it was found that only a small fraction of the total induced charge was mobile, indicating that most of it is trapped in surface states.

It is interesting that a similar scheme, wherein the conductance of a thin semiconductor film is modulated by a capacitatively applied field, had been proposed as a possible amplifier by Heil [18] in 1935. It is quite likely that this proposal was not pursued further at the time because of the detrimental and as yet not understood effect of surface trapping.

An estimate of the density of states on a free surface was also obtained from contact potential measurements. In one experiment Brattain and Shockley [19] studied the variation of contact potential between n- and p-type samples as a function of bulk impurity content. The presence of surface states resulted in a slower increase of contact potential with impurity doping as compared to that calculated on the basis of bulk properties only. In another experiment Brattain [20] measured the change in contact potential upon illumination. The hole–electron pairs produced near the semiconductor surface by the light modify the charge distribution in both the space-charge region and the surface states. This gives rise to a change in the magnitude of the potential barrier and thus to a corresponding change in contact potential.

These and similar studies, combined with careful measurements of point-contact rectification characteristics, brought to light the important role played in surface phenomena by *minority* carriers, even when their concentration in the bulk is very small. It became evident that the space-charge region in germanium and silicon often consisted not only of the immobile donor or acceptor ions constituting the Schottky barrier, but what is more important, of free minority carriers. In an n-type semiconductor, for example, the conductivity type may change from n-type in the bulk to p-type at the surface, giving rise to what is known as an inversion-type space-charge layer. In a classical paper, Brattain and Bardeen [21] showed that it is such an inversion layer that governs the current–voltage characteristics observed on germanium point-contact rectifiers. In the reverse polarity the barrier at the surface is essentially blocking for majority carriers, and the reverse current is composed largely of minority carriers flowing into the metal point. Since minority carriers are almost absent in the bulk, they must be generated thermally in or near the space-charge region. The reverse current is therefore very sensitive to the carrier lifetime near the contact, the lifetime being inversely proportional to the thermal generation rate. In the forward polarity, minority carriers are *injected* from the inversion layer into the bulk semiconductor. If the lifetime is not too short, the excess carriers are able to modulate the spreading resistance under the point contact, giving rise to a forward current considerably larger than that expected from majority-carrier flow alone. This model explained similar results of large forward currents reported previously by Bray [22].

The processes of injection, recombination, and thermal generation which govern the characteristics of point-contact rectifiers, also play a dominant role in the bipolar transistor [16]. The term bipolar refers to the two types of carriers involved in the mode of operation of the transistor: the reverse current of a point contact (the collector) is modulated by the excess minority carriers injected from another point nearby (the emitter). Such modulation is possible only if the lifetime of the injected carriers is sufficiently long to enable them to reach the collector before recombination takes place. This was made possible by the availability of high-grade germanium and silicon obtained after long and intensive materials research.

The development of the junction transistor [23], pioneered principally by Shockley, followed shortly afterwards. This work was undertaken largely with the view of minimizing undesirable surface effects. In the junction transistor the emitter and collector of the point-contact transistor are replaced by two p-n junctions, one on either side of the n-type (or p-type) base material. The theory of p-n junctions has been worked out in detail by Shockley [24], taking into account injection and recombination processes. Davidov [25], as early as 1938, came close to an adequate theory of p-n junctions, but he failed to recognize the process of injection. The characteristics of the junction transistor followed closely the predictions of Shockley's theory. In fact, the p-n junction has become one of the best known semiconductor structures. Most of our detailed understanding of the space-charge region at the free surface has been derived from much the same model.

The discovery of transistor action was a milestone in the history of semiconductor research and technology. The tremendous amount of activity in both areas during the last fifteen years is due largely to the revolutionary importance of transistors and other semiconductor devices in practical applications. Following this discovery, the main effort in semiconductor research shifted from surface phenomena to bulk properties. Major advances were made in materials research leading to purer and more perfect semiconductor crystals, especially those of germanium and silicon. At the same time an intensive and systematic investigation of the fundamental properties of semiconductors was undertaken in many laboratories. This included the study, both experimental and theoretical, of energy band structure, impurity and lattice-defect levels, carrier transport phenomena, and electron–hole recombination processes.

As to the free surface, it was only in the early 1950's that interest was revived and an appreciable effort made in the study of its characteristics. Again, this renewal of interest stemmed largely from practical problems encountered in device fabrication. It was recognized that while transistor action was primarily a bulk phenomenon, bulk properties alone could not account for the anomalous behaviour of diodes and transistors and that many of the spurious effects were associated with the surface. At the same time, the availability of crystals of a high degree of purity and lattice perfection, and the better understanding of bulk properties, made semiconductors an ideal medium for surface studies. Most of the investigations involved germanium and silicon, and it is mainly through work on these materials that surface research has become a quantitative discipline.

2. Recent developments in surface research

In the course of the surface studies of the last decade or so, several demarcation lines have emerged in the classification of the various research activities. In the first place, it has become common to distinguish between two types of surfaces, *real* and *clean*. The former term refers to surfaces obtained by ordinary laboratory procedures, the latter to surfaces prepared under carefully controlled conditions so as to ensure the absence of foreign matter.

A real surface is usually prepared by mechanical polishing followed by chemical etching to remove the outer damaged layers. Such a surface is covered by chemisorbed material, generally an oxide, and by molecules adsorbed from the surrounding ambient. Real surfaces have been studied extensively because they are easily prepared and handled and because they are readily amenable to many types of measurements. Moreover, it is the real surface that is encountered in most practical applications.

Clean surfaces are much more difficult to produce. They can be prepared by cleavage, ion bombardment, or by heating at elevated temperatures. Once obtained, a clean surface must be maintained in ultra-high vacuum (10^{-10} to 10^{-9} mm Hg) to prevent recontamination. It is by now well established that clean surfaces produced in this manner are free of foreign adsorbed matter to better than a few percent of an atomic monolayer. The interest in clean surfaces stems from the fact that they constitute the closest approximation to the *true* crystal surface, and should thus exhibit the fundamental features of the surface *per se*. Due to the many experimental difficulties associated with the handling of clean surfaces, however, such studies have not yet been carried through to the same extent as those on real surfaces.

Another demarcation line divides the recent research activities according to the methods employed in the investigation of real and clean surfaces. Two main categories are apparent, one involving studies of the lattice structure and chemistry of the surface and the other, studies of the various electrical phenomena taking place at and near the surface. Considerable progress has been achieved within each category, but a systematic correlation between the two types of observations is as yet only fragmentary.

In the investigation of surface structure the main emphasis has been placed on clean surfaces. Low-energy electron diffraction, which probes the top few atomic layers of the surface, has proved as powerful a tool in determining the two-dimensional surface lattice parameters as has X-ray diffraction in crystallography. It is generally found that while the surface atoms are displaced from their normal bulk positions, they display long-range order. A common feature of this arrangement is the occurrence along the surface plane of lattice spacings twice as large as those in the bulk. More complex structures are also observed. Considerable attention has been given to the interaction of clean surfaces with gases and vapours, particularly oxygen, and quantitative information is now available on adsorption processes at such surfaces. In the case of real surfaces, studies of chemisorption have proved very useful in elucidating the basic processes involved in catalysis. Solid/liquid interactions such as chemical and electrolytic etching have also been extensively investigated, and numerous etchants have been developed to yield specific surface properties.

The other category of surface studies, and the one with which we shall be mainly concerned in this book, pertains to the various surface phenomena of a predominantly electrical nature. These phenomena involve the free carriers in the space-charge region, the surface states, and the mutual interaction between the two. The most fundamental variable controlling the electronic processes at the surface is the height of the potential barrier — that is, the drop in electrostatic potential between the surface and the underlying bulk. The general approach to the investigation of these processes has been to vary the barrier height, for example by the field-effect technique, and to follow the accompanying changes in electrical characteristics. For any work of a quantitative nature, it is essential to determine the magnitude of the barrier height throughout its variation. This has usually been accomplished by performing measurements on the space-charge region. The measurement of the conductance parallel to the surface, associated with the free carriers in the space-charge layer, has proved particularly useful in this context. In semiconductors, the space-charge region generally extends 10^{-5} to 10^{-4} cm beneath the surface. This distance may be made a sizable fraction of the sample thickness, so that the contribution of the surface conductance to the overall sample conductance can be determined accurately.

Because of the fundamental role played by the space-charge region in surface phenomena, much effort has been devoted to its study. On the theoretical side, the shape of the potential barrier and the density of electrons and holes in the space-charge layer have been calculated as functions of the barrier height and the appropriate bulk parameters. Measurements of surface conductance, space-charge capacitance, contact potential, and infrared carrier absorption, all intimately associated with the characteristics of the space-charge layer, have been found to be in good agreement with the theoretical calculations. Our knowledge of the space-charge region is now quite complete.

This made possible, in turn, a more detailed study of the transport properties of carriers confined to the space-charge layer. In the bulk the carrier mobility is determined by scattering processes arising from the thermal vibrations of the lattice atoms as well as from any impurities and lattice imperfections present in the crystal. One would expect the surface to add its share to these scattering processes, giving rise to a reduced mobility for those carriers moving close to it. The surface conductance would thus be expected to be smaller than that evaluated on the basis of bulk scattering alone. This problem has received considerable attention both because of the importance of surface conductance as an experimental tool and because of the basic information that may be derived from the study of surface scattering on the nature of the surface. The theory of surface transport has been worked out in detail and, in parallel, a substantial amount of experimental data has been accumulated on germanium and silicon.

Perhaps the most concerted effort in the study of the electrical properties of the surface has been directed towards the surface states. This work developed along two different lines. Tamm's theoretical work, which treated a rather simplified model, was extended by Shockley and by other workers to cover more general cases. It was shown that in covalent crystals, surface states may be associated with the unfilled orbitals or dangling bonds of the surface atoms. In germanium and silicon, for example, a bulk atom forms four valence bonds with its nearest neighbours, but a surface atom can form only three. The remaining unfilled orbital may thus be free to trap an electron. On the basis of such considerations, the maximum density of surface states is expected to be of the order of 10^{15} cm^{-2}, one for each surface atom.

The second approach to this problem has been essentially phenomenological. Its aim was to determine the characteristics of the surface states without being particularly concerned with their type or origin. A limited number of parameters was defined for the surface states in a manner completely analogous to that used for bulk impurity states, and theories were developed relating these parameters to measurable quantities. The values of the parameters were deduced from the fit between theory and experiment. This approach has proved extremely fruitful and today one is fairly well acquainted with the characteristics of the surface states in germanium and silicon. The gap is still large, however, between this empirical approach and the fundamental theoretical approach, and a definite relation has yet to be found between the experimentally observed surface states and the theoretically proposed Tamm or Shockley states.

The surface states observed on *real* germanium and silicon surfaces fall into two groups: the "fast states", which are presumably situated at the interface between the crystal and the oxide, and the "slow states", located predominantly at the outer surface of the oxide layer and arising entirely from adsorbed gas atoms. The terms fast and slow refer to the relative speeds with which the respective states interact with the underlying space-charge region of the bulk crystal. The fast states are normally characterized by time constants of the order of microseconds or less, the slow states by time constants ranging from a fraction of a second to hours, depending on the structure of the oxide layer and the nature of the gas ambient. The fast states are usually present with a density of the order of 10^{11} cm^{-2} and their distribution in energy is essentially discrete. The characteristics of those of the fast states that control the recombination of excess carriers at the surface are particularly well known: their energy position and their cross sections for hole and electron capture have been accurately determined. Regarding the slow states, little is known about their energy distribution and other characteristics besides the fact that under normal conditions their density is at least 10^{13} cm^{-2}. This density is sufficient to control the potential barrier at the surface.

The structure of the surface states on *clean* surfaces of germanium and silicon has not yet been studied in as great a detail because of the many experimental difficulties involved. Sufficient data have been accumulated, however, to indicate that their characteristics are similar to those of the fast states on real surfaces, except that their density is considerably higher. The large density points to the possibility that the surface states on a clean surface may be related to the Tamm or Shockley states. This is supported by the observation that oxygen adsorption seems to reduce the number of states, suggesting that the oxide saturates some of the dangling bonds associated with the surface atoms. These are but hypotheses, and the physical origin of the surface states still remains the greatest unsolved problem of surface physics.

The advances in the fundamental understanding of surface phenomena have contributed considerably to device technology. Most of the surface effects detrimental to device performance can be accounted for in terms of the characteristics of the space-charge layer and the surface states, and their known dependence on surface preparation and gas ambient. The conducting channel that often develops across a p–n junction and shorts it out is due to the large surface conductance associated with inversion layers at the surface of the junction, while the excess $1/f$ noise at low frequencies results from slow fluctuations of the potential-barrier height. Both effects are controlled mainly by the slow states and are therefore very sensitive to the surrounding ambient. The large reverse current that is observed sometimes in p–n junctions arises from excessive recombination at the surface resulting from too large a density of fast states. The main problem in device fabrication has not been so much the minimization of such undesirable effects, as the protection against any change with time in device characteristics. Different stabilization methods have been investigated, ranging from hermetic sealing of the device to prevent changes in gas ambients, to shielding by suitable agents put or grown on the exposed surface. Thermal oxidation under controlled conditions following the fabrication of the device has proved particularly effective in silicon. Even today, however, the surface remains a major technological problem, and advances are limited as much by difficulties arising from surface effects as by those originating from device body design and fabrication.

While most of the efforts in device technology are primarily aimed at the elimination of surface effects, there is one notable example where surface phenomena are specifically invoked for device operation. This is the unipolar field-effect transistor, long overshadowed by the bipolar transistor. As has been pointed out above, the field effect brought the important role of surface states into sharp focus, and it was mainly because of such states that its use in device operation was thought unfeasible at the time. Recently, however, means have been devised to nullify the effect of surface states on certain structures involving silicon and cadmium sulphide. Such structures show great promise, and it is quite likely that the field-effect transistor, perhaps the most straightforward of the semiconductor devices, will play an important role in many applications.

References

[1] F. Braun, Pogg. Ann. **153** (1874) 556; Wied. Ann. **1** (1877) 95; **4** (1878) 476; **19** (1883) 340.
[2] A Schuster, Phil. Mag. (4) **48** (1874) 25.
[3] W. G. Adams and R. E. Day, Proc. roy. Soc., London **25** (1876) 113.
[4] A. Sommerfeld, Z. Physik **47** (1928) 1, 43; A. Sommerfeld and H. Bethe, in *Handbuch der Physik* (Edited by H. Geiger and K. Scheel), vol. **24**/2 (Springer, Berlin, 1933), p. 333.
[5] F. Bloch, Z. Physik **52** (1928) 555.
[6] A. H. Wilson, Proc. roy. Soc., London **A133** (1931) 458; **A134** (1931) 277.
[7] J. Frenkel, Phys. Z. Sowjet. **8** (1935) 185.
[8] W. Schottky, Z. Physik **113** (1939) 367; **118** (1942) 539.
[9] N. F. Mott, Proc. roy. Soc., London **A171** (1939) 27.
[10] B. Davidov, J. Phys. (USSR) **1** (1939) 167.
[11] S. Benzer, Phys. Rev. **71** (1947) 141.
[12] W. E. Meyerhof, Phys. Rev. **71** (1947) 727.
[13] J. Bardeen, Phys. Rev. **71** (1947) 717.
[14] I. E. Tamm, Z. Phys. **76** (1932) 849; Phys. Z. Sowjet. **1** (1932) 733.
[15] W. Shockley, Phys. Rev. **56** (1939) 317.
[16] J. Bardeen and W. H. Brattain, Phys. Rev. **74** (1948) 230; **75** (1949) 203.
[17] W. Shockley and G. L. Pearson, Phys. Rev. **74** (1948) 232.
[18] O. Heil, *Improvement in or Relating to Electrical Amplifiers and Other Control Arrangements and Devices*, Brit. Pat. 439 457 (1935).

[19] W. H. Brattain and W. Shockley, Phys. Rev. **72** (1948) 345.
[20] W. H. Brattain, Phys. Rev. **72** (1948) 345.
[21] W. H. Brattain and J. Bardeen, Phys. Rev. **74** (1948) 231.
[22] R. Bray, Bull. Am. phys. Soc. **23** (1948) 21.
[23] W. Shockley, M. Sparks, and G. K. Teal, Phys. Rev. **83** (1951) 151.
[24] W. Shockley, Bell System tech. J. **28** (1949) 435.
[25] B. Davidov, Tech. Phys. (USSR) **5** (1938) 87.

CHAPTER 2

THE SEMICONDUCTOR BULK – A RÉSUMÉ

The physical processes that take place at the surface of a semiconductor are greatly influenced by the properties of the underlying bulk material. One can thus hope to be able to separate surface and bulk effects only by being well acquainted with bulk properties and by possessing a certain measure of control over them. A knowledge of bulk characteristics is also an essential prerequisite to surface studies from the conceptual aspect, and in the present chapter we introduce some of the basic physics of semiconductors. An exhaustive treatment, however, is beyond the scope of this book, and the reader is referred to the existing textbooks [1-6] and review articles [7-9] on the subject. Our main purpose here is to afford the necessary background and provide a convenient reference source for the chapters that follow. Particular attention is given to germanium and silicon which, as a result of the thorough understanding of their bulk properties, have been subjected to the most extensive surface studies.

The first three sections of the present chapter treat the band model and the motion of electrons and holes through a perfect lattice. This is followed by a discussion of lattice vibrations, impurities, and defects normally present in real crystals. In § 5 the occupation statistics of the various energy states is considered, while § 6 deals with transport processes. The chapter concludes with a section on non-equilibrium phenomena and p–n junctions.

1. Wave-mechanical approach to the problem of electrons in solids

An electron in a solid is subject to a field arising from the positive atomic nuclei and from all the other electrons. As in the case of an isolated atom, a full description of the electron behaviour in a crystal can only be given in terms of wave mechanics. Before discussing the various approaches to this problem, it would be instructive to review briefly two simple limiting cases to the actual situation. In the first, the electrons inside the solid are viewed as being completely free except in so far as they are unable to leave the crystal.

This corresponds to the free-electron model applied with considerable success by Sommerfeld [10] to electrons in metals. In the other extreme case the electrons are considered as being tightly bound to individual atoms with no possibility for communication between one atom and the next. This limiting case, less realistic than the first, corresponds closest to molecular crystals wherein the constituent atoms are held together by weak van-der-Waals forces and are far removed from one another.

1.1. FREE AND BOUND ELECTRONS

Consider first an electron in free space, where its potential energy U is zero. For one-dimensional motion, Schrödinger's equation can be written as

$$\frac{d^2\psi(x)}{dx^2} = -\frac{2m}{\hbar^2} E\psi(x). \qquad (2.1)$$

E is the energy of the electron and $\psi(x)$ its wave function; $|\psi(x)|^2 dx$ is the probability of finding the electron in the element dx. The solutions of this equation are:

$$\psi_k(x) = A e^{ikx} \qquad (2.2)$$

and

$$E = (\hbar^2/2m)k^2. \qquad (2.3)$$

Here k is the so-called wave number and can assume any real value; A is a normalizing factor. The time-dependent wave function is

$$\Psi(x, t) = \psi_k e^{-i\omega t} = A e^{i(kx - \omega t)}, \qquad (2.4)$$

where

$$\omega(k) = 2\pi \nu = E_k/\hbar. \qquad (2.5)$$

Ψ represents a travelling wave, moving to the right for positive values and to the left for negative values of k. The wave number k is related to the de Broglie wavelength of the electron by the equation

$$|k| = 2\pi/\lambda. \qquad (2.6)$$

The electron wave is propagated with a velocity ω/k, referred to as the *phase* velocity. A moving electron is represented by a wave packet constructed out of wave functions of the type given by (2.4) and its particle velocity c is the *group* velocity of that wave packet:

$$c = \frac{d\nu}{d(1/\lambda)} = \frac{d\omega}{dk}. \qquad (2.7)$$

From (2.3), (2.5), (2.7) it follows that the electron momentum is

$$p = mc = \hbar k. \tag{2.8}$$

Hence

$$E = p^2/2m = \tfrac{1}{2}mc^2, \tag{2.9}$$

which is the classical expression for the kinetic energy of a free electron. We have thus obtained the well known result that the energy of a free electron is not quantized, *all* values from 0 to ∞ being allowed.

Consider next an electron confined to a box or, in one dimension, to a segment of the x axis. Inside the box the potential energy of the electron has a constant value (which we shall take as zero). The electron can be prevented from leaving the box by erecting impenetrable potential walls at the two ends of the segment. Under these conditions the solutions of the Schrödinger equation are standing waves, since the wave functions must vanish at the boundaries. Alternatively, the confinement of the electron to a one-dimensional box can be effected by taking the segment in the form of a closed loop, resulting in the so-called cyclic or periodic boundary conditions. These conditions are very convenient, since they lead to a running-wave representation as for the case of a free electron. The wave functions must now be periodic in L, the length of the looped segment:

$$\psi(x+L) = \psi(x). \tag{2.10}$$

The energy of the electron is still given in terms of the wave number k by (2.3), but now k is restricted to the values

$$k = \pm n(2\pi/L), \quad n = 0, 1, 2, \ldots. \tag{2.11}$$

If the box is of macroscopic dimensions, corresponding to the case of a free electron in a crystal, the allowed values of k are very closely spaced and the energy levels form in effect a quasi-continuum. If, on the other hand, L is of atomic dimensions, the energy spectrum is discrete with relatively large forbidden regions separating the allowed levels. This represents a hypothetical crystal in which impenetrable walls are interposed between the atoms, so that the electrons are in atomic orbitals.

The above considerations can be extended readily to the three-dimensional case. The wave functions are then of the form

$$\psi_{\mathbf{k}} = A e^{i\mathbf{k}\cdot\mathbf{r}}, \tag{2.12}$$

where r is the position vector and k the wave vector. Furthermore,

$$p = \hbar k = mc; \quad k = 2\pi/\lambda, \qquad (2.13)$$

where k is the absolute magnitude of the vector k.

For cyclic boundary conditions, which may be used to represent an electron in a three-dimensional box as well, the allowed values of k are

$$k_x = \pm n_1(2\pi/L_x); \quad k_y = \pm n_2(2\pi/L_y); \quad k_z = \pm n_3(2\pi/L_z), \qquad (2.14)$$

where n_1, n_2, n_3 are integers.

1.2. ONE-ELECTRON MODELS

The behaviour of electrons in a crystal can be obtained, in principle, by solving Schrödinger's equation involving all the electrons in the solid. The solution of such an equation is well nigh impossible. It is therefore necessary to introduce a number of simplifying approximations. First, the crystal is assumed to be an ideal periodic structure of atoms at rest in their lattice sites. The interaction of the electrons with the thermal vibrations of the nuclei as well as with whatever lattice imperfections are present in a real crystal are treated as perturbations to the ideal case. A further simplification is attained by the use of a one-electron model in which a crystal wave function is approximated by a combination of wave functions each involving the coordinates of one electron. Only the valence electrons need be considered, since the inner-core electrons remain tightly bound to their respective nuclei and play an insignificant role in the electronic processes characterizing the solid as such. Each valence electron is assumed to move under the same potential, consisting of a periodic term due to the fixed positive ions (composed of the nuclei and inner-core electrons) and an average *constant* potential produced by the charge distribution of all the other valence electrons.

In the treatment of this model there are two different approaches, whose starting points are the two limiting cases discussed above. One approach, based on the Heitler-London method of molecular physics [11], pictures the crystal as being composed of separate atoms which interact with one another. It is particularly useful when the atoms are far apart, so that their mutual interaction does not change the atomic properties appreciably. In this case the valence electrons can be considered to be tightly bound to individual atoms, with the crystalline state constituting a relatively weak perturbation.

The other approach, the so-called band model, was advanced by Bloch [12] and extended by Wilson [13]. Here the valence electrons are looked upon as

2. The band model for a one-dimensional periodic lattice

The essential features of the band structure and of electron propagation in a crystal are brought into sharp focus by a consideration of a simple one-dimensional model due to Kronig and Penney [14]. This model can be treated explicitly and will be presented here in some detail. It is of particular interest to us since it underlies the pioneering work of Tamm on surface states, as will be discussed in chapter 5.

2.1. THE KRONIG–PENNEY MODEL

In a one-dimensional crystal lattice the potential energy $U(x)$ of an electron has the form shown in Fig. 2.1a. Kronig and Penney approximated this potential energy by a periodic array of square wells as indicated in Fig. 2.1b. The period of the potential is $(a+b)$; $U = 0$ where $0 < x < a$ and $U = U_0$ where $-b < x < 0$. The Schrödinger equation for an electron moving in this potential is

$$\frac{d^2\psi}{dx^2} + \frac{2m}{\hbar^2}E\psi = 0, \qquad 0 < x < a;$$
$$\frac{d^2\psi}{dx^2} + \frac{2m}{\hbar^2}(E-U_0)\psi = 0, \qquad -b < x < 0. \tag{2.15}$$

There is a general theorem, known as the Bloch theorem [12], which states that the solutions of Schrödinger's equation for a periodic potential are of the form (in one dimension)

$$\psi(x) = u_k(x)e^{ikx}, \tag{2.16}$$

where $u_k(x)$ is a periodic function with the periodicity of the potential:

$$u_k[x+(a+b)] = u_k(x). \tag{2.17}$$

For an infinite lattice, or for a finite lattice associated with cyclic boundary conditions, k can assume only real values. These solutions are called Bloch functions and, when the time dependence is included, they represent running

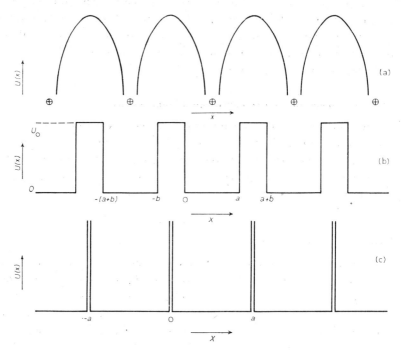

Fig. 2.1. Potential energy of an electron in an infinite one-dimensional lattice. The positions of the positive atomic cores are indicated by the symbol ⊕.
    (a) Schematic representation of actual conditions.
    (b) Kronig–Penney model.
    (c) Periodic delta potential.

waves modulated with the period of the lattice. Substituting (2.16) into (2.15) we obtain the following equations for $u_k(x)$:

$$\frac{d^2u}{dx^2} + 2ik\frac{du}{dx} + (\alpha^2 - k^2)u = 0, \quad 0 < x < a;$$
$$\frac{d^2u}{dx^2} + 2ik\frac{du}{dx} - (\beta^2 + k^2)u = 0, \quad -b < x < 0,$$
(2.18)

where

$$\alpha = \sqrt{\frac{2mE}{\hbar^2}}; \quad \beta = \sqrt{\frac{2m(U_0 - E)}{\hbar^2}}.$$
(2.19)

Ch. 2, § 2.1   BAND MODEL FOR A ONE-DIMENSIONAL LATTICE

The solutions of these equations are readily seen to be

$$u_a(x) = A_1 e^{i(\alpha-k)x} + A_2 e^{-i(\alpha+k)x}, \quad 0 < x < a;$$
$$u_b(x) = B_1 e^{(\beta-ik)x} + B_2 e^{-(\beta+ik)x}, \quad -b < x < 0. \quad (2.20)$$

The constants A_1, A_2, B_1, and B_2 can be determined by the requirements that $u(x)$ and du/dx be continuous at $x = 0$ and at $x = -b$, and by the condition (Bloch's theorem) that $u(x)$ be periodic in $(a+b)$:

$$u_a(0) = u_b(0); \quad u_a(a) = u_b(-b);$$
$$\left.\frac{du_a}{dx}\right|_{x=0} = \left.\frac{du_b}{dx}\right|_{x=0}; \quad \left.\frac{du_a}{dx}\right|_{x=a} = \left.\frac{du_b}{dx}\right|_{x=-b}. \quad (2.21)$$

Equations (2.21) lead to the following linear homogeneous equations:

$$A_1 + A_2 = B_1 + B_2; \quad (2.22)$$

$$e^{i(\alpha-k)a} A_1 + e^{-i(\alpha+k)a} A_2 = e^{-(\beta-ik)b} B_1 + e^{(\beta+ik)b} B_2; \quad (2.23)$$

$$i(\alpha-k)A_1 - i(\alpha+k)A_2 = (\beta-ik)B_1 - (\beta+ik)B_2; \quad (2.24)$$

$$i(\alpha-k)e^{i(\alpha-k)a} A_1 - i(\alpha+k)e^{-i(\alpha+k)a} A_2$$
$$= (\beta-ik)e^{-(\beta-ik)b} B_1 - (\beta+ik)e^{(\beta+ik)b} B_2. \quad (2.25)$$

The condition for a non-trivial solution to these equations is the vanishing of the determinant of the coefficients of A_1, A_2, B_1, B_2. This leads to a rather complicated equation in α, β, and k. To simplify matters Kronig and Penney consider a periodic delta potential (Fig. 2.1c), obtained from that represented in Fig. 2.1b by making U_0 tend to infinity and b tend to zero in such a way that the product $U_0 b$ remains finite. For this potential $E < U_0$ and so β is always real. We define a dimensionless parameter P as

$$P \equiv \lim_{\substack{b \to 0 \\ \beta \to \infty}} \tfrac{1}{2} a(\beta^2 b). \quad (2.26)$$

This parameter is a measure of the binding energy of an electron to an individual core. At the limit, the right-hand side of (2.23) approaches $B_1 + B_2$, since both b and $\beta b = \beta^2 b/\beta$ tend to zero. By subtracting (2.25) from (2.24) and expanding the exponential functions containing β, we see that the right-hand side approaches $(B_1 + B_2) 2P/a$. Thus (2.22)–(2.25) reduce to two linear homogeneous equations in A_1 and A_2:

$$[1-e^{i(\alpha-k)a}]A_1 + [1-e^{-i(\alpha+k)a}]A_2 = 0; \tag{2.27}$$

$$[(2P/a)-i(\alpha-k)(1-e^{i(\alpha-k)a})]A_1 + [(2P/a)+i(\alpha+k)(1-e^{-i(\alpha+k)a})]A_2 = 0.$$

The vanishing of the determinant now yields

$$(P/\alpha a)\sin\alpha a + \cos\alpha a = \cos ka. \tag{2.28}$$

This transcendental equation must have a solution for α in order for wave functions of the form of (2.16) to exist. In Fig. 2.2 the left-hand side of

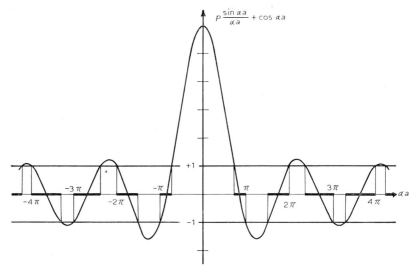

Fig. 2.2. Plot of $(P/\alpha a)\sin\alpha a + \cos\alpha a$ as a function of αa for $P = 5$. The allowed values of αa are indicated by heavy segments.

(2.28) is plotted against αa for an arbitrarily chosen value of P. Since we are seeking here solutions of the Bloch type — that is, running waves representing a uniform probability density — only real values of k are relevant (complex values of k, corresponding to *localized* states at the surface and arising from the abrupt termination of the crystal lattice, will be discussed in chapter 5). Hence the cosine term on the right-hand side of (2.28) can only have values between $+1$ and -1, and consequently only those values of αa are allowed for which the left-hand side falls between these limits. The allowed ranges of αa are indicated by heavy lines in the figure and correspond, through the relation $E = (\hbar^2/2m)\alpha^2$ (eq. (2.19)), to the allowed ranges of the energy E.

The immediate conclusion to be drawn from eq. (2.28) and Fig. 2.2 is that the electron energy spectrum consists of allowed energy bands separated by forbidden zones. The width of a given band decreases with increasing P — that is, with increasing binding energy of an electron to an individual atom. For $P \to \infty$ the energy bands become infinitely narrow and the energy spectrum degenerates into a line spectrum. In this case, (2.28) has solutions only if $\sin \alpha a = 0$, or $\alpha a = \pm n\pi$. This relation is similar to that derived previously (eqs. (2.3), (2.11)), with L replaced by a. It thus corresponds to an electron confined to a box of atomic dimensions. Indeed, $P = \infty$ represents impenetrable potential walls which prevent the tunnelling of the electron from one atom to the next. For $P = 0$ all bands overlap to give a continuum of allowed levels from zero to infinity. This, of course, corresponds to a free electron. The solution of (2.28) reduces to $\alpha = k$, which is identical with (2.3).

From (2.28) it follows that the boundaries of the allowed energy bands (that is, the discontinuities in the E versus k curve) occur at

$$k = \pm n(\pi/a), \quad n = 1, 2, \ldots. \tag{2.29}$$

Furthermore, within any band, the energy is a multivalued function of k and periodic in it, since if one replaces k by $k \pm n(2\pi/a)$ the right-hand side of (2.28) remains unaltered. This conclusion also follows quite generally from the fact that such a substitution preserves the form of the Bloch functions: if $k' = k \pm n(2\pi/a)$ then, from (2.17),

$$\psi = u_k(x) e^{\mp in(2\pi/a)x} e^{ik'x} \equiv u_{k'}(x) e^{ik'x}, \tag{2.30}$$

and since $u_{k'}(x)$ is periodic with the lattice, (2.30) represents a Bloch function.

Bearing these remarks in mind, we now refer the reader to Fig. 2.3, which is a plot of E against k as computed with the aid of Fig. 2.2 (with $P = 5$). The periodic functions represent the first four energy bands that correspond to the heavily drawn segments (on either side of $k = 0$) in Fig. 2.2. The dashed parabolic curve represents the energy of a *free* electron. The modification in the behaviour of such an electron brought about by the crystal lattice can be seen by comparing the dashed curve with those portions of the periodic functions adjacent to it. It is seen that the higher the energy of an electron, the wider is the energy band in which it is located, and the more similar is the E versus k curve to that of a free electron. Physically this is to be expected, since the more energetic the electron the less it is affected by the periodic potential.

Evidently, the wave number k may vary from 0 to $\pm\infty$. The periodicity of the system, however, enables us to express all energies in terms of a "reduced wave number" whose values are confined to a region of length $2\pi/a$, the period of the function $E(k)$. Only values of k within such a region are pertinent to the problem. It is customary to choose this region as

$$-(\pi/a) \leq k \leq (\pi/a), \tag{2.31}$$

the so-called first Brillouin zone. The heavily drawn portions of the E versus k

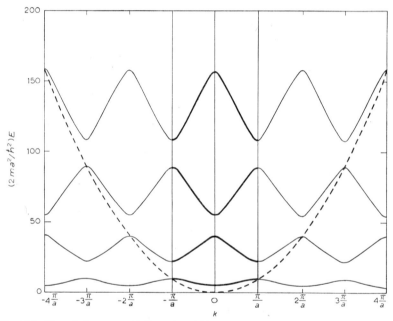

Fig. 2.3. Plot of the energy E (in units of $\hbar^2/2ma^2$) versus wave number k as computed with the aid of Fig. 2.2. The dashed curve represents the energy of a free electron.

curves in Fig. 2.3 represent the band structure in this manner. Due to the periodicity of the function E and its symmetry about $k = 0$, it follows that

$$\left.\frac{dE}{dk}\right|_{k=0,\,\pm\pi/a} = 0. \tag{2.32}$$

In the discussion thus far we have assumed the crystal to be infinite. Consider now a finite crystal of length L containing N atoms ($L = Na$). We

Ch. 2, § 2.2 BAND MODEL FOR A ONE-DIMENSIONAL LATTICE

shall impose here the same cyclic boundary conditions as for the case of a free electron confined to a box (eq. (2.10)). Strictly speaking, these conditions apply only to a closed chain of atoms. If, however, the chain is of macroscopic dimensions ($N \gg 1$), breaking the circular chain at one point in order to form a linear chain cannot alter conditions appreciably except near a few atoms at the ends. In other words, cyclic boundary conditions are a convenient way in which to introduce the finiteness of the crystal, but as long as the crystal is of macroscopic dimensions the behaviour of an electron in the bulk is practically independent of the exact conditions at the boundaries.

Using the condition $\psi(x+Na) = \psi(x)$ (see eq. (2.10)), we obtain that the allowed values of k are integral multiples of $2\pi/L$ (eq. (2.11)). Since, moreover, the pertinent values of k are here further restricted to a Brillouin zone (eq. (2.31)), it follows that the total number of quantum states in each band is N, the number of unit cells. Now it will be recalled that in the one-electron approximation we have been discussing, the effect on any one electron of all the other electrons was taken into account merely by adding a *constant* term to the periodic potential. In this model all electrons are considered equivalent and may occupy any energy level within the allowed bands. We must, however, introduce a restriction on the filling up of these levels in accordance with the Pauli exclusion principle. This principle states that each quantum state can be occupied by no more than two electrons (of opposite spin). Thus the maximum number of electrons that can be accommodated in any band is $2N$, twice the number of unit cells. This conclusion has far-reaching consequences, as will be seen below.

2.2. THE MOTION OF ELECTRONS

An electron in a state k is described by means of an appropriate Bloch function ψ_k. Since ψ_k represents an unattenuated running wave the electron moves unhindered through the lattice, the wave number k being a "constant of motion". As discussed in § 1.2, the particle velocity c in a given state k is given by the group velocity $d\omega/dk$ of a wave packet made up of wave functions of neighbouring states. The angular frequency ω of the de Broglie waves is related to the energy of the particle by the relation $\omega = E/\hbar$, and so

$$c = \frac{1}{\hbar} \frac{dE}{dk}. \tag{2.33}$$

For a free electron c is proportional to k (see eq. (2.8)), whereas for an electron in a periodic potential the relation between the two quantities is more complicated and depends on the energy-band structure. In both cases, however, an electron preserves its velocity indefinitely (k is a constant) provided there are no external forces. A plot of c against k is shown in Fig. 2.4b for the bottom energy band of Fig. 2.3 (redrawn on an expanded scale in Fig. 2.4a). From (2.32) it follows that the velocity vanishes

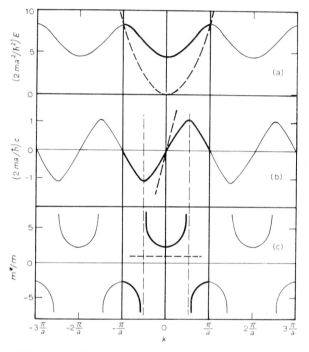

Fig. 2.4. Plot of (a) energy, (b) velocity, and (c) effective mass against the wave number k for an electron in the lowest energy band of Fig. 2.3. The corresponding quantities for a *free* electron are represented by the dashed curves.

at the centre and at the edges of the Brillouin zone. The absolute magnitude of the velocity has a maximum value at the inflection points of the $E(k)$ curve: beyond these points it *decreases* with increasing energy. This behaviour is drastically different from that of a free electron, and more will be said about it later.

Consider now the effect of an electric field on an electron in the crystal. Due to the lattice potential, such an electron is already subject to very intense internal fields ($\approx 10^8$ V/cm), larger than any that could conceivably be applied externally. Thus an external field cannot change significantly the energy-band structure, and we may deduce the behaviour of an electron under such a field on the basis of the foregoing considerations. To avoid complications arising from the Pauli exclusion principle, we shall consider first the case of a single electron occupying an otherwise empty band. The power delivered by an electric field $\mathscr{E}$ to an electron moving with a velocity c is

$$\frac{dE(k)}{dt} = -qc\mathscr{E}, \tag{2.34}$$

where q denotes the absolute magnitude of the electronic charge. Since

$$\frac{dE}{dt} = \frac{dE}{dk}\frac{dk}{dt} = \hbar c \frac{dk}{dt},$$

we have

$$\frac{dk}{dt} = -\frac{q\mathscr{E}}{\hbar}. \tag{2.35}$$

This relation shows that the rate of change of the *wave number* is proportional to the field, whereas in free space it is the rate of change of the *velocity* that is proportional to the field.

The acceleration of the electron is given by

$$a = \frac{dc}{dt} = \frac{1}{\hbar}\left(\frac{d^2E}{dk^2}\right)\frac{dk}{dt},$$

and using (2.35) we have

$$a = -\frac{q\mathscr{E}}{\hbar^2}\frac{d^2E}{dk^2}. \tag{2.36}$$

Comparing this equation with that for a free electron, $a = -q\mathscr{E}/m$, we obtain that the electron in a crystal behaves as though it had an effective mass m^* given by

$$m^* = \hbar^2(d^2E/dk^2)^{-1}. \tag{2.37}$$

The effective mass, as calculated from Fig. 2.4a, is plotted against k in Fig. 2.4c. It is seen that m^* is positive near the minima of $E(k)$ and negative near the maxima: the transitions from positive to negative mass occur at the inflection points of the $E(k)$ curve, where m^* is infinite. Consider, for instance, an electron at $k = 0$. When a constant field is applied in the $-x$ direction, the wave number increases linearly with time (see eq. (2.35)). Initially the electron velocity increases until the velocity attains its maximum value. Beyond this maximum, the same field produces a decrease in c; in other words, the effective mass becomes negative. When k reaches the value π/a, the electron "leaves" the Brillouin zone. From the equivalence of corresponding k values outside and inside the zone (see § 2.1), the electron can then be regarded as "re-entering" the Brillouin zone from the left. This, however, is just a matter of convenience. The important thing to notice is that from this point on the electron velocity becomes negative until, at $k = 0$, we return to the starting point ($c = 0$) and the same cycle is repeated. Thus, under the action of a *constant* field, the electron oscillates to and fro in k-space, as well as in position space, half the time absorbing energy from the external source and half the time giving it back. This behaviour serves to illustrate the difference between a truly free electron in space and a *quasi* free electron in the crystalline lattice. It is a consequence of the fact that an electron in a crystal is subject to both the internal lattice field and the externally applied field. The effective mass has been defined (eq. (2.37)) in such a way as to obtain a formal analogy with the free-electron case. The analogy involves only the external field, the effect of the lattice being implicitly introduced in the form of an energy-dependent effective mass. Although this definition is rather artificial, the effective mass has been found to be a most useful quantity in describing electronic processes in solids.

It should be pointed out that the oscillations described above do not take place in real crystals because the electron undergoes collisions with lattice vibrations and other imperfections so frequently that in ordinary fields k changes on the average very little.

Next we turn to consider the contribution to the electrical conductivity of the whole assembly of electrons in the crystal. At zero absolute temperature the electrons fill the lowest available energy levels, subject to the Pauli exclusion principle. We have seen above that the maximum number of electrons which can be accommodated in any energy band is twice the number of unit cells. If there is an even number of electrons per unit cell, then the electrons fill up completely one or more energy bands, the other bands being entirely empty. This situation is illustrated in Fig. 2.5a, which shows the highest occupied band and lowest unoccupied band. It is easy to see that filled bands cannot carry current. Under an applied field the k values of all electrons vary at the same rate, in accordance with (2.35). Since all the states are occupied, however, the electrons just interchange positions in k space, and the field produces no net change in the occupation of the states. In other words, for every electron accelerated by the field ($m^* > 0$) there is one ($m^* < 0$) that is decelerated. The average velocity of the electrons is therefore the same as in the absence of a field, namely zero. Thus a crystal

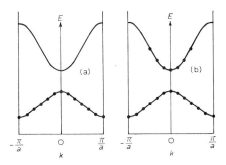

Fig. 2.5. Schematic representation of the highest fully occupied band and the one above it.
(a) Upper band empty.
(b) Upper band partially occupied.

having only bands that are completely filled or entirely empty, is a perfect insulator.

The situation is markedly different in the case of a partially filled band. This is illustrated in Fig. 2.5b where the upper band is shown half filled, the result of an odd number of electrons in each unit cell. Since unoccupied states are now available, an external field would give rise to a net shift in the electron distribution. The new distribution is characterized by a non-vanishing average velocity and therefore results in an electric current.

It will be recalled that the situation depicted by Fig. 2.5a corresponds to absolute zero, where the crystal is in its lowest energy state. Only under such conditions is the crystal truly insulating. At temperatures above zero, some electrons from the highest filled band (usually referred to as the valence band) will be thermally excited into the lowest empty band (the conduction band) and conduction becomes possible. The magnitude of the conductivity depends on the temperature and on the size of the forbidden gap. If the forbidden gap is of the order of several electron volts, the crystal will in effect remain an insulator even at room temperature. For a smaller gap, say 1 eV or less, room-temperature conduction is appreciable and one speaks of an intrinsic semiconductor (see §5.2 below). The distinction between insulators and semiconductors is thus only quantitative: an intrinsic semiconductor is an insulator at sufficiently low temperatures, while an insulator becomes conducting when the temperature is high enough.

2.3. POSITIVE HOLES

It has been mentioned above that in an intrinsic semiconductor a certain number of electrons may be thermally excited into the conduction band. In this process an equal number of vacant states is left in the valence band. Such a situation is illustrated in Fig. 2.6a. Since at thermal equilibrium electrons tend to fill the lowest available energy levels first, the conduction

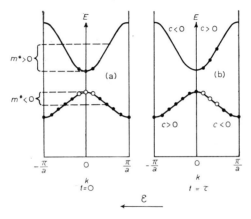

Fig. 2.6. Valence and conduction bands of a one-dimensional intrinsic semiconductor. The dots represent electrons, the circles positive holes.
(a) At thermal equilibrium, prior to the application of an electric field ($t = 0$).
(b) Under an electric field along the $-x$ direction, a short time ($t=\tau$) after its application.

electrons will occupy states near the minimum of the band while the *empty* states in the valence band will be found near the band maximum. The electrons in the conduction band behave essentially as ordinary free electrons in the sense that they are characterized by a positive mass and a negative charge. The valence electrons, on the other hand, may be described in terms of the vacant states or "holes" in the band, characterized by a positive mass and a *positive* charge. This remarkable characteristic of a nearly full band is one of the most interesting features of the band theory. The co-existence of electrons and positive holes in semiconductors is of primary importance

Ch. 2, § 2.3   BAND MODEL FOR A ONE-DIMENSIONAL LATTICE

from both the fundamental and the practical standpoints. The experimental evidence for the physical reality of holes comes from the Hall effect, from transistor physics phenomena, and from cyclotron resonance measurements, as will be discussed below.

Suppose that an electric field is applied along the $-x$ direction. As a result the k values of all electrons are shifted to the right at the *same* rate (eq. (2.35)). In this process, therefore, the holes in the valence band must move along with the electrons in the direction of increasing k. In order to gain an insight into the essential features of the conduction process, it is sufficient to consider the situation after a short time interval τ following the onset of the field (Fig. 2.6b). This situation in fact corresponds to conditions in an actual crystal, where departures from perfect lattice periodicity give rise to a scattering of the electrons (see § 6). Upon being scattered, the electrons "lose memory" of the previous action of the field, and thus the net change Δk in their k values is in effect that taking place during the average time τ between scattering processes. From (2.35) we have

$$\Delta k = +(q/\hbar)|\mathscr{E}|\tau. \qquad (2.38)$$

For the conduction band, the electron velocity c is positive (in the $+x$ direction) for $k > 0$ and negative for $k < 0$, and since $\Delta k > 0$ the velocity of the electrons increases in a direction opposite to that of the field. This is compatible with the assignment of negative charge and positive mass to the conduction electrons. For the electrons in the valence band, on the other hand, c is positive for $k < 0$ and negative for $k > 0$. Under the action of the field there are more electrons on the left side than on the right side of $k = 0$ (see Fig. 2.6b), so that the electrons are again accelerated in the $+x$ direction and negative charge is transported to the right. The holes, however, have an average negative velocity and are therefore accelerated in the direction of the field. Thus *positive* charge is transported in this direction. We shall now express the total electron current in the valence band in terms of the hole motion. Denoting the change in velocity of the electrons by Δc, we can readily see that the total current at time τ is given by $I = -q \sum_\alpha \Delta c_\alpha$, the summation extending over all *occupied* states in the valence band. From (2.36) and (2.37) it follows that

$$I = -q \sum_\alpha \Delta c_\alpha = -q \sum_\alpha (-q\mathscr{E}/m_\alpha^*)\tau = q^2 \tau \mathscr{E} \sum_\alpha (1/m_\alpha^*), \qquad (2.39)$$

where m_α^* is the effective mass corresponding to the α-th occupied state.

We recall now that in a completely filled band the current is zero. For the case being considered, we shall write this condition in the form

$$\sum_\alpha (1/m_\alpha^*) + \sum_\beta (1/m_\beta^*) = 0,$$

where the second sum is to be extended over all the *vacant* states (that is, over all the holes). Introducing this relation into (2.39), we obtain

$$I = q^2 \tau \mathscr{E} \sum_\beta (-m_\beta^*)^{-1}. \qquad (2.40)$$

Since the vacant states are near the top of the band, they are characterized by negative effective masses (see § 2.2). The current is thus just that which would arise from particles with positive charge and positive masses $|m_\beta^*|$ *occupying the vacant states*. In other words, a missing electron in a β-th state of a nearly-filled band behaves as a positive hole of mass $|m_\beta^*|$.

3. Energy bands in a three-dimensional lattice

We shall now indicate how the above considerations can be extended to an actual three-dimensional crystal. A crystal is characterized by the existence of three fundamental vectors a, b, c such that the atomic configuration looks identical when viewed from any two points separated by a vector

$$\boldsymbol{r}_{lmn} = l\boldsymbol{a} + m\boldsymbol{b} + n\boldsymbol{c}, \qquad (2.41)$$

where l, m, n are integers. Thus the physical properties remain unchanged when we make a translation defined by any vector of the type $\boldsymbol{r}_{lmn}$. In particular, the potential energy $U(\boldsymbol{r})$ is a periodic function of the position vector $\boldsymbol{r}$, so that

$$U(\boldsymbol{r} + \boldsymbol{r}_{lmn}) = U(\boldsymbol{r}). \qquad (2.42)$$

The solutions of the Schrödinger equation are now Bloch wave functions of the form

$$\psi_{\boldsymbol{k}}(\boldsymbol{r}) = u_{\boldsymbol{k}}(\boldsymbol{r}) e^{i\boldsymbol{k} \cdot \boldsymbol{r}}, \qquad (2.43)$$

where $\boldsymbol{k}$ is the wave vector characterizing each state and $u_{\boldsymbol{k}}(\boldsymbol{r})$ has the periodicity of the lattice. As before, the allowed energy levels of an electron fall into bands usually separated by forbidden gaps. Within each band the energy is a periodic function of $\boldsymbol{k}$ so that one can introduce a reduced wave vector notation to describe all states in the crystal. The first Brillouin zone is now defined as the smallest *volume* in **k**-space centred at the origin ($k = 0$) that includes all the non-equivalent values of the vector **k**. The number of states in each band is again equal to the number of unit cells in the crystal. The shape of the zone depends on the crystal structure. In a simple cubic crystal (**a**, **b**, **c** orthogonal, and $a = b = c$), for example, the Brillouin zone is a cube, the pertinent values of k_x, k_y, k_z lying between $-\pi/a$ and $+\pi/a$.

3.1. MOTION OF ELECTRONS AND HOLES

We can derive the equations of motion of the electrons in a crystal as before, keeping in mind that the wave number, velocity, and electric field are now vector quantities. Thus, the velocity is given by

$$c = (1/\hbar)\, \text{grad}_k\, E(k) \tag{2.44}$$

and is constant in the absence of external forces. In the presence of an electric field $\mathscr{E}$, the power delivered to an electron is

$$\frac{dE(k)}{dt} = (\text{grad}_k E) \cdot (dk/dt) = -qc \cdot \mathscr{E}; \tag{2.45}$$

and using (2.44), we have

$$(dk/dt) = -(q/\hbar)\mathscr{E}. \tag{2.46}$$

In order to evaluate the acceleration, we differentiate one of the components of (2.44) with respect to time and make use of (2.46). For the x component we get

$$a_x = \frac{dc_x}{dt} = \frac{1}{\hbar}\frac{d}{dt}\left(\frac{\partial E}{\partial k_x}\right) = -\frac{q}{\hbar^2}\left[\frac{\partial^2 E}{\partial k_x^2}\mathscr{E}_x + \frac{\partial^2 E}{\partial k_x \partial k_y}\mathscr{E}_y + \frac{\partial^2 E}{\partial k_x \partial k_z}\mathscr{E}_z\right].$$

By combining this equation with the corresponding ones for a_y and a_z, we can write the vector equation

$$\mathbf{a} = \left(\frac{1}{m_{\eta\zeta}}\right)(-q\mathscr{E}). \tag{2.47}$$

Here $m_{\eta\zeta}$ is the effective mass tensor given by

$$\left(\frac{1}{m_{\eta\zeta}}\right) = \frac{1}{\hbar^2}\left(\frac{\partial^2 E}{\partial k_\eta \partial k_\zeta}\right), \tag{2.48}$$

with η and ζ assuming the values x, y, z. For a coordinate system along the principal axes of the effective-mass tensor (usually the symmetry directions of the crystal), the non-diagonal terms vanish and (2.47) reduces to

$$a_x = -(q/m_x)\mathscr{E}_x; \quad a_y = -(q/m_y)\mathscr{E}_y; \quad a_z = -(q/m_z)\mathscr{E}_z, \qquad (2.49)$$

where m_x, m_y, m_z are the effective masses along the three principal axes:

$$\frac{1}{m_x} = \frac{1}{\hbar^2}\frac{\partial^2 E}{\partial k_x^2}; \quad \frac{1}{m_y} = \frac{1}{\hbar^2}\frac{\partial^2 E}{\partial k_y^2}; \quad \frac{1}{m_z} = \frac{1}{\hbar^2}\frac{\partial^2 E}{\partial k_z^2}. \qquad (2.50)$$

The acceleration in a three-dimensional crystal is in general no longer in the direction of the field, except when the field coincides with one of the principal axes. If the three masses are equal, the mass tensor obviously reduces to a scalar, as for the one-dimensional case.

The current set up by an external field can be expressed in terms of the acceleration, as before. Electrons at the bottom of the conduction band and holes at the top of the valence band are characterized by negative and positive charge, respectively, but they are now generally associated with effective-mass *tensors* (having positive components).

3.2. SHAPE OF THE ENERGY BANDS

The actual calculation of the function $E(\mathbf{k})$ even for the simplest crystal is a formidable problem, and only in a few cases has it been carried out in any detail. Usually the band structure is determined by experimental methods, theoretical considerations serving as a general guide. Since $E(\mathbf{k})$ is a function of three variables, it is rather difficult to represent the shape of the energy bands pictorially. It is customary to draw several diagrams of E as a function of $\mathbf{k}$ along the principal directions of the crystal. This procedure is illustrated in Fig. 2.7a. Here the shape of an energy band near its minimum (taken as zero) is shown, and for simplicity the minimum is assumed to lie at the centre of the Brillouin zone. If x, y, z are along the principal axes of the crystal (assumed orthogonal), we can expand $E(k_x, k_y, k_z)$ about the minimum and for small values of k neglect higher order terms. Thus

$$E = \frac{1}{2}\left[\frac{\partial^2 E}{\partial k_x^2}k_x^2 + \frac{\partial^2 E}{\partial k_y^2}k_y^2 + \frac{\partial^2 E}{\partial k_z^2}k_z^2\right], \qquad (2.51)$$

since the first derivatives vanish at an extremum. Using (2.50) we can write

$$E = \frac{\hbar^2}{2}\left(\frac{k_x^2}{m_x} + \frac{k_y^2}{m_y} + \frac{k_z^2}{m_z}\right), \qquad (2.52)$$

where m_x, m_y, m_z are the diagonal components of the effective-mass tensor corresponding to an electron at the minimum ($E = 0$). In Fig. 2.7a the variation of E along each of the principal axes is shown. Another way of displaying the shape of the band is to plot constant-energy contours in appropriate planes in **k**-space. It is seen from (2.52) that the constant-energy surfaces near the minimum are ellipsoids whose intersection with the $k_x - k_y$, $k_y - k_z$, and $k_z - k_x$ planes are the ellipses shown in Fig. 2.7b.

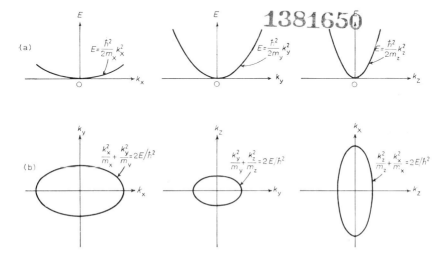

Fig. 2.7. Shape of an energy band near the minimum energy (taken as zero).
(a) Variation of the energy $E(\mathbf{k})$ along the principal axes of the crystal (assumed orthogonal).
(b) Constant-energy contours in the principal planes of the crystal.

Similar conditions hold near the maximum of a band. The derivatives in (2.51) are now obviously negative, corresponding to negative components of the mass tensor. Thus, when the maximum is at $\mathbf{k} = 0$ we may write (taking the zero of energy at the maximum):

$$E = -\frac{\hbar^2}{2}\left(\frac{k_x^2}{|m_x|} + \frac{k_y^2}{|m_y|} + \frac{k_z^2}{|m_z|}\right), \qquad (2.53)$$

and the constant-energy surfaces near the maximum energy are again ellipsoids.

We have discussed conditions near the band edges, where the situation is fairly simple. These regions (the minimum energy for the conduction band and the maximum energy for the valence band) are where the few carriers of a semiconductor will lie, and are therefore frequently of the most interest. In order to determine the carrier densities (see §5), it is necessary to know the density of allowed states in these regions. We shall evaluate this for the simple case in which $m_x = m_y = m_z = m^*$, corresponding to so-called spherical constant-energy surfaces. Equation (2.52) now reduces to

$$E = (\hbar^2/2m^*)k^2, \tag{2.54}$$

and we can adapt the relations derived for free electrons (§1.1) merely by writing m^* for m. From (2.14) it follows that the number of states for which the wave vector has a value in the element $dk_x dk_y dk_z$ of $\mathbf{k}$-space is $(L_x L_y L_z/8\pi^3)dk_x dk_y dk_z$. The *density* of states per unit volume of the crystal (including the two possibilities for the electron spin) is

$$N(k)dk_x dk_y dk_z = \tfrac{1}{4}\pi^{-3} dk_x dk_y dk_z, \tag{2.55}$$

and the density for which the *magnitude* of $\mathbf{k}$ lies between k and $k+dk$ is

$$N(k)dk = \pi^{-2} k^2 dk. \tag{2.56}$$

Combining (2.54) and (2.56), we obtain for the density of states lying in the energy interval dE

$$N(E)dE = (4\pi/h^3)(2m^*)^{\frac{3}{2}} E^{\frac{1}{2}} dE. \tag{2.57}$$

It can be shown that for non-spherical energy surfaces the scalar mass m^* should be replaced by $(m_x m_y m_z)^{\frac{1}{3}}$:

$$N(E)dE = (4\pi/h^3)(8m_x m_y m_z)^{\frac{1}{2}} E^{\frac{1}{2}} dE. \tag{2.58}$$

The same relation holds for holes at the top of the valence band.

In many crystals the extrema of the bands are not at $\mathbf{k} = 0$ but at some other point (k_{x1}, k_{y1}, k_{z1}) in $\mathbf{k}$-space. Equation (2.52) (or eq. (2.53)) should then be replaced by

$$E = \frac{\hbar^2}{2}\left[\frac{(k_x-k_{x1})^2}{m_x} + \frac{(k_y-k_{y1})^2}{m_y} + \frac{(k_z-k_{z1})^2}{m_z}\right]. \tag{2.59}$$

In this case there are often a number of equivalent minima at different points in $\mathbf{k}$-space, in accordance with the requirements of the crystal symmetry.

For many purposes one is not particulary concerned with the shape of the energy bands as a function of $\mathbf{k}$. One can then conveniently represent the energy levels in a semiconductor by the type of diagram shown in Fig. 2.8, where electron energies near the top of the valence band and the bottom of the conduction band are plotted against distance through the crystal. The energy gap (also referred to as the forbidden gap) represents the minimum energetic distance between the two bands. In this diagram *electron* energy increases upwards and the potential, defined as the potential energy of a

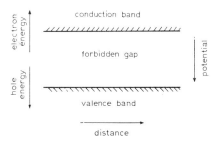

Fig. 2.8. Energy-level diagram for a homogeneous semiconductor. Electron energy increases upwards, hole energy and potential increase downwards.

positive unit charge, increases downwards. The hole energy likewise increases downwards (see § 2.3). For a homogeneous crystal the bands are horizontal, as shown in the figure. In an inhomogeneous crystal or near the surface, the band edges vary with position, as will be discussed below.

3.3. BAND STRUCTURE OF GERMANIUM AND SILICON

Germanium and silicon are in the fourth column of the periodic table and each has four valence electrons, two in s states and two in p states. In the crystalline state every atom forms covalent bonds with four neighbours, each bond consisting of a pair of electrons of opposite spin. The diamond-type lattice to which germanium and silicon belong is shown in Fig. 2.9a. Each atom is surrounded by four nearest-neighbours at the corners of a regular tetrahedron with the atom itself at the centre. The figure represents a unit cell in the sense that it is repeated indefinitely to make up the crystal lattice. Many other choices for such a unit cell are possible, this particular one having the advantage of displaying the cubic symmetry of the diamond lattice. Such a cell, however, is not primitive because it is not the *smallest* parallelepiped which, when repeated, can generate the entire crystal. The primitive cell is a slanting parallelepiped containing two atoms and having a volume one fourth of the cube shown. It is the primitive cell that determines the shape of the Brillouin zone in k-space. The eight valence electrons of the two germanium or silicon atoms in the cell can fill up four energy bands (using both spins).

Let us represent the three-dimensional array of Fig. 2.9a in two dimensions, as shown in Fig. 2.9b. The double links between atoms indicate

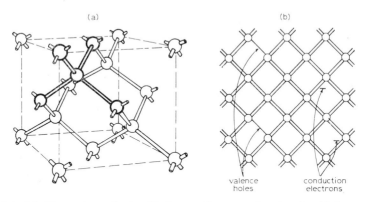

Fig. 2.9. (a) Diamond-type lattice. Each atom forms covalent bonds with four nearest neighbours at the corners of a regular tetrahedron.
(b) Two-dimensional representation of the diamond-type lattice.

the electron-pair bonds. At zero temperature all the valence electrons are used up in forming the covalent bond and are therefore unable to contribute to the conduction process. At temperatures above zero, some of the bonds are broken up: due to thermal agitation, a valence electron may be excited to a higher energy state, leaving a deficit in the valence bond (Fig. 2.9b). Both the excited electrons and the defect sites may move about independently in the crystal, and correspond to electrons in the conduction band and to holes in the valence band. In this simple picture, the forbidden gap represents the energy required to break a bond — that is, to remove a valence electron from one atom and bring it to some other point in the lattice where it is very loosely bound to another atom having saturated bonds.

After this pictorial description of the valence and conduction bands in germanium and silicon, we now turn to their actual structure as determined by a combination of experimental and theoretical work. The two bands are shown rather schematically in Fig. 2.10. The energy plot in Fig. 2.10a is for k in the [100] crystallographic direction in the case of silicon, [111] in the case of germanium. Consider first the conduction band. The fact that the minimum of energy is not at the centre of the Brillouin zone requires the existence of two equivalent minima along k_x ([100] in Si, [111] in Ge) equally spaced on opposite sides of $k_x = 0$. In the directions chosen, these represent the actual minimum energy in the conduction bands. The symmetry of the crystals requires the occurrence of additional equivalent minima along other symmetry axes. Thus the conduction-band edge is multiple, and the electron population must be equally divided among the various minima (or valleys, as they are commonly known). In silicon there are six such valleys,

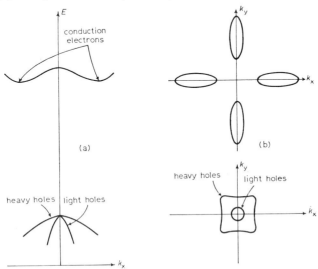

Fig. 2.10. Schematic representation of the valence- and conduction-band structure in germanium and silicon.
(a) E versus k along the [100] crystallographic directions for silicon and along the [111] directions for germanium.
(b) Constant-energy contours in the k_x–k_y plane.

two along each of the [100] crystallographic directions and centred at 0.8 of the distance from $k = 0$ to the edge of the Brillouin zone along these directions. In germanium, eight valleys would be expected, corresponding to the four equivalent [111] directions. Actually there are only four, since the minima occur at the edges of the Brillouin zone so that each minimum can contribute only half a valley.

The constant-energy ellipsoids along two principal directions are shown in Fig. 2.10b. (Actually, in the case of germanium the axes are not perpendicular.) In both substances these are ellipsoids of revolution, so that the energy in each valley can be expressed as

$$E = \frac{\hbar^2}{2}\left(\frac{k_l^2}{m_l} + \frac{k_t^2}{m_t}\right), \qquad (2.60)$$

where the subscripts l and t represent the longitudinal and transverse components of the mass and the wave vector (that is, the components parallel and perpendicular to the major axis of a constant-energy ellipsoid). If we take the x axis along the longitudinal direction, then $k_l = k_x - k_{x1}$ and $k_t^2 = k_y^2 + k_z^2$.

The structure of the valence bands of germanium and silicon is more complicated. In either case the maximum energy is at the centre of the Brillouin zone, but now there are two bands touching at this point, as shown schematically in Fig. 2.10a. This degeneracy arises from the fact that the wave function corresponding to $k = 0$ (which consists of only the periodic, modulating part $u_0(r)$) has within each cell the character of an atomic p state. As such, this state is triply degenerate (apart from spin), giving rise to three bands touching at $k = 0$. Spin-orbit interaction reduces the degeneracy by "pushing down" one of the bands (not shown). This leaves two bands touching at $k = 0$, and the corresponding state is quadruply degenerate. Due to this degeneracy, a simple expansion of E in even powers of k_x, k_y, k_z is no longer possible (even for small k), and a more complex expression must be used. The constant-energy surfaces are now "warped" spheres, as shown in Fig. 2.10b. The anisotropy, however, is not very great, and to a good approximation we can assign scalar masses to the holes in the two bands, as though they were spherical. The band with the greater curvature has of course the smaller mass, and one speaks of heavy and light holes. At normal temperatures the light holes constitute about 5% of the total hole population in germanium and about 20% in silicon.

The effective-mass tensor near the band edges can be determined from many types of measurements. Of these, cyclotron resonance experiments provide the most direct method, and we shall outline the underlying principles. Suppose, for simplicity, that the effective mass of the electron is a scalar m^*. Such an electron, moving in a magnetic field, will rotate about an axis parallel to the field at the cyclotron frequency ω_0. This frequency is determined by balancing the centrifugal force $m^* c^2/r$ and the Lorentz force qcB:

$$\omega_0 = c/r = (q/m^*)B. \tag{2.61}$$

If now a radio-frequency electric field is applied perpendicular to the magnetic field B, the electron will execute a circular oscillatory motion at the frequency of the applied field ω in addition to its rotational motion at the frequency ω_0. If $\omega = \omega_0$, the electron spirals in an increasing orbit and thus continuously absorbs energy. Actually, this process is hampered by scattering, and the absorbed energy is in effect only that gained during the average time τ between scattering processes. If $\omega_0 \gtrsim 1/\tau$, as can be attained, for example, in high-purity germanium and silicon at liquid-helium temperature, this type of interaction gives rise to a detectable resonance absorption of power from the rf field and so can be used for the determination of m^*. In the general case, the components of the effective-mass tensor can be measured by changing the orientation of the magnetic field with respect to the crystal axes.

TABLE 2.1

Values of some parameters characterizing the band structure of germanium and silicon

Quantity	Value	
	Germanium	Silicon
Number and positions of valleys in conduction band	4 minima, [111]	6 minima, [100]
Longitudinal mass for conduction electrons	1.6m	0.97m
Transverse mass for conduction electrons	0.082m	0.19m
Mass of heavy holes (scalar)	$\approx 0.3m$	$\approx 0.5m$
Mass of light holes (scalar)	$\approx 0.04m$	$\approx 0.16m$
Energy gap ($T = 300°$K)	0.67 eV	1.08 eV
Separation of band edges at $k = 0$ ($T = 300$ °K)	0.803 eV	≈ 3.6 eV

The effective-mass components for electrons and holes in germanium and silicon are given in Table 2.1. Also shown are the energy gap (the minimum energetic distance between the valence and conduction bands) and the separation between the band edges at $k = 0$ (the optical gap, as discussed below).

3.4. OPTICAL EXCITATION

At long wavelengths semiconductors are fairly transparent, but as the wavelength decreases a certain threshold is reached, generally in the near or intermediate infrared, where the absorption coefficient rises rapidly and the material becomes opaque (absorption coefficient $\approx 10^5$ cm^{-1}). This threshold, referred to as the fundamental absorption edge, corresponds to the onset of optical excitation of electrons from the valence into the conduction band.

Optical excitation may be considered as a collision between the impinging photons and the electrons in the crystal. In a perfect crystal the transition probability for such a process is zero unless both the momentum and the energy of the colliding particles are conserved. For infrared or visible radiation, the photon momentum h/λ is generally small compared to the electron momentum $\hbar k$. This follows from the fact that the latter is of order h/a, the lattice constant a being much smaller than the wavelength λ of the incident radiation. Hence in an optical transition of this kind the electron wave vector k remains essentially unaltered; in other words, referring to the energy band diagrams in Fig. 2.10a, we may say that transitions can occur only vertically. From the law of energy conservation we can further infer that at any point in k space the photon energy involved in optical excitation must be equal to the *vertical* separation between the bands at that point.

The threshold for optical absorption should thus correspond to the minimum *vertical* distance between the valence and conduction band edges, sometimes referred to as the optical gap. In Ge and Si the optical gap occurs at $k = 0$. It is larger than the energy gap since the maximum of the valence band and the minimum of the conduction band are at different points in k space (see, for example, Fig. 2.10a). This is usually the case, but a few exceptions (such as GaAs) are known where the two gaps are equal.

Experimentally it is found that while a marked change in the absorption coefficient does occur at wavelengths corresponding to the optical gap, absorption is still appreciable at longer wavelengths. The absorption tail extends up to photon energies approximately equal to the energy gap and is more pronounced at higher temperatures. This behaviour indicates that the momentum and energy-conservation laws partially break down in a real crystal and non-vertical or *indirect* transitions become possible. Indeed, because of the interaction of the electrons with the lattice vibrations present in a real crystal (see §§ 4.1, 6.1 below), we have what might be termed a three-body collision: an electron, a photon, and a "phonon", the last being a quantum of vibrational energy. The transition probability for this process increases with the number of phonons or with the extent of excitation of the lattice vibrations — that is, with temperature. The threshold of absorption now corresponds to a phonon-aided transition across the energy gap, the phonon taking up the necessary change in electron momentum without contributing much to the electron energy.

Optical absorption in the infrared, beyond the threshold for excitation across the gap, may take place by free carriers within a single band. In p-type Ge or Si, transitions may also occur among the three energy bands near the top of the valence band. Such absorption will be appreciable only if the electron or hole concentration is sufficiently large.

4. Lattice vibrations, impurities, and defects

Up to now the crystal was assumed to be a perfect periodic structure of atoms at rest. In the present section we review the main types of deviations from ideal conditions encountered in real crystals: thermal vibrations of the atoms, foreign impurities, and structural defects in the crystal lattice. All of these prevent the electrons from maintaining a steady motion as quasi-free particles and are responsible (to varying degrees) for the finite resistance of solids to the flow of current. The effect of the deviations from ideal conditions on electron transport processes will be considered in § 6. Certain impurities and defects introduce localized states having energy levels within the forbidden gap. Such states play a dominant role in controlling the equilibrium electron and hole densities in semiconductors, as will be discussed in § 5. They are also of paramount importance in recombination and trapping processes under non-equilibrium conditions (see § 7). Many other diverse properties — mechanical, optical, chemical — are also greatly influenced by the presence of imperfections, but these will not be considered here.

4.1. LATTICE VIBRATIONS

Due to thermal agitation, the atoms in a crystal execute oscillations about their mean equilibrium positions. The vibrations of the individual atoms are coupled by the strong forces that maintain the lattice. This coupling gives rise to wave propagation in the crystal, and the motion of the atoms can be expressed as

$$D = A \exp i(K \cdot r_{lmn} - \omega t), \quad (2.62)$$

where D is the displacement of an atom having an equilibrium position r_{lmn} and an angular frequency ω. The wave vector is denoted by K and is related to the wavelength by the same type of expression as (2.6). By the use of appropriate boundary conditions, one finds that only certain values of K are allowed, those corresponding to the normal modes. Due to the periodicity of the lattice, we can employ (as in the case of electrons) the reduced wave-vector notation in which all non-equivalent K values are restricted to the first Brillouin zone. The corresponding wavelength ranges from a minimum of twice the lattice constant to a maximum of the order of the size of the crystal. The number of K values is again equal to the number of primitive unit cells in the crystal. The relation between ω and K is determined by the crystal structure and its elastic constants. The general shape of the ω versus **K** curves for lattice vibrations is similar to that of the E versus **k** curves for electrons, the allowed frequencies falling into bands or branches usually separated by forbidden zones. In contrast to the case of electrons in a crystal, however, the number of allowed frequency bands is finite. It is equal to three times the number of atoms per unit cell. This arises from the three possible modes of vibration of the atoms: one longitudinal and two mutually perpendicular transverse modes. In the longitudinal mode the atoms vibrate along the direction of wave propagation and in the transverse modes, perpendicular to it. Thus the total number of modes is equal to three times the number of atoms, or to the total number of degrees of freedom of the system, as expected.

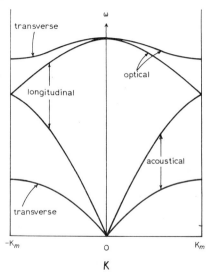

Fig. 2.11. Spectrum of lattice vibrations in germanium. The frequency for the various branches is plotted against K along the [100] directions.

As before, the details of the frequency spectrum can be conveniently displayed by plotting ω as a function of **K** along the principal crystallographic directions. This is illustrated in Fig. 2.11, which is a schematic representation of the various frequency branches in germanium and silicon along the [100] directions. In either crystal there are two atoms per unit cell, so that we should expect altogether six branches. Actually, due to the isotropy of the crystal there are only four, the two transverse branches being each doubly degenerate. For long wavelengths (small K), the vibrations are seen to fall into two distinct classes. In one class, consisting of the two bottom branches, ω is approximately proportional to K. Thus both the group velocity ($d\omega/dK$) and the phase velocity (ω/K) tend to a constant which is the sound velocity in the crystal. The two atoms in each unit cell move approximately in phase. These branches correspond to ordinary sound waves and are referred to as the longitudinal and transverse acoustical branches. The situation is altogether different for the upper longitudinal and transverse branches. Here ω does not go to zero with increasing wavelength, as for the acoustical branches, but on the contrary tends to the maximum value in the spectrum. In this range the two atoms in each unit cell move in approximately *opposite* directions, with the same motion being repeated in the next cell in slightly different phase. Such oscillations are characterized by high frequencies because of the large restoring forces brought into play by the opposing motion of the two atoms in the unit cell. The frequency is then insensitive to K, since the wavelength determines only the relative phase of adjacent cells. This type of vibration is called an optical mode, and the branches are referred to as the longitudinal and transverse optical branches. The term "optical" originates in the fact that if the two atoms in the unit cell are oppositely charged, as is the case in ionic crystals, then the vibrations give rise to oscillating electric dipole moments which can interact with electromagnetic radiation. Due to the small momentum associated with the light waves (see §3.4) the coupling is strong only for vibrations of small K, corresponding to the highest frequencies in the spectrum. In ionic crystals such as NaCl these frequencies generally lie in the far infrared and give rise to the well known absorption in that spectral range. The distinction between the acoustical and optical modes of vibration becomes less sharp as we move towards small wavelengths (large K).

The detailed shape of the various branches can be obtained from a combination of different experiments. These include measurements of sound velocity, cold-neutron diffraction, optical absorption, and recombination radiation.

The vibrational energy contained in each normal mode K of frequency ω_K is quantized, being given by the allowed energy levels of a harmonic oscillator of the same frequency:

$$E(\omega_K) = (n+\tfrac{1}{2})\hbar\omega_K, \qquad n = 0, 1, 2, \ldots. \tag{2.63}$$

The *average* energy associated with each mode at thermal equilibrium is given by Planck's law,

$$\langle E(\omega_K) \rangle = \frac{\hbar\omega_K}{\exp(\hbar\omega_K/kT) - 1}, \qquad (2.64)$$

where k is Boltzmann's constant and T the lattice temperature.

Each mode K can absorb or emit energy only in units of $\hbar\omega_K$. These units are called "phonons" (in analogy with photons of light). The phonon concept is very convenient in describing the interaction of electrons with lattice vibrations, as will be discussed below (§ 6.1). From (2.64) we obtain that the number of phonons at thermal equilibrium in a vibrational mode of frequency ω_K is

$$n_K = \frac{1}{\exp(\hbar\omega_K/kT) - 1}. \qquad (2.65)$$

As can be seen from Fig. 2.11, the number of acoustical phonons (low ω) at any temperature is considerably larger than the number of optical phonons. In germanium, for example, the optical phonons have a frequency of the order of 5×10^{13} sec^{-1}, corresponding to a characteristic temperature ($\hbar\omega/k$) of about 400°K. At room temperature their number in any mode will accordingly be small. For the acoustical branches, on the other hand, $\hbar\omega \ll kT$ and n_K may be approximated by $kT/\hbar\omega$ (see eq. (2.65)). Thus the acoustical phonons will be numerous, especially at the long wavelength limit.

4.2. IMPURITIES

No substance can be made perfectly pure, and even in germanium in its purest form the impurity content is not less than one part in 10^{12}, corresponding to about 10^{11} foreign atoms per cubic centimetre. For many purposes impurities are deliberately introduced into the crystal so as to obtain desired electrical properties. This can be done either in the process of growing the crystal, or by diffusion of impurities into the crystal at elevated temperatures and subsequent cooling (quenching) to "freeze" them in the lattice.

There are two ways in which an impurity atom may enter an otherwise perfect lattice. It can replace an atom of the host crystal, in which case it is called a substitutional impurity, or it can occupy a position between lattice sites — an interstitial impurity. The simplest and best understood impurities are atoms from groups III and V of the periodic table in the group IV semiconductors. Such impurities enter the host lattice substitutionally and are able to form covalent bonds with nearest-neighbour host atoms. Consider first the group V impurities, say arsenic, in germanium. An atom of As introduces into the lattice five valence electrons of which only four are required for the covalent bonds with the neighbouring Ge atoms. The fifth electron is bound to the As$^+$ ion by electrostatic forces. As we shall see in a moment, the binding energy is so small that at room temperature practically all the As atoms lose their extra electrons, which are then free to move about in the crystal. In this condition the As atoms are said to be ionized. Only at very low temperatures are the extra electrons bound in any appreciable numbers to their parent As atoms. In terms of the band picture, we may say that the As atoms introduce localized energy levels just below the bottom of the conduction band. Electrons may be thermally excited out of the levels into the conduction band. Such impurities are therefore referred to as *donor* impurities, and the corresponding levels as donor levels.

Consider next a group III impurity, such as gallium, which has only three valence electrons. In order to complete its four covalent bonds with nearest-neighbour Ge atoms, it tends to pick up an electron from one of the Ge–Ge bonds. This it can do easily at all but very low temperatures, leaving behind a hole which can wander freely throughout the lattice. In terms of the band picture, Ga atoms introduce *acceptor* levels just above the valence-band edge, into which electrons may be excited from the valence band. Alternatively, we may also say that at low temperatures a positive hole is electrostatically bound to the negative Ga ion and is released into the valence band at higher temperatures. The cases of ionized and un-ionized donor and acceptor impurities are illustrated in Fig. 2.12, based on the valence-bond model of Fig. 2.9b. Also shown are the corresponding cases of vacant and occupied donor and acceptor levels in terms of the band picture.

We shall now estimate the binding energy of an electron to a donor. Such an electron is moving under the Coulomb potential arising from a single positive charge located at the ionized donor. The situation is thus similar to that in a hydrogen atom. There are, however, two points of difference. First, the Coulomb force acting on the electron is reduced by a factor $1/\kappa$ where κ is the dielectric constant of the host crystal. This in turn increases the radius of the orbit by a factor κ, so that the binding energy is reduced by κ^2. Secondly, the mass of the electron is different from that of a free electron. If we represent the effective mass tensor of a conduction electron by some sort of average scalar mass m_n, the problem reduces to that of a hydrogen atom having an effective atomic number $Z = 1/\kappa$ and an electron of mass m_n. We then obtain for the allowed energy levels the hydrogen-atom levels scaled down by the factor $(1/\kappa)^2(m_n/m)$. In particular, the ionization energy of hydrogen, 13.6 eV, corresponds to a binding energy of an electron to a donor of $13.6(1/\kappa)^2(m_n/m)$eV. In germanium $\kappa = 16$, $m_n \approx 0.2m$, and so the binding energy is about 0.01 eV. The justification in employing the macroscopic dielectric constant in deriving this value follows from the fact that the first Bohr radius for an electron bound to a donor is κ times as large as in the case of hydrogen. In germanium this radius is 8.5 A, more than three times the distance between nearest neighbours.

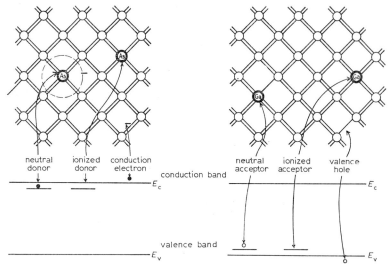

Fig. 2.12. Ionized and un-ionized donor and acceptor impurities. Bottom part shows the corresponding energy-level diagram in terms of the band picture.

Even though the model employed is rather crude, the agreement with experiment is surprisingly good, the donor levels lying very nearly 0.01 eV below the conduction-band edge (0.012, 0.013, 0.0096 eV for P, As, Sb, respectively). The donor levels in silicon are deeper (0.044, 0.049, 0.039 eV for P, As, Sb, respectively) because of the smaller value of the dielectric constant ($\kappa = 11.6$).

Exactly similar arguments can be used to estimate the binding energy of a hole moving in the field of a group III negatively-charged acceptor. Such an estimate gives much the same value for the binding energy as in the case of donors, and the agreement with experiment is again good, with the exception of indium in silicon (0.010, 0.011, 0.011 eV in Ge; 0.045, 0.065, 0.16 eV in Si — for B, Ga, In, respectively).

Ch. 2, § 4.2   LATTICE VIBRATIONS, IMPURITIES, AND DEFECTS   49

The differences in ionization energy between different impurities of the same group in the periodic table arise from the fact that even though the orbits of the bound electron or hole are on the average large, the bound carrier spends some time in the immediate neighbourhood of the impurity ion and is therefore affected to some extent by its properties. This effect is much more pronounced for impurities other than those of group III or V. Such impurities have more than one excess or defect electron, and therefore when they enter the lattice substitutionally they tend to be in a state of higher ionization. This greater net charge of the ionized impurity results in a smaller orbit for a bound electron or hole, which in turn has two consequences: larger binding energies (that is, deeper energy levels) and a greater difference between different impurities of the same column of the periodic table. Unlike the case of the group III or V impurities, there is no simple means of estimating the position of these deeper-lying levels. It is generally possible, however, to predict on the basis of the valence-bond model discussed above whether a given impurity is a donor or an acceptor and also, in each case, the number of levels introduced into the forbidden gap.

A group II substitutional impurity, such as zinc in germanium, tends to pick up two electrons from the valence band in order to complete its four covalent bonds with nearest-neighbour Ge atoms. It is thus a double acceptor, introducing two acceptor levels. In the neutral state of Zn, there exists one level into which an electron can fall. Once this level is occupied, a second electron can fall into another, higher level. By the same arguments, we can infer that group I substitutional impurities should be triple acceptors. These predictions are generally borne out by experiment. For example, Zn in germanium introduces two acceptor levels (at 0.03 and 0.09 eV above the top of the valence band E_v), while Cu is a triple acceptor (two levels at 0.04 and 0.32 eV above E_v, a third at 0.26 eV below the bottom of the conduction band E_c). Gold is the only known exception to this rule; in addition to the three expected acceptor levels (one at 0.15 eV above E_v, two at 0.20 and 0.04 eV below E_c), a fourth *donor* level (0.05 eV above E_v) is also found.

For donor impurities we should look to the other side of the periodic table. Apart from the group V elements, however, very few donor impurities have been found experimentally.

Most of the impurities whose levels have been observed experimentally enter the host lattice substitutionally, as has just been discussed. Quite a few impurities do not seem to introduce any levels in the forbidden gap. Such impurities may occur as interstitial atoms, which usually have ionization energies in the crystal comparable to or greater than the forbidden gap. Hydrogen in Ge and Si may be cited as such an example. Due to the small size of the H atom, the macroscopic dielectric constant does not effect a sufficient reduction in the ionization energy. The electrical inactivity of oxygen in Ge and Si may also be due to its interstitial position. There is one case of an interstitial impurity, however, which does give rise to a donor level: the lithium atom. By releasing its one valence electron, Li attains the noble gas configuration. The ionization energy is 0.0093 eV in Ge and 0.033 eV in Si, and is thus comparable to that of the group V impurities. The reason why Li does provide a donor level while H does not is probably due to the larger size of the former which, as a result, experiences to a greater extent the high dielectric constant of the host crystal.

The energy levels discussed above are characteristic of small concentrations of impurities. At high concentrations these levels broaden into bands. We shall not be concerned in this book with such situations.

4.3. LATTICE DEFECTS

To conclude this section on deviations from ideal conditions in a crystal, we shall mention briefly the various types of structural defects arising from disorder in the atomic arrangement in the lattice. Compared to the case of impurity levels, considerably less is known about the detailed properties of the electronic levels introduced by various defects in semiconductors.

The simplest structural defects are point defects — imperfections centred about single points of the crystal. These may arise from vacant atom sites (vacancies) or from atoms occupying positions in between lattice sites (interstitials). When vacancies and interstitials occur in pairs they are referred to as Frenkel defects. Such a pair is formed when an atom is displaced from its normal position into a near-by interstice, leaving a vacancy behind. Vacancies may also occur unpaired with interstitials when the latter migrate to the surface. These are called Schottky defects.

Vacancies and interstitials occur naturally in crystals due to the constant thermal agitation, their equilibrium concentrations increasing exponentially with temperature. They can be introduced in higher than equilibrium concentrations by heating and subsequent quenching. In this manner excess defects are frozen in. Conversely, the concentration of excess defects can be reduced by annealing: heating for prolonged periods at moderate temperatures, at which the mobility of the defects in the lattice is sufficient to restore equilibrium conditions. Excess vacancies and interstitials can also be introduced by plastic deformation and by bombardment of the crystal by nuclear particles.

Considerations similar to those employed with regard to impurities indicate that vacancies and interstitials in group IV semiconductors should act as acceptors and donors, respectively. A vacancy is surrounded by the four unpaired electrons ordinarily used to bind the missing atom to its neighbours. One, or possibly two electrons could be bound into the vacancy at energies lying in the forbidden gap. An interstitial atom has four valence electrons, one or two of which may be excited into the conduction band. This general picture has been confirmed by studies of neutron bombardment of germanium.

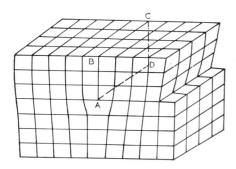 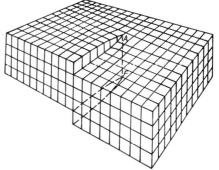

Fig. 2.13. Schematic diagram of an edge dislocation. ABCD is the extra plane in the lattice while AD is the dislocation line. (After Hobstetter, reference 17.)

Fig. 2.14. Schematic diagram of a screw dislocation. AF represents the dislocation line. (After Read, reference 18.)

A second class of imperfections consists of dislocations or line defects. Here the imperfection centres along a line (usually a straight line) in the crystal. When a crystal is subject to stress it may yield by having some planes slip over one another. The dislocation line marks the boundary of a region in the crystal that has slipped with respect to the rest of the crystal. When the direction of slip is at right angles to the line, as illustrated in Fig. 2.13, the defect is called an edge dislocation. The effect of the dislocation is to introduce an extra plane (ABCD) into the crystal. The crystal then has its usual ordered structure everywhere except in the vicinity of the dislocation line (AD). When the direction of slip is parallel to the dislocation line, as shown in Fig. 2.14, we have a screw dislocation. Here again the departure from the regular structure is confined to the vicinity of the dislocation line (AF).

Unless special care is taken, dislocations are produced during the crystal growth. Semiconductor crystals commonly contain about 10^4 dislocation lines per square centimetre of cross section, but some crystals have been grown that contain only a few. Dislocations can be produced in large numbers by plastic deformation. It has been shown that in germanium they introduce deep-lying acceptor levels.

5. Occupation statistics for semiconductors

In the preceding sections we discussed the various energy states that electrons in a semiconductor may occupy. We shall now consider the actual distribution of electrons among these states under conditions of thermal equilibrium. The number of electrons in any small energy interval at a fixed temperature depends on two factors: the Fermi–Dirac distribution function, which represents the probability of the electron having this energy, and the number of available states in the given energy interval.

5.1. THE FERMI–DIRAC DISTRIBUTION LAW

Consider first the quasi-free electrons in the allowed energy bands. The most probable energy distribution of an ensemble of this sort, being subject to the Pauli exclusion principle, is governed by Fermi–Dirac statistics. The probability that an energy level E be occupied at thermal equilibrium by an electron is given by the Fermi–Dirac distribution function

$$f_\mathrm{n}(E) = \frac{1}{1+\exp\left[(E-E_\mathrm{F})/kT\right]}, \qquad (2.66)$$

where E_F, referred to as the *Fermi level*, is determined by the requirement that the total expectation number of electrons be equal to the actual number of electrons involved. For $E = E_\mathrm{F}$, $f = \frac{1}{2}$, there being an equal probability that the level be occupied or vacant. The Fermi level is closely related to the thermodynamic or chemical potential and is a constant (at a given temperature) throughout a system in thermal equilibrium, even when the system is composed of different phases (such as occur in inhomogeneous semiconductors or in different materials in contact). The Fermi–Dirac distribution function is plotted in Fig. 2.15 and is seen to vary from unity at low energies to zero at high energies. The main change in occupation probability occurs in the neighbourhood of E_F, within an energy range $|E-E_\mathrm{F}|$ of a few units of kT.

The probability $f_p(E)$ that a level E be vacant (that a positive hole occupy the level) is obviously given by $f_p(E) = 1 - f_n(E)$; that is,

$$f_p = \frac{1}{1 + \exp\left[(E_F - E)/kT\right]}. \tag{2.67}$$

This function is represented by the dashed curve in Fig. 2.15. Recalling that the energy for holes increases in the downward direction (see § 3.2), we see that the distribution function for holes is identical with that for electrons provided the energy is measured (with respect to E_F) in the opposite direction.

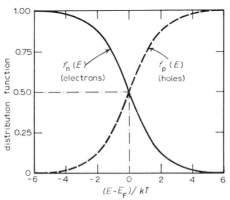

Fig. 2.15. The Fermi–Dirac distribution functions $f_n(E)$ and $f_p(E)$ for electrons and holes, respectively.

In order to obtain the volume density of electrons $n(E)dE$ in an energy interval E, $E + dE$, we must multiply $f_n(E)$ by the volume density of available states $N(E)dE$ in this interval:

$$n(E)dE = f_n(E)N(E)dE. \tag{2.68}$$

The total density of electrons in the conduction band, n_b, is given by

$$n_b = \int_{E_c}^{E_c + A_c} N_c(E)f_n(E)dE, \tag{2.69}$$

where E_c and $E_c + A_c$ are the lower and upper edges of the conduction band, and $N_c(E)$ the density of states per unit volume and unit energy in the band. Similarly the total density of holes p_b in the valence band is given by

$$p_b = \int_{E_v - A_v}^{E_v} N_v(E)f_p(E)dE, \tag{2.70}$$

where $E_v - A_v$ and E_v are the lower and upper edges of the valence band, and $N_v(E)$ the density of states in the band. The subscript b in n_b and p_b stands for "bulk" and signifies that we are dealing with a homogeneous semiconductor.

When the Fermi level is well removed from either of the band edges, the distribution functions (2.66), (2.67) reduce to the classical Maxwell–Boltzmann expressions:

$$f_n(E) \approx e^{-(E-E_F)/kT} = A e^{-E/kT} \quad \text{for} \quad E - E_F \gg kT; \tag{2.71}$$

$$f_p(E) \approx e^{(E-E_F)/kT} = B e^{E/kT} \quad \text{for} \quad E_F - E \gg kT, \tag{2.72}$$

where A and B are normalizing constants. This is only to be expected, since for $|E - E_F| \gg kT$ the electron and hole densities are so low that the restrictions imposed by the Pauli exclusion principle are no longer significant. Under these conditions the semiconductor is said to be non-degenerate.

We now turn to the electrons distributed among the various *localized* states introduced into the forbidden gap by impurity and defect centres. The distribution law for such electrons depends on the characteristics of the centres. In general a centre can capture more than one electron, and in each charge condition there can be several available energy levels (ground and excited states). Moreover, each state may be degenerate (such as when the centre can capture one electron of *either* spin). The occupation statistics for centres of this sort will be discussed in Ch. 5, § 5. Here we shall consider the simple case where, in the charge transfer process between each centre and the energy bands, only *one* electron takes part and the state of this electron is associated with a *single* energy level E_t. The probability that such a centre be occupied by an electron is (see Ch. 5, § 5.2)

$$f_n(E_t) = \frac{1}{1 + (g_0/g_1) \exp\left[(E_t - E_F)/kT\right]}, \tag{2.73}$$

where g_0 and g_1 represent the number of degenerate quantum states of the centre when it is vacant and occupied, respectively. If now we define an effective energy level E_t^f related to the actual level E_t by the equation

$$E_t^f = E_t + kT \ln(g_0/g_1), \tag{2.74}$$

then the distribution function (eq. (2.73)) reduces to that for quasi-free electrons (eq. (2.66)) with E_t replaced by E_t^f:

$$f_n(E_t^f) = \frac{1}{1+\exp\left[(E_t^f-E_F)/kT\right]}. \qquad (2.75)$$

E_t^f is sometimes called the free energy of the centre.

The density of occupied (n_t) and unoccupied (p_t) centres is given by

$$n_t = N_t f_n(E_t^f); \qquad (2.76)$$

$$p_t = N_t f_p(E_t^f), \qquad (2.77)$$

where $f_p(E_t^f)$ is defined similarly to $f_n(E_t^f)$, and $N_t \equiv n_t + p_t$ is the total density of centres having an effective energy E_t^f. It should be noted that now the centres are half occupied ($n_t = p_t$) when the Fermi level coincides with E_t^f, rather than with E_t.

5.2. INTRINSIC SEMICONDUCTORS

In a chemically pure semiconductor having no localized levels, the electron and hole densities must be equal, since every electron excited into the conduction band leaves behind a hole in the valence band. The carrier densities in this case will be denoted by n_i:

$$n_i = n_b = p_b. \qquad (2.78)$$

Applying this condition to (2.69) and (2.70), we can determine the Fermi level and thus obtain an expression for n_i. In a semiconductor the energy gap $E_g (\equiv E_c - E_v)$ is usually large compared to kT so that n_i will be sufficiently small to permit the use of Boltzmann statistics (eqs. (2.71), (2.72)). Furthermore, electrons will be found only near the minimum (E_c) of the conduction band and holes near the maximum (E_v) of the valence band. In these regions the densities of states $N_c(E)$, $N_v(E)$ can be expressed in terms of the components of the effective mass tensor, as we have seen in § 3.2. For the case in which there is a single minimum in the conduction band and a single maximum in the valence band, the densities of states can be written (see eqs. (2.57), (2.58)) in the form

$$N_c(E) = (4\pi/h^3)(2m_n)^{\frac{3}{2}}(E-E_c)^{\frac{1}{2}}; \qquad (2.79)$$

$$N_v(E) = (4\pi/h^3)(2m_p)^{\frac{3}{2}}(E_v-E)^{\frac{1}{2}}. \qquad (2.80)$$

Here m_n and m_p are given by $(m_x m_y m_z)^{\frac{1}{3}}$, where m_x, m_y, m_z represent the diagonal components of the effective-mass tensor for electrons at E_c and for holes at E_v. If there is a number of minima (or maxima), (2.79) and (2.80) should be replaced by appropriate sums over the various valleys. The average masses m_n and m_p appearing in (2.79) and (2.80) are referred to as the "density-of-states" effective masses.

The intrinsic carrier densities are obtained from (2.69) and (2.70) in conjunction with (2.71), (2.72), (2.79), and (2.80):

$$n_b = n_i = (4\pi/h^3)(2m_n)^{\frac{3}{2}} \int_{E_c}^{E_c+A_c} (E-E_c)^{\frac{1}{2}} e^{-(E-E_F)/kT} dE; \quad (2.81)$$

$$p_b = n_i = (4\pi/h^3)(2m_p)^{\frac{3}{2}} \int_{E_v-A_v}^{E_v} (E_v-E)^{\frac{1}{2}} e^{(E-E_F)/kT} dE. \quad (2.82)$$

Because of the rapidly decreasing exponentials under the integral signs, the limits of integration E_c+A_c and E_v-A_v can be replaced by $+\infty$ and $-\infty$, respectively. Consequently

$$n_b = n_i = N_c e^{-(E_c-E_F)/kT}; \quad (2.83)$$

$$p_b = n_i = N_v e^{-(E_F-E_v)/kT}, \quad (2.84)$$

where

$$N_c \equiv 2(2\pi m_n kT/h^2)^{\frac{3}{2}}; \quad (2.85)$$

$$N_v \equiv 2(2\pi m_p kT/h^2)^{\frac{3}{2}}. \quad (2.86)$$

Hence, for the purpose of calculating the carrier densities, the conduction and valence bands can be regarded as two single levels at energies E_c and E_v having effective densities of states N_c and N_v, respectively. From (2.83) and (2.84) we have that

$$n_i^2 = N_c N_v e^{-(E_c-E_v)/kT} = N_c N_v e^{-E_g/kT}. \quad (2.87)$$

The intrinsic Fermi level is given by

$$E_F^{(i)} = \tfrac{1}{2}(E_c+E_v) - \tfrac{1}{2}kT \ln(N_c/N_v). \quad (2.88)$$

If the density-of-states effective masses for electrons and holes are equal, then $N_c = N_v$ and $E_F^{(i)}$ lies exactly midway between the conduction- and valence-band edges. In most semiconductors the deviation from this position is small, since the difference in mass enters only logarithmically into the expression for $E_F^{(i)}$.

Some of the properties of intrinsic germanium and silicon are listed in

Table 2.2. Due to the thermal expansion of the lattice, the energy gap varies with temperature, as indicated in the table. The expression for n_i^2, which has been determined experimentally, includes this variation.

TABLE 2.2

Properties of intrinsic germanium and silicon

Property	Germanium	Silicon
Energy gap at 0°K	0.75 eV	1.153 eV
Energy gap at 300°K	0.67 eV	1.106 eV
n_i^2	$3.1 \times 10^{32} T^3 \exp(-0.785/kT)$	$1.5 \times 10^{33} T^3 \exp(-1.21/kT)$
n_i at 300°K	2.4×10^{13} cm^{-3}	1.5×10^{10} cm^{-3}
Resistivity at 300°K	46 ohm · cm	2.3×10^5 ohm · cm

5.3. SEMICONDUCTORS WITH LOCALIZED LEVELS

We turn now to consider the effect of impurity or defect centres on the carrier densities. Evidently, as long as non-degenerate conditions prevail, the electron and hole densities will still be given by (2.83)–(2.86), so that

$$n_b = N_c e^{-(E_c - E_F)/kT} \quad \text{for} \quad E_c - E_F \gg kT; \tag{2.89}$$

$$p_b = N_v e^{-(E_F - E_v)/kT} \quad \text{for} \quad E_F - E_v \gg kT. \tag{2.90}$$

The Fermi level is no longer given by (2.88) but now depends on the density and energy distribution of the localized levels in the forbidden gap. On the other hand, the product $n_b p_b$ depends, as before, on temperature only:

$$n_b p_b = N_c N_v e^{-E_g/kT} = n_i^2. \tag{2.91}$$

Non-degenerate conditions, for which eqs. (2.89)–(2.91) have been derived, will be assumed to hold throughout the subsequent discussions. The circumstances under which these assumptions are valid will be examined a little further on.

In germanium and silicon the carrier densities are usually controlled by the introduction of shallow impurity levels of group III and V elements. Consider then, a semiconductor having a concentration N_D of shallow donors at energy E_D just below E_c and a concentration N_A of shallow acceptors at energy E_A just above E_v. The density of electrons n_D bound to donors and of holes p_A bound to acceptors can be expressed in terms of the respective effective energies E_D^f and E_A^f (see eqs. (2.74)–(2.77)) in the form

$$n_D = \frac{N_D}{1+\exp\left[(E_D^f-E_F)/kT\right]}; \qquad (2.92)$$

$$p_A = \frac{N_A}{1+\exp\left[(E_F-E_A^f)/kT\right]}. \qquad (2.93)$$

The Fermi level is determined by the condition of overall electrical neutrality. The density of the negative charge is $-q(n_b+N_A-p_A)$ while that of the positive charge is $q(p_b+N_D-n_D)$. Thus

$$n_b+n_D+N_A = p_b+p_A+N_D. \qquad (2.94)$$

Using (2.89), (2.90), (2.92), and (2.93), we obtain from (2.94) a quadratic equation in $\exp(E_F/kT)$ which, in the general case, must be solved numerically (for example, by graphical methods). In order to see the characteristics of the distribution functions, we shall discuss a few limiting cases for which explicit solutions for E_F can be obtained.

Let us first start with an intrinsic semiconductor and introduce into it a few shallow donors whose ionization energy E_c-E_D is small compared to the forbidden gap. (In germanium, for example, $E_c-E_D \approx 0.01$ eV for group V impurities while $E_g \approx 0.7$ eV.) As long as N_D is not too large compared to n_i, E_F will not depart appreciably from its intrinsic position (eq. (2.88)) and will remain well below E_D. Use of (2.89) and (2.92) then yields

$$n_D/n_b = (N_D/N_c)e^{(E_c-E_D^f)/kT}, \qquad (2.95)$$

and thus $n_D \ll n_b$. It therefore follows from the neutrality condition (eq. (2.94)) that $n_b-p_b = N_D$, the donors being in effect completely ionized. Combining this condition with the relation $n_b p_b = n_i^2$ (eq. (2.91)), we obtain

$$\begin{aligned} n_b &= \tfrac{1}{2}N_D[(1+4n_i^2/N_D^2)^{\frac{1}{2}}+1]; \\ p_b &= \tfrac{1}{2}N_D[(1+4n_i^2/N_D^2)^{\frac{1}{2}}-1]. \end{aligned} \qquad (2.96)$$

Obviously n_b is greater than p_b, and we have what is called an n-type semiconductor. (The electrons in this case are referred to as the majority carriers, the holes as the minority carriers.) As more and more donors are added to the semiconductor, p_b decreases while n_b increases, and the Fermi level rises towards the conduction band. For $N_D \gg n_i$ the electrons originating from the donors far exceed those excited across the gap, and $n_b \approx N_D$. Under these conditions the electron density is essentially a constant, independent of temperature, and we have an *extrinsic* n-type semiconductor. The position of the Fermi level is now given by (see eq. (2.89))

$$(E_c - E_F)/kT = \ln(N_c/N_D). \tag{2.97}$$

The assumption of non-degeneracy made above is valid for $N_D \lesssim 0.1 N_c$. For germanium at $300°K$, $N_c \approx 2.5 \times 10^{19}$ cm^{-3} and so we require that $N_D \lesssim 2 \times 10^{18}$ cm^{-3}.

Consider now the effect of the introduction of acceptor levels into an extrinsic n-type sample. For small N_A, the Fermi level still lies above the intrinsic position and the acceptors are completely occupied by the electrons that have dropped down from donor levels. The neutrality condition then yields the relation

$$n_b = N_D - N_A = N_c e^{-W_{bn}/kT} = N_c e^{-w_{bn}} \quad \text{for} \quad N_A \ll N_D, \tag{2.98}$$
$$w_{bn} = \ln[N_c/(N_D - N_A)]$$

where we have introduced the notation, useful in the discussion of surface phenomena,

$$w_{bn} \equiv W_{bn}/kT \equiv (E_c - E_F)/kT. \tag{2.99}$$

The semiconductor thus behaves as if it had an effective density of donors equal to $N_D - N_A$. As N_A approaches N_D, the acceptors will tend to neutralize the donors and E_F will move towards its intrinsic position. When $N_A > N_D$, the Fermi level moves down towards the valence band and the sample becomes a p-type semiconductor with an effective density of acceptors equal to $N_A - N_D$. Under extrinsic conditions we have

$$p_b = N_A - N_D = N_v e^{-W_{bp}/kT} = N_v e^{-w_{bp}} \quad \text{for} \quad N_D \ll N_A, \tag{2.100}$$
$$w_{bp} = \ln[N_v/(N_A - N_D)]$$

where, similarly to the n-type case,

$$w_{bp} \equiv W_{bp}/kT \equiv (E_F - E_v)/kT. \tag{2.101}$$

In future the subscript n or p will be dropped from w_{bn}, W_{bn}, w_{bp}, W_{bp}, the symbols w_b and W_b denoting the energetic distance of the Fermi level from the *majority*-carrier band edge. For an extrinsic semiconductor w_b is seen to be essentially independent of temperature, which enters only logarithmically (see eqs. (2.85), (2.86)) through N_c or N_v.

We have discussed so far the case of complete ionization of impurity levels. Referring to (2.95), we see that in an n-type semiconductor complete ionization prevails ($n_D \ll n_b$) as long as N_D is small compared to N_c and/or

$E_c - E_D^f$ is small compared to kT. Such is usually the case in germanium and silicon except in the range of very low temperatures. Here $E_c - E_D^f$ may no longer be small compared to kT and full ionization will not take place. As the temperature decreases, E_F rises towards the donor level and eventually lies above it. Equations (2.96) are then no longer valid and we must use the Fermi–Dirac distribution function to describe the density of the electrons bound to donors. For sufficiently low temperatures

$$n_b = \sqrt{N_c N_D}\, e^{-(E_c - E_D^f)/2kT} \quad \text{for} \quad E_c - E_D \gg kT \qquad (2.102)$$

and

$$E_F = \tfrac{1}{2}(E_c + E_D^f) - \tfrac{1}{2}kT \ln(N_c/N_D). \qquad (2.103)$$

These equations are identical with (2.87) and (2.88) for an intrinsic semiconductor provided that E_v and N_v are replaced by E_D^f and N_D. Physically this result is obvious, since each electron in the conduction band now originates from a donor which thereby becomes ionized, just as each electron in an intrinsic semiconductor leaves behind a positive hole in the valence band.

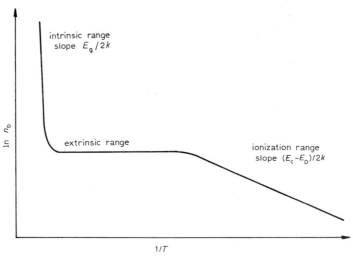

Fig. 2.16. Temperature variation of the electron density in an n-type semiconductor.

We can now see how the free-carrier density in a given semiconductor varies with temperature. This is illustrated in Fig. 2.16 for an n-type semiconductor ($N_D \gg N_A$). Here $\ln n_b$ is plotted against $1/T$, and three ranges are apparent. At very low temperatures (2.102) holds, n_b increasing exponentially with T. The slope in this range yields the ionization energy $E_c - E_D$. As the temperature is raised extrinsic conditions are approached, and thereafter n_b is a constant (equal to $N_D - N_A$) over a considerable temperature range. (In germanium samples, this range usually extends from about 20°K to room temperature.) At still higher temperatures, carrier excitation across the forbidden gap becomes important and (2.96) is applicable. Finally, at sufficiently high temperature such that $n_i \gg N_D - N_A$, intrinsic conditions set in and n_b again varies exponentially with T (eq. (2.87)). The slope now yields the energy gap E_g.

The corresponding variation of the Fermi level with temperature is shown schematically in Fig. 2.17. The position of E_F is above E_D at low temperatures and decreases with increasing temperature until it approaches the intrinsic value (near the mid-gap). The case of a p-type sample is analogous, with E_F approaching the intrinsic value from below as the temperature is raised.

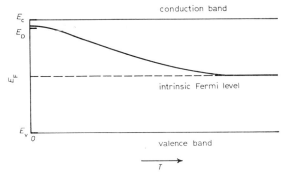

Fig. 2.17. Temperature variation of the Fermi level in an n-type semiconductor.

Throughout the preceding discussion we have considered the case of shallow levels. For deep-lying levels, such as are introduced by gold and copper impurities in Ge and Si, the calculation of the carrier densities is considerably more complicated, especially when several levels are present simultaneously. Most surface studies, however, have been carried out on extrinsic materials and in this book we shall not be concerned much with samples whose carrier densities are controlled by deep-lying levels.

6. Carrier transport

6.1. SCATTERING PROCESSES

As we have seen in §§ 2 and 3, the electron (or hole) flow in a perfectly periodic potential is completely unhampered. Each electron maintains its wave vector $\mathbf{k}$ and velocity $\mathbf{c}$ indefinitely, provided that external forces are absent. In a real crystal, on the other hand, the deviations from ideal conditions caused by lattice vibrations, impurities, and structural defects result in the existence of a non-vanishing transition probability for an electron to be scattered from one state into another. Consequently, the electrons in the crystal change their velocity continually, moving in a random fashion much the same as gas molecules. In the presence of an electric field, a small drift velocity is superimposed on the random motion and a current is set up.

The various scattering processes in a crystal can often be characterized by a relaxation time, a quantity of fundamental importance in transport phenomena. The relaxation time represents the average time required for a disturbance in the equilibrium electron distribution to die out by the randomizing action of the scattering. Suppose that at time $t = 0$ a group of n_0 electrons moves with a velocity $\mathbf{c}$ in a given direction. (This velocity may have been acquired, for instance, by the action of an electric field prior to $t = 0$.) The probability that due to scattering an electron in the group lose its velocity in a small time interval dt may be expressed as dt/τ, where τ is a constant which, as we shall see in a moment, is equal to the relaxation time. The number of electrons $n(t)$ that have *not* been scattered at time t therefore satisfies the equation

$$\frac{dn(t)}{dt} = -\frac{n}{\tau}, \tag{2.104}$$

or

$$n(t) = n_0 e^{-t/\tau}. \tag{2.105}$$

It is readily seen that τ is the average time $\langle t \rangle$ in which an electron in the group loses its velocity:

$$\langle t \rangle = (1/n_0) \int_0^\infty tn(t)dt/\tau = \tau; \tag{2.106}$$

thus τ is just the relaxation time. The mean free path l, defined as the average distance traversed by an electron before it loses its velocity, is given by

$$l = c\tau. \tag{2.107}$$

The orders of magnitude of τ and l in semiconductors at room temperature are 10^{-12} sec and 10^{-5} cm, respectively.

The scattering is *isotropic* when, after a single scattering process, or collision, an electron emerges with a random velocity, having lost all memory of its previous motion. The relaxation time in this case is obviously identical with the mean free time between collisions. If the scattering is not isotropic, several collisions are required to randomize the velocity, and the relaxation time is larger than the mean free time between collisions.

Ohm's law can be derived in an elementary manner for the following simple model. Consider an n-type semiconductor having spherical energy surfaces (scalar mass m_n^*). The current density in any direction, say x, is given by $J_x = -qn_b\langle c_x\rangle$ where n_b is the electron density and $\langle c_x\rangle$ the mean velocity in the x direction. At thermal equilibrium the electron distribution in $\mathbf{k}$ space is spherically symmetrical about $\mathbf{k} = 0$ and $\langle c_x\rangle = 0$. Under an applied electric field $\mathscr{E}$, all electrons change their $\mathbf{k}$ values in accordance with (2.46). For a field in the x direction, we have for the rate of change in the average value of k_x

$$\left(\frac{d\langle k_x\rangle}{dt}\right)_{\text{field}} = -\frac{q}{\hbar}\mathscr{E}_x. \tag{2.108}$$

The field thus tends to shift the electron distribution towards decreasing k_x values. At the same time, scattering tends to restore the random distribution. Steady state is reached when these opposing tendencies balance each other and a net shift of the distribution to the left has occurred. Let us assume that the scattering is characterized by a *constant* relaxation time τ_n, independent of the electron velocity. The rate of change in the average value of k due to scattering is then given by

$$\left(\frac{d\langle k_x\rangle}{dt}\right)_{\text{scattering}} = -\frac{\langle k_x\rangle}{\tau_n}. \tag{2.109}$$

At steady state we obtain from (2.108) and (2.109)

$$\langle k_x\rangle = -(q\tau_n/\hbar)\mathscr{E}_x. \tag{2.110}$$

Recalling that $c_x = \hbar k_x/m_n^*$, $J_x = -qn_b\langle c_x\rangle$, we have

$$\langle c_x\rangle = -(q\tau_n/m_n^*)\mathscr{E}_x; \tag{2.111}$$

$$J_x = n_b(q^2\tau_n/m_n^*)\mathscr{E}_x. \tag{2.112}$$

Thus the average drift velocity $\langle c_x\rangle$ and the current density J_x are proportional to the electric field. This is Ohm's law. We can write (2.111) in the form

$$\langle c_x\rangle = -\mu_n\mathscr{E}_x, \tag{2.113}$$

where μ_n is the *electron mobility*, defined as the absolute magnitude of the drift velocity per unit electric field, and is given by

$$\mu_n = q\tau_n/m_n^*. \tag{2.114}$$

In a similar manner, we obtain for the average drift velocity of holes in a spherical energy band

$$\langle c'_x \rangle = +(q\tau_p/m_p^*)\mathscr{E}_x, \qquad (2.115)$$

the hole mobility being

$$\mu_p = q\tau_p/m_p^*. \qquad (2.116)$$

If both electrons and holes are present, the current density is

$$J_x = -qn_b\langle c_x\rangle + qp_b\langle c'_x\rangle = q(n_b\mu_n + p_b\mu_p)\mathscr{E}_x, \qquad (2.117)$$

or

$$J_x = \sigma\mathscr{E}_x, \qquad (2.118)$$

where σ is the conductivity:

$$\sigma = q(n_b\mu_n + p_b\mu_p). \qquad (2.119)$$

The units commonly used for the mobility are m^2/V·sec or cm^2/V·sec. The corresponding units for the conductivity are ohm^{-1}·m^{-1} or ohm^{-1}·cm^{-1}.

The assumption regarding the relaxation time τ constitutes an oversimplification of the actual situation. Usually τ is a function of the electron velocity, and suitable averaging procedures as prescribed by the Boltzmann transport equation then become necessary. This will be discussed in the next two subsections; for the present, let us note the forms assumed by τ for the various types of scattering processes.

Scattering by lattice vibrations will be considered in some detail since this is the dominant scattering mechanism in chemically pure crystals at ordinary temperatures. By the use of perturbation theory, it can be shown that the transition probability for scattering vanishes unless

$$E(k') - E(k) = \pm\hbar\omega_K; \qquad (2.120)$$

$$k' - k = \pm K, \qquad (2.121)$$

where the unprimed and primed wave vectors refer to the electron before and after scattering; K is the wave vector of a lattice vibration and ω_K the corresponding angular frequency. (To simplify matters, a term has been omitted from (2.121). For a simple cubic lattice, this term is $(2\pi/a)n$, where a is the lattice constant and n a vector with integer components.) These conditions express the conservation of energy and momentum in a two-body system if we look upon the interaction with the lattice vibrations as electron-phonon collisions. The positive sign corresponds to absorption of a phonon by the electron, the negative sign to phonon emission.

It is easy to show that in the interaction with acoustical phonons, while the electron may be scattered over large angles, its energy remains essentially unaltered. For spherical energy surfaces centred at $\boldsymbol{k} = 0$, we have from (2.120)

$$(\hbar^2/2m^*)(k'^2 - k^2) = \pm\hbar\omega_K \approx \pm\hbar c_s K, \qquad (2.122)$$

where c_s is the sound velocity (see § 4.1) and K, k denote the absolute magnitudes of $\boldsymbol{K}$, $\boldsymbol{k}$. Furthermore, the maximum value of K is $k'+k$ (eq. (2.121)). Hence

$$(\hbar^2/2m^*)|k'-k| \leq \hbar c_s.$$

Recalling (eq. (2.44)) that the electron velocity c is equal to $\hbar k/m^*$, we obtain for the fractional change in k

$$\frac{|k'-k|}{k} \leq 2\frac{c_s}{c}. \qquad (2.123)$$

In a non-degenerate semiconductor, the mean-square electron velocity $\langle c^2 \rangle$ is given by

$$\tfrac{1}{2}m^*\langle c^2 \rangle = \tfrac{3}{2}kT,$$

or

$$\langle c \rangle \approx \langle c^2 \rangle^{\frac{1}{2}} = (3kT/m^*)^{\frac{1}{2}}, \qquad (2.124)$$

where $\langle c \rangle$, the thermal velocity, is the mean value of $|c|$. For $m^* = 0.3m$ (electrons in germanium), $\langle c \rangle$ is 10^7 cm/sec at room temperature. On the other hand, c_s is of the order of 10^5 cm/sec, so that $|k'-k|/k$ is very small. It therefore follows from (2.121) and (2.122) that the scattering angle θ and the change in energy δE are given by

$$\sin(\theta/2) \approx K/2k; \qquad (2.125)$$

$$\frac{|\delta E|}{E} \approx \frac{2\hbar c_s k}{E}\sin(\theta/2) = \frac{4c_s}{c}\sin(\theta/2). \qquad (2.126)$$

Since in our case electrons have k values near $k = 0$, only very long-wavelength phonons can be involved in scattering ($K < 2k$, eq.(2.125)). Hence, the phonon energy $\hbar\omega_K$ is much less than the thermal energy kT down to very low temperatures, and a large number of phonons ($n_K = kT/\hbar\omega_K$, see eq. (2.65)) is available for scattering. On physical grounds it is reasonable to expect that the scattering probability $1/\tau_l$ due to phonon emission or absorption be proportional to n_K (that is, to T). Furthermore, there is little variation in the phonon density over the range of energy required for scattering through various angles (eqs. (2.125), (2.126)), so that the scattering is very nearly isotropic. Finally, the probability of scattering of an electron having an energy E, from one state $\boldsymbol{k}$ to another state $\boldsymbol{k}'$ ($k \approx k'$), is also proportional to the density of available states at this energy — that is, it is proportional to $E^{\frac{1}{2}}$ or c (see eq. (2.57)). We may summarize these conclusions as follows:

$$\tau_l \propto T^{-1}c^{-1} \propto T^{-1}E^{-\frac{1}{2}}; \tag{2.127}$$

$$l = c\tau \propto T^{-1}. \tag{2.128}$$

Thus the relaxation time for acoustical-phonon scattering is inversely proportional to the lattice temperature and to the square root of the electron energy. The mean free path, on the other hand, is independent of the electron energy.

The scattering by optical phonons may modify the relations expressed in (2.127) and (2.128). At normal temperatures the optical modes are usually not excited to any appreciable extent (see § 4.1). Optical phonons are therefore not available for absorption, and only the process of phonon emission is important.

A second important scattering mechanism involves impurity centres. In the case of ionized impurities, electrons (or holes) suffer Rutherford scattering due to Coulomb interaction. This type of scattering is highly anisotropic, small angles being strongly favoured. The relaxation time is therefore considerably larger than the mean free time between collisions, and is obtained by averaging over the various scattering angles. A detailed calculation shows that the relaxation time is approximately proportional to the three-halves power of the electron energy E and inversely proportional to the density N_i of ionized impurities:

$$\tau_i \propto c^3/N_i \propto E^{\frac{3}{2}}/N_i. \tag{2.129}$$

Thus, in contrast to the case of lattice scattering, it is the slow electrons that are chiefly involved in scattering by ionized impurities.

The scattering of carriers by neutral impurities is weaker, as expected, and τ is independent of temperature or electron energy.

Scattering by structural defects usually plays a minor role. It has been shown theoretically that for dislocations the relaxation time is inversely proportional to the number of dislocation lines per unit area and to the temperature. Scattering is significant only if the dislocation density is larger than 10^8 cm^{-2}.

In general, all scattering mechanisms are simultaneously present to a greater or lesser extent. If these are assumed to be independent, then the composite relaxation time τ is given by

$$1/\tau = 1/\tau_l + 1/\tau_i + \ldots. \tag{2.130}$$

6.2. THE BOLTZMANN TRANSPORT EQUATION

In most cases of interest the relaxation time is a function of the electron energy, and it is necessary to modify the elementary treatment of transport presented in the preceding subsection. The relaxation time should be suitably averaged over all carriers in the crystal. How to carry out this averaging procedure is shown by the Boltzmann equation, which governs the change in the distribution function due to the application of external forces.

Let $f(\mathbf{k}, \mathbf{r}, t)$ represent the distribution function in momentum and coordinate space. The distribution at thermal equilibrium will be denoted by $f_0(\mathbf{k}, \mathbf{r})$. When expressed as a function of the electron energy E, then f_0 is just the Fermi–Dirac distribution function (2.66).

The steady-state distribution is obtained by equating to zero the total rate of change of f with time, consisting of the rates of change due to external fields and to scattering:

$$\frac{df}{dt} = \left(\frac{\partial f}{\partial t}\right)_{\text{field}} + \left(\frac{\partial f}{\partial t}\right)_{\text{scattering}} = 0. \tag{2.131}$$

Each group of electrons in an element of $(\mathbf{k}, \mathbf{r})$ space moves as an incompressible fluid according to the relations

$$d\mathbf{k} = \hbar^{-1}\mathbf{F}\,dt; \qquad d\mathbf{r} = \mathbf{c}\,dt, \tag{2.132}$$

where $\mathbf{F}$ is the external force acting on the electron (see eq. (2.46)). Thus

$$f(\mathbf{k}+\hbar^{-1}\mathbf{F}\,dt, \mathbf{r}+\mathbf{c}\,dt, t+dt) = f(\mathbf{k}, \mathbf{r}, t).$$

Expansion of the left-hand side yields

$$\hbar^{-1}\mathbf{F} \cdot \text{grad}_k f + \mathbf{c} \cdot \text{grad}_r f + \left(\frac{\partial f}{\partial t}\right)_{\text{field}} = 0,$$

and (2.131) becomes

$$\hbar^{-1}\mathbf{F} \cdot \text{grad}_k f + \mathbf{c} \cdot \text{grad}_r f = \left(\frac{\partial f}{\partial t}\right)_{\text{scattering}}. \tag{2.133}$$

This is the general form of the Boltzmann transport equation.

For sufficiently weak external forces, the deviations from thermal equilibrium are small and one can usually characterize the scattering processes by a relaxation time $\tau(\mathbf{k}, \mathbf{r})$ such that

$$\left(\frac{\partial f}{\partial t}\right)_{\text{scattering}} = -\frac{f(\mathbf{k}, \mathbf{r}, t) - f_0(\mathbf{k}, \mathbf{r})}{\tau(\mathbf{k}, \mathbf{r})}. \tag{2.134}$$

In the general case τ is a function of $\mathbf{k}$ (or E) and $\mathbf{r}$. Equation (2.134) expresses the tendency of scattering to restore the equilibrium distribution. If the external force is suddenly removed (say at time $t = 0$), we obtain from (2.131) and (2.134) that

$$f(\mathbf{k}, \mathbf{r}, t) - f_0(\mathbf{k}, \mathbf{r}) = [f(\mathbf{k}, \mathbf{r}, 0) - f_0(\mathbf{k}, \mathbf{r})]e^{-t/\tau}. \qquad (2.135)$$

This is a generalization of the special case treated in the previous subsection (eq. (2.105)). The distribution at any point $(\mathbf{k}, \mathbf{r})$ relaxes towards its equilibrium value in an average time $\tau(\mathbf{k}, \mathbf{r})$.

We shall now derive the expression for the current in a homogeneous, non-degenerate, n-type semiconductor having spherical energy surfaces centred at $\mathbf{k} = 0$. In this case f is independent of position and, in thermal equilibrium, is given by Boltzmann statistics (eq. (2.71)). The current density along, say, the x direction is obtained by summing the x component of the velocity c_x of all the electrons in the conduction band:

$$J_x = -q \iiint c_x N(\mathbf{k}) f(\mathbf{k}) dk_x dk_y dk_z. \qquad (2.136)$$

$N(\mathbf{k})$ is the density of states in $\mathbf{k}$ space and in the present case is given by $1/4\pi^3$ (see eq. (2.55)). For an electric field along the x direction, $F_x = -q\mathscr{E}_x$, $F_y = F_z = 0$, and (2.133) and (2.134) reduce to

$$f(\mathbf{k}) = f_0(\mathbf{k}) + q\tau \hbar^{-1} \mathscr{E}_x (\partial f/\partial k_x). \qquad (2.137)$$

Obviously, $f_0(\mathbf{k})$ does not contribute to the integral in (2.136). Furthermore, for small fields we may approximate $\partial f/\partial k_x$ by $\partial f_0/\partial k_x$, and then

$$\frac{\partial f}{\partial k_x} \approx \frac{\partial f_0}{\partial k_x} = \frac{\partial f_0}{\partial E} \frac{\partial E}{\partial k_x} = -\frac{f_0}{kT} \hbar c_x,$$

where use has been made of (2.54) and (2.71). We can now write (2.136) in the form

$$J_x = (q^2/4\pi^3 kT)\mathscr{E}_x \iiint c_x^2 \tau f_0 dk_x dk_y dk_z. \qquad (2.138)$$

Since f_0 and τ are functions of $|\mathbf{k}|$ only, the triple integral in (2.138) may be transformed into a single integral by replacing c_x^2 by $\tfrac{1}{3}c^2$ and $dk_x dk_y dk_z$ by $4\pi k^2 dk$:

$$J_x = (q^2/3kT)\mathscr{E}_x \int_0^\infty c^2 \tau(k) f_0(k) k^2 dk/\pi^2. \qquad (2.139)$$

The quantity $f_0 \pi^{-2} k^2 dk$ represents the number of electrons with wave vector in the range dk, so that the total density of electrons in the conduction band is

$$n_b = \int_0^\infty f_0(k) k^2 dk/\pi^2. \qquad (2.140)$$

Thus

$$\langle c^2 \tau \rangle = \frac{\int_0^\infty c^2 \tau(k) f_0(k) k^2 dk/\pi^2}{\int_0^\infty f_0(k) k^2 dk/\pi^2} \qquad (2.141)$$

represents the mean value of $c^2 \tau$ taken over the entire electron distribution. We recall, further, that the mean square velocity $\langle c^2 \rangle$ is given by $3kT/m^*$ (eq. (2.124)). Hence (2.139) can be written as

$$J_x = \frac{q^2 n_b \mathscr{E}_x}{m^*} \frac{\langle c^2 \tau \rangle}{\langle c^2 \rangle} = \sigma \mathscr{E}_x = q \mu_n n_b \mathscr{E}_x, \qquad (2.142)$$

where the mobility μ_n is given by

$$\mu_n = \frac{q}{m^*} \frac{\langle c^2 \tau \rangle}{\langle c^2 \rangle}. \qquad (2.143)$$

A similar expression obviously applies to holes. Equation (2.141) may also be expressed in terms of the electron or hole energy:

$$\frac{\langle c^2 \tau \rangle}{\langle c^2 \rangle} = \frac{\langle E \tau \rangle}{\langle E \rangle} = \frac{\int_0^\infty E^{3/2} \tau(E) e^{-E/kT} dE}{\int_0^\infty E^{3/2} e^{-E/kT} dE}, \qquad (2.144)$$

where use has been made of (2.71).

If τ is independent of E, then (2.143) reduces to $\mu = q\tau/m^*$, as derived for the simplified model. When τ is a function of E, a mean value must be taken. It is important to note that the averaging of τ should be weighted by the factor E (or c^2).

For scattering by acoustical phonons, τ_l is proportional to $E^{-1/2} T^{-1}$ (eq. (2.127)). Evaluation of the integrals in (2.144) for this case yields for the "lattice mobility" the expression

$$\mu_l = \text{const.}\, T^{-3/2}. \qquad (2.145)$$

For ionized-impurity scattering one obtains from (2.129) and (2.144)

$$\mu_i = \text{const}\, T^{3/2}. \qquad (2.146)$$

When both types of scattering are present the mobility μ is approximately given by (see eq. (2.130))

$$1/\mu = 1/\mu_l + 1/\mu_i. \qquad (2.147)$$

It is apparent from (2.145) and (2.146) that lattice scattering predominates at high temperatures and impurity scattering at low temperatures. The transition from one type of scattering to the other depends on the impurity content.

For non-spherical energy surfaces, the conductivity is no longer isotropic and is represented by a tensor. However, for a field along any of the three principal axes, the current is in the direction of the field (see § 3.1) and the respective mobilities are given by

$$\mu_1 = \frac{q}{m_1}\frac{\langle E\tau \rangle}{\langle E \rangle}, \quad \mu_2 = \frac{q}{m_2}\frac{\langle E\tau \rangle}{\langle E \rangle}, \quad \mu_3 = \frac{q}{m_3}\frac{\langle E\tau \rangle}{\langle E \rangle}, \qquad (2.148)$$

where m_1, m_2, m_3 are the effective masses along the principal axes.

In n-type germanium and silicon, where several equivalent minima are present, the mobility is obtained by averaging over the electrons in the various valleys. Although within each valley the mobility is anisotropic, the contribution of all the valleys results in an isotropic mobility. This is particularly easy to see in the case of silicon, where the constant-energy ellipsoids are along three mutually perpendicular directions. In any one of these directions the mobility is

$$\mu = \tfrac{1}{3}(\mu_1 + \mu_2 + \mu_3) = \frac{q}{3}\left(\frac{1}{m_l} + \frac{2}{m_t}\right)\frac{\langle E\tau \rangle}{\langle E \rangle}, \qquad (2.149)$$

where m_l and m_t are the longitudinal and transverse effective masses, respectively (see § 3.3). The mobility is thus independent of direction. This conclusion is quite general for all cubic crystals, as can be shown by symmetry considerations.

For the relatively pure Ge and Si normally employed in surface studies, the mobility is completely determined by lattice scattering from above about liquid nitrogen temperature. The temperature dependence of μ, however, does not exactly follow the $T^{-\frac{3}{2}}$ law derived previously. The deviations are due mainly to the fact that the band structure is more complex than has been assumed and also that optical phonons contribute to the scattering process. Some mobility data for electrons and holes in Ge and Si are summarized in Table 2.3. For holes, the measured values represent weighted averages of the heavy and light hole mobilities.

TABLE 2.3

Electron and hole mobilities in germanium and silicon

	Electrons		Holes	
	μ_n at $T = 300°K$	$\mu_n(T)$	μ_p at $T = 300°K$	$\mu_p(T)$
Ge	3900 cm²/V · sec	$T^{-1.66}$	1900 cm²/V · sec	$T^{-2.3}$
Si	1500 cm²/V · sec	$T^{-2.0}$	400 cm²/V · sec	$T^{-2.7}$

6.3. GALVANOMAGNETIC EFFECTS

Conductivity data, even in an extrinsic semiconductor, yield only the product $\mu_n(T)n_b(T)$ or $\mu_p(T)p_b(T)$. Additional information is required to resolve the two unknowns. This may be provided by the measurement of the Hall effect and magnetoresistance, both associated with the effect of magnetic fields on transport.

The force F exerted on an electron by combined electric ($\mathscr{E}$) and magnetic (B) fields is given by the Lorentz formula

$$F = -q(\mathscr{E} + c \times B). \qquad (2.150)$$

To obtain the current density one should substitute this expression for F in the Boltzmann equation (eq. (2.133)) and evaluate integrals of the form (2.136). This procedure, however, is very involved and will not be undertaken here. Instead we shall write down the equations of motion, solve for the electron velocity, and derive the current by the averaging procedure indicated in the previous subsection[15].

Consider again an extrinsic n-type semiconductor having spherical energy surfaces. The relaxation time is assumed, as before, to be a function of energy only. Let the current (and the electric field) be in the $x-y$ plane, with the magnetic field along the z direction. In the absence of scattering

$$m^* \frac{dc}{dt} = F,$$

and using (2.150) one obtains for the equations of motion of an electron

$$\begin{aligned} \frac{dc_x}{dt} &= -\frac{q}{m^*}\mathscr{E}_x - \omega_0 c_y; \\ \frac{dc_y}{dt} &= -\frac{q}{m^*}\mathscr{E}_y + \omega_0 c_x, \end{aligned} \qquad (2.151)$$

where $\omega_0 \equiv qB/m^*$ is the cyclotron frequency (see eq. (2.61)). The solutions of these equations are

$$
\begin{aligned}
c_x(t) &= c_{x0} \cos \omega_0 t - c_{y0} \sin \omega_0 t \\
&\quad - (q/m^*\omega_0)(\mathscr{E}_x \sin \omega_0 t + \mathscr{E}_y \cos \omega_0 t - \mathscr{E}_y); \\
c_y(t) &= c_{x0} \sin \omega_0 t + c_{y0} \cos \omega_0 t \\
&\quad - (q/m^*\omega_0)(\mathscr{E}_y \sin \omega_0 t - \mathscr{E}_x \cos \omega_0 t + \mathscr{E}_x),
\end{aligned}
\qquad (2.152)
$$

where c_{x0}, c_{y0} are the components of the velocity at $t = 0$.

Scattering hampers the electron motion described by (2.152), each collision acting to randomize the velocity. Under steady-state conditions a non-vanishing mean velocity obtains with components $\langle c_x \rangle$, $\langle c_y \rangle$. For the low-field case under consideration, this acquired velocity is small compared to the mean thermal velocity, so that the energy distribution of the electrons is not appreciably different from that at thermal equilibrium. To determine the mean velocity, we carry out two averaging procedures. First we evaluate the mean-velocity components $c_x^{(E)}$, $c_y^{(E)}$ of a group of electrons having the same energy E (and therefore the same relaxation time $\tau(E)$), and then we average these over the entire energy distribution.

The probability that an electron of energy E be scattered (see § 6.1) after time t is $[dt/\tau(E)] \exp[-t/\tau(E)]$. Hence its contribution to the mean velocity of the group is given by the product of this probability factor and the velocity gained by the action of the fields during the time t. Averaging over all electrons in the group, we have

$$
\begin{aligned}
c_x^{(E)} &= \int_0^\infty c_x(t) e^{-t/\tau(E)} dt/\tau(E); \\
c_y^{(E)} &= \int_0^\infty c_y(t) e^{-t/\tau(E)} dt/\tau(E),
\end{aligned}
\qquad (2.153)
$$

where $c_x(t)$, $c_y(t)$ are given by (2.152). The contribution of the terms containing c_{x0}, c_{y0} is zero, since they correspond to the random equilibrium distribution. The integration yields

$$
\begin{aligned}
c_x^{(E)} &= -\frac{q}{m^*}\left(\mathscr{E}_x \frac{\tau}{1+\omega_0^2\tau^2} - \omega_0 \mathscr{E}_y \frac{\tau^2}{1+\omega_0^2\tau^2}\right); \\
c_y^{(E)} &= -\frac{q}{m^*}\left(\mathscr{E}_y \frac{\tau}{1+\omega_0^2\tau^2} + \omega_0 \mathscr{E}_x \frac{\tau^2}{1+\omega_0^2\tau^2}\right).
\end{aligned}
\qquad (2.154)
$$

For sufficiently low magnetic fields such that $\omega_0^2 \tau^2 \ll 1$ and for the case $|\mathscr{E}_y| \ll |\mathscr{E}_x|$ (see below), the above equations reduce to

$$c_x^{(E)} = -(q/m^*)(\mathscr{E}_x \tau - \omega_0 \mathscr{E}_y \tau^2 - \omega_0^2 \mathscr{E}_x \tau^3);$$
$$c_y^{(E)} = -(q/m^*)(\mathscr{E}_y \tau + \omega_0 \mathscr{E}_x \tau^2). \tag{2.155}$$

Next we average these expressions over the entire electron distribution, using the weighting factor indicated by the analysis of the preceding subsection. The mean acquired velocity can then be expressed as

$$\langle c_x \rangle = \frac{\langle E c_x^{(E)} \rangle}{\langle E \rangle} = -\frac{q}{m^*}\left(\mathscr{E}_x \frac{\langle E\tau \rangle}{\langle E \rangle} - \omega_0 \mathscr{E}_y \frac{\langle E\tau^2 \rangle}{\langle E \rangle} - \omega_0^2 \mathscr{E}_x \frac{\langle E\tau^3 \rangle}{\langle E \rangle}\right);$$
$$\langle c_y \rangle = \frac{\langle E c_y^{(E)} \rangle}{\langle E \rangle} = -\frac{q}{m^*}\left(\mathscr{E}_y \frac{\langle E\tau \rangle}{\langle E \rangle} + \omega_0 \mathscr{E}_x \frac{\langle E\tau^2 \rangle}{\langle E \rangle}\right), \tag{2.156}$$

where $\langle E\tau^2 \rangle$ and $\langle E\tau^3 \rangle$ represent integrals of the type appearing in the numerator of the right-hand side of (2.144) (with τ replaced by τ^2 and τ^3, respectively).

The components of the current density are

$$J_x = -qn_b \langle c_x \rangle; \quad J_y = -qn_b \langle c_y \rangle. \tag{2.157}$$

Measurements are usually carried out on rectangular samples whose cross sectional dimensions are small compared to their length. The current is confined to the long axis (x), so that $J_y = 0$. Under these conditions a field $\mathscr{E}_y$ is set up along the y axis to counterbalance the Lorentz force. Equations (2.156), (2.157) then reduce to

$$\theta \approx \tan\theta \equiv \mathscr{E}_y/\mathscr{E}_x = -\omega_0 \frac{\langle E\tau^2 \rangle}{\langle E\tau \rangle};$$
$$J_x = (q^2/m^*) n_b \mathscr{E}_x \frac{\langle E\tau \rangle}{\langle E \rangle}\left[1 - \omega_0^2 \frac{\langle E\tau^3 \rangle \langle E\tau \rangle - \langle E\tau^2 \rangle^2}{\langle E\tau \rangle^2}\right], \tag{2.158}$$

where θ is the so-called Hall angle. The first equation expresses the Hall effect, which consists in the appearance across the sample of a transverse voltage following the application of a magnetic field. The second equation expresses the transverse magnetoresistance — that is, the change in sample resistance due to the magnetic field. While the Hall effect is a first-order effect (involving the first power of ω_0), the magnetoresistance is a second-order effect and may be neglected for sufficiently weak magnetic fields.

The *Hall coefficient* is

$$R_\mathrm{H} = \frac{\mathscr{E}_y}{J_x B} \approx -\frac{1}{qn_b} \frac{\langle E\tau^2 \rangle \langle E \rangle}{\langle E\tau \rangle^2}. \tag{2.159}$$

It follows from (2.119) and (2.159) that

$$|R_\mathrm{H}|\sigma = \mu_n \frac{\langle E\tau^2 \rangle \langle E \rangle}{\langle E\tau \rangle^2} \equiv \mu_{n\mathrm{H}}, \tag{2.160}$$

where $\mu_{n\mathrm{H}}$ has the dimensions of mobility and is referred to as the electron Hall mobility. The *transverse magnetoresistance* $\Delta\rho/\rho B^2$ is given by

$$\frac{\Delta\rho}{\rho B^2} \approx -\frac{\Delta\sigma}{\sigma B^2} = \frac{q^2}{m^{*2}} \frac{\langle E\tau^3 \rangle \langle E\tau \rangle - \langle E\tau^2 \rangle^2}{\langle E\tau \rangle^2}, \tag{2.161}$$

where $\rho \equiv 1/\sigma$ is the resistivity of the semiconductor and $\Delta\rho$ is the change in resistivity in a magnetic field; $\Delta\rho$ is always positive. The term *transverse* indicates that the magnetic field is perpendicular to the direction of the current. The change in resistance due to a magnetic field *along* the current (not treated here) is referred to as the *longitudinal* magnetoresistance. For the case of spherical energy surfaces under consideration, the longitudinal magnetoresistance is always zero.

If τ is independent of energy, $\mu_{n\mathrm{H}} = \mu_n$ and the Hall coefficient becomes a direct measure of the carrier concentration: $R_\mathrm{H} = -1/qn_b$. The transverse magnetoresistance vanishes in this case.

In the general case the integrals $\langle E\tau^2 \rangle$, $\langle E\tau^3 \rangle$ should be calculated using the appropriate energy dependence of τ. For lattice scattering ($\tau \propto E^{-\frac{1}{2}}$), integration yields for the Hall mobility, Hall coefficient, and transverse magnetoresistance the following expressions:

$$\mu_{n\mathrm{H}} = \tfrac{3}{8}\pi\mu_n; \qquad R_\mathrm{H} = -\tfrac{3}{8}\pi(1/qn_b); \tag{2.162}$$

$$\Delta\rho/\rho B^2 = (\tfrac{3}{8}\pi)^2(4/\pi - 1)\mu_n^2 \approx 0.4\mu_n^2. \tag{2.163}$$

Identical expressions hold for holes, with the sign of the Hall angle and Hall coefficient reversed, and n_b, μ_n, $\mu_{n\mathrm{H}}$ replaced by p_b, μ_p, $\mu_{p\mathrm{H}}$, respectively. The observation of a *positive* Hall coefficient in certain materials constituted one of the earliest and most striking pieces of evidence for the existence of holes in solids.

If both electrons and holes are present, (2.157) must be summed over the two types of carriers. When lattice scattering prevails the Hall coefficient is given by

$$R_H = \frac{3\pi}{8q} \frac{p_b \mu_p^2 - n_b \mu_n^2}{(p_b \mu_p + n_b \mu_n)^2}. \tag{2.164}$$

The sign of R_H depends on both the conductivity-type and the mobility ratio.

For non-spherical energy bands, the analysis of galvanomagnetic effects is considerably more complicated and involves details of the band structure. In germanium and silicon (as in any cubic crystal), the Hall effect is isotropic. The magnetoresistance, being a second-order effect, is anisotropic. Measurements of Hall effect and magnetoresistance have proved to be powerful tools in the study of the band structure of these and many other semiconductors.

7. Non-equilibrium phenomena

So far we have been concerned with situations in which the carrier densities were constant at their equilibrium values. In this section we consider departures from thermal equilibrium and the electronic processes taking place in the semiconductor under these conditions.

7.1. LIFETIME OF EXCESS CARRIERS

Thermal equilibrium may be upset by many methods. Photons of energy greater than the forbidden gap can excite electrons from the valance into the conduction band, producing electron–hole pairs (see § 3.4). X-rays, energetic electrons, or nuclear particles are similarly effective. As we shall see in § 7.4, minority carriers may be injected into the bulk from p–n junctions or metal–semiconductor contacts. Through all these processes, the densities of electrons and of holes are generally altered both in the valence and conduction bands and on localized imperfection levels in the forbidden gap.

In order to characterize the processes that tend to restore thermal equilibrium, let us consider a homogeneous semiconductor in which a flash of uniform excitation has initially produced an equal excess of free holes and electrons. Subsequently these carriers may disappear in pairs by transitions across the gap, or they may fall into localized levels. Preservation of electrical neutrality requires that the total of free and bound hole and electron excess densities be equal. If the net rates of fall-in are equal for both types of carriers, then the overall occupation of the levels remains unchanged and the decay of excess electrons and of excess holes proceeds at the same rate. The decay can then be characterized by a common lifetime τ defined by the relation [†]

$$U = -\frac{d(\delta p_b)}{dt} = -\frac{d(\delta n_b)}{dt} = \frac{\delta p_b}{\tau} = \frac{\delta n_b}{\tau}, \qquad (2.165)$$

where U is the net rate of recombination per unit volume and $\delta p_b \equiv p_b^* - p_b = \delta n_b \equiv n_b^* - n_b$ is the excess density of holes and electrons. Throughout this book the starred notation will be employed to designate carrier densities under non-equilibrium conditions.

For sufficiently small δp_b, the lifetime may be considered a constant. In this case the decay of excess carriers following the cessation of external excitation is exponential with a time constant equal to τ:

$$\delta p_b = \delta n_b = (\delta p_b)_0 e^{-t/\tau}. \qquad (2.166)$$

Under *constant* external excitation, steady-state conditions are reached when the net rates of volume excitation $\mathscr{L}$ and recombination U are equal. The steady-state excess densities are given by

$$\delta p_b = \delta n_b = \mathscr{L}\tau \quad \text{for} \quad \mathscr{L} = U. \qquad (2.167)$$

When holes and electrons fall into localized levels at unequal rates, carriers of one type are preferentially captured or *trapped* in bound states. This may lead to an appreciable inequality in the free hole and electron densities and thus to different hole and electron lifetimes. Trapping effects, however, are negligible even for unequal rates of capture, provided the density of localized levels is sufficiently small. Conditions of "no trapping" will be assumed to hold throughout the subsequent considerations.

7.2. RECOMBINATION PROCESSES

The various processes of recombination fall into two classes, depending on whether the electrons and holes recombine directly by band-to-band transitions, or indirectly via intermediate localized energy levels in the forbidden gap. Each process can be characterized by an average capture cross section A for recombination. In terms of this cross section, the average probability K that in unit time a free carrier make a transition to a localized level or across the gap is given by $\langle c \rangle A$ per localized level or per free carrier of the opposite sign ($\langle c \rangle$ is the thermal velocity). Measured values of cross section

---

[†] It should be noted that the same symbol τ is used for both the lifetime and the relaxation time (see § 6). There is little danger of confusion, however, since the two quantities relate to altogether different processes.

have a vast range, from 10^{-22} to 10^{-12} cm^2. A commonly reported value for localized levels is 10^{-15} cm^2, which is the magnitude to be expected from the physical dimensions of an atom or an ion.

The cross section for any given process is influenced greatly by the manner in which the carriers are able to release energy when making transitions to lower energy states. In direct recombination, the electrons must release an amount of energy approximately equal to the full energy gap, while an equal amount of energy must be disposed of in two stages in the indirect process. The following physical mechanisms have been envisaged for energy release: (1) phonon emission, (2) radiation of photons, and (3) transfer to another free carrier (Auger process). For direct recombination light emission is generally the important mechanism, while for indirect transitions phonon emission predominates.

Carrier lifetime is usually found to be an extremely structure-sensitive quantity. In different samples of the same material, variations in lifetime of many orders of magnitude have been found with but minor differences in other electrical properties. For example, lifetimes in germanium have been reported ranging from 10^{-8} to 10^{-2} sec, depending on the growth conditions and subsequent treatment of the crystal. Even the highest reported value of lifetime is still much lower than that estimated from theoretical considerations for radiative band-to-band transitions. It is thus apparent that indirect transitions via localized imperfection levels generally play the dominant role in recombination even in the most perfect crystals available.

As has been mentioned above, the dominant mechanism of energy release involved in a transition via a localized state is phonon emission. This mechanism, however, is by no means a simple one. The transfer of energy to the lattice requires the excitation of many phonons, because even the highest phonon energy is usually much smaller than the energy that has to be transferred (at least half the forbidden gap). Such a multi-phonon emission is an extremely unlikely process. It has therefore been proposed that the electron cascades into the ground state of an imperfection centre via its excited states, emitting a single phonon in each stage. A similar process is suggested for hole capture by the centre.

The statistical treatment of indirect recombination, as given by Shockley and Read[16], will be outlined briefly in the present section. A more detailed discussion is deferred to Ch. 5, §§ 7, 8, where the similar problem of surface recombination is treated. Consider a non-degenerate homogeneous semiconductor under uniform and steady excitation. The recombination centres, with density N_t, are all assumed to be at an energy E_t (and effective energy E_t^f) in the forbidden gap; each centre is capable of capturing one electron when empty and one hole when occupied by an electron, the capture probabilities for these processes being K_n and K_p, respectively. Assuming that the thermal velocity of electrons and holes is the same, we may express the capture probabilities in terms of the capture cross sections in the form $K_n = \langle c \rangle A_n$, $K_p = \langle c \rangle A_p$. The rate (per unit volume) at which free holes (p_b^*) are captured by occupied centres (n_t^*) is given by

$$R_{tv} = K_p p_b^* n_t^*. \tag{2.168}$$

Similarly the rate of capture of free electrons (n_b^*) by empty centres ($N_t - n_t^* = p_t^*$) is

$$R_{ct} = K_n n_b^* p_t^*. \tag{2.169}$$

Simultaneously with the capture of free carriers, bound carriers are continuously re-emitted thermally from the centres back into the bands. The rates of these reverse processes for holes (G_{vt}) and for electrons (G_{tc}) are proportional to the densities of vacant (p_t^*) and occupied (n_t^*) centres, respectively. The proportionality factors can be obtained in terms of K_p, K_n, and E_t^f from the principle of detailed balance, which states that at thermal equilibrium $R_{tv} = G_{vt}$, $R_{ct} = G_{tc}$. A calculation based on these equations — for which the free and bound carrier densities assume their *equilibrium* values p_b, n_b, p_t, n_t — yields for the generation rates

$$G_{vt} = K_p p_1 p_t^*; \tag{2.170}$$

$$G_{tc} = K_n n_1 n_t^*, \tag{2.171}$$

where

$$p_1 \equiv p_b n_t/p_t = N_v e^{-(E_t^f - E_v)/kT}; \tag{2.172}$$

$$n_1 \equiv n_b p_t/n_t = N_c e^{-(E_c - E_t^f)/kT}. \tag{2.173}$$

It is seen that p_1 and n_1 are equal to the free hole and electron densities for the case in which the Fermi level coincides with the effective energy E_t^f.

The net transition rates from the valence and conduction bands to the recombination centres are

$$U_p = R_{tv} - G_{vt} = K_p(p_b^* n_t^* - p_1 p_t^*); \tag{2.174}$$

$$U_n = R_{ct} - G_{tc} = K_n(n_b^* p_t^* - n_1 n_t^*). \tag{2.175}$$

Under steady-state conditions $U_p = U_n$, a requirement used to evaluate the

Ch. 2, § 7.3 NON-EQUILIBRIUM PHENOMENA 79

fraction n_t^*/N_t of centres occupied by electrons. Substituting this value for n_t^* in (2.174) or (2.175), we obtain for the net steady-state hole-electron recombination rate U the expression

$$U = U_p = U_n = N_t K_p K_n \frac{p_b^* n_b^* - p_b n_b}{K_p(p_b^* + p_1) + K_n(n_b^* + n_1)}. \quad (2.176)$$

For the case we are considering — no trapping and low-level excitation —

$$\delta p_b = \delta n_b \ll p_b + n_b, \quad (2.177)$$

and the common lifetime is given by (see eq. (2.167))

$$\tau = \frac{\delta p_b}{U} \approx \frac{1}{K_p N_t} \frac{n_b + n_1}{n_b + p_b} + \frac{1}{K_n N_t} \frac{p_b + p_1}{n_b + p_b}. \quad (2.178)$$

It can readily be seen that the lifetime reduces to $1/K_p N_t$ or $1/K_n N_t$ for extrinsic n- or p-type samples, respectively. In either case the rate limiting process in recombination is the capture of *minority* carriers; the majority carriers are so numerous that they will immediately recombine with the captured minority carriers before the latter can be re-emitted into the appropriate band. The lifetime for less extrinsic samples is larger, since thermal re-emission hampers the recombination.

7.3. DRIFT AND DIFFUSION OF EXCESS CARRIERS

In the preceding discussion of the physical processes involved in recombination, it was sufficient to consider a homogeneous semiconductor with uniform volume excitation. In practice, the carrier densities are often functions of position (either because of the non-homogeniety of the semiconductor or because of non-uniform excitation) and, in addition, an external electric field is usually present. It is therefore necessary to formulate the appropriate equations governing the behaviour of carriers under these more general conditions. The discussions that follow will be confined to non-degenerate semiconductors.

When the densities are not uniform, carriers tend to diffuse from regions of high to regions of low concentration. Thus, in addition to the *drift* current due to the external field, a *diffusion* current is also established. If the latter is assumed to be directly proportional to the concentration gradient, an assumption amply borne out by experiment, we may write the hole and electron current densities as

$$J_p = q\mu_p p^* \mathscr{E} - qD_p \operatorname{grad} p^*;$$
$$J_n = q\mu_n n^* \mathscr{E} + qD_n \operatorname{grad} n^*; \qquad (2.179)$$
$$J = J_p + J_n.$$

Here p^* and n^* are the position-dependent hole and electron densities (the omission of the subscript b indicates that the carrier densities in the bulk material are not necessarily uniform), and D_p and D_n are the so-called diffusion constants for holes and electrons; J is the total current density. The signs in front of the diffusion terms are in accordance with the convention regarding the current polarity.

The relation between the diffusion constant and the mobility of either carrier type is obtained by considering an inhomogeneous semiconductor under thermal equilibrium. For this case p^* and n^* in (2.179) are replaced by their respective equilibrium values p and n. According to the principle of detailed balance, in the absence of external fields the hole and electron currents must vanish separately. Let us consider first the electrons and assume that a concentration gradient exists only in the x direction. Using (2.179), we have

$$\mu_n n \mathscr{E}_x = -D_n (dn/dx). \qquad (2.180)$$

The field $\mathscr{E}_x$ is established internally to counteract the diffusion flow. If V denotes the electrostatic potential set up in this manner, then

$$\mathscr{E} = -\operatorname{grad} V; \quad \mathscr{E}_x = -dV/dx. \qquad (2.181)$$

Combining (2.180) and (2.181) we obtain

$$n = \operatorname{const.} e^{\mu_n V / D_n}. \qquad (2.182)$$

Since $-qV$ is the potential energy of an electron, we may express the Maxwell–Boltzmann distribution law (see eq. (2.71)) in the form

$$n = A\, e^{qV/kT}, \qquad (2.183)$$

where A is a normalizing constant.

Thus, it follows from (2.182) and (2.183) that

$$D_n = (kT/q)\mu_n. \qquad (2.184)$$

In a similar manner one obtains for holes

$$D_p = (kT/q)\mu_p. \qquad (2.185)$$

These are the well-known Einstein relations between mobility and diffusion constant. The units of D_n and D_p are m^2/sec if μ is in $m^2/V\cdot\text{sec}$ and cm^2/sec if μ is in $cm^2/V\cdot\text{sec}$.

Due to the dependence on position of the potential V in an inhomogeneous semiconductor, an electron in, say, the lowest state of the conduction band will have different energies in different parts of the crystal. On the other hand, at thermal equilibrium the Fermi level E_F must be constant throughout the crystal (see § 5.1). Using (2.89) and (2.183) we have

$$n = N_c e^{(E_F - E_c)/kT} = A e^{(qV/kT)}.$$

Thus the edge of the conduction band must be related to the potential by

$$E_c = \text{const} - qV. \qquad (2.186)$$

The edge of the valence band, as well as all other levels in the crystal, follows the variation of the potential just as E_c does. This behaviour is a consequence of the fact that the separations between these various energy levels are determined by the large, short-range crystal fields and therefore remain practically unaffected by the much weaker fields associated with the inhomogeneity of the crystal. The energy-band diagram of an inhomogeneous semiconductor is schematically represented in Fig. 2.18 and should be contrasted with that for a homogeneous semiconductor (Fig. 2.8). Also shown in Fig. 2.18 is the corresponding variation of the electrostatic potential V. As is well known, the zero of the electrostatic potential may be defined arbitrarily. It will be convenient here to choose this zero in such a manner that

$$-qV = E_i = \tfrac{1}{2}[E_c + E_v - kT \ln(N_c/N_v)]. \qquad (2.187)$$

The energy E_i lies approximately midway between E_v and E_c and corresponds, in a homogeneous sample, to the intrinsic Fermi level (eq. (2.88)). With this choice, we may express the carrier densities at any point in the crystal as

$$\begin{aligned} p &= n_i e^{-(qV + E_F)/kT}; \\ n &= n_i e^{(qV + E_F)/kT}, \end{aligned} \qquad (2.188)$$

where n_i is the intrinsic carrier density (eq. (2.87)).

It should be noted that in the foregoing discussion, thermal equilibrium conditions apply only as far as the carrier densities are concerned. The impurities and other imperfection centers giving rise to the inhomogeneity of the crystal do not of course have an equilibrium distribution – they maintain their position in the lattice only because of their vanishingly small mobility at the temperatures considered.

We now return to non-equilibrium situations. The combined action of diffusion and drift may be conveniently expressed by means of the *quasi* Fermi levels F_p and F_n for holes and electrons. These are defined in terms of the carrier densities p^*, n^* and the potential V by

$$p^* = n_i e^{-(qV+F_p)/kT};$$
$$n^* = n_i e^{(qV+F_n)/kT}.$$
(2.189)

Thus, the hole and electron current densities (eqs. (179)) may be written as

$$\mathbf{J}_p = -q\mu_p p^* \text{ grad } V + (qD_p/kT)p^* \text{ grad }(qV+F_p);$$
$$\mathbf{J}_n = -q\mu_n n^* \text{ grad } V + (qD_n/kT)n^* \text{ grad }(qV+F_n).$$

Using the Einstein relation we obtain

$$\mathbf{J}_p = \mu_p p^* \text{ grad } F_p;$$
$$\mathbf{J}_n = \mu_n n^* \text{ grad } F_n.$$
(2.190)

The quasi Fermi levels F_p and F_n are not to be confused with the actual

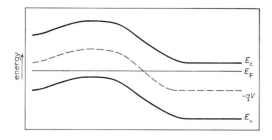

Fig. 2.18. Energy-level and potential diagrams in an inhomogeneous semiconductor.

Fermi level, defined only under equilibrium conditions. At thermal equilibrium, of course, F_p and F_n reduce to E_F.

The behaviour of carriers in a semiconductor under conditions in which the densities are functions of time and position is governed by the continuity equation. This equation expresses the principle of particle conservation, taking into account recombination and external excitation. The general form of the continuity equation is

$$\partial p^*/\partial t = \mathscr{L} - U_p - (1/q) \operatorname{div} \mathbf{J}_p;$$
$$\partial n^*/\partial t = \mathscr{L} - U_n + (1/q) \operatorname{div} \mathbf{J}_n. \tag{2.191}$$

Here $\mathscr{L}$ is the external volume excitation rate and U_p and U_n are the net rates of hole and electron decay due to recombination and trapping. The third term in each equation represents the divergence of carriers by diffusion and drift (eqs. (2.179)).

We shall discuss a few limiting cases for which the solution of the continuity equation is not too involved. Consider a homogeneous semiconductor under *non-uniform* volume excitation and an applied electric field. For low-level excitation and no trapping, carrier recombination is characterized by a common constant lifetime τ. Under these conditions the continuity equations may be written as

$$\frac{\partial(\delta p)}{\partial t} = \mathscr{L} - \frac{\delta p}{\tau} - \mu_p p^* \operatorname{div} \mathscr{E} - \mu_p \mathscr{E} \cdot \operatorname{grad}(\delta p) + D_p \operatorname{div} \operatorname{grad}(\delta p);$$
$$\frac{\partial(\delta n)}{\partial t} = \mathscr{L} - \frac{\delta n}{\tau} + \mu_n n^* \operatorname{div} \mathscr{E} + \mu_n \mathscr{E} \cdot \operatorname{grad}(\delta n) + D_n \operatorname{div} \operatorname{grad}(\delta n), \tag{2.192}$$

where use has been made of (2.179).

We shall first show that the excess hole and electron densities δp and δn must be nearly equal. Unless this is the case, a space charge will be set up, and we have from Poisson's equation

$$\operatorname{div} \mathscr{E} = (q/\varepsilon_0 \kappa)(\delta p - \delta n), \tag{2.193}$$

where $\varepsilon_0 \kappa$ is the permittivity. In a typical n-type germanium sample $n_b \approx 10^{14}$ cm^{-3} and $\delta p \approx 10^{12}$ cm^{-3}. A departure of 1% from the equality $\delta p = \delta n$ results in $d\mathscr{E}_x/dx \approx 10^3$ V/cm^2, a very large field gradient. Fields of this magnitude are hardly ever encountered in a homogeneous semiconductor. (We shall see below, however, that strong fields are quite common at the surface and in inhomogeneous structures such as p–n junctions.)

The neutrality condition greatly simplifies matters. By multiplying the first line of (2.192) by $\mu_n n^*$, the second line by $\mu_p p^*$ and adding, we obtain

$$\frac{\partial(\delta p)}{\partial t} = \mathscr{L} - \frac{\delta p}{\tau} + D \text{ div grad }(\delta p) - \mu \mathscr{E} \cdot \text{grad }(\delta p) \quad \text{for} \quad \delta p = \delta n, \quad (2.194)$$

where

$$D = \frac{p^* + n^*}{p^*/D_n + n^*/D_p}, \quad \mu = \frac{n^* - p^*}{p^*/\mu_n + n^*/\mu_p}. \quad (2.195)$$

D and μ are called the *ambipolar diffusivity* and *ambipolar mobility*. For small δp (eq. (2.177)), μ and D are approximately independent of excess density:

$$D = \frac{p_b + n_b}{p_b/D_n + n_b/D_p}; \quad \mu = \frac{n_b - p_b}{p_b/\mu_n + n_b/\mu_p}. \quad (2.196)$$

The ambipolar diffusivity and mobility characterize the motion of both the excess minority and the excess majority carrier densities. For extrinsic material ($n_b \gg p_b$ or $p_b \gg n_b$), D and μ reduce to the *minority* carrier diffusion constant and mobility. In this case the motion of minority carriers is unaffected by the presence of the majority carriers, the latter being able to follow the former by means of a comparatively small distortion in their spatial distribution. For less extrinsic material, the relative change in the majority carrier distribution is no longer negligible; both carrier excesses — which, from the neutrality requirement, must always move at the same rate — are characterized by a diffusivity and mobility given by (2.196).

For planar geometry ($\mathscr{E}$ along the x direction, δp a function of x only), we have from (2.194) that

$$\frac{\partial(\delta p)}{\partial t} = \mathscr{L} - \frac{\delta p}{\tau} + D \frac{\partial^2(\delta p)}{\partial x^2} - \mu \mathscr{E} \frac{\partial(\delta p)}{\partial x}. \quad (2.197)$$

Suppose that carrier excitation takes place only very close to the plane $x = 0$, so that elsewhere $\mathscr{L} = 0$. (This may be accomplished in practice by illuminating a filament in a narrow strip.) Three cases will be considered, and in all of these low-level excitation (eq. (2.177)) is assumed.

Case a: Steady excitation in the absence of a field. Here $\partial(\delta p)/\partial t = 0$ under steady-state conditions, and the solution of (2.197) is

$$\delta p = \delta p(0) e^{-|x|/(D\tau)^{\frac{1}{2}}}, \quad (2.198)$$

where $\delta p(0)$ is the excess carrier density at $x = 0$. Thus the carriers diffuse symmetrically in both directions, away from the plane of excitation. The length $(D\tau)^{\frac{1}{2}}$, referred to as the *diffusion length*, represents the average distance an excess carrier diffuses before it recombines. One of the earliest methods of measuring lifetime employed a configuration based on such considerations.

Case b: Steady excitation under strong fields. For sufficiently strong fields, the diffusion term in (2.197) may be neglected. The solution then becomes

$$\begin{aligned} \delta p(x) &= \delta p(0) e^{-x/\mu \mathscr{E} \tau} \quad \text{for } x/\mu > 0; \\ \delta p(x) &= 0 \quad \text{for } x/\mu < 0. \end{aligned} \quad (2.199)$$

The so-called *drift length* $\mu \mathscr{E} \tau$ is the average distance an excess carrier drifts under the action of the field before recombining. It is evident that neglecting the diffusion term in (2.197) is permissible as long as $\mu \mathscr{E} \tau \gg (D\tau)^{\frac{1}{2}}$.

For an extrinsic n-type sample $\mu = \mu_p$ (see eqs. (2.196)), and both holes and electrons drift to the *right* (in the direction of the field). For an extrinsic p-type sample $\mu = -\mu_n$, and both carriers move to the *left*; the drift length is now that of the electrons, the minority carriers in this case. For an intrinsic semiconductor ($p_b = n_b$) $\mu = 0$, and (2.199) do not apply. The solution is then identical to that of case *a*, with D equal to $D_p D_n/(D_p + D_n)$. Thus the excess carrier distribution is not affected by the field, diffusion being the only controlling process.

Case c: Flash excitation under strong fields. Here a short flash of excitation is applied at $t = 0$, producing a narrow distribution of carriers (of density $\delta p(0, 0)$) at the plane $x = 0$. Equation (2.197) now becomes

$$\frac{\partial(\delta p)}{\partial t} + \mu \mathscr{E} \frac{\partial(\delta p)}{\partial x} = -\frac{\delta p}{\tau}. \quad (2.200)$$

In a frame of reference moving with a velocity $\mu \mathscr{E}$, so that $x = \mu \mathscr{E} t$, (2.200) reduces to

$$\frac{d(\delta p)}{dt} = -\frac{\delta p}{\tau}.$$

The solution is given by

$$\begin{aligned} \delta p(x, t) &= \delta p(0, 0) e^{-t/\tau} \quad \text{for } x = \mu \mathscr{E} t; \\ \delta p(x, t) &= 0 \quad \text{otherwise.} \end{aligned} \quad (2.201)$$

Thus, the disturbance in carrier density produced by the flash excitation moves with a drift velocity $\mu \mathscr{E}$, decaying exponentially as it advances. In an extrinsic n-type sample the disturbance moves to the right with a drift velocity $\mu_p \mathscr{E}$, in a p-type sample it moves to the left with a drift velocity $\mu_n \mathscr{E}$. The drift velocity in an intrinsic sample is zero.

Actually, due to diffusion, the pulse of excess carriers broadens as it advances. The peak, however, moves with the velocity $\mu \mathscr{E}$ and is attenuated exponentially, as in the absence of diffusion. By detecting the pulse at a subsequent time and observing its shape, it is possible to measure μ, D, and τ. This method, known as the Haynes-Shockley technique, has been widely used for the direct measurement of minority-carrier mobilities in semiconductors.

7.4. P–N JUNCTIONS

Perhaps the most common inhomogeneous structure in semiconductors is the p–n junction. Such a junction consists of the transition zone between two regions of the crystal of opposite conductivity type. It may be produced by converting one part of the crystal from p to n type (or from n to p type) by diffusing-in donor (or acceptor) impurities. Alternatively, a p–n junction can be formed during crystal growth by proper control of impurity segregation.

In general the transition in conductivity type is not abrupt, but in order to simplify matters we shall assume that such is the case. At the junction there is a large concentration gradient of holes and electrons. As a result, holes tend to diffuse from the p to the n region and electrons from the n to the p region. The space charge formed in this manner establishes a potential barrier V_0 that is just sufficient to counteract the diffusion. Figure 2.19a is a schematic representation of a p–n junction under thermal equilibrium. Except near the junction, the p and n regions are assumed uniform. (The electron and hole densities are denoted by n_p and p_p in the p region and by n_n and p_n in the n region.)

We shall now derive the rectification characteristics of a p–n junction provided with two (non-rectifying) metal contacts at its two ends X_p and X_n. Suppose that a forward bias V_a is applied across the contacts (p side positive), as shown in Fig. 2.19b. The applied voltage reduces the potential barrier at the junction and, as a result, "injection" of holes from the p into the n region and of electrons from the n into the p region takes places. We assume that the uniform sections are sufficiently wide so that excess carriers can recombine before reaching the end contacts. Hence the quasi Fermi levels F_p and F_n at each end coincide and differ from E_i by the same amount as in the equilibrium case. Referring to Fig. 2.19b, we thus have

$$F_p(X_p) = F_n(X_p) = F_p(X_n) - qV_a = F_n(X_n) - qV_a. \tag{2.202}$$

Next we shall show that F_p is nearly constant in the p region and increases by qV_a in the n region while F_n is nearly constant in the n region and drops by qV_a in the p region, as indicated in Fig. 2.19b. To do this we consider

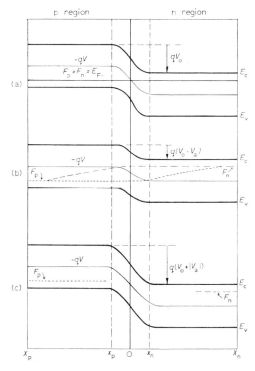

Fig. 2.19. Energy-level diagram for a p–n junction (a) at thermal equilibrium, (b) under a forward bias ($V_a > 0$), and (c) under a reverse bias ($V_a < 0$). F_p and F_n are the quasi Fermi levels for holes and electrons, $-qV$ is the potential energy of an electron.

the points x_p and x_n just outside the space-charge region. At x_p we have from (2.190)

$$J_p(x_p) = \mu_p p_p^*(x_p) \frac{dF_p}{dx}\bigg|_{x_p}. \tag{2.203}$$

The transition region will be assumed small compared to the diffusion lengths for holes and electrons (see § 7.3). Under these conditions, recombination in the junction is negligible and the same hole current flows at x_n:

$$J_p(x_p) = J_p(x_n) = \mu_p\, p_n^*(x_n)\left.\frac{dF_p}{dx}\right|_{x_n}. \qquad (2.204)$$

Thus

$$\left.\frac{dF_p}{dx}\right|_{x_n} \div \left.\frac{dF_p}{dx}\right|_{x_p} = \frac{p_p^*(x_p)}{p_n^*(x_n)}. \qquad (2.205)$$

For sufficiently low applied voltage such that the injected minority density p_n^* is small compared to the majority density n_n, this ratio is very large, since the two regions are assumed to be extrinsic. Similar arguments apply to F_n. We thus conclude that the major changes in F_p and F_n occur in the regions where the corresponding carriers are *minority* carriers, as was maintained above.

The total current $J = J_p + J_n$ may be conveniently obtained by evaluating J_p at x_n and J_n at x_p. This procedure, rather than summing up the hole and electron currents at the same point, is permissible in view of the negligible recombination in the junction. As compared to its equilibrium position, F_p at x_n is nearer E_v by the amount qV_a (see Fig. 2.19b). Use of (2.188) and (2.189) therefore leads to

$$p_n^*(x_n) = p_n\, e^{qV_a/kT};$$
$$\delta p_n(x_n) = p_n^*(x_n) - p_n = p_n(e^{qV_a/kT} - 1). \qquad (2.206)$$

Similarly,

$$n_p^*(x_p) = n_p\, e^{qV_a/kT};$$
$$\delta n_p(x_p) = n_p^*(x_p) - n_p = n_p(e^{qV_a/kT} - 1). \qquad (2.207)$$

Thus the excess minority carrier densities just outside the junction increase exponentially with applied voltage. This is the justification of our usage of the term *injection*: the applied voltage introduces excess minority carriers into two regions. From the points x_p, x_n outwards the hole and electron currents are carried by diffusion, since the field outside the junction is very small.

Using (2.198) for diffusion away from a plane source of excess carriers, we obtain

$$J_p(x_n) = -qD_p\left.\frac{\partial(\delta p_n)}{\partial x}\right|_{x_n} = \frac{qp_n D_p}{\sqrt{D_p \tau_p}}(e^{qV_a/kT} - 1), \qquad (2.208)$$

where τ_p is the hole lifetime in the n region. A similar expression holds for $J_n(x_p)$. The total current density is given by

$$J = J_p(x_n) + J_n(x_p) = (J_{ps} + J_{ns})(e^{qV_a/kT} - 1), \qquad (2.209)$$

where

$$J_{ps} = \frac{qp_n D_p}{\sqrt{D_p \tau_p}} \; ; \qquad J_{ns} = \frac{qn_p D_n}{\sqrt{D_n \tau_n}} . \qquad (2.210)$$

The same arguments apply to the case of reverse bias ($V_a < 0$) shown in Fig. 2.19c, and eqs. (2.209), (2.210) hold for this range as well.

Equations (2.209), (2.210) express the rectification characteristics of a p–n junction. In the forward direction ($V_a > 0$) the current increases exponentially with applied voltage; in the reverse direction the exponential term vanishes and the current approaches the saturation current $-(J_{ps} + J_{ns})$.

Similar rectification and injection phenomena may occur at a metal-semiconductor contact. A potential barrier analogous to that in a p–n junction often exists at the semiconductor side of the contact, as will be discussed in Ch. 4, § 1.

References

[1] W. Shockley, *Electrons and Holes in Semiconductors* (Van Nostrand, New York, 1950).
[2] W. C. Dunlap, Jr., *An Introduction to Semiconductors* (John Wiley, New York, 1957).
[3] W. Ehrenberg, *Electric Conduction in Semiconductors and Metals* (Clarendon Press, Oxford, 1958).
[4] E. Spenke, *Electronic Semiconductors* (McGraw-Hill, New York, 1958).
[5] A. F. Ioffe, *Physics of Semiconductors* (Infosearch, London, 1960).
[6] R. A. Smith, *Semiconductors* (University Press, Cambridge, 1961).
[7] E. M. Conwell, Proc. Inst. Radio Engrs. **46** (1958) 1281.
[8] N. B. Hannay (Editor), *Semiconductors* (Reinhold, New York, 1959).
[9] A. F. Gibson, P. Aigrain, and R. E. Burgess (Editors), *Progress in Semiconductors* (Heywood, London) vol. I (1956), vol. II (1957), vol. III (1958);
A. F. Gibson, F. A. Kröger, and R. E. Burgess (Editors), *Progress in Semiconductors* (Heywood, London) vol. IV (1960), vol. V (1960), vol. VI (1962), vol. VII (1964).
[10] A. Sommerfeld, Z. Physik **47** (1928) 1, 43; A. Sommerfeld and H. Bethe in *Handbuch der Physik* (Edited by H. Geiger and K. Scheel), vol. **24/2** (Springer-Verlag, Berlin, 1933), p. 333.
[11] W. Heitler and F. London, Z. Physik **44** (1927) 455.
[12] F. Bloch, Z. Physik **52** (1928) 555.
[13] A. H. Wilson, Proc. roy. Soc. London A **133** (1931) 458; A**134** (1931) 277.
[14] R. de L. Kronig and W. G. Penney, Proc. roy. Soc. London A**130** (1931) 499.
[15] H. Brooks in *Advances in Electronics and Electron Physics* (Edited by L. Marton), vol. 7 (Academic Press, New York, 1955), p. 85.
[16] W. Shockley and W. T. Read, Phys. Rev. **87** (1952) 835.
[17] J. N. Hobstetter in *Semiconductors* (Edited by N. B. Hannay), (Reinhold, New York, 1959), p. 511.
[18] W. T. Read, *Dislocations in Crystals* (McGraw-Hill, New York, 1953), p. 16.

CHAPTER 3

THE LATTICE STRUCTURE AND CHEMICAL REACTIVITY OF THE SURFACE

In the previous chapter we summarized the bulk properties of a semiconductor viewed as an infinite periodic medium. We now turn to the surface of a finite crystal and begin by reviewing in this chapter the main structural and chemical characteristics associated with the abrupt termination of the crystal lattice. The different environment of the topmost atomic layers as compared to that of parallel layers within the bulk usually gives rise to a rearrangement of the surface atoms, while the special position of such outer atoms and the unsaturated bonds associated with them promote a strong interaction with various species in the surrounding media.

The most fundamental information on both the structure and reactivity of the surface has been obtained from studies of clean surfaces. Such studies were made possible by the advent of ultra-high vacuum techniques by Alpert [1]. Working pressures as low as 10^{-10} mm Hg are now quite common[2], and they enable one to maintain an atomically clean surface for hours before a monolayer of gas is adsorbed. Two approaches have been used in obtaining clean surfaces. In one, a single crystal is cleaved in high vacuum along one of its crystallographic planes or simply crushed into tiny fragments. In either case the surfaces produced are initially clean. In the other, more widely used approach, one starts with a real surface and removes the adsorbed foreign matter by ion bombardment, by heating to elevated temperatures, or by applying high electric fields normal to the surface. Several methods have been developed for determining the degree of cleanliness of the surface. These usually rely on the reproducibility of results following successive cleaning cycles. The exact criterion of a clean surface thus depends on the sensitivity of the method employed. Very direct evidence for the cleanliness of ion bombarded surfaces has been provided by comparison with data obtained [3] on cleaved surfaces, the latter being of course atomically clean.

A considerable effort has been devoted to the study of the lattice structure of clean surfaces using low-energy electron diffraction. Clean surfaces, properly annealed, display long-range order, a condition that is essential for obtaining the sharp diffraction patterns used in the determination of the surface lattice spacings. This long-range order is no doubt one of the most striking features of clean surfaces. Other methods, such as field emission and field ion emission microscopy, which are not sensitive to the ordering of the lattice, have also proved very useful. All of these methods are also employed in observing changes in surface structure brought about by the interaction of clean surfaces with different gases and vapours. Clean surfaces are highly reactive and adsorb foreign matter readily. The adsorption is customarily divided into two categories, chemisorption and physical adsorption. The former is characterized by strong, short-range chemical bonding, accompanied by the liberation of large quantities of heat ($\gtrsim 10$ kcal/mole). Oxidation is a typical example of adsorption of this type. Physical adsorption, on the other hand, is characterized by long-range, van der Waals interaction and low heat of adsorption. Noble-gas adsorption on solid surfaces may be cited as an example of physical adsorption. (By its very nature, however, the distinction between chemisorption and physical adsorption is not sharp.) The coverage of the surface by adsorbed matter may be inferred from measurements of the changes in sample weight or ambient gas pressure. The binding energy, if sufficiently large, can be derived from heat of adsorption measurements.

The first two sections of the present chapter discuss the various techniques employed for producing clean surfaces and for studying their physical and chemical characteristics. Section 3 summarizes the available data on the surface structure and adsorption properties of the diamond-type semiconductors. The last section is concerned with real surfaces — that is, with surfaces prepared by chemical etching and then exposed to the atmosphere or to other gaseous ambients. Here as well, both the reactivity and structural characteristics have been investigated. In contrast to the case of clean surfaces, however, the chemical studies were concerned mainly with the interactions between the surface and the liquid phase. Structure studies were confined mostly to the gross features, particular emphasis having been placed on those surface defects from which information can be obtained on the imperfections in the underlying bulk. Due to the practical importance of real surfaces and the ease with which they can be prepared and handled, a large body of experimental data has been accumulated on their chemical and structural characteristics. It is beyond the scope of this book to give a comprehensive review of the existing literature, and the reader is referred to recent books on the subject [4-6] and to the papers cited therein. What we shall attempt to do here is to give a general picture of the physical state of a real surface and to point out some of the important chemical reactions between the surface and various surrounding media. The subject of surface catalysis, however, will not be dealt with at all since it constitutes an entire field by itself.

1. Preparation of clean surfaces

1.1. CLEAVAGE

This is perhaps the most direct method of preparing clean surfaces [7,8]. It is used mainly for structural and electrical studies, the areas obtained being too small for adsorption measurements. The samples are mounted on a rod and placed inside a high vacuum system. For crystals that cleave easily, such as gallium arsenide and indium antimonide, an externally manipulated hammer or blade is allowed to fall on the sample [9]. In the case of germanium and silicon, however, a different procedure has been found more suitable [7,9,10]. A small scratch is scribed where the cleavage is to commence. The sample is then clamped at one end and a bending force is applied at the other. For silicon the result of this method is a true cleavage in the upper half of the surface (the part under tension during the bending) and a rough fracture in the lower half (under compression) [10]. Electron diffraction studies indicate a high degree of perfection in the top surface layers and the absence of amorphous and foreign material. Electron micrographs show relatively flat regions several thousand angstroms in size separated by steps 50–100 Å high [10]. More sensitive measurements, using special decoration techniques (electroplating of small amounts of metal), indicate that the flat regions actually contain many small steps, presumably monatomic [11].

The preparation of clean surfaces by cleavage is restricted to the cleavage planes of the crystal. In germanium and silicon, as in diamond, these are the (111) planes, whereas in the intermetallic compounds they are generally the (110) planes [12].

1.2. CRUSHING

This procedure results in an exceedingly large surface-to-volume ratio and is very useful for measuring rates or heats of adsorption [13]. Crushing consists in pulverizing (in vacuum) thin wafers of the crystal by repeated blows from a glass hammer. Prior to crushing, the wafers are heated in vacuum or in an atmosphere of hydrogen in order to remove gases (mostly oxygen) from the bulk.

Microscopic examination of a freshly crushed germanium powder shows [14] the existence of particles ranging in size from below a micron to above a millimetre, covered with pits much less than a micron in diameter. The surface area has been evaluated before and after prolonged heating (50 hr at 400–600°C), and a decrease of about one half was found. This is apparently due to surface diffusion of germanium, which promotes a thermal polish [14].

The distribution of the principal crystallographic planes in a crushed powder cannot be measured directly. It is usually estimated [13, 15, 16] from surface energy considerations [12]. The weighted average of the planes is used to calculate the surface density of adsorption sites (assumed equal to the number of broken or dangling bonds).

In order to determine the total surface area one usually uses the BET method, based on the theory of physical adsorption developed by Brunauer, Emmett, and Teller [17]. This theory provides an expression connecting the quantity of gas sufficient for a monolayer coverage with the rate of gas adsorption at a given temperature. Krypton is usually used as the adsorbate [18]. Once the number of molecules that may be accommodated in a monolayer is obtained, one must still determine the area covered by each molecule. Here, again, only theoretical approaches are available [19]. Although the basic assumptions of the BET theory have been questioned, the results obtained using weakly cohesive molecules on moderately heterogeneous surfaces agree well with those arrived at by other, more involved methods [20]. Green [15] estimates that the accuracy in such a surface area determination is 10 %.

1.3. HYDROGEN REDUCTION

Reduction by hydrogen is similar to crushing in that a large surface-to-volume ratio is obtained. Dehydrated germanium dioxide powder is heated at about 600°C in purified hydrogen for several hours [21] and is reduced by the hydrogen to pure germanium and water. Oxygen adsorption data obtained by Dell [22] on a surface prepared in this manner correspond closely to those on crushed germanium. This is the only criterion of the cleanliness of the surface, and on that basis an upper limit of 20 % surface contamination has been proposed [23].

1.4. HEATING IN HIGH VACUUM

In contrast to the methods described above, those discussed in this and the following subsection aim at the cleaning of the *real* surface of etched single-crystal samples. Clean surfaces prepared in this manner are amenable to both structural and adsorption studies because there is no restriction on the choice of the crystallographic plane, and the shape and area of the surface can be completely controlled. Moreover, the ease with which contacts can be applied to regularly-shaped samples makes such surfaces very suitable for electrical measurements.

Thermal cleaning consists in inserting a freshly etched sample into a high vacuum system and heating it to elevated temperatures until all the adsorbed gases and other foreign matter are driven from the surface. In silicon a clean surface can normally be obtained by heating at 1280°C for two minutes, but if a thick oxide layer ($\approx$ 300 Å) is present, heating at 1380°C for five minutes is found to be necessary [24]. For germanium, heating at 690°C for two to four hours is sufficient provided a special etching procedure is employed[25]. The etchant must be continuously diluted with deionized water in such a way that the germanium sample is never exposed to air during the flooding-out process. The purpose of this is to avoid contamination of the surface by remnants of the etchant. Germanium surfaces prepared without flooding cannot be cleaned by heating alone.

Heating at lower temperatures is generally sufficient for the regeneration of initially clean surfaces. For germanium several minutes at 500°C is usually able to restore the clean surface even after several months of exposure to the atmosphere, while a few seconds at 800–900°C is all that is required for silicon[26]. Following exposure to oxygen, initially clean surfaces of gallium antimonide and indium antimonide can be regenerated[27] by heating at 350–400°C. The exact time required (of the order of minutes) depends on the extent of oxygen exposure. It is not possible, however, to regenerate the clean surface of gallium antimonide after the adsorption of carbon dioxide even by raising the temperature to 450°C (close to the melting point) for a period of several hours.

It appears that the diffusion of oxygen into the crystal bulk during the heating process plays an important role in the removal of oxygen from the surfaces of gallium and indium antimonide [27]. Wolsky, on the other hand, finds that the regeneration of germanium is always accompanied by a decrease in sample weight [28], so that in this case desorption, rather than the inward diffusion of oxygen or the outward diffusion of germanium atoms, seems to be the dominant mechanism. Such a conclusion is supported by the work of Rosenberg et al. [14], who propose that the chemisorbed oxygen is liberated by GeO distillation.

1.5. ION BOMBARDMENT AND ANNEALING

Ion bombardment is the most widely used method for producing clean surfaces [29-33]. The sample is placed in a noble-gas atmosphere, usually argon, which is characterized by weak adsorption properties. A convenient ionization technique uses an incandescent filament as a source of electrons [34]. These are accelerated towards a positive grid, some managing to pass through and ionize the gas beyond. The positive ions are accelerated towards the negatively-biased sample.

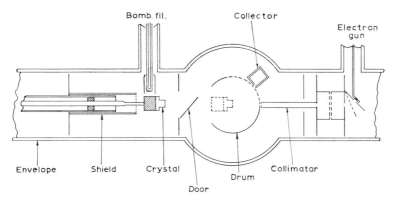

Fig. 3.1. Schematic outline of experimental arrangement used for ion bombardment and electron diffraction studies. The sample is first cleaned by ion bombardment in the left compartment and then introduced through the door into the central chamber where diffraction measurements are carried out. (After Farnsworth et al., reference 37.)

Farnsworth and his co-workers [29] were the first to use this method systematically for obtaining clean germanium and silicon surfaces. A schematic diagram of their apparatus is shown in Fig. 3.1. The system is pumped and baked until a pressure of 10^{-9} mm Hg or less is attained. The sample, mounted on a molybdenum support, is then thoroughly outgassed at elevated temperatures. Finally, very pure argon is introduced and ion bombardment is carried out for several minutes. Typically, the argon pressure is 10^{-3}–10^{-4} mm Hg and the accelerating potential is a few hundred volts, producing an ion current density of 10–100 $\mu a/cm^2$ at the bombarded surface.

During the bombardment, lattice atoms are displaced from their normal sites and argon ions are adsorbed at the surface. The damage can be removed by annealing at elevated temperatures, where the mobility of the defects at the surface is sufficient to restore the ordered lattice. At the same time, such heating (carried out in high vacuum) serves to desorb the argon atoms. The optimum conditions for obtaining a clean surface consist of a series of bombardments of a few minutes each followed by annealing for one or two minutes at a temperature of 500°C for Ge, 700°C for Si [35], 350–400°C for GaSb and InSb [27]. Alternatively, cleaning and annealing can be carried out simultaneously by keeping the sample hot during the bombardment [28]. The total bombarding time necessary under such conditions is between half an hour and an hour. The sputtering rates of argon bombarded surfaces have been determined by measurements of weight changes [31]. The threshold energy for atom ejection from a (100) germanium surface by argon ions is found to be 46 eV.

Electron diffraction measurements (see § 2.1 below) indicate that surfaces prepared by ion bombardment are covered by less than 5 % of a monolayer of foreign atoms.

2. Measurements on clean surfaces

2.1. LOW-ENERGY ELECTRON DIFFRACTION

The first experiments utilizing the diffraction of low-energy electrons for determining the structure of the topmost atomic layers on a crystal surface were carried out by Davisson and Germer [36]. Farnsworth and his co-workers [29,30] developed this technique and applied it extensively to the study of semiconductor surfaces. In their experimental arrangement, a mono-energetic electron beam from an electron gun impinges on the surface of the sample. The surface structure is deduced from the diffraction patterns of the reflected electrons. Because of the low energy of the electrons (10–150 eV), penetration is not significant and the reflection is controlled by the top few atomic layers. Typically, for 50 eV electrons, about 75 % of the reflected beams originate in the outermost surface layer. This method is thus very sensitive to the structure of the surface whether in the clean state or in the presence of adsorbed foreign matter.

For the purpose of analysing the diffraction patterns, one can look upon the surface as consisting of rows of scattering centres — the atoms — which diffract as the lines of a diffraction grating. Usually the direction of the impinging electron beam is chosen normal to the surface. If, in addition, the plane formed by the incident and reflected beams is perpendicular to the atomic rows, then the diffraction maxima are given by the relation $n\lambda = d \sin \alpha$. Here d is the separation of the rows, α is the angle between the incident and reflected beams, and n is the diffraction order. The electron wavelength λ is equal to h/mc, where c is the electron velocity. The latter is determined by the accelerating voltage V according to the expression $qV = \frac{1}{2}mc^2$. Thus

$$V = \frac{h^2}{2qm} \frac{n^2}{d^2 \sin^2 \alpha}. \tag{3.1}$$

Actually, because of the influence of the crystal field, the electron wavelength inside the crystal is less than that in vacuum. As a result, the positions of the maxima are displaced from those indicated by (3.1) and their intensities are modified. Although the displacement cannot be calculated exactly because of the lack of knowledge of the precise form of the internal field, the *separation* of the maxima is still given by (3.1). It is thus possible to obtain accurate data on d by measuring the angular and voltage dependence of the reflected beam intensities.

A typical diffraction apparatus employed by Farnsworth et al.[37] is shown in Fig. 3.1. The same apparatus is used for cleaning the surface, as described in §1.5. An electron gun furnishes a narrow beam of electrons which is accelerated and collimated to impinge upon the crystal surface at normal incidence. The diffracted electrons are detected by a Faraday collector which can be rotated by remote magnetic control. By applying a retarding potential difference between the inner and outer walls of the collector, only the elastically-scattered electrons are recorded. The accelerating voltage is varied while the angle is held fixed, and this is then repeated for other angles (usually at half-degree intervals). In order to obtain the diffraction pattern for the different atomic rows of the surface, the crystal is rotated about the incident beam.

This method of detection is of necessity slow and tedious, even though a semi-automatic operation has been introduced[30] to record the diffracted beam intensities. As a result, only a relatively limited amount of information on surface structure was accumulated. Scheibner, Germer, and Hartman[38] improved the diffraction apparatus considerably by employing a fluorescent screen to detect the diffracted beams[39]. Figure 3.2 shows a schematic diagram of one of the most recent instruments[40]. The electrons emitted from the cathode (the electron gun) are accelerated, collimated, and decelerated to the desired energy. The reflected beams pass through grids where they are first retarded (to eliminate inelastically-scattered electrons) and then accelerated by a high voltage towards a semispherical fluorescent screen. In this manner the entire diffraction pattern is displayed at the same time. A typical pattern from a clean (100) germanium surface is shown in Fig. 3.3. This method, which is both fast and productive, provided a great stimulus to the study of surface structure and is presently in a state of rapid development.

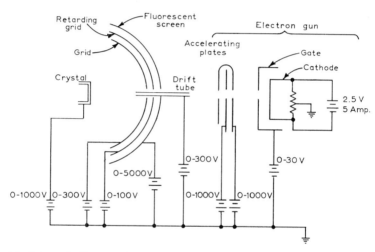

Fig. 3.2. Schematic diagram of electron diffraction apparatus that employs a fluorescent screen for the detection of the diffracted beams. (After Lander et al., reference 40.)

The diffraction patterns obtained from a surface cleaned by ion bombardment and annealing consist of sharp intensity maxima characteristic of both the bulk and the surface structure. Following gas adsorption, the maxima show a marked decrease in intensity, or else disappear entirely [30]. By assuming the amplitude of the diffracted beam (or the square root of its intensity) to be proportional to the amount of uncovered surface, one obtains the following relation for the extent of coverage by an adsorbate:

$$\theta = 1 - \sqrt{I/I_0}, \tag{3.2}$$

where I is the intensity of a diffraction maximum produced by a surface of fractional coverage θ, and I_0 is the value of I for a clean surface. Until

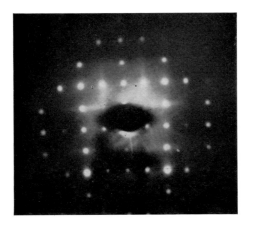

Fig. 3.3. Typical low-energy electron diffraction pattern as displayed on a fluorescent screen (clean (100) germanium surface). (After Lander et al., reference 40.)

recently, the main criterion that I_0 correspond to a clean surface has been the reproducibility of the data. Diffraction measurements carried out lately [3] on *cleaved* (111) germanium and silicon surfaces (which are obviously clean), however, were found to be in very good agreement with those obtained on similar faces cleaned by ion bombardment.

In some cases the intensities of the maxima do not decrease to zero at monolayer coverage. Equation (3.2) is then evidently not valid, and a more complicated expression is required [41].

In order to measure the sticking coefficient S of the adsorbate, defined as the rate at which atoms are adsorbed on the surface divided by the rate they impinge on it, a small quantity of the gas is introduced into the system and the pressure p is monitored continuously. After a short while the system is again pumped out and the diffraction maxima measured. In this way one can determine I as a function of $\int p dt$. The value of S is evaluated from the relation

$$S = \frac{N}{v} \frac{\delta \theta}{\delta \int p \, dt}, \qquad (3.3)$$

where N is the number of atoms per unit surface area at monolayer coverage ($\theta = 1$), and v is the rate at which the atoms impinge on a unit surface area (at unit pressure). The advantage of this method lies in the fact that it permits determination of the adsorption even for small coverages. Its disadvantage is that the measurement cannot be carried out during the adsorption process itself (when the gas is in the system), and so is applicable only to gases that do not desorb on pumping.

2.2. PHOTOELECTRIC EMISSION AND CONTACT POTENTIAL

The photoelectric effect consists of the emission of electrons from the surface of a solid upon irradiation with light of suitable wavelength. The photoelectric current as a function of wavelength has a threshold at the minimum photon energy capable of ejecting electrons. This corresponds to the energy separation between the electron energy in vacuum and the highest level in the crystal that is sufficiently populated by electrons (usually the top of the valence band). Photoelectric emission will be discussed in some detail in chapters 7 and 9. For the present it is enough to note that the emission threshold is a sensitive function of the conditions at the surface. In particular, adsorbed atoms alter the surface dipole layer which, in turn, changes the threshold. This method can thus be used as a detector of surface cleanliness.

Measurements on silicon show that the photoelectric threshold attains a reproducible minimum value for surfaces found clean by other methods [24]. A slight contamination results in a sharp increase in the threshold, 10 % of a monolayer being readily detectable.

The work function is another quantity that is sensitive to conditions at the surface, as will be discussed in subsequent chapters. It is determined from measurements of contact potential (see Ch. 7, § 2) between the semiconductor sample and a reference electrode whose work function is known and not expected to be affected by the gaseous ambient. It is found [42] that the value of the work function of germanium attains a reproducible minimum for clean surfaces (obtained by ion bombardment and annealing). This may be used, similarly to the photoelectric-emission threshold, as a measure of surface cleanliness.

2.3. FIELD EMISSION AND FIELD ION MICROSCOPY

In the field-emission microscope, first devised by Müller [43], the electron emissivity of the surface under high electric fields is measured. The sample is shaped, usually by electrolytic etching, to a fine hemispherical point (radius about 1000 Å) and is affixed to the cathode of an electron tube. A potential of a few thousand volts applied to the anode produces a field of $2-4 \times 10^7$ V/cm at the surface of the point. This is sufficient to extract electrons from the surface by means of a tunnelling process (see Ch. 7, § 4). The emission current is projected onto a fluorescent screen where it portrays a magnified image of the point. The magnification, given by the ratio of the sample–screen separation to the radius of curvature of the point, can be made as high as 10^6, corresponding to a resolution along the surface of 20 Å. The emission current from any small area of the point (and thus the brightness of its image on the screen) is an exponential function of both the field intensity and the potential barrier for tunnelling in that area. The emissivity is thus very sensitive to steps and other gross structural features at the surface through their effect on the field in their vicinity, as well as to adsorbed foreign atoms through their influence on the barrier for tunnelling.

Silicon surfaces cleaned by heating in high vacuum yield reproducible field emission patterns having the same symmetry as the crystal substrate [24]. Adsorbed gases obliterate the patterns, and it is possible to detect a few percent of a monolayer coverage [44].

If the dc voltage is replaced by an ac voltage (60 cps), electrons are emitted as before in the negative cycle, but now desorption of positive ions

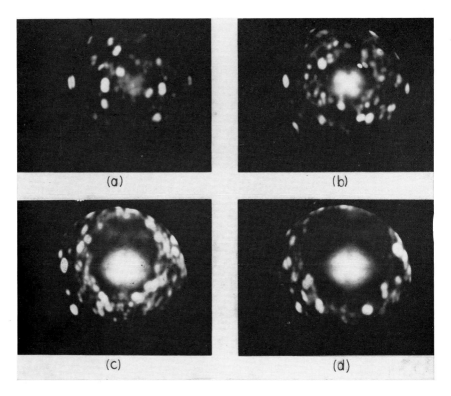

Fig. 3.4. Field emission patterns obtained from a germanium tip during cleaning by field desorption (at room temperature).
(a) Most of the surface is covered by oxide.
(b), (c), (d) Through progressive cleaning, the four-fold symmetry of the germanium crystal emerges at the centre while the oxide recedes further towards the periphery.
(After Allen, reference 44.)

is possible during the positive cycle provided that the voltage is sufficiently high (about five times that required for electron emission). In this manner silicon and germanium surfaces have been cleaned and studied simultaneously [44]. A dc biasing voltage must be used to limit the electron emission current. In the case of silicon, cleaning by field desorption is possible at a temperature of 1000°C, which is considerably lower than that needed to clean the surface by heating alone (see § 1.4). As for germanium, field desorption produces atomically clean surfaces even at room temperature and without recourse to the special etching procedure mentioned previously (§ 1.4). Figure 3.4 shows several stages of the surface following successive field-desorption cleaning cycles.

In principle it is possible to use the field desorbed ions themselves in order to obtain the magnified image of the surface being cleaned. The ion current, however, is weak and of short duration [45]. In order to overcome this difficulty, an inert gas is introduced into the system [46]. As a result of the high field in the vicinity of the point, the gas molecules are ionized just outside the surface. They are then drawn to the fluorescent screen where they reproduce on an enlarged scale the field distribution at the surface of the point. To improve the resolution, the measurements are carried out at low temperatures so as to reduce the thermal velocity of the ions. With this technique it is possible to resolve monatomic steps and even single atoms. So far it has been applied mainly to metals, but there are already some preliminary results on silicon [47]. Field ion microscopy seems to be the most promising tool for studying local surface structure on a microscopic scale.

2.4. AUGER EJECTION OF ELECTRONS BY NOBLE GAS IONS

While the primary objective of this experiment is to elicit information on the Auger ejection phenomenon and on the energy band structure in solids, it has also proved very useful for determining the state of cleanliness of the surface. The experimental procedure has been developed by Hagstrum first for metals [48] and later for silicon [49] and germanium [50]. The phenomenon studied is the release of electrons from the solid under the incidence of slowly moving noble gas ions (most commonly, helium ions). When the positive ion approaches very close (≈ 2 Å) to the surface of a semiconductor, an electron from the valence band can tunnel through the potential barrier between the solid and the ion and be captured at the ground state of the atom. The energy released by this neutralization process (of the order of the ionization energy of the atom) is transferred by an Auger process to a second electron in the valence band. Because of the large ionization energy of the noble gas atom, the excess energy released is sufficient to enable an appreciable fraction of the secondary electrons to escape from the solid. The experiment consists of measuring the total yield and the kinetic energy distribution of the secondary electrons (as a function of the incident ion energy). These characteristics are very sensitive to conditions at the surface. Atomically clean surfaces of silicon and germanium yield reproducible electron emission whose characteristics are those to be expected from the known general form of the density-of-states function of the valence band of these materials. The results are in good agreement with the theory of the Auger neutralization process worked out by Hagstrum [51]. The fit of the experimental data is achieved by varying the parameters characterizing the valence band structure, and it enables the determination of the numerical values of these parameters. The presence of adsorbed impurities at the surface, on the other hand, alters the characteristics of the electron emission. A monolayer of adsorbed oxygen or carbon monoxide is sufficient to cause a drastic change [24, 52].

The results on surface cleanliness deduced from Auger ejection measurements [50, 52, 53] agree with the findings of Farnsworth and co-workers [35] that ion bombardment and annealing produce atomically clean surfaces of germanium and silicon. They also agree with the conclusion of the latter workers that prolonged heating of germanium just below the melting point (without using the flooding-out technique discussed in § 1.4) does not produce a completely clean surface. From the observed change in secondary emission characteristics following the bombardment of such surfaces, Hagstrum concludes [50] that a substantial fraction of a monolayer remains on germanium surfaces subjected to heating alone.

As opposed to the electron diffraction patterns, the secondary emission characteristics are apparently uninfluenced by annealing of the surface. No essential difference is observed between surfaces that are bombarded only and those that are annealed following bombardment. This is attributed [50] to the fact that the Auger ejection process is insensitive to disorder among the atoms of an atomically clean surface, being affected only by the small number of surface atoms in the immediate vicinity of the approaching noble gas ion. Electron diffraction, on the other hand, depends on the structure of the entire surface area covered by the electron beam. In fact, clear diffraction patterns are obtained only if a substantial amount of long-range order exists at the surface.

2.5. ADSORPTION MEASUREMENTS

The experimental methods discussed so far deal, inasmuch as the reactivity of the surface is concerned, with changes in surface *structure* introduced by adsorption. A number of techniques are available for measuring the *amount* of material adsorbed when the surface is exposed to various ambients and under different surface conditions. The most direct method of this type is that employed by Wolsky[31]. Here the weight changes (as small as 0.15 μg) that the sample undergoes in a gaseous ambient can be measured by an ultra-sensitive quartz microbalance. Although this technique seems straightforward, a considerable amount of effort has gone into its development in order to achieve the necessary sensitivity.

The amount of adsorption on the surface can also be measured indirectly by following the pressure changes of the ambient. Care must be taken, however, to minimize and control the adsorption on all other exposed surfaces (such as the walls of the chamber). This is conveniently done in the flash filament technique [54−56]. A filamentary sample is used which can be rapidly heated by passing an electric current through it. The sample is inserted into a high vacuum system and, following cleaning by heating at elevated temperatures (see § 1.4), the gas being studied is introduced through a porous diaphragm interposed between the gas container and the continuously pumped system. The rate at which the gas enters the system soon equals the pumping rate, and a constant pressure is established. After a time t has elapsed, a high current is passed through the filament, heating it quickly to a temperature where all the gas previously adsorbed (during the time interval t) is released. The resulting change in pressure Δp is measured, and if the desorption rate is made large compared to the pumping rate, Δp is a direct measure of the amount of adsorbed gas. The number of atoms N released per unit surface is given by

$$N = 2KV\Delta p/A, \qquad (3.4)$$

where K is the number of gas molecules per unit volume at the temperature of the surroundings and at unit pressure, V is the volume of the system, and A is the surface area of the filament. The factor 2 appears because the gas is assumed to be diatomic. By repeating this flashing procedure for different exposure times, it is possible to obtain N as a function of the time t elapsed between the entry of the gas and the flash. The sticking coefficient $S(t)$ (see definition in § 2.1) can then be evaluated from the expression

$$S(t) = \frac{1}{2vp_0} \frac{dN}{dt} = \frac{KV}{vp_0 A} \frac{d(\Delta p)}{dt}, \tag{3.5}$$

where v is the rate at which gas molecules impinge on a unit surface area at unit pressure, and p_0 is the constant pressure prevailing in the system during the adsorption process. The rate v is easily calculated from Avogadro's number, the molecular weight of the gas, and the temperature of the system.

The flash filament technique is not suitable for gases that desorb in the form of compounds (such as oxygen on germanium, which desorbs under heat as GeO — see §1.4). In such cases the measurement must be carried out *during* the adsorption process itself [56]. To do this, the system is continuously pumped while the clean sample is heated to a sufficiently high temperature where no adsorption can take place. The gas is then introduced through the porous diaphragm and the system allowed to reach steady-state. Under these conditions, the rates of gas inflow through the diaphragm and of gas outflow through the pumping hole are equal and a constant pressure p_0 is maintained in the system. These rates can be expressed as $vp_0 a$, where a is the area of the pumping hole. At a particular instant $t = 0$, the sample is allowed to cool rapidly. As a result, the clean surface immediately begins to adsorb and the pressure $p(t)$ decreases with time. The pressure in the gas container is much larger than the pressure $p(t)$ in the system, so that the rate of gas inflow through the diaphragm remains constant at its initial value $vp_0 a$. The continuity equation for this process is thus

$$v[(p_0 - p)a - pSA] = KV(dp/dt). \tag{3.6}$$

By recording the pressure as a function of time, one can obtain the sticking coefficient from the equation

$$S(t) = \frac{a}{A}\left(\frac{p_0}{p} - 1\right) - \frac{KV}{vpA}\frac{dp}{dt}, \tag{3.7}$$

and the surface density of adsorbed atoms from

$$N(t) = 2v\frac{a}{A}\int_0^t (p_0 - p)dt + \frac{2KV}{A}(p_0 - p). \tag{3.8}$$

The pressure can be measured by an Alpert ionization gauge [1,2] or, preferably, by a partial-pressure gauge such as an omegatron [57-61]. These gauges are suitable down to pressures of 10^{-10} mm Hg and can follow pressure changes quickly. Special difficulties are encountered in oxygen adsorption measurements because of the reactivity of the ionization gauge filament towards the oxygen. It is then necessary to use either well de-carbonized filaments [59], or a magnetron gauge [62] which operates without an incadescent filament.

2.6. HEAT OF ADSORPTION

This type of experiment provides information concerning the binding energies of chemisorbed gases as a function of surface coverage. It is applicable only for chemisorption, where large quantities of heat are liberated, and even here only the initial, rapid adsorption can be measured. Satisfactory results have been obtained with evaporated films deposited on a glass tube having very thin walls [63] to ensure fast response and maximum sensitivity. A platinum wire, precision-wound around the glass tube, serves as one arm of a Wheatstone bridge, and the heat liberated by the adsorption process is detected as a change in resistance. The gas is admitted in small known quantities and the heat of adsorption measured for each dose. The fractional coverage is calculated from the amount of gas introduced and the known surface area (usually measured by the BET method, see § 1.2). The heats of adsorption derived in this manner are averages for each dose introduced, since the heat is liberated faster than the response of the apparatus. Data have been obtained only up to about one monolayer coverage, the heat liberated upon further adsorption being too small to be detected with the available sensitivity.

3. Clean surface structure of diamond-type semiconductors

The experimental data on the structure of clean semiconductor surfaces consist mainly of results on the ordering of the surface atoms and on their interaction with different gases. The latter is concerned chiefly with chemisorption, physical adsorption having been studied only in a few cases such as the germanium–krypton [64] and germanium–argon [65] systems.

3.1. LATTICE STRUCTURE OF CLEAN SURFACES

Low-energy electron diffraction measurements on the clean surface of diamond-type semiconductors display two distinct patterns. One consists of diffraction spots (or maxima) in cubic array for (100) faces and hexagonal array for (111) faces, characteristic of the diamond-type lattice. The lattice spacings calculated from this pattern agree well with the known unit-cell dimensions of the crystal being studied. Superimposed on this "bulk" pattern there is an array of "fractional-order" diffraction maxima due to a two-dimensional surface lattice whose spacings are larger than those in the substrate. The term fractional order refers to the fractional-integral values that n assumes in (3.1) for these maxima when the appropriate bulk spacing is substituted for d. Most commonly, the observed diffraction patterns are characterized by half-order maxima, corresponding to a spacing ($2d$) twice that in the bulk (d).

Detailed results are available for the elemental semiconductors germanium and silicon [30, 41, 66, 67] and for the group III–V compounds gallium antimonide and indium antimonide [27, 32]. In all these materials, half-order diffraction maxima have been observed in almost every crystallographic plane investigated, and their intensities are usually comparable to those of the integral-order maxima. In some cases, fractional-order maxima have also been observed. In the following paragraphs we shall summarize the available data on surface structure. Since different crystals display similar characteristics along the same crystallographic face, it will be more convenient to classify the data according to the crystallographic plane rather than the material studied. Where available, structure models proposed to account for the observed diffraction patterns will be described. It should be pointed out that a $(1/n)$th-order diffraction pattern may be obtained from *any* surface structure having a regularity n times that in the bulk. The preference of one particular model over many other possible ones in any given instance is usually made on the basis of diffraction-maxima intensity measurements [66]. In some cases physical considerations regarding the nature of the chemical bonds at the surface serve as a guide.

(100) *crystallographic face.* In Ge and Si [41, 66], and in InSb [27], half-order maxima are observed for diffraction from atomic rows along crystallographic directions (within the (100) surface) the sums of whose indices are *even*. Such is the case, for example, for rows along the [011] directions. This feature is independent of the energy of the argon ions used in the bombardment and of the precise annealing procedure employed in producing the clean surface. After careful annealing of Ge and Si surfaces, quarter-order maxima begin to appear as well [66], indicating that spacings *four* times as large as in the bulk develop on such surfaces. The diffraction from atomic rows along directions the sums of whose indices are odd consists of only integral-order maxima, corresponding to normal bulk spacings.

To account for the observed double spacings, Schlier and Farnsworth [30, 41] proposed the model of a (100) surface shown in Fig. 3.5a. The dots in the figure represent surface atoms, the open circles atoms immediately

Ch. 3, § 3.1   STRUCTURE OF CLEAN SURFACES   107

below the surface. The spacing of atoms within each row in the [01$\bar{1}$] direction is assumed to be the same as that within parallel rows in the bulk. Due to the interaction of the dangling bonds, however, atoms in adjacent rows are displaced in opposite directions to form pairs in alternate rows, each pair being associated with one lattice point. Since the [011] and [01$\bar{1}$] directions are completely equivalent, a pairing of atoms resulting from a

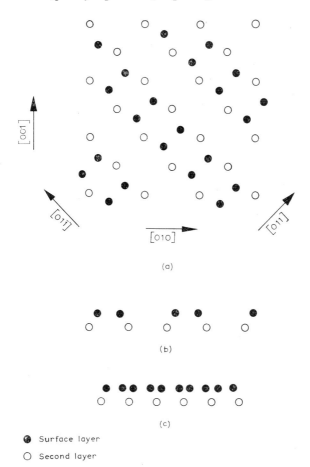

Fig. 3.5. Schematic diagram illustrating the pairing model of a (100) surface. The atoms in the surface layer are designated by dots, those in the second layer by open circles. (a) View from the top. (The arrows indicate the principal crystallographic directions.) (b) Projection of the atoms in the two layers onto the (01$\bar{1}$) plane. (c) Projection of the atoms in the two layers onto the (001) plane.
(After Schlier and Farnsworth, reference 30.)

displacement of [011] rows (corresponding to a rotation of the figure by 90°) is just as likely. An actual surface is thus expected to contain regions of both orientations. The projections of the atoms in the two layers of Fig. 3.5a in the (01$\bar{1}$) and the (001) planes (both normal to the surface plane) are shown in Figs. 3.5b and 3.5c, respectively. It is seen that whereas the separation of like surface configurations along the [011] direction is twice the bulk separation, the two separations are the same along the [010] direction. (This situation is not altered if the [011] rows are displaced.) The model thus explains the occurrence of half-order diffraction maxima in the "even" directions and only integral-order maxima in the "odd" directions. Calculations of surface energy [12] suggest that such a pairing of surface atoms is energetically favourable. A somewhat different model was suggested by Haneman[27, 68], in which every second row is raised normal to the surface rather than displaced sideways.

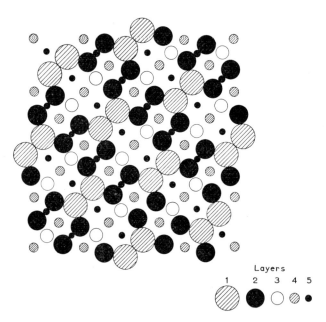

Fig. 3.6. Model of a carefully annealed (100) surface capable of accounting for both the half and quarter-order diffraction maxima observed. The positions of the atoms in the first five layers are indicated by circles of different sizes. The arrangement of the atoms in the second layer is the same as that of the surface atoms in Fig. 3.5, while the first layer contains only half the number of atoms normally present in a monolayer. (After Lander and Morrison, reference 66.)

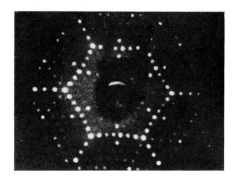

Fig. 3.7. Low-energy electron diffraction pattern obtained from a (111) silicon surface following heating at 800°C. The spots at the corners of the hexagon are integral-order maxima, those along the sides are the seventh-order maxima. (After Lander and Morrison, reference 66.)

To account for the observed quarter-order maxima in Ge and Si, Lander and Morrison [66] propose that as a result of careful annealing, an additional layer appears on the surface. This outermost layer contains only half the number of atoms in parallel bulk layers and, as in the previously discussed model, the surface atoms associate in pairs. Lander's model is shown in Fig. 3.6. The third, fourth, and fifth layers display the atomic arrangement in the bulk. The second layer (full circles) has the same pair structure proposed by Schlier and Farnsworth (Fig. 3.5), while the topmost layer (large shadowed circles) represents the additional atom pairs appearing on prolonged annealing. The lattice spacing along the [011] direction is four times that in parallel rows within the bulk. (This is seen most clearly by comparing the shadowed circles with the smallest circles, corresponding to the fifth layer.) It should be noted that the rectangular unit cell of the topmost layer has only two-fold symmetry, whereas the observed quarter-order diffraction patterns have four-fold symmetry. This is only to be expected [66], since the surface area covered by the impinging electron beam (about 0.01 cm^2) is large enough to have considerable rotational symmetry disorder. As has been pointed out above, the two crystallographic directions [011] and [01$\bar{1}$] are equivalent, and the target area contains many regions of either orientation.

(111) *crystallographic face.* Half-order maxima have been observed for Ge [41, 66], GaSb [27], and InSb [32]. In contrast to the case of the (100) face, these are not confined to the even directions and arise from all sets of parallel atomic rows. Silicon shows an anomaly in that no half-order maxima are observed, only fractional order ones (except in a cleaved surface, as will be discussed in a moment). The exact fractional order depends on both the ion bombardment and the annealing treatments [41, 66]. Such fractional-order, treatment-dependent maxima are also found for Ge [41, 66] and InSb [32].

In the case of silicon, heating at about 800°C gives rise [41, 66] to seventh-order maxima, as shown by the diffraction pattern in Fig. 3.7. The array of spots exhibits the six-fold symmetry characteristic of the atomic arrangement on a (111) face. The spots at the corners of the hexagon correspond to the integral-order maxima, those along the sides to the seventh-order maxima. Annealing at 600°C, on the other hand, gives rise to fifth-order maxima which presumably correspond to the stable configuration of the surface.

For Ge, Lander *et al.* [67] find that the half-order spots consist actually of doublets. At 200° C the separation of each doublet is one-twelfth that of the integral-order spots, indicating that surface spacings twelve times those in the bulk are also involved. After the sample is slowly cooled to room temperature, the twelfth-order spots give way to eighth-order ones.

Cleaved surfaces exhibit quite different features. Farnsworth et al.[3] reported half-order diffraction maxima in the case of silicon. More detailed measurements by Lander et al.[67] show the presence of a surface layer with two-fold rotational symmetry, placed on the three-fold symmetrical substrate. In both germanium and silicon, integral-order maxima are observed along one crystallographic direction and approximately half-order ones at right angles to this direction. These arise from a rectangular unit cell which

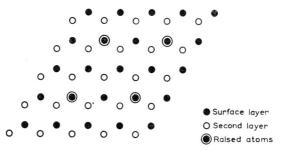

Fig. 3.8. Model of a (111) surface, as proposed by Haneman (references 27, 68). The atoms in the surface layer are designated by dots, those in the second layer by open circles. The surface atoms indicated by the circled dots are assumed to be raised.

measures $\sqrt{3} \times 1$ with respect to the substrate cell. When the cleaved surface is annealed, the diffraction pattern is always transformed into that obtained from a surface prepared by bombardment and annealing, the two structures being completely indistinguishable.

Several models have been proposed to account for the observed diffraction patterns. In one, due to Haneman [27, 68], every second surface atom, counting along alternate (close-spaced) rows, is assumed to be raised with respect to its neighbours. This is illustrated in Fig. 3.8, where the surface atoms are designated by dots and the atoms in the layer underneath by circles. The raised atoms, one fourth of the top layer, are denoted by circles drawn around the respective dots. The spacing between like surface configurations is thus twice the bulk separation irrespective of the crystallographic orientation, in accordance with the observed diffraction data. Haneman uses the following arguments to justify his model. The asymmetric restraining forces tend to pull the surface atoms down. This, in turn, exerts strong lateral forces on the atoms of the second layer. To relieve the pressure, the central atom within each hexagonal ring of atoms (the circled dots surrounded by six dots in Fig. 3.8) is raised, thus allowing more space for the accomodation of the laterally displaced nearest neighbours in the second layer. The bonding angles of the raised atoms are expected to be decreased from the bulk value of 109° to about 90°. This implies that the raised atoms each have three p-like covalent bonds and one s-like dangling bond (instead of the tetragonal sp^3 bonds for bulk atoms). Those atoms that are not raised, on the other hand, have three nearly coplanar trigonal sp^2 covalent bonds and one p-like dangling bond. Thus, three quarters of the surface atoms are associated with p-like dangling bonds, in accordance with Haneman's supposition that the dangling bonds tend to be predominantly p-like. The extent of the elevation of the surface atoms should depend, according to this model, on the strength of the bonding between atoms: the weaker the bonding forces, the greater the elevation and therefore the larger the intensity of the half-order maxima. This seems to be the observed trend; the half-order intensities from InSb (weak bonding, energy gap 0.25 eV) are comparable to the integral-order intensities [32], whereas in Ge, Si, and GaSb (stronger bonding, energy gaps 0.7, 1.1, and 0.7 eV, respectively), they are on the average only about a quarter as strong.

Lander and co-workers contend that the fractional-order maxima obtained from annealed Ge and Si surfaces are due to missing atoms rather than to raised or paired atoms. The vacancies are presumably associated with the stable configuration of the surface, and they can be formed at the elevated temperatures used in the annealing treatment. An example of the "warped benzene-type" structure proposed by these authors [66] is illustrated in Fig. 3.9. The hexagonal rings consist each of three surface atoms (dots) and three atoms in the second layer (circles). Three warped rings are disrupted around each vacant site. The figure represents a unit cell, each side of which extends over seven bulk spacings. Such a cell can account for the six-fold symmetrical array of seventh-order maxima obtained from Si at elevated temperatures (see Fig. 3.7). This type of structure can be arranged in many symmetrical ways, including those actually observed: the 5×5, 8×8, and 12×12 patterns mentioned above. (The latter two patterns may arise from rotational disorder involving three-fold domains of 2×8 and 2×12 structure, respectively.)

Perhaps the strongest evidence in support of the warped benzene structure is provided by comparison of annealed and cleaved surfaces. Obviously, surfaces cleaved at room temperature cannot be associated with missing surface atoms, at least not on a long-range order basis. The half-order maxima must therefore arise from short-range displacements, consisting probably of the pairing of neighbouring atoms. (The lack of six-fold symmetry in the diffraction patterns is most likely related to the presence of a preferred direction introduced by the manner in which the crystals have been cleaved.) Indeed [67], upon adsorption of iodine or cesium (at room temperature), the half-order maxima immediately disappear, indicating that the adsorbate restores (by saturating the dangling bonds) the paired atoms to their normal bulk positions. On annealed surfaces, on the other hand, such adsorption does not obliterate the fractional-order maxima, the iodine

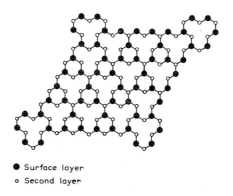

● Surface layer
○ Second layer

Fig. 3.9. Warped benzene-type model of a (111) surface capable of accounting for the observed seventh-order maxima on Si. (After Lander and Morrison, reference 66.)

or cesium layer assuming the structure of the substrate. If the fractional-order maxima were due to paired atoms rather than to vacancies, then upon adsorption the structure would have converted to that in the bulk, as in the case of cleaved surfaces. Further support for the warped benzene-ring structure is provided by calculations [67] based on maxima intensity measurements.

(110) *crystallographic face*. This face has been investigated only in the case of germanium [41]. Both half-order and treatment-dependent, fractional-order diffraction maxima were observed. No model for the surface structure of such faces is available as yet.

Adsorption. The adsorption of oxygen and other gases usually results in a general weakening of both the fractional and the integral-order maxima. The adsorbed layer in such cases must thus be highly disordered or even amorphous. There are, however, quite a few exceptions. For example, oxygen adsorbed on germanium weakens the fractional-order maxima (characteristic of the surface structure) much more than the integral-order maxima [41]. This implies that the adsorbed layer is not amorphous and that the oxygen atoms assume the ordered bulk lattice positions of the substrate. For the (111) plane there is a displacement of the integral-order maxima, suggesting that the spacing in a direction normal to the surface is altered [41]. Phosphorus and iodine adsorbed on silicon [66], iodine on germanium [67], and carbon dioxide on gallium antimonide [27] (all on the (111) surface) also show an ordered surface structure.

3.2. CHEMISORPTION ON CLEAN SURFACES

Most of the studies in this field have been carried out on the semiconductor/oxygen system, oxygen being the active gas normally in contact with the surface. As soon as a clean surface is exposed to oxygen, a rapid adsorption takes place up to a coverage of about one monolayer: approximately 0.7 of a monolayer in germanium and silicon [13, 15, 26, 41, 69], 1–1.5 in the arsenides [16], and 2–2.5 in the antimonides [16, 27, 32]. The time required for such coverage (at the pressures studied, usually 10^{-3}–10^{-1} mm Hg) is of the order of seconds to minutes, so that information in this range has been obtained mostly from flash filament techniques [26, 69] and from indirect measurements using electron diffraction [27, 30, 32, 41]. The orders of magnitude of the sticking coefficients obtained from such measurements (and extrapolated to zero surface coverage) are listed in Table 3.1. It is interesting that the sticking coefficient in most cases is very small, each oxygen molecule striking the surface on the average many times before being finally adsorbed.

TABLE 3.1.
Sticking coefficients of oxygen for zero surface coverage

Crystal plane	(111)	(100)	($\bar{1}\bar{1}\bar{1}$) *
Germanium	10^{-3}–10^{-4}	10^{-3}	—
Silicon	10^{-1}	10^{-2}	—
Gallium antimonide	10^{-5}		10^{-4}
Indium antimonide	10^{-5}		10^{-5}

* The (111) plane of the group III–V compounds consists of either three atoms of group V or three atoms of group III. The latter plane is denoted by ($\bar{1}\bar{1}\bar{1}$).

The sticking coefficients decrease rapidly with increasing surface coverage. The slow adsorption range has been investigated mainly by crushing [15, 16, 70] and by weighing [71] techniques. In germanium, the adsorption process from about one monolayer to two or three monolayers is logarithmic, following the Elovitch rate equation: the number of atoms N adsorbed per square centimetre may be expressed as

$$N = a + b \log_{10} t, \tag{3.9}$$

where t is the time in minutes, and a and b are temperature-dependent parameters (a is also pressure dependent). This equation holds (in the pressure range studied) from about 1 to 10^3 min. The parameters a and b apparently vary with crystal orientation and with the amount of strain present in the surface. Different measurements [13, 22] have yielded values for the ratio a/b ranging from 5 to 25, all for about the same temperatures and pressures. The orders of magnitude of a and b are 10^{15} and 10^{14} cm^{-2}, respectively. The Elovitch equation describes a general phenomenon occurring in many substances, and it can be derived from quite general considerations [72]. It is therefore difficult to draw from its occurrence any definite conclusion about the adsorption processes at the surface. After a coverage of about two to three monolayers, oxygen adsorption (at room temperature) comes to a halt, or at least becomes considerably slower than logarithmic.

For silicon, opinion is divided [15, 73] as to the existence of a logarithmic adsorption range. In the case of crushed powders it is found [15] that adsorption ceases after a coverage of nearly one and a half monolayers (or becomes too slow to be detected). The situation is similar in the group III–V compounds: the initial rapid oxygen adsorption is followed by a much slower process [16].

The differential heat of adsorption of oxygen on germanium and silicon surfaces is found [63] to remain essentially constant up to about 0.7 of a monolayer, when it decreases sharply. The integral heat of adsorption (122 kcal/mole for Ge, 218 kcal/mole for Si) agrees well with the heat of oxidation, suggesting that there should not be a great deal of difference between the chemisorbed oxygen monolayer and the ordinary oxide.

The main conclusion to be drawn from the data presented above is the existence of three well-defined ranges: a rapid adsorption up to a coverage of about one monolayer, a much slower subsequent adsorption up to a coverage ranging from one and a half monolayers in Si to two or three monolayers in the other crystals studied and, finally, a region where adsorption usually ceases or is too small to be detected. Each range must be characterized by an entirely different adsorption mechanism. The initial rapid adsorption is most likely associated with the saturation by oxygen of the dangling bonds of the substrate surface atoms. Another monolayer or two may be adsorbed by some sort of short-range rearrangement of the substrate and oxygen layers so as to expose the former for additional uptake of oxygen. This process is slow and characterized by low heat of adsorption. Beyond this point any further uptake can occur only by ordinary oxidation, involving the migration of ions and defects through the previously adsorbed layers. At room temperatures this process is extremely slow. (At elevated temperatures, on the other hand, the atomic mobility may be considerably larger, with a correspondingly faster rate of oxide formation. In germanium, the rate observed [74] at elevated temperatures is faster than logarithmic.)

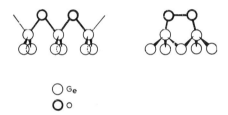

Fig. 3.10. Proposed model for the initial oxygen adsorption on germanium. (After Green and Liberman, reference 75.)

Green and co-workers have proposed specific models for the atomic arrangement on germanium and silicon surfaces following oxygen adsorption. These models assume that the normal covalencies of the substrate and adsorbate atoms are satisfied. The saturation of the dangling bonds on the (100) and (111) faces of germanium (following the rapid initial adsorption) is envisaged [75] as taking place in a manner illustrated in Fig. 3.10. In the (100) face each Ge atom has two dangling bonds, and these are used to link two oxygen atoms, so that each oxygen is bound to two neighbouring Ge atoms. In the (111) face (as well as in the (110) face not shown here) only one dangling bond is available and two neighbouring oxygen atoms pair up. On all faces, full saturation of the dangling bonds would lead to a monolayer coverage with one adsorbed atom for each germanium atom at the surface. It is suggested, however, that after about 70 % coverage, additional oxygen atoms find it energetically more favourable to interact with the strained

Ge–Ge bonds. As a consequence, some of these bonds break up and Ge atoms interchange positions with adsorbed oxygen atoms so as to reach the outer surface where further adsorption can take place. This process is much slower and would account for the observed drop in the rate of adsorption following about one monolayer coverage. Green and Liberman [75] assume that the more oxygen atoms are already adsorbed in the vicinity of a certain point, the less is the strain energy and consequently the slower is the adsorption. Such a process can account for the observed logarithmic rate of adsorption.

A similar model has been proposed by Green and Maxwell [15] for the (111) and (110) faces of silicon. This model accounts very elegantly for the observed termination of oxygen adsorption at a coverage of 1.5 monolayers,

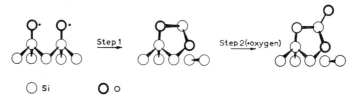

Fig. 3.11. Proposed model for chemisorption on (111) silicon surfaces that accounts for 1.5 monolayer coverage. (After Green and Maxwell, reference 15.)

and is illustrated in Fig. 3.11. After the rapid initial adsorption, an inversion of alternate Si–O pairs occurs due to the repulsion between oxygen neighbours (Step 1). To satisfy the valencies of the three underlying Si atoms previously attached to the Si atom that has been moved to the surface, one of these Si atoms is assumed to bond to the oxygen atom now in the second layer while the other two Si atoms bond to one another. A silicon atom that has reached the top layer bonds to the oxygen atom already attached to the neighbouring surface Si, forming a cyclic structure. The last stage is assumed to be the adsorption of oxygen with the still unsaturated silicon atoms in the top layer (Step 2), giving a structure corresponding to a coverage of 1.5 monolayers. According to this model, the first adsorbed 1.5 monolayers are attached to the surface by essentially covalent bonding. Further oxidation must take place by a different mechanism, involving the diffusion of Si ions to the surface or the diffusion of oxygen ions into the substrate.

Several measurements have been carried out on the adsorption of gases other than oxygen, mostly on germanium surfaces. Those gases that have been found to chemisorb [76], do so rapidly and attain a coverage of about one monolayer. Such is the case, for example, with water vapour on germanium. Both germanium and silicon chemisorb iodine, while silicon has been found very reactive to phosphorus [66, 67]. The initial sticking coefficients of both iodine and phosphorus are close to unity, and in suitable temperature and pressure ranges the adsorbates exhibit an ordered surface structure [66, 67]. On the other hand, a number of gases and vapours such as N_2, CO_2, CO, CCl_4, and $CHCl_3$ are not chemisorbed [76]. (CO_2, however, is chemisorbed [27] on GaSb, the sticking coefficient being 10^{-6}–10^{-5}.) Molecular hydrogen does not chemisorb [76] but atomic hydrogen does [30, 77], and a monolayer coverage is apparently attained. Green and Maxwell [76] attempt to explain the preference of germanium for the adsorption of certain gases on the basis of the electrostatic interaction between the adsorbed molecules and the germanium atoms. In those adsorbates where one can expect a form of hydrogen bonding, numerical calculations give binding energies (12–15 kcal/mole) that are probably sufficient to account for chemisorption, while in other cases the calculations give much lower binding energies ($\approx$ 200 cal/mole). While the results are in agreement with experiment, they cannot serve as a general criterion for the adsorption or non-adsorption of gases on Ge, since electrostatic interaction may not be the only important mechanism.

4. Real surfaces

4.1. SAMPLE PREPARATION

In general the sample to be studied is cut from a single-crystal ingot and then ground (or sandblasted) and polished to the desired shape and size. This shaping produces a dense pattern of conchoidal fractures of the dimensions of the abrasive particles (usually of the order of one micron) used in polishing [78, 79]. Excessive heating of the sample during the polishing must be prevented since at elevated temperatures the defects may migrate deep into the underlying bulk. In order to remove the damaged layer and attain a surface as smooth (on an atomic scale) as possible, the sample is etched in suitable chemical etchants. Quite frequently electrolytic etching is employed, the sample serving as the anode. The portions of the sample not to be etched (such as the contact regions) are masked with organic compounds such as wax or polystyrene. To insure an even etching, the sample must first be carefully cleaned by organic solvents and deionized water to remove grease and other foreign matter. The best procedure is to insert the washed sample immediately (without drying) into the etchant so as to minimize the residual deposit. After etching, care must be taken to wash away the remnants of the etch. A common practice is to remove the sample from the etchant and rinse it in deionized water. It is found, however, that this procedure leaves a considerable residue on the surface [80]. Moreover, it causes inhomogeneities due to the continuation of the etching process on parts of the surface before they are reached by the water. A better procedure is to flood-out the etchant by adding successive amounts of deionized water to the etching solution, as has been discussed in § 1.4. In this manner, when the sample is eventually exposed to the air, the etchant in contact with the surface is infinitely diluted, and in drying a minimum deposit is left.

The physical and chemical state of a freshly etched surface depends to a large extent on the etchant employed. Germanium and silicon surfaces produced by etching in strong acid mixtures (based on HNO_3 and HF) give clear diffraction patterns when medium-energy electron beams are used [81], as shown in Fig. 3.12. This indicates that such surfaces are smooth on an atomic scale (although they are sometimes found to be undulating on a macroscopic scale). Any surface layer of disordered structure or of foreign matter (such as an oxide) that is present must be thinner than the limit of detectability obtainable with medium-energy electrons, which is about 4Å for Ge and 8Å for Si. A thin oxide layer (1–2 monolayers) has actually been detected (by cathodic reduction measurements) in the case of germanium [82]. When germanium is etched in weaker solutions (based on H_2O_2), medium-energy electron diffraction reveals a thin amorphous layer which is very likely an oxide or a hydrated oxide [81]. When cleanliness of the surface is of importance, care should be taken to eliminate the presence of metal ions in the etchant, either as impurities or in the form of deliberately added compounds. Metals that are close to the semiconductor in the electrochemical series tend to deposit as compounds from acid solutions, and may thus contaminate the etched surface [81]. This is illustrated in Fig. 3.13, which represents a diffraction pattern obtained from a silicon sample etched in the presence of cobalt. The ring pattern arises from a surface layer consisting of a cobalt compound, apparently an oxide.

After freshly etched germanium and silicon samples have been exposed to the air for some time, the surface becomes covered with additional oxide as well as with gases and greases adsorbed from the atmosphere. It has been found that heating such surfaces to 800°C results in the evolution of nitrogen, carbon monoxide and dioxide, and hydrogen, equivalent to a total of two to four monolayers coverage [83]. (Mechanically polished surfaces evolve about three times as much gas.) Thick oxide films can be obtained by thermal oxidation. The sample is heated to elevated temperatures ($\approx 1000°C$ for silicon, above $500°C$ for germanium) either in air or in some controlled environment (usually oxygen and/or water vapour). The process of thermal oxidation was studied in both germanium [84-86] and silicon [87]. In the latter the oxide layer is amorphous and if special precautions are taken in cleaning the surface prior to oxidation, it is quite continuous and acts as a good insulator (see also chapter 9).

To reveal the gross structure of a real surface, optical or electron microscopy is used [88], depending on the resolution desired. In the former, various illumination techniques are employed, such as direct reflection, dark-field illumination, and polarized light. In electron microscopy, transmission techniques using replicas are the most common. A thin film is deposited on the surface in order to reproduce the surface topography; the film is then stripped off and its electron transmissivity examined. The resolution obtained in this manner is normally about 100 Å. A considerable improvement in contrast and resolution can be achieved by preparing the replica by vacuum deposition on the surface of an extremely thin metal layer at an oblique angle [89]. Any projections or steps receive a heavier deposit (on the side facing the evaporating source) while, at the same time, they cast shadows in which no metal is deposited. With this shadowing procedure, a resolution down to 20–30 Å is attainable. The multiple-beam interferometry techniques developed by Tolansky [90] may also be advantageously applied to semiconductor surface research. The choice of the particular technique employed by various workers is determined, to a large extent, by the apparatus available to them. Standard optical microscopy is no doubt the one most commonly used.

4.2. CHEMICAL ETCHING

Etching, both chemical and electrolytic, can perform a number of important tasks. Besides the obvious function of dissolving the damaged layers introduced by the mechanical polish, it can be employed for the controlled removal of material to obtain samples of desired shape and dimensions. Uniform wafers as thin as 1000 Å can be produced in this manner. Many etching solutions attack different crystallographic faces preferentially and can be used to develop specific planes at the surface of the crystal. With a suitable arrangement, it is possible to delineate p–n junctions and impurity concentration gradients. Very wide use is made of special etching procedures for revealing dislocation lines and grain boundaries that extend up to the surface.

Most etchants act through an intermediate oxide layer which is continually formed by an oxidizing agent and removed by dissolution. When the oxide is insoluble or its rate of dissolution is slower than that of formation, a suitable reagent must be added to convert the oxide into a soluble anion. (In germanium and silicon, hydrofluoric acid is widely used for this purpose.) In the absence of such a reagent, passivation of the surface occurs, the oxide layer eventually blocking the etching [91, 92].

The etchant generally attacks rough spots and irregularities more than smooth areas and thus tends to reduce the surface to a uniform level. Atoms in a disordered region are bound more loosely to the solid and are therefore more easily dissolved away. A mechanically polished surface represents an extreme case of lattice disorder and is indeed dissolved much faster than a smooth etched surface [78]. In the latter, lattice defects such as dislocations and grain boundaries may become important factors in the etching process. Due to the enhanced rate of dissolution in the vicinity of such defects, pits tend to form around them. Etch pits of this sort are particularly pronounced with slow etchants. When, on the other hand, the rate of dissolution is sufficiently fast that the rate limiting process in the etching becomes the thermal and chemical diffusion in the solution, lattice defects are of secondary significance and a much smoother surface (on an atomic scale) results.

A very large number of etchants have been developed, each intended to perform some specific function or to emphasize a particular feature of the surface. Detailed lists of the various etchants available for semiconductors and their characteristic properties have been compiled by Holmes [80] and by Faust [93]. Only a few of the important etchants will be mentioned here, together with a brief description of the physical state of the resulting surface. The effect of various etchants on the electrical properties of the surface will be discussed in chapter 9. One of the best known and most widely used etchants is the so-called CP–4 etchant [94]. One form (designated by the suffix A) consists of 3 parts by volume of HF, 5 of HNO_3, and 3 of acetic acid. (The liquid components here and subsequently are taken to be in their usual concentrated commercial forms.) It is used to give smooth surfaces on Ge, Si [95], and InSb [96]. Another form contains, in addition, 0.3 mg of bromine (dissolved in the acetic acid) for every 55 ml of CP-4A solution. This results in more reproducible surface properties. Many other etchants are based on HF and HNO_3 in varying proportions. The so-called white

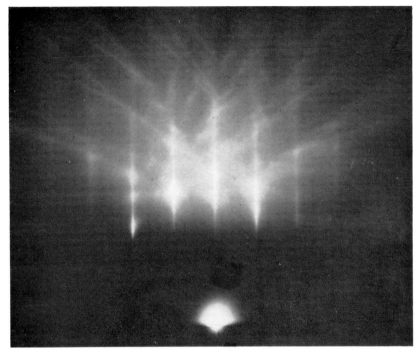

Fig. 3.12. Medium-energy electron diffraction pattern obtained from a germanium surface etched in CP-4, the ordered array of spots indicating the absence of any detectable deposit. (After Holmes and Newman, reference 81.)

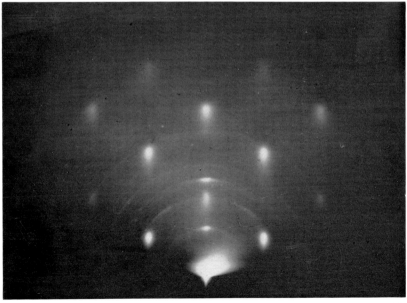

Fig. 3.13. Medium-energy electron diffraction pattern of a silicon surface etched in HF/HNO$_3$ in the presence of cobalt. The deposited compound is detected by the appearance of a ring pattern. (After Holmes and Newman, reference 81.)

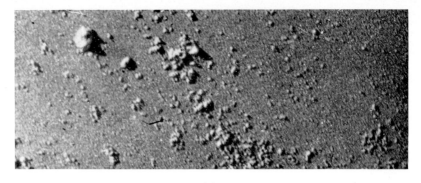

(a)

(b)

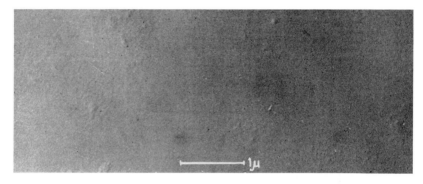

(c)

Fig. 3.14. Electron micrographs of etched germanium surfaces.
        (a) Superoxol etchant.
        (b) CP–4 etchant.
        (c) Electrolytic polish.
(After Ellis, reference 105.)

etchant [80] (1 part by volume of HF, 3 of HNO_3), for example, serves as a chemical polish for silicon, while the Dash etchant [97] (1 HF, 3 HNO_3, 8 to 12 acetic acid), when applied for several hours, can be used to reveal dislocation lines terminating on any of the crystallographic planes of silicon. Several etchants contain hydrogen peroxide. Hot concentrated H_2O_2 gives clean mat surfaces on germanium and is particularly useful for revealing p–n junctions and grain boundaries [98]. The Superoxol etchant (1 H_2O_2, 1 HF, 4 H_2O) is used for (100) and (111) planes of germanium [99, 100]. When greatly diluted with water, it produces (after etching for hours) etch pits on all crystallographic planes of germanium [101]. A hot mixture of aqueous NaOH and KOH (in varying proportions) is useful for revealing structural details on silicon [81, 102]. The ferricyanide etchant [103] (12 g KOH, 8 g $K_3Fe(CN)_6$, 100 g H_2O — one minute boiling) produces clear triangular etch pits on (111) faces of germanium.

4.3. ELECTROLYTIC ETCHING

Similar oxidation and dissolution processes are involved in electrolytic etching. The sample is immersed in a suitable electrolyte and serves as the anode in the current flow. Germanium can be dissolved anodically in most electrolytes and in many non-aqueous solutions [104]. Aqueous solutions of NaOH or KOH are the most common electrolytes used, however, the concentrations not being critical. During the dissolution process, about a monolayer of oxide is always present. In silicon, where the oxide is insoluble in water, special electrolytes must be employed. Satisfactory results are obtained using a largely non-aqueous mixture of hydrofluoric acid and hydrophilic organic compounds such as alcohols, glycols, or acetic acid. If an aqueous HF solution is used, electropolishing starts only after a critical current density is exceeded[104]. Several electrolytes are available for etching group III–V compounds [93].

The chemical smoothing action is considerably enhanced by the very nature of the electrolytic etching process. Since the conductance of the electrolyte is lower than that of the semiconductor sample, the electric field is largest at raised areas and other irregularities on the surface. The current density at such regions is therefore larger and a correspondingly faster rate of dissolution takes place. It is indeed found that by a suitable choice of the operating voltage range, a smooth surface on the atomic scale can be obtained, provided that the crystal is sufficiently uniform [104]. The electron micrographs [105] in Fig. 3.14 (taken by replica techniques) show the contrast between a germanium surface etched chemically (a and b) and electrolytically (c). Mirror-like smooth surfaces have been obtained by using a rapidly rotating gold-plated cathode placed very close to the surface being etched [106]. The intense stirring enhances the polishing action.

The presence of free holes in the semiconductor was found to promote anodic dissolution [107]. (The same effect is sometimes observed in chemical etching.) The free holes represent missing electronic bonds (see Ch. 2, § 3.3), where the atoms are tied more loosely to the crystal lattice. P-type samples are therefore etched more readily, while in n-type samples etching can be enhanced by supplying holes by external means. For polishing action the best procedure is to illuminate the surface uniformly, whereas for shaping purposes localized illumination or hole injection is very useful.

To reveal p–n junctions, a reverse bias is applied across the ends of the sample while it is immersed in the electrolyte [108]. In this manner the p side becomes the cathode and the n side the anode. Only the latter can be etched, and a step will appear at the interface between the p and n regions. An ac modification of this method can be used for revealing multiple junctions [109]. Similar techniques are available for tracing resistivity gradients [108].

Following electrolytic etching, the surface is normally covered with a few monolayers of oxide. At large current densities, however, the rate of oxide formation can exceed that of dissolution, giving rise to a much heavier oxide growth. Such anodic oxidation yields oxide films 200–1200 Å thick on germanium [110, 111], and up to 2000 Å thick on silicon [110, 112].

4.4. PREFERENTIAL ETCHING AND ETCH PITS

In many cases the etchant attacks certain crystallographic faces faster than others. As a result, a crystal undergoing dissolution eventually reaches a stable shape, known as the dissolution figure, which is not altered upon further etching except for a proportionate reduction in size. The dissolution figure depends on both the sample material and the etchant employed. Several such figures have been reported for germanium [113], silicon [114], and gallium arsenide [115], all in the form of polyhedra with slightly curved faces. The faces of the polyhedra obviously correspond to orientations of fast etching rates, the corners to slow rates. Conversely, if a hollow is etched within the crystal, the faces developed would correspond to orientations of slow etching rates. Figure 3.15 shows three dissolution figures obtained

(a) (b) (c)

Fig. 3.15. Silicon spheres after prolonged etching.
(a) 10 % NaOH solution ((100) faces are developed).
(b) 30 % NaOH solution ((100) and (110) faces developed).
(c) CP–4 etchant (non-preferential).
(After Faust, reference 114.)

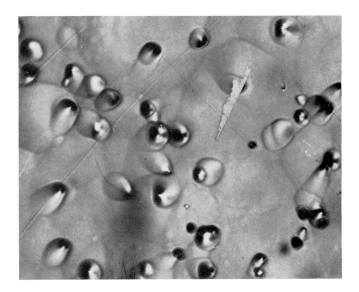

Fig. 3.16. Micrograph of a germanium surface showing conical etch pits obtained by a non-preferential etchant (CP–4). (After Bardsley *et al.*, reference 116.)

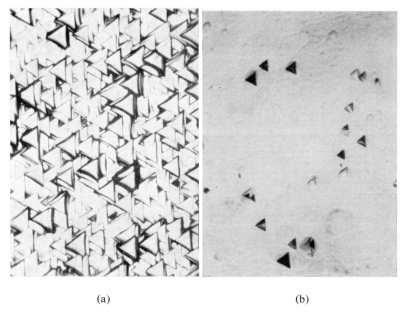

(a) (b)

Fig. 3.17. (111) germanium surfaces preferentially etched following different modes of preparation.

    (a) Background pitting on abraded surface.
    (b) Relatively few dislocation pits on chemically polished surface.

        (After Faust, reference 114.)

from spherical silicon samples using different etchants [114]. The cube was obtained with a 10 % NaOH solution, in which the (100) faces etch fastest. With a 30 % NaOH solution, a polyhedron results having faces along the (100) and (110) planes, the etching rates at such planes being equal and faster than that at the (111) face. The spherical figure was obtained using a CP–4 etch which, due to its fast action, is essentially non-preferential.

Dislocation etch pits appear whenever the etching rate associated with the strained lattice around the dislocation lines is faster than that of the crystallographic face on which the lines emerge. Microscopic examination at a magnification of about 100 is normally sufficient to observe etch pits and to provide sufficient resolution for their counting. The shape of the etch pits depends on whether a preferential or non-preferential etch (with respect to crystallographic orientation) is employed. The latter gives rise to conically shaped pits, symmetrical if the dislocation line is normal to the surface and asymmetrical if the line is oblique. This is shown in Fig. 3.16 by the round form of the etch pits obtained [116] on a (111) germanium face following a non-preferential etch (CP–4). Most of the pits arise from oblique dislocation lines. With a preferential etch, on the other hand, the same crystallographic face develops triangular etch pits characteristic of the crystal symmetry, as shown in Fig. 3.17.

It should be pointed out that not all etch pits originate from bulk dislocation lines. Quite often an array of pits develops on the surface, identical in appearance to the dislocation pits and ten or a hundred time denser (10^5–10^6 cm^{-2}). Such background pitting is probably associated with irregularities or shallow dislocations introduced by abrasion and not entirely removed by the treatment used prior to the dislocation etching[114]. An extreme case of background pitting is shown in Fig. 3.17a. A considerable reduction in the number of pits can be obtained if the surface is first polished either by a fast chemical etchant or electrolytically, as illustrated in Fig. 3.17b.

Several tests may be employed to ascertain whether an observed pattern of etch pits actually does originate from dislocation lines in the underlying bulk. The sample can be mechanically and chemically polished and etched several times in succession to reveal whether the pattern of pits is reproducible in depth. Bardsley et al. [116] examined opposite surfaces of a thin sample and found a systematic correlation between the two etch pit patterns, indicating that the observed pits were indeed the termination of dislocation lines traversing the sample. Another check was provided by the comparison of etch pit data with infrared examination of dislocation lines along which copper had been precipitated [117].

Nearly all etchants show an orientation dependence in the formation of dislocation pits [118]. In the diamond-type semiconductors, etch pits are confined mainly to surfaces close to the (111) plane. Several etchants are available for producing dislocation pits on both (111) and (110) surfaces, but only a few (such as the Dash etchant on silicon) give pits on all faces.

A grain boundary consists of a planar array of dislocation lines and can therefore be traced by etching solutions that reveal dislocations. Such boundaries, however, are usually visible following almost any etch, and they can often be seen by the unaided eye.

Etch pits are also used to align a sample accurately along a particular crystallographic direction. The background pitting, whose density can be made much larger than that of the dislocation pits (for example by Superoxol etch for Ge, warm caustic soda for Si), is very convenient for this purpose [119-122]. A collimated beam of light is directed at the surface, and the reflected pattern is observed on a screen. Surfaces that are accurately cut parallel to crystallographic planes of low indices — such as (100), (110), (111) — give symmetrical patterns about a point at which the image would have been formed had there been a mirror in place of the surface. Deviations from the crystal plane cause the reflected image to be asymmetrical, and the sample can be adjusted accordingly. The accuracy obtained is about $\frac{1}{2}°$. Apparatus for this type of alignment is now available commercially.

References

[1] D. Alpert, J. appl. Phys. **24** (1953) 860.

[2] P. A. Redhead, J. P. Hobson, and E. V. Kornelsen in *Advances in Electronics and Electron Physics* (Edited by L. Marton), vol. **17** (Academic Press, New York, 1962), p. 323; M. Pirani and J. Yarwood, *Principles of Vacuum Engineering* (Chapman & Hall, London, 1961), ch. 8; S. Dushman, *Scientific Foundations of Vacuum Technique*, 2nd ed. (Edited by J. M. Lafferty), (John Wiley, New York, 1962), ch. 9.

[3] H. E. Farnsworth, J. B. Marsh, and J. Toots, *Proc. Int. Conf. Semiconductor Physics*, Exeter, 1962 (Institute of Physics and Physical Society, London, 1962), p. 836. See also reference 67.

[4] H. C. Gatos (Editor), *The Surface Chemistry of Metals and Semiconductors* (John Wiley, New York, 1960).

[5] P. J. Holmes (Editor), *The Electrochemistry of Semiconductors* (Academic Press, New York, 1962).

[6] R. K. Willardson and H. L. Goering (Editors), *Preparation of III–V Compounds* (Reinhold, New York, 1962).

[7] G. A. Barnes and P. C. Banbury, Proc. phys. Soc. **71** (1958) 1020.

[8] D. R. Palmer and C. E. Dauenbaugh, Bull. Am. phys. Soc., Ser. II **3** (1958) 138.
[9] P. C. Banbury, G. A. Barnes, D. Haneman, and E. W. J. Mitchell, Vacuum **9** (1959–60) 126.
[10] G. W. Gobeli and F. G. Allen, J. Phys. Chem. Solids **14** (1960) 23.
[11] W. Mehl and M. D. Coutts, J. appl. Phys. **34** (1963) 2120.
[12] G. A. Wolff and J. D. Broder, Acta Cryst. **12** (1959) 313. (References).
[13] M. Green, A. J. Kafalas, and P. H. Robinson in *Semiconductor Surface Physics* (Edited by R. H. Kingston), (University of Pennsylvania Press, Philadelphia, 1957), p. 349.
[14] A. J. Rosenberg, P. H. Robinson, and H. C. Gatos, J. appl. Phys. **29** (1958) 771.
[15] M. Green and K. H. Maxwell, J. Phys. Chem. Solids **13** (1960) 145.
[16] A. J. Rosenberg, J. Phys. Chem. Solids **14** (1960) 175.
[17] S. Brunauer, P. H. Emmett, and E. Teller, J. Am. chem. Soc. **60** (1938) 309.
[18] A. J. Rosenberg, J. Am. chem. Soc. **78** (1956) 2929.
[19] P. H. Emmett and S. Brunauer, J. Am. chem. Soc. **59** (1937) 1553.
[20] P. H. Emmett in *Advances in Catalysis* (Edited by W. G. Frankenburg, V. I. Komarewsky, and E. K. Rideal), (Academic Press, New York), vol. 1 (1948), p. 65.
[21] L. M. Dennis, K. M. Tressler, and F. E. Hance, J. Am. chem. Soc. **45** (1923) 2033.
[22] R. M. Dell, J. phys. Chem. **61** (1957) 1584.
[23] M. Green in *Progress in Semiconductors* (Edited by A. F. Gibson, F. A. Kröger, and R. E. Burgess), (Heywood, London), vol. IV (1960), p. 45. (References).
[24] F. G. Allen, J. Eisinger, H. D. Hagstrum, and J. T. Law, J. appl. Phys. **30** (1959) 1563.
[25] M. Green and I. A. Liberman, *Proc. Int. Conf. Semiconductor Physics, Prague*, 1960 (Czechoslovak Academy of Sciences, Prague, 1961), p. 536.
[26] H. D. Hagstrum, J. appl. Phys. **32** (1961) 1020. (References).
[27] D. Haneman, Phys. Rev. **121** (1961) 1093.
[28] S. P. Wolsky, J. appl. Phys. **29** (1958) 1132.
[29] H. E. Farnsworth, R. E. Schlier, T. H. George, and R. M. Burger, J. appl. Phys. **26** (1955) 252.
[30] R. E. Schlier and H. E. Farnsworth in *Semiconductor Surface Physics* (Edited by R. H. Kingston), (University of Pennsylvania Press, Philadelphia, 1957), p. 3.
[31] S. P. Wolsky, Phys. Rev. **108** (1957) 1131; S. P. Wolsky, E. J. Zdanuk, and D. Shooter, Surface Science **1** (1964) 110.
[32] D. Haneman, J. Phys. Chem. Solids **14** (1960) 162.
[33] H. D. Hagstrum, Phys. Rev. **119** (1960) 940.
[34] R. S. Ohl, Bell. System tech. J. **31** (1952) 104.
[35] H. E. Farnsworth in reference 4, p. 21. (References).
[36] C. J. Davison and L. H. Germer, Phys. Rev. **30** (1927) 705; L. H. Germer, Z. Physik **54** (1929) 408.
[37] H. E. Farnsworth, R. E. Schlier, T. H. George, and R. M. Burger, J. appl. Phys. **29** (1958) 1150.
[38] E. J. Scheibner, L. H. Germer, and C. D. Hartman, Rev. sci. Instr. **31** (1960) 112; L. H. Germer and C. D. Hartman, *ibid.* **31** (1960) 784.
[39] W. Ehrenberg, Phil. Mag. **18** (1934) 878.
[40] J. J. Lander, J. Morrison, and F. Unterwald, Rev. sci. Instr. **33** (1962) 782.
[41] R. E. Schlier and H. E. Farnsworth, J. chem. Phys. **30** (1959) 917.
[42] J. A. Dillon and H. E. Farnsworth, J. appl. Phys. **28** (1957) 174.
[43] E. W. Müller, Z. Physik **106** (1937) 541.
[44] F. G. Allen, J. Phys. Chem. Solids **19** (1961) 87. (References).
[45] E. W. Müller, Phys. Rev. **102** (1956) 618.

[46] E. W. Müller, Z. Physik **136** (1951) 131; E. W. Müller and K. Bahadin, Phys. Rev. **102** (1956) 624; E. W. Müller, Ann. N.Y. Acad. Sci. **101** (1963) 585. (References).
[47] E. W. Müller. Private communication.
[48] H. D. Hagstrum, Rev. sci. Instr. **24** (1953) 1122.
[49] H. D. Hagstrum, J. Phys. Chem. Solids **8** (1959) 211.
[50] H. D. Hagstrum, J. Phys. Chem. Solids **14** (1960) 33.
[51] H. D. Hagstrum, Phys. Rev. **122** (1961) 83.
[52] H. D. Hagstrum, J. appl. Phys. **32** (1961) 1015.
[53] H. D. Hagstrum, Phys. Rev. **119** (1960) 940.
[54] J. A. Becker and C. D. Hartman, J. phys. Chem. **57** (1953) 153.
[55] J. A. Becker in *Advances in Catalysis* (Edited by W. G. Frankenburg, V. I. Komarewsky, and E. K. Rideal), (Academic Press, New York), vol. VII (1955), p. 163.
[56] J. Eisinger and J. T. Law, J. chem. Phys. **30** (1959) 410.
[57] H. Sommer, H. A. Thomas, and J. A. Hipple, Phys. Rev. **82** (1951) 697.
[58] D. Alpert and R. S. Buritz, J. appl. Phys. **25** (1954) 202.
[59] R. E. Schlier, J. appl. Phys. **29** (1958) 1162.
[60] E. J. Zdanuk, R. Bierig, L. G. Rubin, and S. P. Wolsky, Report on 19th Annual Conference on Physical Electronics, M.I.T., Cambridge, 1959, p. 162.
[61] S. P. Wolsky and E. J. Zdanuk, Proc. 6th National Symposium on Vacuum Technology of the American Vacuum Soc., Philadelphia, 1959 (Pergamon Press, London, 1960), p. 6.
[62] J. P. Hobson and P. A. Redhead, Canadian J. of Phys. **36** (1958) 271.
[63] D. Brennan, D. V. Hayward, and B. M. W. Trapnell, J. Phys. Chem. Solids **14** (1960) 117.
[64] A. J. Rosenberg, J. phys. Chem. **62** (1958) 1112.
[65] M. J. Sparnaij, J. Phys. Chem. Solids **14** (1960) 111. (References).
[66] J. J. Lander and J. Morrison, Ann. N.Y. Acad. Sci. **101** (1963) 605; J. appl. Phys. **34** (1963) 1403.
[67] J. J. Lander, G. W. Gobeli, and J. Morrison, J. appl. Phys. **34** (1963) 2298; J. J. Lander and J. Morrison, *ibid.* **34** (1963) 1411.
[68] D. Haneman, *Proc. Int. Conf. Semiconductor Physics*, Exeter, 1962 (Institute of Physics and Physical Society, London, 1962), p. 842.
[69] J. Eisinger and J. T. Law, J. chem. Phys. **30** (1959) 410.
[70] M. Green, J. Phys. Chem. Solids **14** (1960) 77. (Review).
[71] S. P. Wolsky and E. J. Zdanuk, J. Phys. Chem. Solids **14** (1960) 124.
[72] P. T. Landsberg, J. chem. Phys. **23** (1955) 1079.
[73] J. T. Law, J. Phys. Chem. Solids **4** (1958) 91.
[74] J. R. Ligenza, J. phys. Chem. **64** (1960) 1017.
[75] M. Green and I. A. Liberman, J. Phys. Chem. Solids **23** (1962) 1407.
[76] M. Green and K. H. Maxwell, J. Phys. Chem. Solids **11** (1959) 195. (References).
[77] G. Heiland and P. Handler, J. appl. Phys. **30** (1959) 446.
[78] T. M. Buck in reference 4, p. 107. (Review).
[79] H. C. Gatos, M. C. Lavine, and E. P. Warekois, J. electrochem. Soc. **108** (1961) 645.
[80] P. J. Holmes in reference 5, p. 329.
[81] P. J. Holmes and R. C. Newman, Proc. Inst. elec. Engrs. B**106** (Suppl. 15) (1959) 287.
[82] D. R. Turner, J. electrochem. Soc. **106** (1959) 786.
[83] V. M. Kozlovskaya, Fiz. tver Tela **1** (1959) 1027 [translation: Soviet Phys. — Solid State **1** (1959) 940].
[84] J. T. Law and P. S. Meigs in *Semiconductor Surface Physics* (Edited by R. H. Kingston), (University of Pennsylvania Press, Philadelphia, 1957), p. 383.
[85] R. B. Bernstein and D. Cubicciotti, J. Am. chem. Soc. **73** (1951) 4112.

REFERENCES

[86] H. C. Gatos and M. C. Lavine, Ann. N.Y. Acad. Sci. **101** (1963) 983.
[87] M. M. Atalla, E. Tannenbaum, and E. J. Scheibner, Bell System tech. J. **38** (1959) 749.
[88] See, for example, A. E. J. Vickers (Editor), *Modern Methods of Microscopy* (Butterworth, London, 1956).
[89] See, for example, R. B. Fisher, *Applied Electron Microscopy* (Indiana University Press, Bloomington, 1953).
[90] S. Tolansky, *Multiple Beam Interferometry* (Clarendon Press, Oxford, 1948); J. Elect. Control **4** (1958) 63.
[91] M. C. Cretella and H. C. Gatos, J. electrochem. Soc. **105** (1958) 487.
[92] H. Robins and B. Schwartz, J. electrochem. Soc. **107** (1960) 108.
[93] J. W. Faust, Jr. in reference 6, p. 445.
[94] R. D. Heidenreich, U.S. Pat. 2 619 414 (1952).
[95] P. J. Holmes, Proc. Inst. elec. Engrs. **B106** (Suppl. 17) (1959) 861.
[96] J. W. Allen, Phil. Mag. [8] **2** (1957) 1475.
[97] W. C. Dash, J. appl. Phys. **29** (1958) 705.
[98] O. Rosner, Z. Metallk. **46** (1955) 225.
[99] B. W. Batterman, J. appl. Phys. **28** (1957) 1236.
[100] P. R. Camp, J. electrochem. Soc. **102** (1955) 586.
[101] S. G. Ellis, J. appl. Phys. **26** (1955) 1140.
[102] J. Franks, G. A. Geach, and A. T. Churchman, Proc. phys. Soc. **B68** (1955) 111.
[103] E. Billig, Proc. roy. Soc. **A235** (1956) 37.
[104] D. R. Turner in reference 4, p. 285 and reference 5, p. 155. (Reviews).
[105] S. G. Ellis, J. appl. Phys. **28** (1957) 1262.
[106] D. L. Klein, G. A. Kolb, L. A. Pompliano, and M. V. Sullivan, J. electrochem. Soc. **108** (1961) 60C.
[107] W. H. Brattain and C. G. B. Garrett, Bell System tech. J. **34** (1955) 129; Physica **20** (1954) 885.
[108] J. I. Pankove in reference 5, p. 312.
[109] M. Sparks, U.S. Pat. 2 656 496 (1953).
[110] P. F. Schmidt and W. Michel, J. electrochem. Soc. **104** (1957) 230.
[111] S. Zwerdling and S. Sheff, J. electrochem. Soc. **107** (1960) 338.
[112] A. Politychi and E. Fuchs, Z. Naturf. **A14** (1959) 271.
[113] R. C. Ellis, J. appl. Phys. **25** (1954) 1497; **28** (1957) 1068.
[114] J. W. Faust, Jr., in reference 4, p. 151.
[115] J. L. Richards and A. J. Crocker, J. appl. Phys. **31** (1960) 611.
[116] W. Bardsley, R. L. Bell, and W. B. Straugham, J. Elect. Control **5** (1958) 19.
[117] W. C. Dash, J. appl. Phys. **27** (1956) 1193.
[118] P. J. Holmes, Acta Met. **7** (1959) 283.
[119] R. D. Hancock and S. Edelman, Rev. sci. Instr. **27** (1956) 1082.
[120] G. A. Wolff, J. M. Wilbur, and J. C. Clark, Z. Electrochem. **61** (1957) 101.
[121] G. W. Schwuttke, J. electrochem. Soc. **106** (1959) 315.
[122] P. L. Ostapkovich and M. J. Ronaldi, J. appl. Phys. **30** (1959) 2019.

CHAPTER 4

THE SURFACE SPACE-CHARGE REGION

The concepts of a space-charge region at a semiconductor surface and the potential barrier associated with it evolved initially from studies of rectification processes in metal–semiconductor contacts [1-3]. During the last decade or so, such contacts — as well as their near relatives, the p–n junctions [4] — have been the subject of extensive investigation, and we refer the reader to the literature on the subject [5,6]. What we are interested in here is the space-charge region at a *free* surface. A detailed knowledge of the characteristics of this region is an essential prerequisite to any electrical surface studies of a quantitative nature.

The shape of the potential barrier can be determined from a solution of Poisson's equation. The excess carrier densities at the surface are then obtained by suitable integration procedures. This problem was first treated quantitatively by Schrieffer [7], Kingston and Neustadter [8], and Garrett and Brattain [9]. Since then, several workers [10-17] have succeeded in extending the range of the numerical calculations and in introducing various convenient analytical approximations. In the present chapter these calculations are considered in detail. After a discussion of the origin of the space-charge region, Poisson's equation is solved under conditions of thermal equilibrium and non-degenerate statistics, and the space-charge density and the shape of the potential barrier are obtained. Next the excess hole and electron surface densities are calculated. In the regions of interest, exact solutions are presented in graphical form and are extended to include surface degeneracy. Approximate solutions are also derived and their physical significance discussed. Following this there is a section on steady-state non-equilibrium conditions, and the problem is shown to reduce to that at thermal equilibrium. The impurity and other localized centres are always taken to be fully ionized, except in the concluding section. There the case is considered in which the occupation of the localized centres in the space-charge region is a function of barrier height. This situation is quite common in wide-gap semiconductors, and obtains even in germanium and silicon at very low temperatures.

Throughout the chapter it is assumed that we are dealing with a homogeneous single crystal whose dimensions are large compared to the width of the space-charge region.

1. The origin of the space-charge region

The space-charge region near the surface of a conducting solid may be produced by an electric field outside the solid, such field being established by an external source or by the proximity of another solid with a different work function. Alternatively, the space-charge region may result from the presence of a localized charged layer at the surface proper, due usually to surface states. These possibilities will now be discussed.

1.1. EXTERNAL FIELD

Consider a parallel-plate capacitor having one electrode a metal and the other a semiconductor. The application of a voltage across the capacitor results in the establishment of an electric field $\mathscr{E}$ between the plates. A displacement of mobile charge-carriers near the surface of each plate takes place, thus giving rise to two space-charge regions. The density Q_s of the induced charge on each plate is given by Gauss's law: $Q_s = \varepsilon_0 \mathscr{E}$. Since the free carrier density in a semiconductor is much smaller than that in a metal, the space-charge region will extend much farther into the bulk of the former. This situation is illustrated schematically in Fig. 4.1 for the case of an n-type semiconductor devoid of surface states, for various values of the bulk mobile carrier density n_b. The polarity of the applied voltage V_0 has been chosen so as to make the metal plate negative. Note that because the electrons have a negative charge, the electron *energy* increases in the direction of decreasing potential. To simplify the presentation in Fig. 4.1, the discontinuity in potential caused by the electron affinity has been omitted (see below). Also, in order to avoid the discontinuity of the field at the surface, the position coordinate z has been changed to z/κ, where κ is the dielectric constant in the medium concerned. The penetration of the field into the semiconductor produces a potential barrier beneath the surface. In order to obtain an idea of the magnitude of the total potential drop V_s between the surface and the underlying bulk, we shall approximate the potential in the space-charge region by a linear extension (dashed) of that in the vacuum.

From the figure we see that

$$\frac{V_s}{V_0 - V_s} = \frac{z_s/\kappa}{z_0}, \tag{4.1}$$

where z_0 is the separation of the capacitor plates and z_s the approximate distance (within the semiconductor) over which the potential drops to zero.

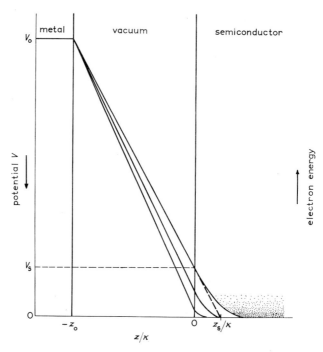

Fig. 4.1. Space-charge region in an n-type semiconductor devoid of surface states as produced by an applied voltage between a metal and the semiconductor. The potential and electron energy are shown as functions of the normalized distance z/κ. The three curves are for different values of the bulk electron density n_b, the uppermost curve corresponding to the smallest n_b. The dots illustrate the electron energy distribution in the semiconductor.

An estimate of Q_s is readily obtained by assuming that all the electrons in the space-charge region up to the distance z_s are removed by the field, thus leaving behind an equal static positive charge. By Gauss's law

$$-\frac{\varepsilon_0(V_0 - V_s)}{z_0} \approx q n_b z_s. \tag{4.2}$$

Combining (4.1) and (4.2) we have

$$z_s \approx \sqrt{\frac{\kappa \varepsilon_0 |V_s|}{n_b q}} \; ; \qquad \frac{V_s}{V_0 - V_s} \approx -\frac{\varepsilon_0(V_0 - V_s)}{\kappa z_0^2 n_b q}. \qquad (4.3)$$

Despite the fact that the approximations used in deriving equation (4.3) appear rather crude, nevertheless the results yield the correct order of magnitude (see below, eq. (4.27)). It is seen from (4.3) that as n_b increases, $|V_s|$ and z_s decrease. This is illustrated in Table 4.1, where values of V_s and z_s are given for various values of n_b in germanium ($\kappa = 16$). The applied voltage V_0 is -100 V and the plate separation z_0 is 10^{-2} cm.

TABLE 4.1

Values of V_s and z_s for different values of n_b in germanium ($\kappa = 16$).
$V_0 = -100$V, $z_0 = 10^{-2}$ cm

n_b(cm^{-3})	V_s(V)	z_s(cm)
10^{13}	-3.5×10^{-1}	5.6×10^{-4}
10^{15}	-3.5×10^{-3}	5.6×10^{-6}
10^{17}	-3.5×10^{-5}	5.6×10^{-8}

Obviously, although a space-charge region exists in metals as well, the corresponding values of V_s and z_s are so small as to be negligible for most purposes.

In the construction of Fig. 4.1 and the derivation of equations (4.1)–(4.3), the effect of surface states has been omitted. As will be discussed below, such states can give rise to a charged layer at the surface. A space-charge region will then be formed below the surface to compensate this surface charge. If now an external field is applied, charge will be displaced from both the surface states and the already-present space-charge region. For a sufficiently large density of occupied surface states, the external field will be terminated almost entirely at the surface proper, leaving the space-charge region essentially undisturbed. Thus surface states can screen the underlying region from external effects.

1.2. CONTACT POTENTIAL AND THE METAL–SEMICONDUCTOR CONTACT

Although electrons in a crystal are free to move about, they are not able to leave the crystal surface easily. This implies that their energy is lower inside the solid than outside. The work required to remove an electron at the

Fermi level to a point in free space just outside the solid is defined as the work function W_ϕ of the solid. In semiconductors it is found useful to define a second quantity, the electron affinity χ, as being the work required to remove an electron from the bottom edge of the conduction band at the surface to a point in free space just outside the semiconductor. The corresponding changes in potential as one moves from the solid into free space are very steep since they arise from the short-range binding forces exerted by the crystal lattice on the electron (typically, 10^8 V/cm over an interatomic

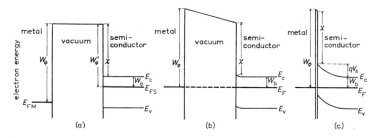

Fig. 4.2. Energy-level diagram for an electron in a metal and an n-type semiconductor devoid of surface states. The two media are (a) isolated from one another, (b) in thermal equilibrium, and (c) in close proximity.

spacing). Such potential steps are sometimes referred to as micropotentials [18, 19], to stress their microscopic character. The work function and affinity are illustrated in Fig. 4.2a, which represents an energy-level diagram for an electron in a system composed of a metal and an n-type semiconductor separated by free space (vacuum). We assume that thermal equilibrum exists *within* each solid but not yet *between* them. In other words, although the Fermi level is horizontal within each medium, it is not necessarily at the same height in the metal and in the semiconductor. The separation between the Fermi levels is not a well-defined quantity; it is determined by the potential distribution in free space which, in turn, is largely dependent on the previous history of the system. For simplicity, the free-space potential has been taken horizontal in Fig. 4.2a (zero field), implying that an electron in vacuum has the same energy everywhere.

We now make an electrical contact between the solids in such a way as to allow the flow of electrons from one medium to the other but without disturbing the surfaces presented in the figure. This can be done, for example, by connecting a metal wire between the back faces of the solids. (In principle, even without such a connection, charge exchange will take place by thermionic emission from the two solids into the vacuum, but this process is extremely slow at normal temperatures.) In the case illustrated, the work function of the semiconductor W'_ϕ is lower than that of the metal W_ϕ. Electrons in the semiconductor thus tend to flow into the metal, and a space charge builds up at the two surfaces. This flow continues until the Fermi levels are adjusted to the same height, corresponding to thermal equilibrium conditions, as shown in Fig. 4.2b. The electrostatic field established by the accumulated space charge is then just sufficient to stop any further charge transfer. This field is usually much weaker than the crystal field, so that the macroscopic potential from which it is derived leaves the micropotential associated with W_ϕ and χ practically unaffected. The overall drop in the macropotential between the metal and semiconductor bulks is referred to as the contact potential and is given by the difference in work functions. As far as this potential is concerned, the situation here is analogous to that discussed in the previous subsection. Thus if one bears in mind that the macropotential is superimposed on the short-range micropotential, it is seen that practically the whole contact potential distributes itself between the vacuum and the semiconductor. In the latter region the energy-band edges assume the shape of the macropotential. This follows from the fact that the carrier density at any point can be expressed in equivalent forms in terms of either the macropotential or the band edges, as discussed in Ch. 2, §7.3. In order that these expressions yield the same value for the carrier densities, the band edges at any point must differ from the electrostatic energy $-qV$ by at most a constant. Since the zero of electrostatic potential is arbitrary, it is possible to identify the macropotential energy with either of the band edges (or with any other energy level at a fixed distance from them).

Because the separation of the metal and the semiconductor is large, the contact potential falls mostly across the vacuum (eq. (4.3)), as shown in Fig. 4.2b. Accordingly, the energy bands of the semiconductor bend only slightly upwards, corresponding to a small space charge. As the separation decreases, the bending of the bands increases until, when the surfaces are at atomic distances (Fig. 4.2c), practically the whole contact potential falls across the space-charge region and V_s is given by

$$-qV_s = W_\phi - W'_\phi = W_\phi - \chi - W_b. \tag{4.4}$$

Here W_b represents the energy difference between the Fermi level E_F and conduction-band edge E_c in the semiconductor bulk. At such separations, despite the large barrier height between the metal and the semiconductor due to the electron affinity on the semiconductor side and the work function on the metal side, the barrier width is so small that the electrons are easily able to penetrate it by quantum-mechanical tunnelling. It is therefore customary to omit this barrier in conventional diagrams.

It follows from (4.4) that V_s depends on the work function of the metal and on the affinity and conductivity of the semiconductor. Contrary to these theoretical expectations, experimental evidence accumulated on germanium and silicon has shown that V_s is usually practically independent of the metal and of the conductivity of the semiconductor. To explain the discrepancy, Bardeen[20] proposed the existence of surface states — energy levels in the forbidden gap at the semiconductor surface. (The states actually responsible for these experimental results are the so-called slow surface states, as will be discussed in subsequent chapters.)

1.3. SURFACE STATES

In the absence of surface states, the energy bands of a semiconductor continue straight up to the surface, provided there is no external field. When acceptor-like surface states are introduced below the Fermi level, they will not be in equilibrium with the energy bands as long as they remain unoccupied. This situation is illustrated in Fig. 4.3a, where the surface states

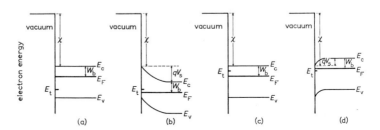

Fig. 4.3. Energy-level diagram for an n-type semiconductor in the presence of acceptor-like ((a) and (b)) and donor-like ((c) and (d)) surface states at an energy level E_t.
(a) and (c) represent conditions immediately following the introduction of the surface states; (b) and (d), after thermal equilibrium has been reached.

have been introduced at an energy level E_t. Since the states are empty and below the Fermi level, some of the electrons in the conduction band fall into them. In this process, the surface becomes negatively charged while a positive space-charge layer forms below it. Consequently, the energy bands at the surface bend upwards with respect to the Fermi level. Since the energy position of the surface states in the forbidden gap is determined by short-range atomic forces and is thus unaffected by the presence of the macropotential, the surface levels rise together with the band edges. The process of charge transfer continues until equilibrium is reached (Fig. 4.3b). In the particular case shown, the surface states at thermal equilibrium are somewhat above the Fermi level and so are only partially filled. Actually, the final position of the surface states with respect to the Fermi level is determined by the condition that the positive charge in the space-charge region just balance the negative equilibrium charge in the surface states. This is made necessary by the requirement that overall neutrality prevails. Thus the larger the surface-state density, the higher the bending of the bands at the surface. The situation for donor-like surface

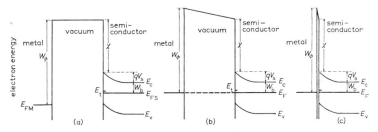

Fig. 4.4. The effect of (acceptor-like) surface states on the metal-semiconductor contact.
        (a) Prior to electrical contact.
        (b) After thermal equilibrium has been reached.
        (c) At close proximity.

states placed above the Fermi level is completely analogous (Figs. 4.3c, 4.3d). The introduction of acceptor-like states above the Fermi level or of donor-like states below the Fermi level has of course no influence on the shape of the energy bands.

Let us now consider the modification that the presence of surface states introduces into the treatment of the metal–semiconductor contact. If such states are appropriately located with respect to the Fermi level, the energy bands will be bent even before any electrical contact is made with the metal. Figure 4.4a illustrates this situation for the case of a large density of charged acceptor-like surface states. When thermal equilibrium is established and the two Fermi levels coincide, the energy bands of the semiconductor bend only slightly higher (Fig. 4.4b). This slight additional bending is sufficient to cause the required number of electrons to leave the surface states and enter the metal. Thus, the field caused by the contact potential is almost entirely terminated by the surface states rather than by the space charge, and nearly the whole contact potential falls across the vacuum. This situation continues as the metal–semiconductor separation is decreased to a few interatomic distances (Fig. 4.4c), provided the surface-state density is sufficiently large (see Ch. 9, §1 for a more quantitative discussion). The barrier height V_s then remains practically independent of the work functions of the solids, and the system behaves as though the semiconductor surface were a thin metallic film screening the bulk from external fields.

2. The space-charge density and the shape of the potential barrier

In the previous section we discussed possible origins of the space-charge region near the surface. Once such a region exists in a given semiconductor, its detailed characteristics are uniquely determined by the height of the potential barrier at the surface proper. We shall now calculate the magnitude of the space-charge density and the shape of the potential in the space-charge region as a function of barrier height.

2.1. CONCEPTS AND DEFINITIONS

It will be useful at this stage to define the various parameters associated with the space-charge region near the surface.

The potential ϕ is defined by the equation

$$q\phi \equiv E_F - E_i, \qquad (4.5)$$

where E_i is parallel to the band edges and in the bulk coincides with the intrinsic Fermi level (Fig. 4.5). Its exact position is given by (2.88), and is usually close to the mid-gap. The value of ϕ in the bulk is called the bulk potential ϕ_b and its value at the surface, the surface potential ϕ_s.

The potential barrier V is defined as

$$V \equiv \phi - \phi_b \qquad (4.6)$$

and represents the potential at any point in the space-charge region with respect to its value in the bulk. In particular, the barrier height $V_s (\equiv \phi_s - \phi_b)$ is the total potential difference between the surface and the bulk.

It is convenient to define dimensionless potentials u, v by the equations

$$u \equiv q\phi/kT; \qquad v \equiv qV/kT. \qquad (4.7)$$

Ch. 4, § 2.1 SPACE-CHARGE DENSITY AND POTENTIAL BARRIER SHAPE 137

By means of the intrinsic carrier concentration n_i (eq. (2.87)) and the above definitions, the electron density n and hole density p in a non-degenerate semiconductor are given at every point by

$$n = n_i e^u = n_b e^v; \quad p = n_i e^{-u} = p_b e^{-v}. \tag{4.8}$$

The condition $u > 0$ signifies that at that point $n > p$, and conversely; $v > 0$ implies $n > n_b$ and $p < p_b$, and conversely. At the point where

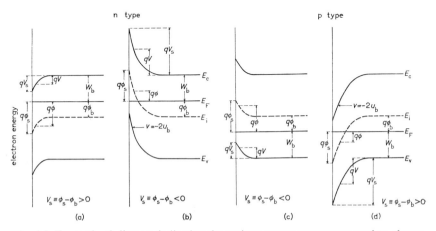

Fig. 4.5. Energy-level diagram indicating the various energy parameters used to characterize the space-charge region, as defined in § 2.1.
    (a) n-type semiconductor, accumulation layer.
    (b) Inversion layer.
    (c) p-type semiconductor, accumulation layer.
    (d) Inversion layer.

$u = 0$ the carrier densities are intrinsic; in particular $u_b = 0$ corresponds to an intrinsic bulk, $u_s = 0$ to an intrinsic surface. When $v_s = 0$ there is no bending of the energy bands and they continue straight from the bulk to the surface. This is known as the flat-band condition.

When the majority-carrier density in the space-charge region is greater than that in the bulk, the space-charge region is termed an accumulation layer. This condition obtains when the sign of v_s is the same as that of u_b (eq. (4.8)): positive for n type and negative for p type. When the sign of v_s is opposite to that of u_b, we have either a depletion or an inversion layer. The space-charge region up to the point where the minority-carrier density equals the majority-carrier bulk density ($v = -2u_b$) is called the depletion region. Between this point and the surface the minority-carrier density exceeds the majority-carrier bulk density, and this region is called the inversion region. (Of course the situation can exist where the depletion region continues right up to the surface.) Figure 4.5 illustrates the above definitions for an n- and a p-type semiconductor.

2.2. POISSON'S EQUATION

We shall treat a semi-infinite, homogeneous crystal under thermal equilibrium. The surface is represented by the plane $z = 0$ and the bulk, by positive values of z. The potential V at any point is then a function of z only. It is determined by Poisson's equation which, for the planar geometry assumed, has the form

$$\frac{d^2 V}{dz^2} = -\frac{\rho}{\kappa \varepsilon_0}, \qquad (4.9)$$

where ρ is the charge density (coul/m^3) and is composed of positive and negative static charge and the charge due to mobile electrons and holes. The appropriate boundary conditions at the surface ($z = 0$) and in the bulk ($z \to \infty$) are $V = V_s$ and $V = 0$, respectively. For the case of constant impurity concentration and complete ionization, the static charge density is the difference between the donor and acceptor concentrations ($N_D - N_A$) and, from the condition of charge neutrality in the bulk, is equal to the difference in electron and hole bulk densities. The mobile electron and hole densities in the case of non-degenerate statistics are given by (4.8). With the help of the reduced potential v, Poisson's equation can be rewritten as

$$\frac{d^2 v}{dz^2} = -\frac{q^2}{\kappa \varepsilon_0 kT} (n_b - p_b + p_b e^{-v} - n_b e^v). \qquad (4.10)$$

In order to obtain a feel for the problem, we shall first consider the case of small disturbances, corresponding to $|v| \lesssim \frac{1}{2}$. Under these conditions it is permissible to neglect the nonlinear terms in the series expansion of the exponential functions. We then have

$$\frac{d^2 v}{dz^2} \approx \frac{v}{L^2}, \qquad (4.11)$$

where

$$L \equiv \sqrt{\frac{\kappa \varepsilon_0 kT}{q^2 (n_b + p_b)}}. \qquad (4.12)$$

Ch. 4, §2.2 SPACE-CHARGE DENSITY AND POTENTIAL BARRIER SHAPE

Equation (4.11) can be integrated directly, and yields the result

$$v = v_s e^{-z/L}. \tag{4.13}$$

It thus follows that in terms of z/L the potential barrier for small disturbances is independent of the bulk potential u_b. The value of $|v|$ decreases exponentially with characteristic length L, the *effective* Debye length. The so-called Debye length L_D was defined by Shockley [21] as $(\kappa\varepsilon_0 kT/2q^2 n_i)^{\frac{1}{2}}$, and is the value of L for an intrinsic semiconductor. As will be seen below, the effective Debye length L, and not L_D, characterizes the width of the space-charge region in the general case as well [22]. Accordingly, the parameter L will be used in all future discussions.

It can be seen from the approximate solution (eq. (4.13)) that the derivative dv/dz vanishes in the bulk and that the boundary condition $v = 0$ for $z \to \infty$ is equivalent to $v = 0$ for $z \gg L$. In the case of thin slabs, where the condition $z \gg L$ no longer holds, the situation is more complicated and will not be dealt with here [23].

With the help of (4.8) and (4.12), Poisson's equation (eq. (4.10)) can be written as

$$\frac{d^2 v}{dz^2} = \frac{1}{L^2}\left[\frac{\sinh(u_b+v)}{\cosh u_b} - \tanh u_b\right]. \tag{4.14}$$

Multiplying both sides by $2dv/dz$ we can integrate once and, using the condition $dv/dz = 0$ for $v = 0$, we obtain

$$\frac{dv}{dz} = \mp\frac{F}{L}; \quad \frac{z}{L} = \int_{v_s}^{v}\frac{dv}{\mp F}, \tag{4.15}$$

where

$$F(u_b, v) = \sqrt{2}\left[\frac{\cosh(u_b+v)}{\cosh u_b} - v\tanh u_b - 1\right]^{\frac{1}{2}}. \tag{4.16}$$

The upper sign (minus) refers to $v > 0$, the lower sign (plus) to $v < 0$; this convention will be adhered to throughout the book. Equation (4.15) cannot be solved in closed form for the general case, and numerical solutions are given below. For the particular case of an intrinsic semiconductor, however, this equation can be solved analytically [10] to give

$$F(0, v) = \pm 2\sinh(v/2); \quad \frac{z}{L} = \ln\frac{\tanh(v_s/4)}{\tanh(v/4)}. \tag{4.17}$$

The space-charge density Q_{sc} is defined as the total net charge in the space-charge region per unit surface area. As such, Q_{sc} satisfies Gauss's law:

$$Q_{sc} = \kappa \varepsilon_0 \mathscr{E}_s, \qquad (4.18)$$

where $\mathscr{E}_s$ is the electrostatic field inside the semiconductor just below the surface and is positive when directed from the surface outwards. The value and sign of $\mathscr{E}_s$ are given by the derivative of the potential V at the surface. With the help of (4.7), (4.12), (4.15), we can express Q_{sc} in the form

$$Q_{sc} = \mp q(n_b + p_b) L F_s, \qquad (4.19)$$

where F_s is the value of F at the surface ($v = v_s$). The function F_s is plotted in Figs. 4.7, 4.8 below. The space-charge density Q_{sc} arises from charge that is unevenly distributed throughout the space-charge region. We shall find it useful to define an effective charge distance L_c as the position (measured from the surface) of the centre of the space charge:

$$L_c \equiv \frac{\int_0^\infty \rho z \, dz}{\int_0^\infty \rho \, dz} \equiv \frac{\int_0^\infty \rho z \, dz}{Q_{sc}}. \qquad (4.20)$$

Integrating by parts and using (4.9), (4.15), (4.19), we obtain

$$L_c = -\frac{[zF]_0^\infty}{F_s} + \frac{\int_0^\infty F \, dz}{F_s}. \qquad (4.21)$$

The first term vanishes at the upper limit as well as at the lower, because $F \to 0$ exponentially as $z \to \infty$ (eqs. (4.13), (4.15)). The numerator of the second term is just $|v_s| L$, as can be seen from (4.15). Thus

$$L_c = \frac{|v_s|}{F_s} L. \qquad (4.22)$$

This equation has a very simple physical interpretation. If we define the space-charge capacitance C_{sc} as the ratio of the space-charge density Q_{sc} to the barrier height V_s, we obtain (using (4.7), (4.12), (4.19), (4.22))

$$C_{sc} \equiv \left| \frac{Q_{sc}}{V_s} \right| = \frac{\kappa \varepsilon_0}{L_c}. \qquad (4.23)$$

The space-charge capacitance is thus seen to be the capacitance of a parallel-plate capacitor of unit area having one plate situated at the surface and the other at the centre of the space charge. It is easily seen (eqs. (4.13), (4.15), (4.22)) that for small values of $|v_s|$, L_c approaches L. The behaviour of L_c as a function of $|v_s|$ is shown for various values of $|u_b|$ in Figs. 4.7 and 4.9.

It should be noted that two basic simplifications were made in solving Poisson's equation. First, the donor and acceptor ions were represented by a uniform charge distribution. In reality the impurity ions are distributed randomly and, since the spacing between neighbours is usually of the same order of magnitude as the thickness of the space-charge layer, the potential will fluctuate along the surface on that scale. This effect would be particularly pronounced in depletion layers, where most of the space charge consists of ionized impurities. Experimentally measured quantities usually involve large surface areas, however, so that the fluctuations about the average value may be neglected for most purposes. A second assumption concerns the use of classical statistics in the expression for the free carrier densities in terms of the electrostatic potential (eq. (4.8)). This is permissible as long as the variation in potential over a distance equal to the electron wavelength is small compared to kT/q. For steeper barriers (such as are encountered in very strong inversion or accumulation layers), a modification of the treatment may be necessary in order to include possible quantization effects. Both of these limitations are discussed in some detail by Garrett and Brattain[9].

2.3. APPROXIMATE SOLUTIONS FOR EXTRINSIC SEMICONDUCTORS

The space-charge density Q_{sc} is determined by both the electrons and the holes that are either accumulated or depleted in the space-charge region. But when one type of carrier becomes dominant, it is this one that will effectively determine Q_{sc}. One can then derive simple approximations[11,13-15] for the function F and the shape of the potential barrier $v(z)$.

The majority-carrier bulk density exceeds that of the minority carrier by the factor $\exp|2u_b|$, so that for an extrinsic semiconductor ($|u_b| \gtrsim 2$), minority carriers in the bulk constitute at most 2% of the total number of carriers (eq. (4.8)). The situation in accumulation layers (v positive for n type, negative for p type) is even more favourable, so that here Q_{sc} is determined solely by the majority carriers. Thus in accumulation layers the function F can be approximated by

$$F(u_b, v) \approx \sqrt{2}[e^{|v|} - |v| - 1]^{\frac{1}{2}} \equiv F^+(v) \qquad (4.24)$$

and is independent of the bulk potential u_b. This approximation is equivalent to neglecting the minority-carrier density in Poisson's equation.

For strong accumulation layers ($|v| \gtrsim 3$), the function $F^+(v)$ increases according to $\exp \tfrac{1}{2}|v|$ and the centre of charge rapidly approaches the surface (see eq. (4.22)). The shape of the potential barrier in this region is obtained from an approximate solution to (4.15):

$$z/L \approx \sqrt{2}[e^{-\frac{1}{2}|v|} - e^{-\frac{1}{2}|v_s|}]. \tag{4.25}$$

In depletion and inversion layers of an extrinsic semiconductor, the majority carriers are depleted. In the region where the minority-carrier density remains small ($0 > v > -2u_b$ for n type, $0 < v < -2u_b$ for p type), Q_{sc} is determined by the static space charge, and F is given by

$$F(u_b, v) \approx \sqrt{2}[e^{-|v|} + |v| - 1]^{\frac{1}{2}} \equiv F^-(v). \tag{4.26}$$

Here F increases slowly with $|v|$, approximately parabolically. The reason for this is that the space charge consists of static charge only, and this remains uncompensated by the depleted majority carriers. When $|v_s|$ increases, the space charge can become larger only to the extent that it penetrates farther into the bulk, and the centre of the space charge (eq. (4.22)) recedes from the surface approximately as the square root of $|v_s|$. For $|v| \gtrsim 3$, the exponential term can be neglected, and (4.15) can be integrated directly:

$$|v| \approx \tfrac{1}{2}[\sqrt{2|v_s| - 2} - z/L]^2 + 1. \tag{4.27}$$

This corresponds to the so-called Schottky barrier[1], which is characterized by a quadratic dependence of the potential on z.

For weak inversion layers, Q_{sc} is determined by both the static space charge and the minority carriers, and F can be approximated by

$$F(u_b, v) \approx \sqrt{2}[e^{|v| - 2|u_b|} + |v| - 1]^{\frac{1}{2}}. \tag{4.28}$$

The parabolic region continues up to where the exponential term can be neglected. Thus for large $|u_b|$, the parabolic dependence extends slightly into the inversion region. For larger values of $|v|$ (strong inversion region) the linear term can be neglected, and near the surface it is the minority carriers which are dominant. (Even though, farther removed from the surface, the space charge region is still determined by the static charge.) Here F can be written approximately as

$$F(u_b, v) \approx \sqrt{2}\, e^{\frac{1}{2}(|v| - 2|u_b|)}, \tag{4.29}$$

Ch. 4, §2.4 SPACE-CHARGE DENSITY AND POTENTIAL BARRIER SHAPE

which is equivalent to neglecting both the static space charge and the majority carriers in Poisson's equation. The function F increases exponentially with $\frac{1}{2}|v|$, as in the accumulation-layer case, but now its absolute magnitude is determined by $\exp\frac{1}{2}(|v|-2|u_b|)$. Equation (4.15) can be integrated directly to yield

$$z/L \approx \sqrt{2}\,[e^{\frac{1}{2}(2|u_b|-|v|)} - e^{\frac{1}{2}(2|u_b|-|v_s|)}]. \qquad (4.30)$$

This result becomes identical with that for accumulation layers if v is measured from the point where $v = -2u_b$; that is, from the point where the minority-carrier density is equal to that of the bulk majority carriers. Physically, this follows from the fact that in both cases the slope of the potential barrier is determined only by the carriers dominant at the surface. In an inversion layer, the value of L_c decreases with increasing $|v_s|$ after having reached its maximum value at $v_s = -2u_b$.

For near-intrinsic semiconductors ($|u_b| < 2$), the situation is qualitatively the same, except for the absence of the parabolic region between the inversion region and the bulk.

2.4. NUMERICAL SOLUTIONS

In Fig. 4.6 results are shown of the numerical integration of (4.15). The potential barrier $|v|$ is plotted as a function of z/L, the normalized distance from the surface, for various values of the bulk potential $|u_b|$. In accumulation layers the shape is independent of the bulk potential for $|u_b| \gtrsim 2$, as can be seen from (4.24) and (4.25). As the accumulation layer becomes more pronounced, $|v|$ varies very rapidly with z/L. The curve for an intrinsic semiconductor ($u_b = 0$) has been calculated from (4.17) and is seen to lie quite close to that for $|u_b| \gtrsim 2$. The curve for $|u_b| = 0.5$, not shown in the figure, falls approximately midway between the two curves.

The shape in inversion and depletion layers is also shown in Fig. 4.6. For strong inversion layers, the shape is similar to that in accumulation layers (eq. (4.30)). This is shown by the plots for $|u_b| < \frac{1}{2}|v_s|$ at small values of z/L. At somewhat larger values of z/L, we enter a transition region. Following this, the value of $|v|$ decreases much more slowly with z/L (the parabolic region, eq. (4.27)) and the curves are approximately parallel. All the curves for $|u_b| \gtrsim 10$ coincide because, for the value of v_s chosen here ($|v_s| = 20$), these curves correspond to the case where the depletion region extends right up to the surface (depletion layer). The plot for $u_b = 0$

144   THE SPACE-CHARGE REGION   Ch. 4, § 2.4

is of course the same as the one shown in the accumulation-layer case.

Although the surface in Fig. 4.6 is represented by $z/L = 0$, nevertheless the shape for any barrier height $|v'_s|$ smaller than that shown here can be obtained from the figure simply by a translation along the z/L axis. This follows from the fact that (4.15) can be rewritten as

$$\frac{z'}{L} + \frac{(z-z')}{L} = \int_{v_s}^{v'_s} \frac{dv}{\mp F} + \int_{v'_s}^{v} \frac{dv}{\mp F}, \tag{4.31}$$

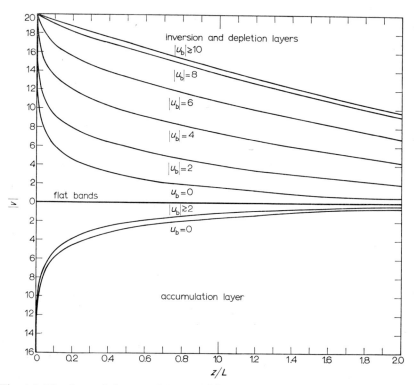

Fig. 4.6. The shape of the potential barrier $|v|$ as a function of the normalized distance from the surface z/L for various values of the bulk potential $|u_b|$. (The value of $|v_s|$ has been taken equal to 20.)

where z' is the point at which $v = v'_s$. Thus for the new barrier height v'_s, one merely measures z/L from the point where the curve for the appropriate $|u_b|$ intersects the horizontal line $|v| = |v'_s|$.

2.5. EXTENSION FOR DEGENERATE SURFACE CONDITIONS

The case of a degenerate semiconductor has been treated by Seiwatz and Green [17] who were able to integrate Poisson's equation once and obtain an expression for the function F (as defined in eq. (4.15)). In practice, the semiconductor bulk is usually not degenerate, and only the space-charge region near the surface becomes at times degenerate. In such a case, for extrinsic semiconductors and complete ionization in the bulk, the function F assumes a simpler form. In the regions where only one carrier is dominant at the surface, similar approximations can be made for F as in the non-degenerate case above. For accumulation layers

$$F(w_b, v_s) \approx \sqrt{2}\,[\tfrac{4}{3}\pi^{-\frac{1}{2}} e^{w_b} F_{\frac{3}{2}}(|v_s| - w_b) - |v_s| - 1]^{\frac{1}{2}}, \qquad (4.32)$$

where

$$F_{\frac{3}{2}}(\eta) \equiv \int_0^\infty \frac{x^{\frac{3}{2}}\,dx}{1+\exp(x-\eta)} \qquad (4.33)$$

and has been tabulated [24]. Here, as in the non-degenerate case, F is independent of u_b. But now an additional parameter w_b enters, the energy distance (in units of kT) between the Fermi level and the majority-carrier band edge in the bulk (see Fig. 4.5 and eqs. (2.99), (2.101)). This is to be expected, since $w_b - |v|$ determines where the degenerate conditions set in.

For inversion layers

$$F(u_b, e_g - w_b, v_s) \approx$$
$$\sqrt{2}\,[\tfrac{4}{3}\pi^{-\frac{1}{2}} e^{(e_g - w_b) - 2|u_b|} F_{\frac{3}{2}}(|v_s| - e_g + w_b) + |v_s| - 1]^{\frac{1}{2}}. \qquad (4.34)$$

Here, too, there is an additional parameter, $(e_g - w_b)$, the energy distance (in units of kT) between the Fermi level and the minority-carrier band edge in the bulk.

If the separation between the Fermi level and the pertinent band edge in the space-charge region is large ($w_b - |v_s| \gtrsim 4$ for accumulation layers, $(e_g - w_b) - |v_s| \gtrsim 4$ for inversion layers), then (4.32) and (4.34) reduce [24] to (4.24) and (4.28), respectively, as of course they should. It is to be noted that when the space-charge region becomes degenerate, the space-charge density Q_{sc} increases relatively more slowly than in the non-degenerate case. This follows from the fact that now the filling of the energy states in the pertinent band obeys Fermi–Dirac rather than Boltzmann statistics.

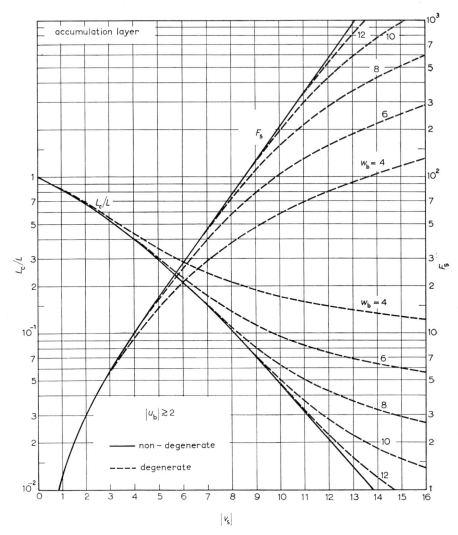

Fig. 4.7. The functions F_s and L_c/L plotted against the barrier height $|v_s|$ in accumulation layers. The dashed curves represent degenerate surface conditions for various values of w_b.

In Fig. 4.7 the functions F_s and L_c/L for extrinsic semiconductors are plotted against $|v_s|$ for accumulation layers. The dashed curves represent F_s and L_c/L in the degenerate case for several values of the parameter w_b. The derivation of (4.22) for L_c was not restricted to the non-degenerate

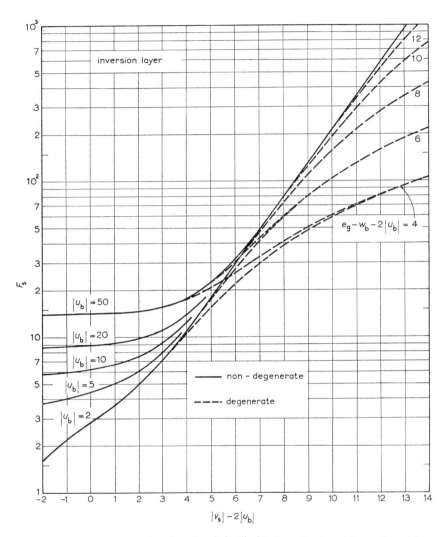

Fig. 4.8. The function F_s plotted against $|v_s|-2|u_b|$ in inversion layers for various values of $|u_b|$. The dashed curves represent degenerate surface conditions for different values of $e_g-w_b-2|u_b|$.

case and so is applicable here as well, provided the appropriate F_s is used. In Fig. 4.8, F_s in inversion layers is plotted as a function of $|v_s|-2|u_b|$ for different values of $|u_b|$. The abscissa was so chosen because in this region F_s depends mainly on $|v_s|-2|u_b|$, as can be seen from (4.28) and

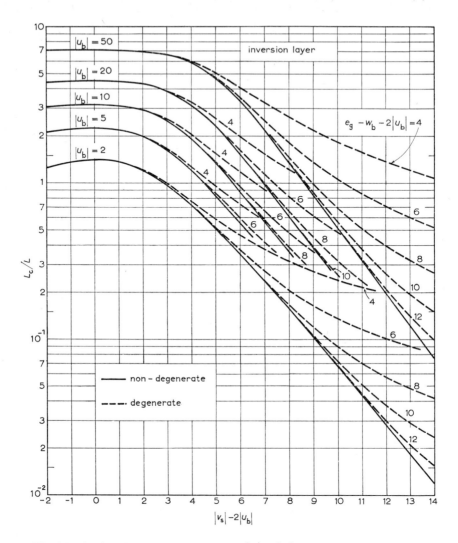

Fig. 4.9. The function L_c/L plotted against $|v_s|-2|u_b|$ in inversion layers for various values of $|u_b|$. The dashed curves represent degenerate surface conditions for different values of $e_g - w_b - 2|u_b|$.

(4.34). The additional dependence on $e_g - w_b - 2|u_b|$ for degenerate conditions is also shown. The corresponding values of L_c/L are presented in Fig. 4.9.

3. The excess surface-carrier densities ΔN and ΔP

3.1. DEFINITIONS

The excess surface-carrier densities are defined as the number (per unit surface area) of mobile electrons ΔN and holes ΔP in the space-charge layer with respect to their numbers at flat bands ($v_s = 0$). It should be noted that if ΔN is positive then ΔP is negative, and conversely. According to this definition, the densities can be expressed as

$$\Delta N = \int_0^\infty (n-n_b)\mathrm{d}z; \quad \Delta P = \int_0^\infty (p-p_b)\mathrm{d}z. \tag{4.35}$$

In terms of these quantities, the space-charge density Q_{sc} is given by

$$Q_{sc} = q(\Delta P - \Delta N). \tag{4.36}$$

By changing the variable of integration to v and using (4.8) and (4.15), one obtains for the non-degenerate case

$$\Delta N = n_b L \int_{v_s}^0 \frac{e^v - 1}{\mp F(u_b, v)} \mathrm{d}v; \tag{4.37}$$

$$\Delta P = p_b L \int_{v_s}^0 \frac{e^{-v} - 1}{\mp F(u_b, v)} \mathrm{d}v. \tag{4.38}$$

Substituting $y = -v$ in the second equation and using the identity

$$F(-u_b, v) \equiv F(u_b, -v), \tag{4.39}$$

one can write (4.37) and (4.38) for the case of majority carriers in accumulation layers as

$$\Delta N = n_b L G^+(u_b, v_s) \quad \text{for} \quad v_s \geqq 0 \text{ in n type } (u_b \geqq 0); \tag{4.40}$$

$$\Delta P = p_b L G^+(u_b, v_s) \quad \text{for} \quad v_s \leqq 0 \text{ in p type } (u_b \leqq 0). \tag{4.41}$$

Here

$$G^+(u_b, v_s) \equiv \int_0^{|v_s|} \frac{e^v - 1}{F(|u_b|, v)} \mathrm{d}v, \tag{4.42}$$

and is always positive.

For majority carriers in depletion and inversion layers, (4.37) and (4.38) can be rewritten (again by substituting $y = -v$, this time in eq. (4.37)) in the form

$$\Delta N = n_b L G^-(u_b, v_s) \quad \text{for} \quad v_s \leqq 0 \text{ in n type } (u_b \geqq 0); \quad (4.43)$$

$$\Delta P = p_b L G^-(u_b, v_s) \quad \text{for} \quad v_s \geqq 0 \text{ in p type } (u_b \leqq 0), \quad (4.44)$$

where

$$G^-(u_b, v_s) \equiv \int_0^{|v_s|} \frac{e^{-v}-1}{F(-|u_b|, v)} \, dv, \quad (4.45)$$

and is always negative.

In the case of minority carriers in accumulation layers, (4.37) and (4.38) can be expressed with the help of (4.8) as

$$\Delta N = p_b L g^-(u_b, v_s) \quad \text{for} \quad v_s \leqq 0 \text{ in p type } (u_b \leqq 0); \quad (4.46)$$

$$\Delta P = n_b L g^-(u_b, v_s) \quad \text{for} \quad v_s \geqq 0 \text{ in n type } (u_b \geqq 0). \quad (4.47)$$

Here

$$g^-(u_b, v_s) \equiv e^{-2|u_b|} \int_0^{|v_s|} \frac{e^{-v}-1}{F(|u_b|, v)} \, dv, \quad (4.48)$$

and is always negative.

For minority carriers in depletion and inversion layers, (4.37) and (4.38) can be rewritten in the form

$$\Delta N = p_b L g^+(u_b, v_s) \quad \text{for} \quad v_s \geqq 0 \text{ in p type } (u_b \leqq 0); \quad (4.49)$$

$$\Delta P = n_b L g^+(u_b, v_s) \quad \text{for} \quad v_s \leqq 0 \text{ in n type } (u_b \geqq 0), \quad (4.50)$$

where

$$g^+(u_b, v_s) \equiv e^{-2|u_b|} \int_0^{|v_s|} \frac{e^v-1}{F(-|u_b|, v)} \, dv, \quad (4.51)$$

and is always positive.

Keeping in mind that ΔN and ΔP are always opposite in sign, we have from (4.36) that $|Q_{sc}|$ is equal to the sum of the absolute values of $q\Delta N$ and $q\Delta P$. Using (4.8), (4.19), and (4.40)–(4.51), we obtain from (4.36)

$$G^+ + |g^-| = (1+e^{-2|u_b|})F(|u_b|, |v_s|); \quad (4.52)$$

$$|G^-| + g^+ = (1+e^{-2|u_b|})F(-|u_b|, |v_s|). \quad (4.53)$$

3.2. Graphical representation and approximate expressions

The functions G^+, G^-, g^+, g^- defined above cannot be obtained in closed form in the general case and must therefore be evaluated numerically.

For the particular case of an intrinsic semiconductor ($u_b = 0$), however, (4.42), (4.45), (4.48), and (4.51) can be integrated directly. Using (4.17), we obtain

$$G^+ = g^+ = 2(e^{\frac{1}{2}|v_s|} - 1); \qquad (4.54)$$

$$G^- = g^- = 2(e^{-\frac{1}{2}|v_s|} - 1). \qquad (4.55)$$

It is seen that in this case the density of the accumulated carriers (electrons for $v_s > 0$, holes for $v_s < 0$) increases rapidly (approximately exponentially) with $|v_s|$. The carriers of the opposite sign, on the other hand, are depleted near the surface and their excess density tends to a negative saturation value. This behaviour is quite general, being exhibited by extrinsic and near intrinsic samples as well.

Before discussing in detail the general case, the reader may perhaps find it helpful to look at the main features of the four functions G^+, G^-, g^+, g^-. For this purpose they are presented in compact form in Fig. 4.10 as a

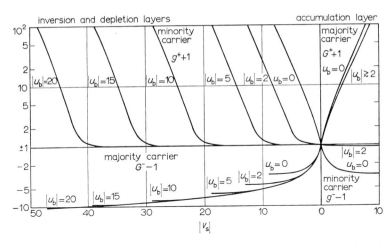

Fig. 4.10. The four functions representing the excess surface carrier densities plotted against the barrier height $|v_s|$ for various values of the bulk potential $|u_b|$.

function of $|v_s|$ for various values of $|u_b|$. Because of their practical usefulness, these functions are displayed again on a much larger scale in Figs. 4.11 and 4.12. To obtain ΔN or ΔP, one must multiply the appropriate function by the product of L and the *majority*-carrier bulk density. Thus, for a given value of u_b, the functions represent the *relative* magnitudes of the excess carrier densities. Obviously, all functions approach zero as v_s

tends to zero. This situation is awkward to represent on a logarithmic scale, and so $G^+ + 1$ rather than G^+ (and similarly for the other functions) has been plotted in Fig. 4.10. For large values of $|v_s|$, G^+ and g^+ increase rapidly with $|v_s|$ whereas G^- and g^- tend to a (negative) saturation value similarly to the intrinsic-sample case discussed above. As G^+ represents the accumulated majority carriers, it starts increasing rapidly even at low values of $|v_s|$. The function $g^+ + 1$, on the other hand, corresponds to minority carriers; hence its sharp increase with $|v_s|$ sets in at $v_s \approx -2u_b$, when the depletion layer gives place to an inversion layer. The functions G^- and g^- correspond to a depletion of majority and minority carriers, respectively, and are thus insensitive to $|v_s|$. These characteristics, as well as a more detailed insight into the nature of the space-charge region, can be obtained by considering a few limiting cases. In those ranges where the excess surface density of one type of carrier can be neglected with respect to that of the other, the dominant carrier density is obtainable from Q_{sc}. Such is the case in extrinsic semiconductors ($|u_b| \gtrsim 2$) for majority carriers in accumulation and depletion layers, where the situation is that of a single-carrier system. Here we neglect $\exp(-2|u_b|)$ with respect to unity and drop g^-, g^+ in (4.52) and (4.53). By using the approximate expression for F (eq. (4.24) for accumulation layers, eq. (4.26) for depletion layers), we then obtain

$$G^+(u_b, v_s) \approx F^+(v_s); \tag{4.56}$$

$$G^-(u_b, v_s) \approx F^-(v_s). \tag{4.57}$$

The approximations are better than 1 %, as can be verified by direct integration.

In the case of degenerate conditions at the surface, (4.36) is still valid and G^+ is definable by (4.40) and (4.41). The function G^+ can be approximated in accumulation layers by (4.32) and obtained from Fig. 4.7.

In the transition between depletion and strong inversion layers, Q_{sc} is determined by both majority and minority carriers, so that here there is no convenient approximation to ΔN or ΔP. In strong inversion layers, however, the minority-carrier concentration near the surface is very large and every increment in $|v_s|$ falls across that portion of the space-charge region just below the surface (Fig. 4.6). The space-charge region does not penetrate farther into the bulk, so that G^- must tend to saturation. This can best be seen by approximating F by its expression in strong inversion layers (eq. (4.29)) and integrating G^- from the point $|v_0|$ where (4.29) still holds and up to the surface:

$$\delta G^- \equiv \int_{|v_0|}^{|v_s|} \frac{e^{-v}-1}{F(-|u_b|,v)} dv \approx -\sqrt{2}\left[e^{\frac{1}{2}(2|u_b|-|v_0|)} - e^{\frac{1}{2}(2|u_b|-|v_s|)}\right]. \quad (4.58)$$

Here δG^- represents the increment (negative) in the function G^-. The minority-carrier excess surface density, on the other hand, becomes large in this region. Because the majority-carrier density reaches saturation, any increase in Q_{sc} must be due entirely to minority carriers. Thus g^+ can be approximated by (see eq. (4.53))

$$g^+(u_b, v_s) \approx F(-|u_b|, |v_s|) - |G^-(u_b, \infty)|, \quad (4.59)$$

where $G^-(u_b, \infty)$ represents the saturation value of G^- for large $|v_s|$. As shown above (eq. (4.29)), F here is exponential in $\frac{1}{2}(|v_s|-2|u_b|)$ and so g^+ depends mainly on this parameter.

In the case of surface degeneracy, g^+ can be defined by (4.49) and (4.50), so that (4.59) still gives the correct value of g^+ provided the appropriate expression (eq. (4.34)) is used for F (see also Fig. 4.8).

For minority carriers in accumulation and depletion layers, convenient approximations are again not possible. In the case of an extrinsic semiconductor, however, such approximations are in any case of no interest because of the small minority-carrier density.

In Fig. 4.11, numerically computed values of G^+ and g^- (majority and minority carriers in accumulation layers) are shown as functions of $|v_s|$ for various values of $|u_b|$. As explained above, the curves of G^+ coincide in the range $|u_b| \gtrsim 2$. It is also seen that g^- for $|u_b| = 2$ can be safely neglected with respect to G^+; for that matter, so can it for any u_b (including zero) provided $|v_s|$ is not too small.

Computed values of g^+ and G^- (minority and majority carriers) are shown for inversion layers in Fig. 4.12 as a function of $|v_s|-2|u_b|$ for various values of $|u_b|$. The advantages in the choice of abscissa are obvious in the case of g^+: as can be seen from (4.28) and (4.59), this function in inversion layers depends mainly on $|v_s|-2|u_b|$. For $|v_s| < 2|u_b|-2$, the value of g^+ is so small compared to G^- that this range is of no practical interest. For inversion layers that are stronger than those shown in the figure, the value of g^+ is easily obtained from (4.28), (4.59), and the plot of G^-. In the case of G^-, it is mainly the saturation region that is shown, because for depletion layers a very convenient approximation is available (eq. (4.57)).

Under conditions of surface degeneracy, the dominant carrier density is obtainable from the function F_s (Figs. 4.7, 4.8) as discussed above.

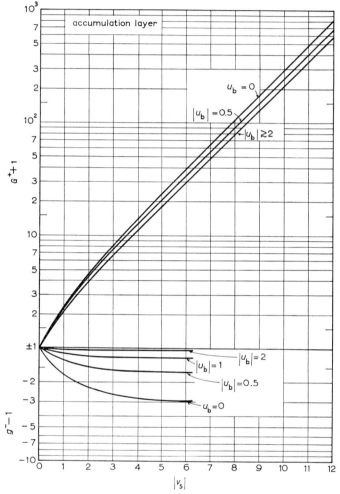

Fig. 4.11. An expanded plot of the accumulation-layer region of Fig. 4.10 showing the function $G^+ + 1$ (majority carriers) and $g^- - 1$ (minority carriers) *versus* the barrier height $|v_s|$ for various values of the bulk potential $|u_b|$.

The reader should be cautioned that the definitions and parameters used here in characterizing the space-charge region are different from those employed in the literature [8, 9, 16]. In particular, in the latter the excess surface-carrier densities ΔN and ΔP are expressed as functions of the surface potential $u_s = v_s + u_b$. This presentation is cumbersome, and

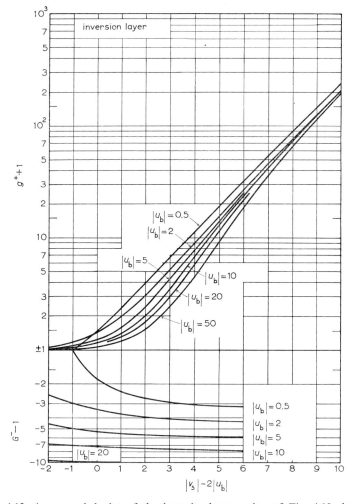

Fig. 4.12. An expanded plot of the inversion-layer region of Fig. 4.10 showing the functions $g^+ + 1$ (minority carriers) and $G^- - 1$ (majority carriers) for various values of the bulk potential $|u_b|$.

consequently graphical results have been published only for a restricted range of u_b and v_s. With the introduction of the variable $|v_s|$ for accumulation layers and $|v_s| - 2|u_b|$ for depletion and inversion layers, a considerable simplification is achieved as regards both range of graphical presentation and facility of interpolation in u_b. In fact, in this system the functions G^+ and g^+, which control the space charge, are insensitive to the value of u_b over a wide range, as can be seen from Figs. 4.11 and 4.12.

4. The space-charge region under non-equilibrium conditions

In the preceding sections the semiconductor was assumed to be at thermal equilibrium. Any disturbance that results in carrier excitation generally modifies the potential barrier and charge distribution at the surface. In order to obtain the characteristics of the space-charge region under such non-equilibrium conditions, we consider the case of uniform excitation of carriers throughout the semiconductor sample (produced, for example, by illumination in the absorption edge). If the carrier diffusion length (see Ch. 2, § 7.3) is large compared to the width of the space-charge layer and if no recombination takes place at the surface, then there can be no significant diffusion current of either carrier type towards the surface. Under these conditions the quasi-Fermi levels for electrons F_n and for holes F_p are position independent and continue in straight horizontal lines up to the surface (see eqs. (2.190)). Actually, to a good approximation, the quasi-Fermi levels are horizontal in the space-charge region even in the presence of weak surface recombination [25], as is normally encountered at etched germanium surfaces. Precise conditions for the range of validity of this approximation are given in Ch. 5, § 7.

Under the above assumptions, the steady-state density of electrons n^* and of holes p^* at any point in the semiconductor is given by (eqs. (2.187), (2.189))

$$n^* = n_i e^{(F_n - E_i)/kT}; \qquad p^* = n_i e^{(E_i - F_p)/kT}. \tag{4.60}$$

Let us define new parameters pertinent to the problem at hand — the "steady-state mean Fermi level" E_F^*, the steady-state potential u^*, and the "steady-state intrinsic carrier density" n_i^* —

$$E_F^* \equiv \tfrac{1}{2}(F_n + F_p); \tag{4.61}$$

$$u^* \equiv (E_F^* - E_i)/kT; \tag{4.62}$$

$$n_i^* \equiv \sqrt{n_b^* p_b^*} = n_i e^{(F_n - F_p)/2kT}. \tag{4.63}$$

The steady-state carrier densities (eq. (4.60)) can then be rewritten as

$$n^* = n_i^* e^{u^*} = n_b^* e^{v^*}; \qquad p^* = n_i^* e^{-u^*} = p_b^* e^{-v^*}, \tag{4.8*}$$

where n_b^*, p_b^* are the bulk values of n^*, p^*. Here v^* is the steady-state potential barrier and is defined as

$$v^* \equiv u^* - u_b^*, \tag{4.64}$$

where u_b^* is the "steady-state bulk potential". Keeping in mind that neutrality prevails in the bulk and neglecting bulk trapping, we have $N_D - N_A = n_b^* - p_b^*$. Poisson's equation (eq. (4.14)) can then be written in terms of the new variables as

$$\frac{d^2 v^*}{dz^2} = \left(\frac{1}{L^*}\right)^2 \left[\frac{\sinh(u_b^* + v^*)}{\cosh u_b^*} - \tanh u_b^*\right], \tag{4.14}*$$

where L^* is the steady-state effective Debye length and is given by

$$L^* \equiv \sqrt{\frac{\kappa \varepsilon_0 kT}{q^2(n_b^* + p_b^*)}}. \tag{4.12}*$$

The boundary conditions are as before: $v^* = v_s^*$ at $z = 0$, and $v^* \to 0$ in the bulk ($z \to \infty$). It is thus seen that the whole problem has been reduced to that at thermal equilibrium, and the considerations of this chapter are therefore valid for non-equilibrium steady-state conditions as well. Hence all the solutions (numerical and approximate) obtained for thermal equilibrium can be used for the present case provided that the parameters are replaced by their parallel, steady-state values (starred).

As a demonstration of the usefulness of the above treatment, we shall consider the case of surface photovoltage. This is defined as the change in barrier height due to illumination: $(v_s^* - v_s)(kT/q)$. The situation in the general case, when surface states are involved, is quite complicated and will be discussed in Ch. 7, § 2. Here we shall consider the simpler case in which surface states are absent or are inoperative due to poor communication with the space-charge region (as, for example, when only slow states are involved). It then follows that the space-charge density Q_{sc} is unchanged upon illumination. Hence, by the use of (4.12), (4.19), and (4.12)*,

$$F(u_b^*, v_s^*) = \sqrt{\frac{n_b + p_b}{n_b^* + p_b^*}} F(u_b, v_s). \tag{4.65}$$

Thus v_s^* can be calculated from the values of u_b and v_s in the dark and the values of the bulk carrier densities n_b^*, p_b^* under illumination. Since F_s is a monotonic function of v_s^*, it is seen that $|v_s^*|$ decreases as the steady-state bulk densities increase (increasing illumination). In the limit of very strong illumination, v_s^* approaches zero. In other words, the effect of illumination is always such as to decrease the absolute magnitude of the barrier height. As will be shown in Ch. 7, § 2, this conclusion generally holds even when surface states are present. Such a behaviour is just that to be expected from physical considerations. If the quasi-Fermi level is to remain flat up to the surface, the only way a constant space charge can be maintained is by a readjustment of the barrier resulting in a decrease in the magnitude of the barrier height.

As an example, we shall calculate the surface photovoltage for strong accumulation layers and weak light intensities such that the function $F^+(v_s)$ can be approximated by the exponential term (§ 2.3) both in the dark and under illumination. In this case it follows from (4.65) that

$$|v_s^*| - |v_s| = -\ln\left(1 + \frac{\delta n_b + \delta p_b}{n_b + p_b}\right), \qquad (4.66)$$

where δn_b, δp_b are the excess carrier densities in the bulk and vary monotonically with the light intensity.

5. The space-charge region in the presence of deep traps

So far we have considered a semiconductor in which the state of occupation of the various localized levels in the space-charge region is independent of the potential barrier. Such a situation applies quite well to the germanium and silicon samples normally employed in surface studies. Here the only significant localized levels arise from shallow donor and acceptor impurities, and these are always fully ionized except at very low temperatures. In most wide-gap semiconductors (or insulators), however, deep-lying levels are the rule rather than the exception, and it is such traps that usually dominate the characteristics of the space-charge region. Even in germanium and silicon, when doped with deep-lying impurities such as copper or gold, the space-charge region may become trap dominated.

A general solution of Poisson's equation under these conditions along the lines discussed above is not possible, and each distribution of traps has to be treated individually. In order to bring out the main features of the space-charge region in the presence of traps, we shall consider two relatively simple cases; one corresponds to a discrete set of traps[26], the other to a continuous trap distribution[27]. The energy bands will be taken to be non-degenerate, so that the free-carrier densities are given by Boltzmann statistics.

5.1. DISCRETE SET OF LOCALIZED LEVELS

The semiconductor will be assumed to contain only one type of deep trap, lying at an energy E_t and with a density N_t. These traps may be present in addition to the ionized donor (or acceptor) impurities, or they may represent the impurities when they are not fully ionized (at low temperatures). The charge density ρ at any point in the space-charge region is given by the excess (or deficit) free and trapped carrier densities with respect to their bulk values (where electrical neutrality prevails):

$$\rho = q[(p-p_b)-(n-n_b)-(n_t-n_{tb})]. \qquad (4.67)$$

Here n_b and p_b are the free electron and hole densities in the bulk, as before, while n_t and n_{tb} denote the density of occupied traps in the space-charge region and in the bulk, respectively. The latter are given by the appropriate Fermi factors (compare with eqs. (2.92), (2.93)), taking into account the displacement of the trap energy (with respect to the Fermi level) when the traps are in the potential barrier of the space-charge region,

$$n_{tb} = \frac{N_t}{1+\exp e_{tF}}; \quad n_t = \frac{N_t}{1+\exp(e_{tF}-v)}. \qquad (4.68)$$

Here e_{tF} is the energy separation in the bulk between the effective trap energy and the Fermi level, expressed in units of kT,

$$e_{tF} \equiv (E_t^f - E_F)/kT. \qquad (4.69)$$

The integration of Poisson's equation (eq. (4.9)), using the charge distribution given by (4.67)–(4.69), yields for the field (compare the derivation in § 2)

$$\frac{dv}{dz} = \mp \frac{F_t}{L}, \qquad (4.70)$$

where L is the effective Debye length defined by (4.12), and F_t is given by

$$F_t = \sqrt{2} \left\{ \frac{\cosh(u_b+v)}{\cosh u_b} - v \tanh u_b - 1 \right. \\ \left. + \frac{N_t}{n_b+p_b} \left[\ln\left(1 + \frac{e^v-1}{1+\exp e_{tF}}\right) - \frac{v}{1+\exp e_{tF}} \right] \right\}^{\frac{1}{2}}. \qquad (4.71)$$

The function F_t is analogous to F (eq. (4.16)), but in the present case it is governed by both the free and the trapped carriers. It is for this reason that the effective Debye length L is no longer a measure of the width of the space-charge layer. To get an idea of what the actual width should be in the presence of traps, we consider first the case of small barriers, $|v_s| \ll 1$. Under these conditions we can use a series expansion of the functions under the square root sign in (4.71) and, by neglecting higher order terms, we obtain

$$F_t \approx \left\{ v^2 \left[1 + \frac{N_t/(n_b + p_b)}{1 + \exp|e_{tF}|} \right] \right\}^{\frac{1}{2}}. \qquad (4.72)$$

The integration of (4.70) for this case is straightforward; the resulting potential barrier decreases exponentially with the distance from the surface, just as when no traps are present (eq. (4.13)). The effective width of the space-charge layer, however, is smaller, being given by

$$L_t = \sqrt{\frac{\kappa \varepsilon_0 kT}{q^2 [n_b + p_b + N_t/(1 + \exp|e_{tF}|)]}}. \qquad (4.73)$$

The situation is thus equivalent to that in a semiconductor with fully ionized levels but having an additional density n_{tb} of free electrons, if the traps are above the Fermi level, or $p_{tb} (\equiv N_t - n_{tb})$ of free holes, if the traps are below.

For large potential barriers we shall consider only a few limiting cases for which simple approximations can be applied.

Case a. The traps are well removed from the Fermi level both in the bulk and in the space-charge layer. Under these conditions Boltzmann statistics apply for the trapped carriers as well. If the traps are above the Fermi level, then (compare with the derivation of eq. (2.95))

$$\frac{n}{n_t} = \frac{n_b}{n_{tb}} = \frac{N_c}{N_t} e^{-(E_c - E_t^f)/kT} \equiv \theta_n \quad \text{for} \quad e_{tF} > 0, \qquad (4.74)$$

while if they lie below the Fermi level

$$\frac{p}{p_t} = \frac{p_b}{p_{tb}} = \frac{N_v}{N_t} e^{-(E_t^f - E_v)/kT} \equiv \theta_p \quad \text{for} \quad e_{tF} < 0, \qquad (4.75)$$

the constants θ_n and θ_p being independent of the potential barrier v. Poisson's equation now reduces to (4.10) corresponding to a semiconductor with fully ionized levels, provided n_b is replaced by $n_b(1 + 1/\theta_n)$ or p_b by $p_b(1 + 1/\theta_p)$, as the case may be. The shape of the potential barrier, the surface charge density, and the excess surface-carrier densities can then be obtained from the results of §§ 2, 3 with u_b replaced by

$$u_b + \tfrac{1}{2} \ln (1 + 1/\theta_n) \quad \text{for} \quad e_{tF} > 0, \qquad (4.76)$$

or by

$$u_b - \tfrac{1}{2}\ln(1+1/\theta_p) \quad \text{for} \quad e_{tF} < 0. \tag{4.77}$$

It should be noted that the excess carrier densities (as derived for example, from the plots of the G, g functions in Figs. 4.11, 4.12) will now include both the free and the trapped carrier densities. These can be separated by the use of (4.74) or (4.75), which determines their ratio.

Case b. The traps in the bulk are located at the Fermi level ($e_{tF} = 0$). The discussion will be limited to large barriers in an extrinsic semiconductor (taken, for the sake of concreteness, as n type). For depletion layers such that $-2u_b \leq v_s \ll 1$, the logarithmic term in (4.71) may be expanded to yield $\exp v - \ln 2$ over most of the extent of the space-charge region, and F_t can be approximated by

$$F_t \approx \sqrt{2}\{(|v|-1)[1+\tfrac{1}{2}N_t/(n_b+p_b)]\}^{\frac{1}{2}}. \tag{4.78}$$

The barrier is nearly parabolic as for the case of no traps (eq. (4.27)), but with a reduced effective Debye length given by

$$L_t = \sqrt{\frac{\kappa\varepsilon_0 kT}{q^2(n_b+p_b+\tfrac{1}{2}N_t)}}. \tag{4.79}$$

For accumulation layers ($v_s \gtrsim 4$), F_t assumes the approximate form

$$F_t \approx \sqrt{2}\left[e^v - v - 1 + \frac{\tfrac{1}{2}N_t}{n_b+p_b}(v-\ln 4)\right]^{\frac{1}{2}}. \tag{4.80}$$

As v_s increases, the last term under the square root (representing the trapped carriers) becomes increasingly less significant compared to the exponential term (representing the free carriers), and F_t approaches F. Thus for strong accumulation layers and moderate trap densities, the characteristics of the space-charge region are insensitive to the presence of deep traps. Physically, this follows from the fact that the width of the barrier is very small, and the change in occupation of the relatively few traps lying in the space-charge region is negligible compared to the large surface density of free carriers.

Case c. The traps cross the Fermi level in the space-charge region but are well removed from it in the bulk. An n-type semiconductor will again be considered. Obviously such a crossing will occur in a depletion layer if the traps in the bulk are below the Fermi level ($e_{tF} < 0$), and in an accumulation layer if they are above ($e_{tF} > 0$). For $|v_s| < |e_{tF}|$, *case a* is applicable. As $|v_s|$ increases beyond $|e_{tF}|$, more and more traps cross the Fermi level and conditions become similar to those in *case b*. For strong depletion layers, F_t can be approximated by

$$F_t \approx \sqrt{2}\left[|v|-1+\frac{N_t}{n_b+p_b}(|v|-|e_{tF}|)\right]^{\frac{1}{2}}, \qquad (4.81)$$

while the effective Debye length approaches

$$L_t \approx \sqrt{\frac{\kappa\varepsilon_0 kT}{q^2(n_b+p_b+N_t)}}. \qquad (4.82)$$

Here again the effect of the traps is to decrease the width of the nearly parabolic potential barrier. The space-charge region now consists not only of ionized shallow donors (if present) but also of ionized deep traps. The reduction in width is particularly marked in insulators, where the density of free carriers is usually much smaller than that of deep traps. The effect of traps on accumulation layers is far less significant. As v_s is increased, the free excess density of electrons increases rapidly and the traps gradually lose their influence on the characteristics of the space-charge region.

The effective charge distance L_{ct} for the various cases discussed above is obtained in an analogous manner to that employed in the derivation of (4.22) and is given by

$$L_{ct} = L\frac{|v_s|}{F_{ts}}, \qquad (4.83)$$

where L is the effective Debye length of the same crystal but devoid of deep traps (eq. (4.12)), and F_{ts} is the value of F_t at the surface ($v = v_s$).

5.2. CONTINUOUS DISTRIBUTION OF TRAPS

A particularly simple solution can be obtained in the case of an insulator in which the distribution in energy of the traps is continuous and uniform[27]. Let $\bar{N}_t dE$, where $\bar{N}_t$ is a constant, represent the volume density of traps in the energy interval dE. The traps are assumed to extend over the entire energy gap. If we neglect the free carrier density, the charge density ρ at any point in the space-charge region at which the potential is V can be written in the form

$$\rho = -q\overline{N}_t \left\{ \int_{E_{v0}-qV}^{E_{c0}-qV} \frac{dE}{1+\exp[(E-E_F)/kT]} - \int_{E_{v0}}^{E_{c0}} \frac{dE}{1+\exp[(E-E_F)/kT]} \right\}$$

$$= -q\overline{N}_t \left\{ \int_{E_{v0}-qV}^{E_{v0}} \frac{dE}{1+\exp[(E-E_F)/kT]} - \int_{E_{c0}-qV}^{E_{c0}} \frac{dE}{1+\exp[(E-E_F)/kT]} \right\},$$

(4.84)

where E_{v0} and E_{c0} denote the valence- and conduction-band edges in the bulk. We shall assume that the Fermi level is well removed from the band edges throughout the space-charge region. Under these conditions the Fermi function under the third integral sign is very close to unity while that under the fourth is negligibly small. Thus the charge density reduces to

$$\rho = -q^2 \overline{N}_t V, \quad (4.85)$$

and the solution of Poisson's equation assumes the simple form

$$v = v_s e^{-z/L_t}. \quad (4.86)$$

The potential barrier decreases exponentially with a characteristic length given by

$$L_t = \sqrt{\frac{\kappa \varepsilon_0 kT}{q^2(kT\overline{N}_t)}}. \quad (4.87)$$

The characteristic length has the same form as the effective Debye length appropriate to an insulator having a discrete set of traps (eq. (4.82), with $n_b = p_b = 0$), but now N_t is replaced by $\overline{N}_t kT$, the volume density of traps in an energy interval kT.

References

[1] W. Schottky, Z. Physik **113** (1939) 367; **118** (1942) 539; W. Schottky and E. Spenke, Wiss. Veröffentl. Siemens-Werken **18** (1939) 3.
[2] N. F. Mott, Proc. roy. Soc. London **A11** (1939) 27.
[3] B. Davidov, J. Phys. (U.S.S.R.) **1** (1939) 167.
[4] W. Shockley, *Electrons and Holes in Semiconductors* (Van Nostrand, New York, 1950).
[5] E. Spenke, *Electronic Semiconductors* (McGraw-Hill, New York, 1958).
[6] H. K. Henisch, *Rectifying Semiconductor Contacts* (Clarendon Press, Oxford, 1957). (References)
[7] J. R. Schrieffer, Phys. Rev. **97** (1955) 641.
[8] R. H. Kingston and S. F. Neustadter, J. appl. Phys. **26** (1955) 718.
[9] C. G. B. Garrett and W. H. Brattain, Phys. Rev. **99** (1955) 376.
[10] G. C. Dousmanis and R. C. Duncan, Jr., J. appl. Phys. **29** (1958) 1627.
[11] H. Flietner, Ann. Physik Leipzig **3** (1959) 396.

[12] E. Groschwitz and R. Ebhardt, Z. angew. Physik **11** (1959) 9; E. Groschwitz, E. Hofmeister, and R. Ebhardt, *ibid.* **12** (1960) 544.
[13] D. R. Frankl, J. appl. Phys. **31** (1960) 1752.
[14] N. B. Grover, Y. Goldstein, and A. Many, J. appl. Phys. **32** (1961) 2538.
[15] Y. Goldstein, N. B. Grover, A. Many, and R. F. Greene, J. appl. Phys. **32** (1961) 2540.
[16] C. E. Young, J. appl. Phys. **32** (1961) 329.
[17] R. Seiwatz and M. Green, J. appl. Phys. **29** (1958) 1034.
[18] J. Bardeen, Phys. Rev. **49** (1936) 653.
[19] E. Callen, Am. J. Phys. **25** (1957) 138.
[20] J. Bardeen, Phys. Rev. **71** (1947) 717.
[21] W. Shockley, Bell System tech. J. **28** (1949) 435.
[22] A. Many, N. B. Grover, Y. Goldstein, and E. Harnik, J. Phys. Chem. Solids **14** (1960) JD5 (facing p. 298).
[23] This case has been considered by R. F. Greene, D. R. Frankl, and J. Zemel, Phys. Rev. **118** (1960) 967.
[24] J. McDougall and E. C. Stoner, Trans. roy. Soc. London **A237** (1938) 67.
[25] W. H. Brattain and J. Bardeen, Bell System tech. J. **32** (1953) 1.
[26] J. R. Macdonald, J. chem. Phys. **29** (1958) 1346.
[27] A. Rose, Helv. Phys. Acta **29** (1956) 199.

CHAPTER 5

SURFACE STATES

1. Introduction

In chapter 3 we considered the structural and chemical properties of the surface by virtue of its being the outer boundary of the crystal lattice. In the present chapter we shall be concerned with the effect of the lattice discontinuity on the electrical processes at the surface. The most important of these arise from the presence of surface states and their interaction with the underlying space-charge region.

The concept of surface states was introduced by Tamm [1] who showed from fundamental quantum-mechanical considerations that, in contrast to the situation in the bulk, *localized* states can appear at the surface. In other words, whereas in the bulk a valence electron belongs to the entire crystal (the Bloch approach), at the surface it may be confined to an atomic layer. Tamm's theoretical work, which treated a one-dimensional Kronig–Penney type model, was later extended to include more general cases, both one- and three-dimensional [2]. It became apparent that two types of surface states originating from the termination of the crystal lattice were possible, the Tamm states and the Shockley states [3]. Formal conditions were derived for the existence or non-existence of either type of states on ideal surfaces (that is, surfaces whose lattice structure is identical to that of parallel crystallographic planes in the bulk). This work will be discussed in the first two sections of the present chapter, including two examples of a detailed calculation. It should be pointed out that so far, except for the indication that surface states can exist in principle, no attempt has been made to calculate accurately their structures in an actual semiconductor even in the case of a *clean* surface. The main difficulty underlying such a calculation arises from the displacement of the surface atoms from their ideal lattice sites. As far as a real surface is concerned, covered as it is with an adsorbed layer of foreign matter, the problem is of course even more formidable. Several studies along these lines were reported by Koutecký [4] and Grimley [5].

In parallel with this theoretical work, a considerable effort was devoted to the study of the surface states actually observed experimentally. Most of the studies were concerned with those processes able to shed light on the physical characteristics of such states. The underlying phenomenological theories will be developed in the second part of the chapter. In that part we assume the existence of surface states characterized by a few operational parameters, without being concerned with their type or origin. Quantitative relationships are then developed between these parameters and quantities that can be measured experimentally. In §§4, 5 we calculate the occupational statistics of surface states at thermal equilibrium as a function of the surface potential. Next we formulate the kinetics of interactions for non-equilibrium processes between the surface states and the free carriers in a single energy band. The final sections consider the theory of surface recombination, which involves the interaction of the states with both bands. In all the calculations a homogeneous, non-degenerate semiconductor will be considered in which the surface states are associated with discrete energy levels. This last restriction leads to a considerable simplification in the theory. Its justification lies in the fact that the dominant surface states found to date (on germanium and silicon) are characterized by an essentially discrete distribution.

2. Tamm and Shockley states

2.1. ONE-DIMENSIONAL PERIODIC POTENTIAL

In order to calculate the wave functions and energy levels of electrons in a solid, it is customary to consider an ideal crystal of macroscopic dimensions with cyclic boundary conditions (see Ch. 2, §§ 1–3). The solutions of Schrödinger's equation for this case are in the form of Bloch functions: $\psi_k(r) = u_k(r) \exp(i\mathbf{k} \cdot \mathbf{r})$, where $u_k(r)$ has the periodicity of the lattice. Since the function ψ must describe an electron having equal probability of being found in any unit cell of the crystal, all the components of $\mathbf{k}$ have to be real. This restriction leads to the existence of allowed energy bands corresponding to real values of $\mathbf{k}$ and to forbidden energy zones corresponding to complex values of $\mathbf{k}$. Whereas such a treatment provides a suitable description of the bulk properties, it inherently eliminates the effect of the surface.

Tamm was the first to realize that in the presence of the surface, $\psi_k(r)$ functions having complex values of $\mathbf{k}$ may describe quantum states that are physically possible. A function of this type decreases exponentially as one moves away from the surface into the bulk, and thus corresponds to a localized state. Tamm treated a modified Kronig–Penney model (Ch. 2, § 2.1) appropriate to a semi-infinite one-dimensional crystal. The actual potential in such a structure is schematically illustrated in Fig. 5.1a. The abrupt termination of the lattice at the surface results in a potential wall corresponding to the binding energy of an electron to the crystal. The potential in Fig. 5.1a is approximated by square walls inside the crystal as for the case of an infinite lattice (see Fig. 2.1). Now, however, there is a square wall at the

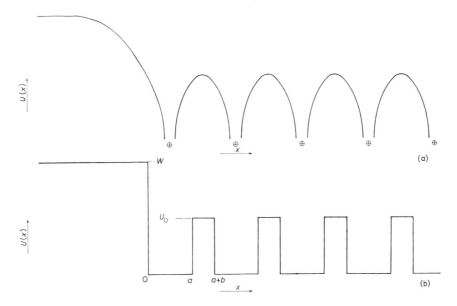

Fig. 5.1. Potential energy of an electron in a semi-infinite one-dimensional lattice.
(a) Schematic representation of actual conditions.
(b) Square-well approximation employed by Tamm.

surface ($x = 0$), as shown in Fig. 5.1b. The electron energy E is assumed to be less than the potential energy W in free space and also less than the peaks U_0 of the periodic potential. As in Ch. 2, § 2.1 we let U_0 tend to infinity and b to zero, and obtain the condition for the existence of Bloch solutions for positive x given by (2.28). The solution of Schrödinger's equation for free space ($x \leq 0$), in which the potential energy is a constant ($U = W$), is easily seen to be given by an exponentially decaying function of the form

$$\psi_n(x) = Ce^{\eta x} \quad \text{for} \quad x \leq 0, \tag{5.1}$$

where

$$\eta = \frac{\sqrt{2m(W-E)}}{\hbar}. \tag{5.2}$$

The boundary conditions are obtained from the requirements that every wave function from within the crystal ($x > 0$) join a corresponding ψ_n function at $x = 0$ and that the slope at this point be continuous. These boundary conditions can always be met for real values of k since for every allowed energy $E(k)$ there exist two Bloch functions corresponding to $\pm k$. By taking a linear combination of these functions, the two coefficients can be determined so as to meet both boundary conditions at $x = 0$. Thus the quasi-continuous spectrum of allowed energy levels in each band is not affected by the termination of the crystal lattice by the surface. At the same time, one must also consider solutions corresponding to complex values of k, provided they satisfy (2.28). Because the left-hand side of this equation is real, only those complex solutions are possible which have the form $k' = i\xi$ or $k' = \pi + i\xi$. Since, moreover, the wave function must remain finite as x tends to infinity, ξ has to be positive. In this case, the condition expressed in (2.28) is transformed into

$$(P/\alpha a) \sin \alpha a + \cos \alpha a = \pm \cosh \xi a, \tag{5.3}$$

the plus sign corresponding to $k' = i\xi$, the minus sign to $k' = \pi + i\xi$. We have from the boundary conditions at $x = 0$, using (2.16), (2.20), and (5.1), that

$$A_1 + A_2 = C; \quad i\alpha(A_1 - A_2) = C\eta. \tag{5.4}$$

Inserting in these last equations the value of A_1/A_2 from (2.27), we obtain

$$\eta \frac{\sin \alpha a}{\alpha} + \cos \alpha a = \pm e^{-\xi a}, \tag{5.5}$$

where the plus and minus signs again refer to $k' = i\xi$ and $k' = \pi + i\xi$, respectively. The elimination of ξ from (5.3) and (5.5) gives for αa the relation

$$\alpha a \cot \alpha a = \frac{a^2 \eta^2}{2P} - \sqrt{a^2 \eta^2 - a^2 \alpha^2}, \tag{5.6}$$

where

$$\gamma^2 = \alpha^2 + \eta^2 = \frac{2mW}{\hbar^2}. \tag{5.7}$$

Equation (5.6) has a single solution for the energy (through the relation $E = (\hbar^2/2m)\alpha^2$) in every interval $n\pi$, $(n+1)\pi$. But in each such interval there is one allowed energy band corresponding to k real (compare with Fig. 2.2). It therefore follows that there is an allowed energy level between every two allowed energy bands. The termination of the lattice thus introduces localized surface states whose energies lie in the forbidden zones. Similarly, for a potential representing a chain of atoms finite in both directions, two sets of surface states are obtained — one at each end of the chain. These are the so-called Tamm surface states.

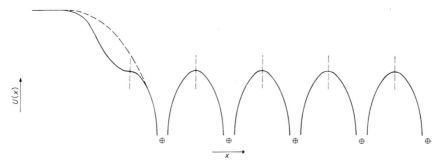

Fig. 5.2. Potential energy of an electron in a semi-infinite one-dimensional lattice, corresponding to a symmetrical termination at the surface (the Shockley model). Also shown (dashed curve) is the potential energy corresponding to asymmetrical termination (compare Fig. 5.1a).

Shockley [3] has pointed out that the Tamm states are a direct consequence of the *asymmetrical* termination of the periodic potential at the surface. Shockley, too, considered a one-dimensional lattice but he employed a more general shape for the crystal potential, as illustrated in Fig. 5.2. The underlying assumption in this model is that the potential maintains perfect periodicity *right up to the surface*, at which point it is terminated *symmetrically*. This is to be contrasted with the *asymmetrical* termination of the potential (see Fig. 5.1a, redrawn as a dashed curve in Fig. 5.2 for purposes of comparison) which served as the starting point for the modified Kronig–Penney model employed by Tamm. With his model Shockley finds that for large interatomic distances no surface states are possible. The analysis is based on the form that the energy bands assume as a function of the lattice constant a. This is illustrated in Fig. 5.3 for two neighbouring bands. For large values of a, the bands degenerate into discrete energy levels corresponding to isolated free atoms, as discussed in Ch. 2, § 2.1. As the lattice constant is decreased and the atoms draw closer together, the allowed energy bands broaden and the edges of the neighbouring bands approach. When a particular value of the lattice constant is reached, the boundary curves of the bands (the dashed lines in Fig. 5.3) intersect. As a is decreased further, the boundary curves interchange their roles and two energy levels are split off, one from each band. These correspond to two localized wave functions, one on each of the two surfaces, the so-called

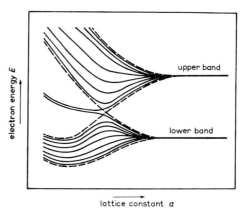

Fig. 5.3. The form of the energy bands for a one-dimensional crystal (eight atoms) as a function of the lattice constant. (After Shockley, reference 3.)

Shockley states. Thus, the condition for the existence of surface states in this model is that the actual lattice constant be smaller than a certain minimum value.

Lipmann [6] showed that for a one-dimensional crystal, this type of intersection of boundary curves can take place only for an electron energy E that is *larger* than the potential energy at every point of the unit cell. (This is not a necessary condition for a three-dimensional crystal.) Such a situation cannot, of course, occur in Tamm's model, where U_0 tends to infinity. Thus the existence of Tamm states must be due entirely to the *asymmetrical* termination of the periodic potential. Shockley states, on the other hand, are obtained for a symmetrical termination, subject to the conditions discussed above.

2.2. EXTENSION TO THREE-DIMENSIONAL CRYSTALS

The one-dimensional models considered in the previous subsection have been extended by various workers to cover three-dimensional crystals. It is beyond the scope of the book to discuss this work in detail and the reader is referred to the comprehensive review given by Koutecký [2]. In the present subsection we shall point out the main features resulting from the calculations, without going into the mathematical detail, and in the following section we present examples of an actual calculation using one of the accepted procedures.

The general trend in the literature is to distinguish between Tamm-like and Shockley-like states in a three dimensional crystal on the basis of much the same criteria that characterize the two types of states in the one dimensional models. Thus, Shockley states are identified with those states obtained when the periodic potential is symmetrically terminated at the surface and the atoms are close together (strong interaction between neighbours). Tamm states, on the other hand, are associated with an asymmetric termination of the potential and a large separation between atoms (weak interaction). Accordingly, a convenient method for the calculation of Tamm states is the Heitler–London or tight-binding approximation mentioned in Ch. 2, §1.2. One starts by considering an electron moving in the field of a single atom and then expresses the presence of the remaining atoms of the crystal in the form of a perturbation. The crystal potential is taken as the sum of the potentials of the individual atoms. To a zeroth approximation, a wave function in the crystal is written as a linear combination with undetermined coefficients of atomic (unperturbed) functions, one for each atom in the crystal. One thus assumes that each energy band originates from *one* atomic level only, independently of the other bands.

Following perturbation theory, we substitute this approximation for ψ into Schrödinger's equation and obtain a system of equations for the calculation of the coefficients. The condition for non-trivial solutions determines the energies of the allowed levels of the crystal. The expression for these energies (eq. (5.20)) contains three terms: the energy of the unperturbed atomic level, the Coulomb integral α (which expresses the perturbation introduced by the potential of all other atoms), and a term dependent on the exchange integral γ between wave functions of neighbouring atoms (see eq. (5.13)). It is γ that gives rise to a splitting of the atomic level into an energy band. So much for bulk states; we now turn our attention to surface states.

The termination of the periodic potential by the surface is introduced via the assumption that the Coulomb integral α is different for surface atoms. This leads to boundary conditions which, in turn, determine whether there will or will not be surface states. The greater the deformation of the potential caused by the surface, the more favourable become the conditions for the existence of Tamm states. These levels constitute an energy band at the surface that is either in the forbidden zone above the allowed band from which it originates or else partially overlaps that band. The number of surface states is the same as the number of surface atoms if we neglect the effect of spin (which, of course, introduces a two-fold degeneracy in each state).

The situation is quite different in the case of Shockley states, which are more likely to occur when the atoms in the crystal are close together and interact strongly. The free-electron model, having proved very successful in determining the band structure under these conditions, would be expected to describe adequately the surface states as well. And indeed, several workers [7-11] have employed this approach, assuming a symmetrical termination of the crystal potential at the surface and using as a starting point the solutions of Schrödinger's equation in the free-electron approximation. For a surface perpendicular to the x direction, each wave function within the crystal is taken as a linear combination of all such solutions having a given set of values of the energy E and the components k_y, k_z of the wave vector $\mathbf{k}$. The energy of course is always real, and k_y, k_z can likewise assume only real values because of the fact that the crystal is infinite in the y and z directions. The x component k_x, however, may assume both real and complex values. In the former case the solutions of Schrödinger's equation are clearly unattenuated Bloch waves, while in the latter case they represent exponential functions either increasing or decreasing with distance from the surface. Only the decreasing functions are admissible as physical solutions. If for a given set E, k_y, k_z, all the solutions have complex values of k_x, then the resulting crystal function is damped and therefore represents a localized surface state. The energy is determined, as before, by the condition that each crystal function join smoothly at the surface onto an exponentially decaying function outside the crystal.

All band-structure calculations made have been concerned with real values of $\mathbf{k}$, and by now the corresponding Bloch functions are known for a fair number of crystals. For the purpose of studying surface states, on the other hand, one is primarily interested in solutions having complex values of k_x. A detailed study of such solutions has been made by Heine[11], who was able to show quite generally which solutions can be of interest in the construction of surface-state wave functions. More important still, he indicated how such solutions can be derived from the known Bloch functions. This work substantiates previous approximations[10] made mostly on an intuitive basis.

Shockley states can also be treated by the tight-binding approximation used for Tamm states, appropriately modified. Because of the strong interaction between neighbouring atoms, the allowed energy bands broaden and approach one another or even overlap. It is therefore no longer sufficient to treat each energy band as originating from a single atomic level as in the case of Tamm states. Now the function ψ should be approximated by a linear combination involving more than one wave function of each atom. Of most interest are the s and p functions, since these give rise to the valence and conduction bands in germanium and silicon (as well as in many other semiconductors). The subsequent discussions will accordingly be confined to such states and to the two neighbouring bands that originate from them. One proceeds as in the case of Tamm states, but the calculations are much more complicated. The main difference is the existence now of non-vanishing exchange integrals between s and p functions of two neighbouring atoms (γ_{sp} in eq. (5.28)). These integrals are assumed to be the same for both bulk and surface atoms. The deformation of the potential at the surface results in a change in the values of the Coulomb integrals (α_s and α_p). Moreover, due to this deformation, the Coulomb integrals between wave functions of the same atom (α'_{sp} in eq. (5.28)), taken to be zero in the bulk, are assumed not to vanish for surface atoms. From these boundary conditions one can again determine whether or not there exist surface states[12]. When the separation between the two energy bands is large, the solution reduces to that for the non-degenerate case considered above, as it should. If the deformation of the potential at the surface is sufficiently large, Tamm-like states are obtained between the two bands (and also above the upper band). When the bands approach one another, the energy of these states decreases and they gradually sink into the lower band and disappear. For an interaction between the energy bands such that Tamm states disappear, Shockley-like states are obtained provided that the deformation of the potential at the surface is small. These levels also make up a surface band which can overlap partially or completely the bulk bands.

The Shockley states are believed to be associated with the unsaturated or dangling bonds of the surface atoms [2,13,14]. Such a conclusion, however, is based on general physical considerations rather than on rigorous analysis. This problem has been examined on a more quantitative basis by Koutecký [15,16], using a one-dimensional model.

3. The tight-binding approximation

3.1. TAMM STATES

In order to obtain the characteristic features of Tamm states, it is sufficient to restrict ourselves to a crystal having a simple cubic lattice and to an energy band originating from the s-states of the individual atoms. Following Goodwin [17] and Artman [12], we shall consider a crystal that has N atomic layers in the x direction and is infinite in the other two directions. The position of the atoms will be represented by the vector r_{lmn}, where $l = 1, 2, \ldots, N$ and the integers m, n extend from $-\infty$ to $+\infty$. The potential energy of an electron moving in the potential of an *isolated* atom is of the form $U_{lmn} = U_1(|r - r_{lmn}|)$, where r is the position vector of the electron. Schrödinger's equation for such a bound electron is

$$\frac{\hbar^2}{2m} \Delta\phi + [E - U_{lmn}]\phi = 0. \tag{5.8}$$

The solution appropriate to an s state (spherical symmetry) of energy E_s will be indicated by $\phi_s(|r - r_{lmn}|)$. We shall approximate the electron potential energy U due to all the atoms in the crystal by the sum

$$U = \sum_{lmn} U_{lmn}. \tag{5.9}$$

The wave function ψ of an electron in the crystal must satisfy Schrödinger's equation

$$\frac{\hbar^2}{2m} \Delta\psi + (E - U)\psi = 0. \tag{5.10}$$

As a zeroth approximation, ψ is taken to be a linear combination of the wave functions ϕ_s of the individual atoms:

$$\psi = \sum_{lmn} a_{lmn} \phi_s(|r - r_{lmn}|). \tag{5.11}$$

In order to determine the coefficients a_{lmn} and the energy E, we substitute this expression for ψ into (5.10) and obtain, using (5.8),

$$\sum_{lmn} a_{lmn}[(E-E_s)-(U-U_{lmn})]\phi_s(|\mathbf{r}-\mathbf{r}_{lmn}|) = 0. \tag{5.12}$$

Here $E-E_s$ is the change in the atomic level (E_s) brought about by the crystal state, and $U-U_{lmn}$ is the potential energy of an electron due to all the atoms in the crystal except the lmn-th one. In order to calculate the allowed energy levels E, we multiply this equation by the complex conjugate function $\phi_s^*(|\mathbf{r}-\mathbf{r}_{l'm'n'}|)$ and integrate over all space. (Actually, since the s-states are non-degenerate, ϕ_s can always be taken as real.) Only three types of terms will be of interest, and these we denote as follows:

$$\begin{aligned} -\int \phi_s^*(|\mathbf{r}-\mathbf{r}_{l'm'n'}|)[U-U_{lmn}]\phi_s(|\mathbf{r}-\mathbf{r}_{lmn}|) d\tau & \\ = \alpha_s \text{ for } l, m, n = l', m', n' \text{ and } l \neq 1, N & \\ = \alpha_s' \text{ for } l, m, n = l', m', n' \text{ and } l = 1, N & \\ = \gamma_s \text{ for } l, m, n = l', m', n' \pm 1 \text{ or } l', m' \pm 1, n' \text{ or } & \\ l' \pm 1, m', n' \text{ (nearest neighbours)} & \end{aligned} \right\} . \tag{5.13}$$

The minus sign before the integral was introduced in order to make α_s, α_s', γ_s positive, the potential energy $U-U_{lmn}$ being negative. The Coulomb integrals α_s and α_s' represent the change in energy of an electron bound to a bulk and a surface atom, respectively, as a result of the force field arising from all *other* atoms. Clearly, $\alpha_s \geqq \alpha_s'$. The exchange integral γ_s represents an additional contribution to the energy arising from the overlap of the wave functions of nearest-neighbour atoms and is assumed to be the same for both bulk and surface atoms. The tight-binding approximation inherently implies little overlap between different wave functions, so that γ_s is small. We cannot neglect it compared to α_s and α_s', however, because the latter are themselves small due to the absence from the potential term of the contribution of the atom in question. On the other hand, the exchange integral between wave functions of atoms that are not nearest neighbours may justifiably be neglected. Finally, we assume that all wave functions are orthonormal:

$$\int \phi_s^*(|\mathbf{r}-\mathbf{r}_{l'm'n'}|)\phi_s(|\mathbf{r}-\mathbf{r}_{lmn}|) d\tau$$
$$\begin{aligned} = 1 \text{ for } l, m, n, = l', m', n' & \\ = 0 \text{ otherwise} & \end{aligned} \right\}. \tag{5.14}$$

Note that the influence of the surface is expressed only in the fact that the Coulomb integral over the surface atoms α'_s differs from that over the bulk atoms α_s. The bulk atoms, including those lying next to the surface atoms, are assumed to be characterized by the same value of α_s — that is, the distortion of the potential at the surface penetrates only one atomic layer. The short-range distortion in potential also justifies our taking the exchange integral γ_s to be the same for all atoms in a *row* (m, n constant), including the two surface atoms at the ends of the row. The assumption that two nearest-neighbour *surface* atoms are again associated with the same γ_s is harder to justify even for the case of an ideal surface being considered. Its introduction, however, greatly facilitates the calculations since it enables the reduction of the treatment to that of a one-dimensional problem.

Using (5.12)–(5.14), we obtain the following equations in the coefficients a_{lmn}:

$$(E-E_s+\alpha_s)a_{lmn}+\gamma_s(a_{lm(n+1)}+a_{lm(n-1)}$$
$$+a_{l(m+1)n}+a_{l(m-1)n}+a_{(l+1)mn}+a_{(l-1)mn}) = 0 \quad \text{for} \quad l \neq 1, N, \quad (5.15)$$

with the boundary conditions

$$(E-E_s+\alpha'_s)a_{1mn}+\gamma_s(a_{1m(n+1)}+a_{1m(n-1)}$$
$$+a_{1(m+1)n}+a_{1(m-1)n}+a_{2mn}) = 0; \quad (5.16\text{a})$$

$$(E-E_s+\alpha'_s)a_{Nmn}+\gamma_s(a_{Nm(n+1)}+a_{Nm(n-1)}$$
$$+a_{N(m+1)n}+a_{N(m-1)n}+a_{(N-1)mn}) = 0. \quad (5.16\text{b})$$

As all the atoms are equivalent with respect to a transformation in the y and z directions and so must contribute equally to the wave function ψ, it is easy to see that the a_{lmn} are of the form $a_l \exp[ia(k_y m + k_z n)]$, where k_y, k_z are real (a is the lattice constant). Using this expression for a_{lmn}, (5.15) and (5.16) reduce to

$$[E-E_s+\alpha_s+2\gamma_s(\cos k_y a + \cos k_z a)]a_l$$
$$+\gamma_s(a_{l+1}+a_{l-1}) = 0 \quad \text{for} \quad l \neq 1, N; \quad (5.17)$$

$$[E-E_s+\alpha'_s+2\gamma_s(\cos k_y a + \cos k_z a)]a_1 + \gamma_s a_2 = 0; \quad (5.18\text{a})$$

$$[E-E_s+\alpha'_s+2\gamma_s(\cos k_y a + \cos k_z a)]a_N + \gamma_s a_{N-1} = 0. \quad (5.18\text{b})$$

Equations (5.17), (5.18) constitute a system of N homogeneous equations in the coefficients $a_1, a_2, \ldots, a_N$. The allowed energies E can be obtained from the requirement that there exist non-trivial solutions — that is, from the vanishing of the determinant. Alternatively, the $N-2$ equations (5.17) can be solved by the substitution

$$a_l = A e^{iak_x l} + B e^{-iak_x l}, \tag{5.19}$$

which immediately yields for the allowed energies the expression

$$E(k_x, k_y, k_z) = E_s - \alpha_s - 2\gamma_s(\cos k_x a + \cos k_y a + \cos k_z a). \tag{5.20}$$

The constants A and B must now be determined in such a way that the solution (eq. (5.19)) also satisfy the boundary conditions (eqs. (5.18)). From this requirement we obtain two homogeneous equations in A and B. The vanishing of the corresponding determinant and the use of (5.20) gives for k_x the relation

$$\sin\left[(N+1)(\pi - k_x a)\right] - 2(\varepsilon_s/\gamma_s) \sin\left[N(\pi - k_x a)\right]$$
$$+ (\varepsilon_s/\gamma_s)^2 \sin\left[(N-1)(\pi - k_x a)\right] = 0, \tag{5.21}$$

where $\varepsilon_s \equiv \alpha_s - \alpha_s'$ and is positive. The solution $k_x a = \pi$ has no physical meaning, as it results in the vanishing of all the a_{lmn}. Besides this trivial one, (5.21) has N solutions for $k_x a$. It therefore follows from (5.20) that there are N allowed energy levels for each (k_y, k_z) pair. Goodwin [17] solved (5.21) graphically and showed that for N large, there exist N *real* solutions for k_x provided that $\varepsilon_s/\gamma_s < 1$. For a larger deformation at the surface such that $\varepsilon_s/\gamma_s > 1$, on the other hand, two of these real solutions, those corresponding to the largest energies, disappear from the bulk energy band. The value of ε_s/γ_s cannot affect the number of solutions, however, so that these two solutions have to be complex. The energy must be real, and so for complex k_x one can write $k_x a$ in the form $k_x a = \pi + i\xi$, where $\xi > 0$. By substituting this expression into (5.21), we obtain two solutions which, for large N, are approximately equal and given by

$$e^\xi = \varepsilon_s/\gamma_s. \tag{5.22}$$

Inserting this value into (5.20), we obtain for the energy levels of the surface states

$$E(\xi, k_y, k_z) = E_s - \alpha_s + 2\gamma_s(\cosh \xi - \cos k_y a - \cos k_z a). \tag{5.23}$$

It can easily be seen that all the solutions with k_x real correspond to Bloch functions. For k_x complex, ψ takes the form [17]

$$\psi_{\xi k_y k_z} \propto \sum_{mn} e^{ia(k_y m + k_z n)} \{ [\phi_s(|\mathbf{r}-\mathbf{r}_{1mn}|)$$
$$\pm \phi_s(|\mathbf{r}-\mathbf{r}_{Nmn}|)] + e^{-\xi}[\phi_s(|\mathbf{r}-\mathbf{r}_{2mn}|) \pm \phi_s(|\mathbf{r}-\mathbf{r}_{(N-1)mn}|)] \quad (5.24)$$
$$+ e^{-2\xi}[\phi_s(|\mathbf{r}-\mathbf{r}_{3mn}|) \pm \phi_s(|\mathbf{r}-\mathbf{r}_{(N-2)mn}|)] + \ldots \},$$

where either the upper sign or the lower sign is to be taken throughout. For every (k_y, k_z) pair these functions (one symmetric, one antisymmetric) describe two surface states, one on each surface. This follows from the fact that $\psi_{\xi k_y k_z}$ is composed of atomic wave functions whose contribution decreases by the factor $\exp(-\nu\xi)$ with increasing ν — that is, with increasing distance from the surfaces. The number of different values that (k_y, k_z) can assume is equal to the number of atoms in a layer perpendicular to x, so that there is one surface state per surface atom. By comparing (5.23) with (5.20), we see that the energy of the surface states for fixed (k_y, k_z) is larger than the other energies for k_x real. Because $\cos k_y a$, $\cos k_z a$ independently assume values between $+1$ and -1, the surface-state band can either be completely above the bulk band or can overlap it partially. This is determined by ξ — that is, by the value of ε_s/γ_s.

3.2. SHOCKLEY STATES

When the width of the allowed bulk bands is comparable to the energy difference between the atomic levels from which the bands originate, it is necessary to construct the function ψ as a linear combination of more than one atomic function of each atom. As has been pointed out above, the s and p_x, p_y, p_z states are usually the ones of most interest. In order to simplify the calculations, we shall restrict ourselves to the s and p_x states only. The wave functions corresponding to these states will be denoted by ϕ_s and ϕ_p, respectively. (The function ϕ_p is characterized by cylindrical symmetry about the x direction.) In addition, we shall deal only with a semi-infinite crystal.

As in the preceding subsection, we take as an approximate solution to Schrödinger's equation the function ψ, composed of linear combinations of atomic functions. Now, however, the p_x functions are also included:

$$\psi = \sum_{lmn} a_{lmn} \phi_s(|\mathbf{r}-\mathbf{r}_{lmn}|) + \sum_{lmn} b_{lmn} \phi_p(\mathbf{r}-\mathbf{r}_{lmn}). \quad (5.25)$$

(Here l, m, n are integers, and $l \geq 1$.) The coefficients in (5.25) can be written as

$$a_{lmn} = a_l e^{ia(k_y m + k_z n)}; \quad b_{lmn} = b_l e^{ia(k_y m + k_z n)}. \quad (5.26)$$

Ch. 5, § 3.2   TIGHT-BINDING APPROXIMATION   179

We introduce the following notation (in addition to eq. (5.13)):

$$
\begin{aligned}
-\int \phi_p^*(r - r_{l'm'n'})[U - U_{lmn}]\phi_p(r - r_{lmn})\,d\tau & \\
= \alpha_p \quad \text{for } l, m, n = l', m', n' \text{ and } l \neq 1 & \\
= \alpha_p' \quad \text{for } l, m, n = l', m', n' \text{ and } l = 1 & \\
= -\gamma_p \quad \text{for } m, n = m', n' \text{ and } l = l' \pm 1 & \\
\text{(nearest neighbours in the } x \text{ direction)} & \\
= \delta_p \quad \text{for } m, n = m', n' \pm 1 \text{ or } m' \pm 1, n' \text{ and } l = l' & \\
\text{(nearest neighbours in the } y \text{ or } z \text{ direction)} &
\end{aligned} \quad (5.27)
$$

$$
\begin{aligned}
\int \phi_s^*(|r - r_{l'm'n'}|)[U - U_{lmn}]\phi_p(r - r_{lmn})\,d\tau & \\
= \alpha_{sp} = 0 \quad \text{for } l, m, n = l', m', n' \text{ and } l \neq 1 & \\
= -\alpha_{sp}' \quad \text{for } l, m, n = l', m', n' \text{ and } l = 1 & \\
= \pm \gamma_{sp} \quad \text{for } m, n = m', n' \text{ and } l = l' \pm 1 &
\end{aligned} \quad (5.28)
$$

$$
\begin{aligned}
\eta_s &= E_s - \alpha_s - 2\gamma_s(\cos k_y a + \cos k_z a) \\
\eta_p &= E_p - \alpha_p - 2\delta_p(\cos k_y a + \cos k_z a) \\
\eta_s' &= E_s - \alpha_s' - 2\gamma_s(\cos k_y a + \cos k_z a) \\
\eta_p' &= E_p - \alpha_p' - 2\delta_p(\cos k_y a + \cos k_z a).
\end{aligned} \quad (5.29)
$$

From (5.13) and (5.29) it follows that

$$\eta_s' - \eta_s = \varepsilon_s; \quad \eta_p' - \eta_p = \varepsilon_p, \quad (5.30)$$

where $\varepsilon_p \equiv \alpha_p - \alpha_p'$ and is positive.

By substituting ψ from (5.25) into Schrödinger's equation (5.10) we obtain, in a manner completely analogous to the derivation in the previous subsection, the following homogeneous equations for the coefficients a_l, b_l:

$$
\begin{aligned}
(E - \eta_s)a_l &= -\gamma_s(a_{l+1} + a_{l-1}) + \gamma_{sp}(b_{l+1} - b_{l-1}) \quad \text{for } l > 1; \\
(E - \eta_p)b_l &= -\gamma_{sp}(a_{l+1} - a_{l-1}) + \gamma_p(b_{l+1} + b_{l-1}) \quad \text{for } l > 1,
\end{aligned} \quad (5.31)
$$

with the boundary conditions

$$
\begin{aligned}
(E - \eta_s')a_1 &= -\gamma_s a_2 + \gamma_{sp} b_2 - \alpha_{sp}' b_1; \\
(E - \eta_p')b_1 &= -\gamma_{sp} a_2 + \gamma_p b_2 - \alpha_{sp}' a_1.
\end{aligned} \quad (5.32)
$$

The solution of (5.31) and (5.32) for the general case with arbitrary values for the α and γ parameters is most complicated and not very much can be learned from it. It therefore becomes necessary to introduce a number of simplifying assumptions. Because γ_s, γ_p, γ_{sp} have the same sign and are of comparable magnitude we shall consider, following Artman[12], the special case

$$\gamma_s = \gamma_p = \gamma_{sp} = 1. \tag{5.33}$$

As γ has the dimensions of energy, this equation implies that from now on all energies are expressed in units of γ and can be treated as dimensionless quantities. (The value of γ is of the order of 1 eV). In addition, because the parameters relevant to the surface are likewise of the same sign and are of comparable magnitude, we assume that

$$\varepsilon_s = \varepsilon_p = \alpha'_{sp} \equiv \beta > 0. \tag{5.34}$$

With these assumptions, (5.31) and (5.32) can be rewritten as

$$\begin{aligned}(E-\eta_s)a_l &= -a_{l+1}-a_{l-1}+b_{l+1}-b_{l-1} \quad \text{for} \quad l > 1; \\ (E-\eta_p)b_l &= -a_{l+1}+a_{l-1}+b_{l+1}+b_{l-1} \quad \text{for} \quad l > 1;\end{aligned} \tag{5.35}$$

$$\begin{aligned}(E-\eta_s-\beta)a_1 &= -a_2+b_2-\beta b_1; \\ (E-\eta_p-\beta)b_1 &= -a_2+b_2-\beta a_1.\end{aligned} \tag{5.36}$$

Consider first the case of a crystal infinite in the x direction as well. Equations (5.35) are then solved by the substitutions

$$a_l = Ae^{iak_xl}; \quad b_l = \Lambda A e^{iak_xl}. \tag{5.37}$$

If we define h as

$$h \equiv \tfrac{1}{4}(\eta_p-\eta_s) = \tfrac{1}{4}(E_p-E_s)-\tfrac{1}{4}(\alpha_p-\alpha_s)+\tfrac{1}{2}(1-\delta_p)(\cos k_y a+\cos k_z a), \tag{5.38}$$

then the expression for the allowed energies is given by

$$E = \tfrac{1}{2}(\eta_s+\eta_p)\pm 2\sqrt{(1+he^{iak_x})(1+he^{-iak_x})}. \tag{5.39}$$

The upper sign (plus) refers to the upper band and the lower sign to the lower band. Similarly, the solution for Λ is

$$\Lambda = -i\frac{h+\cos k_x a}{\sin k_x a} \mp \frac{i}{\sin k_x a}\sqrt{(1+he^{iak_x})(1+he^{-iak_x})}. \tag{5.40}$$

The pertinent values of $k_x a$ lie in the interval $0, \pm\pi$, so that the minimum of the upper band and the maximum of the lower band for fixed (k_y, k_z) are given by

$$E = \tfrac{1}{2}(\eta_s+\eta_p) \pm 2|1-|h||. \tag{5.41}$$

So much for the energy bands in the bulk. For a finite crystal (5.35) and (5.36) are solved, in a manner analogous to that in the previous subsection, by the substitutions

$$a_l = Ae^{iak_xl} + Be^{-iak_xl}; \quad b_l = \Lambda(Ae^{iak_xl} - Be^{-iak_xl}), \tag{5.42}$$

with Λ given by (5.40). The calculation shows that the condition for obtaining complex solutions for $k_x a$ (corresponding to surface states) depends on the magnitudes of both $|h|$ and β. For $|h| \gg 1$ — that is, for a large energy separation between the two bands — the problem reduces to that of the previous subsection. For $\beta > 1$, corresponding to a large perturbation at the surface, surface states are obtained above both the upper and lower bands (Tamm states). When $|h|$ decreases the interaction between the two bands increases, and for $|h| < 1$ those Tamm states lying between the two bands disappear. In this case ($|h| < 1$), on the other hand, Shockley states appear, provided that now $\beta < 1$ (small perturbation at the surface). In the other two cases ($|h| > 1$, $\beta < 1$ and $|h| < 1$, $\beta > 1$) no surface states can appear between the bands. It should be noted that h depends also on the values of k_y and k_z (see eq. (5.38)). If the energy difference $E_p - E_s$ between the two atomic levels from which the two bands originate is such that $|h|$ can have values both above and below unity, then the surface states between the bands will be Tamm-like when $\beta > 1$ and Shockley-like when $\beta < 1$. The number of states is determined by the number of pairs (k_y, k_z) for which $|h| > 1$ in the first case or $|h| < 1$ in the second. In either case there will generally be fewer surface states than surfaces atoms (if the effect of spin is disregarded).

One should bear in mind that the conclusions reached concerning the characteristics of the surface states have been drawn from a consideration of a relatively simple model and rather special assumptions (eqs. (5.33), (5.34)). Essentially the same results have been obtained from calculations using modified atomic functions (Wannier functions) and based on somewhat different assumptions [13].

For an actual crystal it is necessary in each case to evaluate the exchange and Coulomb integrals, to insert the values obtained into a system of homogeneous equations of the type (5.31) and (5.32), and to solve for the wave functions and energy levels. Such calculations (including also the p_y and

p_z states) have in fact been carried out for ideal surfaces of diamond [14, 18]. An estimate of the integrals shows that bands of Shockley states should exist on such surfaces both between and overlapping the conduction and valence bands.

4. Occupation statistics of surface states

In the previous sections we saw that under certain conditions there can exist surface states. From now on we assume the presence of localized states at the surface without going into the question of their origin; they can be of the Tamm or Shockley type, or they may be associated with defects or impurities. In a manner analogous to the case of localized bulk states, we shall look upon the surface states as arising from centres located at the surface and consider their effect on the electronic processes in this region. In the present section we calculate the occupation of the surface states as a function of surface potential. (The latter can be varied, for example, by an electrostatic field applied normal to the surface.) Another topic to be considered is the effect of the surface states on the variation of surface potential with external field.

4.1. SINGLE-CHARGE SURFACE STATES

Consider first the simplest case, that in which each centre at the surface can capture or release only one electron, thereby introducing a single allowed energy level E_t in the forbidden gap. All centres are assumed to have identical characteristics and to be present with a density (N_t) that is sufficiently low to eliminate any mutual interaction. The equilibrium densities of occupied (n_t) and unoccupied ($p_t \equiv N_t - n_t$) centres are given by Fermi–Dirac statistics in terms of the energy position of the surface states with respect to the Fermi level at the surface, in a manner analogous to that in Ch. 2, § 5.3. Now, however, N_t, n_t, p_t refer to densities per unit area. As in the case of bulk states, we introduce an *effective* energy level E_t^f (eq. (2.74)) to allow for possible multiplicity of the surface states. Recalling, further, that at the surface (see Ch. 4, § 2.1)

$$(E_t^f - E_F)/kT = (E_t^f - E_i)/kT - u_s \tag{5.43}$$

(u_s being the surface potential), we have that

$$n_t/N_t = f_n(E_t^f); \qquad p_t/N_t = f_p(E_t^f), \tag{5.44}$$

where the Fermi distribution functions $f_n(E_t^f)$ and $f_p(E_t^f)$ are given by

$$f_n(E_t^f) = \frac{1}{1+\exp[(E_t^f - E_i)/kT - u_s]};$$

$$f_p(E_t^f) = \frac{1}{1+\exp[u_s - (E_t^f - E_i)/kT]}.$$
(5.45)

From (5.43)–(5.45) we see that when $u_s = (E_t^f - E_i)/kT$ (that is, when the Fermi level passes through E_t^f), then $n_t = p_t = \tfrac{1}{2}N_t$. The main change in occupation takes place within a few units of u_s; the steepest slope occurs at $f_n(E_t^f) = \tfrac{1}{2}$ and is equal to $\tfrac{1}{4}$.

In general, we can expect the surface to have several such independent levels at energies E_{tj} and of densities N_{tj}. In this case the density of electrons captured in *all* surface states will be given by the sum $\sum_j N_{tj} f_n(E_{tj}^f)$. For purposes of illustration, the occupation function for two surface levels of equal density $(N_{t1} = N_{t2})$ has been plotted in Fig. 5.4 against

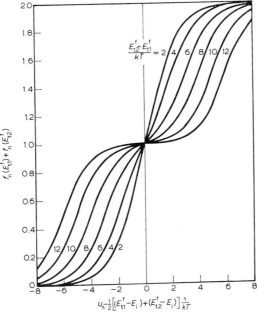

Fig. 5.4. Overall fractional occupation of two independent centres present with equal densities as a function of the distance of the Fermi level at the surface from the mean value of the two effective energies E_{t1}^f and E_{t2}^f. The different curves are for various values of $(E_{t2}^f - E_{t1}^f)/kT$. The same curves are equally applicable to a double-charge centre having the same values of E_{t1}^f and E_{t2}^f (see § 5.2).

$u_s - \frac{1}{2}[(E_{t1}^f - E_i) + (E_{t2}^f - E_i)]/kT$, the distance of the Fermi level from the average of the effective energies E_{t1}^f, E_{t2}^f. The different curves are for various values of $(E_{t2}^f - E_{t1}^f)/kT$ and represent, in each case, the overall fractional occupation of the states.

4.2. THE VARIATION OF SURFACE POTENTIAL WITH EXTERNAL FIELD

The value of the surface potential in the absence of an external electric field is determined by the density and energy distribution of the surface states. The neutrality condition requires that $Q_{ss}^0 + Q_{sc}^0 = 0$; Q_{ss} is the charge density (per unit area) in the surface states and is equal to $-qn_t$ for acceptor-like states and $+qp_t$ for donor-like states, and Q_{sc} is the space-charge density as defined in Ch. 4, § 2.2. The superscript o indicates the absence of an external field. By substituting for n_t, p_t, Q_{sc} from (5.44) and (4.19), we obtain

$$(n_b + p_b)LF_s = N_t f(E_t^f), \tag{5.46}$$

where the distribution function $f(E_t^f)$ represents $f_n(E_t^f)$ for acceptor-like states and $f_p(E_t^f)$ for donor-like states. In the derivation of (5.46), the sign of Q_{sc} (see eq. (4.19)) was determined by the condition that for acceptor-like states the energy bands bend upwards ($v_s < 0$) while for donor-like states the opposite is true (compare Figs. 4.3b and 4.3d). Equation (5.46) can be solved graphically (see Figs. 4.7, 4.8, 5.4) for the surface potential u_s or the barrier height $v_s (\equiv u_s - u_b)$. It is easily seen that for large surface-state densities ($N_t \gg L(n_b + p_b)$), E_t^f will be well removed from E_F and donor-like states will remain relatively full and acceptor-like states relatively empty. In general, u_s (or v_s) will be determined by several levels of this type, and then in place of $N_t f(E_t^f)$ there will appear in (5.46) the sum $\sum_j N_{tj} f(E_{tj}^f)$ over all surface levels present.

We shall now see how the surface potential changes under the effect of an external electrostatic field applied normal to the surface. For such a case we obtain that $Q_{ss} + Q_{sc} = Q_s$, where Q_s is the total charge induced by the field. This equation expresses the fact that the induced charge Q_s is distributed between the surface states and the space-charge region. Under equilibrium conditions we have, similarly to (5.46), that

$$\mp (n_b + p_b)LF_s \mp N_t f(E_t^f) = Q_s/q. \tag{5.47}$$

Here the upper sign before the second term (minus) refers to acceptor-like states and the lower sign (plus) to donor-like states. As for the first term, its sign is actually that of Q_{sc} (negative for $v_s > 0$, positive for $v_s < 0$). Equation (5.47), which determines the new value of the surface potential, is difficult to solve in the general case. Experimentally, one usually measures Q_{sc} and Q_s and in this way determines the surface-state density (see Ch. 6, § 4). What we are interested in here is the rate of change of v_s (or u_s) with respect to the total induced charge (dv_s/dQ_s), this derivative being a measure of the extent to which an external field is able to swing the surface potential. Differentiation of (5.47) yields

$$\frac{dv_s}{d(Q_s/q)} = -\frac{1}{(n_b+p_b)L|dF_s/dv_s| + N_t|df(E_t^f)/dv_s|}. \tag{5.48}$$

It follows that the larger the density N_t, the more difficult it is to vary the potential barrier. A very large density "anchors" the barrier height at its field-free value, the surface states effectively screening the space-charge region from the influence of the external field. In the more general case when several sets of surface states are present, the second term in the denominator of (5.48) must be replaced by a sum of terms corresponding to the different levels. The derivative $|df(E_t^f)/dv_s|$ is greatest when the Fermi level passes through E_t^f, so that if the densities of the various sets of surface states are comparable, then the most effective level will be the one nearest the Fermi level. As far as the space-charge region itself is concerned, there will be similar anchoring for strong accumulation or inversion layers even without surface states, since here $|dF_s/dv_s|$ varies exponentially with v_s (see eqs. (4.24), (4.29)).

5. Occupation statistics for complex surface states

5.1. GENERAL CASE

In the preceding section we used eqs. (5.44), which are the analogues of eqs. (2.76), (2.77), in order to express the densities of the occupied and unoccupied surface states. These equations are valid only for the simple case where, in the charge transfer processes between the surface states and the energy bands, only one electron per centre takes part and the state of this electron is associated with a single energy. We now extend the treatment to centres capable of capturing more than one electron. For each condition of charge, these centres can exist in several quantum states associated with different energies (ground and excited states). Centres of this sort have not yet been found on the surfaces but are known to exist in the bulk [19-22].

Consider the situation in which several electrons are bound to the centre and, because of energy considerations, cannot become free. When in such a charge condition, the centre will be called unoccupied. Let us assume that the unoccupied centre can exist in different quantum states and designate the energy of the m-th state by E_{0m}. When the centre has captured j additional electrons, we say that it is charged j times and denote the corresponding energies by E_{jm}. The densities of the centres in the different charge conditions can be derived by finding the distribution that maximizes the probability. Several authors [23] have performed calculations of this type, and here we present only the results. Shockley and Last [23] find that the density of centres charged j times and corresponding to the m-th state is given by

$$M_{jm} \propto N_t \exp\left[(jE_F - E_{jm})/kT\right]. \tag{5.49}$$

The proportionality constant is determined from the normalization requirement $\sum_{jm} M_{jm} = N_t$, where N_t is the total density of the centres. In order to obtain the density of the centres charged j times, it is necessary to perform the summation

$$M_j = \sum_m M_{jm} \propto N_t Z_j \exp(jE_F/kT), \tag{5.50}$$

where

$$Z_j \equiv \sum_m \exp(-E_{jm}/kT). \tag{5.51}$$

The ratio of the density of centres charged j times to that of centres charged $(j-1)$ times can be written as

$$M_j/M_{j-1} = \exp\left[(E_F - E_{tj}^f)/kT\right], \tag{5.52}$$

where E_{tj}^f is given by

$$E_{tj}^f = kT \ln(Z_{j-1}/Z_j). \tag{5.53}$$

It can be shown [23] that in order to transform a centre charged j times into one charged $(j-1)$ times and thereby raise an electron from the centre into the conduction-band edge, the ionization energy $E_c - E_{tj}^f$ must be added to it on the average. (The averaging procedure takes into consideration the various states in the two conditions of charge.) For this reason we can look upon the centre as though the j-th electron is effectively at an energy E_{tj}^f. Because the centre becomes more negatively charged with the capture of each additional electron, the ionization energy $E_c - E_{tj}^f$ decreases with increasing j. The effective energies E_{tj} thus constitute a sequence of increasing energies.

Obviously, only those conditions of charge in which $E^f_{tj} < E_c$ can be stable.

We now choose a few relatively simple special cases to treat in the following subsections.

5.2. SINGLE- AND DOUBLE-CHARGE CENTRES WITHOUT EXCITED LEVELS

In this subsection it will be assumed that in every condition of charge the centres can be associated with a single energy only. The number of degenerate quantum states of a centre that is empty, charged once, and charged twice will be denoted by g_0, g_1, and g_2.

We first illustrate how the relationships governing the occupation of *single*-charge centres (Ch. 2, § 5.1 and Ch. 5, § 4.1) can be derived from the generalized statistics just discussed. From (5.51) we obtain for this case that $Z_0 = g_0 \exp(-E_0/kT)$ and $Z_1 = g_1 \exp(-E_1/kT)$. The substitution of these values into (5.53) yields

$$E^f_t = (E_1 - E_0) - kT \ln(g_1/g_0), \qquad (5.54)$$

where $E^f_t = E^f_{t1}$, the subscript 1 being superfluous in the case of single-charge centres. (This notation will always be used in future for single-charge centres whenever there is no danger of ambiguity.) Equations (2.75)–(2.77), (5.44), (5.45) follow directly from (5.52) since now M_j and M_{j-1} are simply n_t and p_t, respectively, with $n_t + p_t = N_t$. Thus a centre having an energy E_0 (g_0-times degenerate) when empty and E_1 (g_1-times degenerate) when occupied introduces an energy level $E_t = E_1 - E_0$ for electron occupation. The corresponding effective energy is given by (5.54). This case serves to illustrate in a simple way the relation between the energies of the centre under different charge conditions (E_0 and E_1) and the electronic energy level (E^f_t) introduced by the centre in the conventional band scheme. As has been pointed out in the previous subsection, the connecting link between the two representations is the ionization energy $E_c - E^f_t$, which is the average energy required to raise an electron from the charged centre into the conduction-band edge.

Next we proceed to calculate the relevant densities for double-charge centres. For this case Z_0, Z_1, and E^f_{t1} are the same as for the single-charge centre. The expression for Z_2 is again obtained from (5.51) as $g_2 \exp(-E_2/kT)$, and the effective energy of the centre charged twice is $E^f_{t2} = (E_2 - E_1) - kT \ln(g_2/g_1)$. By using the condition that $M_0 + M_1 + M_2 = N_t$ (that is, that the sum of the densities of the centres which are empty, charged once, and charged twice is equal to the total density of the centres), we obtain from (5.43) and (5.52) that

$$\frac{M_0}{N_t} = \frac{1}{1+\exp[u_s-(E_{t1}^f-E_i)/kT]+\exp[2u_s-\{(E_{t1}^f-E_i)+(E_{t2}^f-E_i)\}/kT]};$$

$$M_1 = M_0 \exp[u_s-(E_{t1}^f-E_i)/kT]; \tag{5.55}$$

$$M_2 = M_0 \exp[2u_s-\{(E_{t1}^f-E_i)+(E_{t2}^f-E_i)\}/kT].$$

The total density of electrons captured by the centres is now given by M_1+2M_2.

It turns out that to a good approximation M_1+2M_2 is equal to the total density of electrons occupying two *independent* centres, provided that $E_{t2}^f-E_{t1}^f \gtrsim kT$. This situation has already been illustrated in Fig. 5.4, where the occupation is shown of two independent levels each of equal density N_t. Identical curves are obtained for double-charge centres having the same density N_t and the same effective energies E_{t1}^f and E_{t2}^f. Thus, as far as the overall electron occupation is concerned, a multicharge centre is indistinguishable from a corresponding set of independent centres, provided that the energies E_{tj}^f are not too close to one another.

5.3. SINGLE-CHARGE CENTRE WITH EXCITED LEVELS

We now consider centres that can capture and release only one electron, but which in both charge conditions (vacant and occupied) may possess more than one energy. For simplicity we assume that when the centre is vacant there are only two quantum states: the ground state at energy E_{00} and a single excited state at energy E_{01}. Similarly, the occupied centre has only two quantum states, at energies E_{10} and E_{11}. All four states are thus assumed to be non-degenerate. The densities of the unoccupied and occupied centres in the ground and excited states will be denoted by p_t, p_t', n_t, and n_t'. Here and subsequently, primed quantities refer to excited states. For such centres, (5.51) yields $Z_0 = \exp(-E_{00}/kT) + \exp(-E_{01}/kT)$ and $Z_1 = \exp(-E_{10}/kT) + \exp(-E_{11}/kT)$. From (5.53) it follows that the effective energy E_t^f is equal to $kT \ln(Z_0/Z_1)$. The densities of the centres in the various states can be obtained from (5.50), (5.52), and the condition that $n_t+n_t'+p_t+p_t' = N_t$:

$$(n_t+n_t')/N_t = f_n(E_t^f); \qquad (p_t+p_t')/N_t = f_p(E_t^f);$$
$$n_t' = n_t \exp[(E_{10}-E_{11})/kT]; \qquad p_t' = p_t \exp[(E_{00}-E_{01})/kT], \tag{5.56}$$

where the functions f_n and f_p are as defined in (5.45). We thus see that from the point of view of the overall electron occupation $(n_t + n'_t)$, the centres behave as simple centres having no excited states (§ 4.1), provided we take for their effective energy E_t^f the value just determined.

6. The interaction of surface states with a single band

In § 4.2 we considered the steady-state distribution of charge between the space-charge region and the surface states as induced by a constant electrostatic field normal to the surface. In this section we shall look into the transient behaviour of the induced charge and the processes leading up to steady-state conditions. Such processes involve carrier exchange between the surface states and the conducting bands, and through their experimental investigation considerable light can be shed on the characteristics of the surface states involved. The simplest case from both the theoretical and the experimental points of view corresponds to the interaction of the surface states with a single band. This can easily be realized by choosing an extrinsic semiconductor and restricting the surface potential to values corresponding to accumulation and depletion layers only. Under these conditions minority carriers are effectively absent, and any charge exchange takes place exclusively between the surface states and the majority-carrier band. Such a situation will be assumed to hold throughout the discussion in this section. A more general discussion will be given in Ch. 6, §§ 6, 7.

Consider for example an n-type semiconductor having an accumulation layer at the surface, as shown in Fig. 5.5a. Suppose that at $t = 0$ an electric field is abruptly produced at the surface by applying a voltage pulse between the semiconductor and a metal electrode parallel to the surface. If the electrode is made positive then the field, which momentarily penetrates into the semiconductor, will cause a flow of electrons from the contacts to the surface until the metal–semiconductor capacitor is charged up. The time constant associated with this process — the charge relaxation time (see Ch. 6, § 6.1) — depends on the experimental configuration employed, and in most cases is very small (of the order of 10^{-8} sec). Accordingly, a very short time ($t = 0_+$) after the application of the field, the induced charge in the conduction band attains an equilibrium distribution, the Fermi level being well defined in the space-charge region as well as in the bulk. The barrier height becomes more positive (bands bending downwards) to permit the induced charge to be accomodated in the conduction band. This situation is shown schematically in Fig. 5.5b by the termination of the lines of force at free electrons in the accumulation region. At this stage the surface states are not as yet in equilibrium with the induced charge. Excess free electrons tend to drop into them and a relaxation of the barrier height v_s takes place. When equilibrium conditions are reached, part of the induced charge resides in the space-charge region and the rest appears in the surface states, as indicated in Fig. 5.5c.

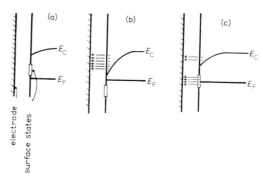

Fig. 5.5. Shape of the conduction-band edge in an n-type semiconductor under an electrostatic field in the form of a step function applied normal to the surface.
   (a) Prior to the application of the field ($t < 0$).
   (b) Just following the onset of the field ($t = 0_+$).
   (c) Under steady-state conditions ($t \to \infty$).

In order to derive the main features of the interaction process between the conduction band and the surface states, we consider a relatively simple case. The surface states are assumed to arise from single-charge centres associated with a discrete energy level E_t (and effective energy E_t^f), present with a density N_t. The rate of change in the density of occupied centres (n_t) can be expressed as

$$\frac{dn_t}{dt} = K_n[n_s(N_t - n_t) - n_1 n_t], \tag{5.57}$$

where K_n is the probability per unit time that an electron be captured by a vacant state, and n_s is the free electron density at the surface. The former is related to the capture cross section A_n and the average thermal velocity of the electrons $\langle c \rangle$ by the expression $K_n = \langle c \rangle A_n$ (see Ch. 2, §7.2). Equation (5.57) is completely analogous to (2.175), except that now n_t and N_t are densities per unit *area* and n_s is a function of the surface potential. The value of the emission constant n_1 is given by the principle of detailed balance, as in Ch. 2, § 7.2. At thermal equilibrium, $dn_t/dt = 0$, and (5.57) gives

$$n_1 = n_s(N_t - n_t)/n_t, \qquad (5.58)$$

where n_s and n_t now refer to the equilibrium values of free and trapped electrons at the surface. By substitution of the values of n_s and n_t from (4.8) and (5.44), the dependence on u_s cancels out and (5.58) reduces to (2.173). Thus n_1 is equal to the electron density at the surface (n_s) when the Fermi level coincides with E_t^f.

The kinetics of the filling and emptying of the surface states is determined by (5.57) together with the appropriate boundary conditions. The boundary conditions most commonly used are those corresponding to the experimental layout illustrated in Fig. 5.5. Here, at time $t = 0_+$, n_s abruptly assumes a new value n_{s0}, governed by the appearance of all the induced charge Q_s as a change in $q\Delta N$, while the value of n_t remains as yet unaltered and equal to its value n_{t0} prior to the application of the field. (An n-type sample is again being considered.) As n_t can change only by capturing from or emitting electrons into the conduction band, we have

$$\delta(\Delta N) + \delta n_t = 0 \quad \text{for} \quad t \geq 0_+, \qquad (5.59)$$

where $\delta(\Delta N)$ and δn_t represent changes in the values of ΔN and n_t. The general solution of (5.57) is quite complicated, and only a few limiting cases will be considered.

Case a. Small disturbances. This case has been analysed by Low[24]. We introduce the following notation:

$$\begin{aligned} n_t &= n_{t\infty} + \delta n_t; \\ n_s &= n_{s\infty} + \delta n_s; \\ \Delta N &= \Delta N_\infty + \delta(\Delta N), \end{aligned} \qquad (5.60)$$

where $n_{t\infty}$, $n_{s\infty}$, ΔN_∞ are the final equilibrium values of n_t, n_s, ΔN. The amplitude of the applied pulse is assumed to be sufficiently low to ensure that the increments δn_t, δn_s, $\delta(\Delta N)$ are all small. By expanding ΔN and n_s (as functions of the barrier height v_s) about their final values and neglecting the nonlinear terms, we obtain

$$\delta n_s \approx \left.\frac{dn_s}{dv_s}\right|_\infty \delta v_s; \qquad \delta(\Delta N) \approx \left.\frac{d(\Delta N)}{dv_s}\right|_\infty \delta v_s. \qquad (5.61)$$

Eliminating δv_s from these equations and using (5.59) and (5.60), we have from (5.57) that

$$-\frac{\mathrm{d}}{\mathrm{d}t}[\delta(\varDelta N)] = K_n \left[(N_t - n_{t\infty})\frac{\mathrm{d}n_s}{\mathrm{d}(\varDelta N)}\bigg|_\infty + n_{s\infty} + n_1\right]\delta(\varDelta N), \quad (5.62)$$

where we have neglected the product $(\delta n_t)(\delta n_s)$ and made use of the fact that $\mathrm{d}n_t/\mathrm{d}t$ vanishes when equilibrium conditions are reached. The solution of (5.62) yields an exponential decay:

$$\delta(\varDelta N) = \delta_0(\varDelta N)e^{-t/\tau_0}, \quad (5.63)$$

where $\delta_0(\varDelta N)$ is the value of $\delta(\varDelta N)$ immediately following the application of the field ($t = 0_+$), and the time constant τ_0 is given by

$$\tau_0 = \left\{K_n\left[(N_t - n_{t\infty})\frac{\mathrm{d}n_s}{\mathrm{d}(\varDelta N)}\bigg|_\infty + n_{s\infty} + n_1\right]\right\}^{-1}. \quad (5.64)$$

By substituting the values of n_s and $\varDelta N$ from (4.8) and (4.37), we obtain for strong accumulation layers that

$$\frac{\mathrm{d}n_s}{\mathrm{d}(\varDelta N)} \approx \frac{F_s}{L} \approx \frac{\sqrt{2}e^{\frac{1}{2}v_s}}{L}, \quad (5.65)$$

while $n_{s\infty} = n_b \exp v_s$. Thus the time constant τ_0 is short and the system approaches its final state rapidly. In depletion layers, on the other hand, there results approximately

$$\frac{\mathrm{d}n_s}{\mathrm{d}(\varDelta N)} \approx \frac{F_s e^{-|v_s|}}{L} \approx \frac{\sqrt{2|v_s|}e^{-|v_s|}}{L}, \quad (5.66)$$

and $n_{s\infty} = n_b \exp(-|v_s|)$. Both terms are small for large $|v_s|$, and it is mainly n_1 that governs the decay. For deep-lying levels and low temperatures, this decay may be slower by many orders of magnitude than that in accumulation layers.

Case b. Large disturbances. Let us now assume that the amplitude of the pulsed field is sufficiently large so that the final value of the electron density in the surface states $n_{t\infty}$ differs substantially from the density n_{t0} at the application of the pulse. In this case either the surface states were initially almost fully occupied and so they empty out (negative pulse) when final equilibrium conditions have been reached, or else they were initially unoccupied and so they fill up (positive pulse). The first case has been treated by Rupprecht [25]. Since the polarity of the pulse here is such as to decrease n_s, and the disturbance is assumed sufficiently large so that n_{s0} is very small, we can neglect initially $n_s(N_t - n_t)$ with respect to $n_1 n_t$. Equation (5.57) then reduces to

$$\frac{dn_t}{dt} = -K_n n_1 n_t, \qquad (5.67)$$

which again represents an exponential decay with time constant τ_1:

$$\tau_1 = \frac{1}{K_n n_1}. \qquad (5.68)$$

This solution is valid for the range of t in which (5.67) still remains a good approximation; for larger values of t the decay will gradually be transformed into that for small disturbances. It is clear from (5.68) that the emptying of the surface states is a strictly thermal process, n_1 being a measure of the rate of thermal emission. Consequently, τ_1 increases rapidly with decreasing temperature and with increasing energy separation between the conduction-band edge and the surface states. This method (large disturbances) can thus become a powerful tool for the determination of the energy level associated with the surface states. One should be cautioned, however, that the use of such large disturbances (which is the essence of this technique) might be complicated by the appearance of field-enhanced emission processes that may completely mask the thermal release, as will be discussed in Ch. 7, § 4 and Ch. 9, § 2.

In the case of the filling of the surface states (positive pulse), the polarity of the pulse is such as to increase n_s. Since we have assumed that prior to the disturbance the surface states are essentially unoccupied, we may neglect initially the terms involving n_t in (5.57):

$$\frac{dn_t}{dt} = K_n n_s N_t. \qquad (5.69)$$

The solution of this equation is no longer exponential. As the amplitude of the decay is approximately equal to N_t, we define a "time constant" τ_2 as

$$\tau_2 \equiv N_t \bigg/ \left(\frac{dn_t}{dt}\right)_{t=0} = \frac{1}{K_n n_{s0}}. \qquad (5.70)$$

This time constant characterizes the decay directly following the application of the field, and depends on the magnitude of n_{s0}. Thus for accumulation layers it is very short and difficult to determine experimentally.

In the above treatments we have expressed the kinetics in the limiting cases of small and large disturbances in terms of time constants. The reader should bear in mind that in an exponential process τ determines only the relative decay rate whereas the absolute value is given by the derivative dn_t/dt. This explains what at first glance may seem somewhat surprising — that the time constant τ_1 of the exponential decay under large-signal conditions (eq. (5.68)) is greater than the time constant τ_0 of the tail of this decay (corresponding to a small disturbance, eq. (5.64)).

7. Surface recombination

When the density of free carriers exceeds the thermal equilibrium value (as a result of optical generation or the like), the excess carriers recombine both in the bulk and at the surface. It has been found experimentally that in both regions the recombination proceeds mainly through localized levels in the forbidden gap. Using a model that assumes the presence of such levels, Shockley and Read [26] have analysed the statistics of recombination in the semiconductor bulk, as discussed in Ch. 2, § 7.2. A similar treatment has been carried out by Brattain and Bardeen [27] for surface recombination via surface states. This was subsequently extended [28, 29] to yield an explicit expression for the surface recombination in terms of the surface potential and the pertinent parameters of the surface states involved. The surface states were assumed to be associated with single-charge centres having a single, discrete energy level in the forbidden gap. Because of the considerable success this analysis has had in accounting for the experimental results on germanium and silicon, we shall present it here in some detail. The case of surface recombination involving more complex centres will be treated in the next section.

Consider a homogeneous semiconductor under uniform and steady excitation. Bulk trapping will be assumed to be negligible, so that the excess electron and hole densities in the bulk are equal. The presence of surface recombination in addition to bulk recombination results in a net flow of carriers towards the surface ($z=0$). Let qU_n and $-qU_p$ denote the electron and hole current densities corresponding to this flow. We then obtain, similarly to the previous section, that

$$U_n = K_n[n_s^*(N_t - n_t^*) - n_1 n_t^*]; \tag{5.71}$$

$$U_p = K_p[p_s^* n_t^* - p_1(N_t - n_t^*)]. \tag{5.72}$$

Here n_s^*, p_s^* are the mobile electron and hole densities at the surface, and n_t^* is the new (steady-state) density of occupied surface states. The constants n_1 and p_1 express the thermal emission rate of electrons and holes from the states into the respective bands and are given by (2.172) and (2.173). Under steady-state conditions, the electron and hole currents are equal. Equating U_n and U_p from (5.71) and (5.72), we obtain for n_t^*

$$\frac{n_t^*}{N_t} = \frac{K_n n_s^* + K_p p_1}{K_n(n_s^* + n_1) + K_p(p_s^* + p_1)}. \tag{5.73}$$

The substitution of this value of n_t^* into (5.71), (5.72) and the use of the relation $n_1 p_1 = n_b p_b$ (Ch. 2, § 5.3) yield

$$U_n = U_p = U = \frac{K_n K_p N_t (n_s^* p_s^* - n_b p_b)}{K_n(n_s^* + n_1) + K_p(p_s^* + p_1)}. \tag{5.74}$$

Before we proceed further, we shall make some assumptions regarding the steady-state surface densities n_s^* and p_s^*. These can be very simply related to the respective bulk densities n_b^* and p_b^* *just outside the space-charge region*, provided that the quasi Fermi levels F_n and F_p are horizontal throughout the space-charge region. (The reader should note that in the presence of surface recombination, the steady-state bulk densities will in general not be uniform as we move away from the surface. The symbols n_b^* and p_b^* are therefore used here to denote the carrier densities at a point immediately below the space-charge region. This question will be taken up again in Ch. 7, § 1.) We then have (see Ch. 4, § 4) $n_s^* = n_b^* \exp v_s^*$ and $p_s^* = p_b^* \exp(-v_s^*)$. To determine the range of validity of such an assumption [27], we write the current density from the bulk to the surface due to either of the carrier types in the form (see eq. (2.190))

$$J_n = |J_p| = \mu_p p^*(z) \left|\frac{dF_p}{dz}\right|. \tag{5.75}$$

Usually one is interested in the case in which the bulk lifetime is sufficiently long for the diffusion length to be large compared to the width of the space-charge layer. Under these conditions, recombination in the space-charge region is negligible and $|J_p| = qU_p$ throughout the region, independently of the distance z from the surface. Thus, for a given rate of recombination at the surface, the smaller the density p^*, the larger is the change in F_p. This change will be greatest for barriers corresponding to depletion of holes ($p_s^* < p_b^*$), and in this range (5.75) yields

$$\left|\frac{dF_p}{dz}\right| \lesssim \frac{|J_p|}{\mu_p p_s^*} = \frac{qU_p}{\mu_p p_s^*}. \tag{5.76}$$

The value of U_p is given by (5.72) and is seen to be always smaller than $K_p p_s^* N_t$. Thus

$$\left|\frac{d(F_p/kT)}{dz}\right| \leq \frac{qK_p N_t}{\mu_p kT} = \frac{K_p N_t}{D_p}, \quad (5.77)$$

where we have made use of the Einstein relationship (2.185) between the mobility μ_p and the diffusion constant D_p. The width of the space-charge region is of the order of the effective Debye length L, so that the maximum change δF_p in the value of F_p at the surface (with respect to its bulk value just outside the space-charge region) can be estimated by the inequality

$$\delta|F_p/kT| \leq \frac{K_p N_t L}{D_p}. \quad (5.78)$$

A completely analogous estimate can be made for δF_n. Thus in order for the quasi Fermi level of the depleted carrier to be straight, it is sufficient that $KN_t L/D$ be small compared to unity. As an example, consider germanium at room temperature and with an impurity concentration $N_D - N_A = 10^{14}$ cm^{-3}, so that $L \approx 5 \times 10^{-5}$ cm. For a surface-state density N_t of 10^{12} cm^{-2} and a capture probability K equal to 10^{-8} cm^3/sec, the ratio $KN_t L/D$ is of the order of 10^{-2}.

We now return to (5.74). From the assumptions that the quasi Fermi levels are flat and that $\delta n_b = \delta p_b$, we obtain

$$n_s^* p_s^* = n_b^* p_b^* = (n_b + \delta p_b)(p_b + \delta p_b). \quad (5.79)$$

We shall consider here only the case of small disturbances, where the density of excess carriers can be neglected with respect to that of *all* the carriers in the crystal at thermal equilibrium — that is,

$$\delta n_b = \delta p_b \ll n_b + p_b. \quad (5.80)$$

Calculations have also been made [30] for the case of larger disturbances. These of course are more complicated, and will not be dealt with here.

By substituting for n_1, p_1 from (2.172), (2.173) and for n_s^*, p_s^* from Ch. 4, § 4, we obtain from (5.74)

$$s \equiv \frac{U}{\delta n_b} = \frac{U}{\delta p_b} = \frac{\sqrt{K_n K_p} N_t(n_b + p_b)}{2n_i^*\{(n_i/n_i^*)\cosh[(E_t^f - E_i)/kT - u_0] + \cosh(u_s^* - u_0)\}}, \quad (5.81)$$

where u_0 is given by

$$u_0 = \ln\sqrt{K_p/K_n}. \quad (5.82)$$

The *surface recombination velocity* s, defined in (5.81), is the ratio of the rate of electron (or hole) flow into a unit surface area to the excess carrier density in the bulk just beneath the surface.

If the deviation δp_b from the thermal equilibrium density is small *even with respect to the minority-carrier bulk density*, then $n_i^* = n_i$, $u_s^* \approx u_s$, and s can be expressed in terms of equilibrium variables only:

$$s = \frac{\sqrt{K_n K_p}\, N_t(n_b + p_b)}{2n_i\{\cosh[(E_t^f - E_i)/kT - u_0] + \cosh(u_s - u_0)\}}. \quad (5.83)$$

The maximum value of s is obtained at $u_s = u_0$ and is given by

$$s_{\max} = \frac{\sqrt{K_n K_p}\, N_t(n_b + p_b)}{2n_i\{\cosh[(E_t^f - E_i)/kT - u_0] + 1\}}. \quad (5.84)$$

It is often convenient to express the surface recombination velocity relative to its maximum value:

$$\frac{s}{s_{\max}} = \frac{\cosh[(E_t^f - E_i)/kT - u_0] + 1}{\cosh[(E_t^f - E_i)/kT - u_0] + \cosh(u_s - u_0)}. \quad (5.85)$$

This ratio is plotted in Fig. 5.6 as a function of $(u_s - u_0)$ for various values of the parameter $|(E_t^f - E_i)/kT - u_0|$. The surface recombination velocity is seen to be symmetrical about $u_s = u_0$. The value of u_0 (eq. (5.82)) is a measure of the inequality between the capture probability for holes and that for electrons. It is positive or negative according as K_p is larger or smaller than K_n. The function s is constant in the region about $u_s = u_0$, but falls off rapidly for large $|u_s - u_0|$ as the second term in the denominator of (5.83) (or (5.85)) becomes predominant. Physically, this behaviour can be understood from the following arguments. The rate of recombination depends on the cooperation of both electrons and holes in being captured by the surface states. The slower of the two processes will of course be the rate-limiting one. Recombination is a maximum ($s = s_{\max}$) when conditions are favourable for the capture of *both* types of carriers. This occurs when $K_n n_s = K_p p_s$, which is equivalent to $u_s = u_0$ (eqs. (4.8), (5.82)). On either side of u_0 the rate of recombination decreases. Consider for the sake of concreteness an n-type sample. In strong accumulation layers, the hole capture is limited by the small number of holes (p_s) able to surmount the potential barrier and reach the surface from the bulk. The surface states fill up and the rate of electron capture is slowed down to match the weak rate of hole capture. For strong inversion layers, on the other hand, the rate-limiting process is that of electron capture; even though the states are essentially vacant, very few electrons are able to reach the surface (small n_s).

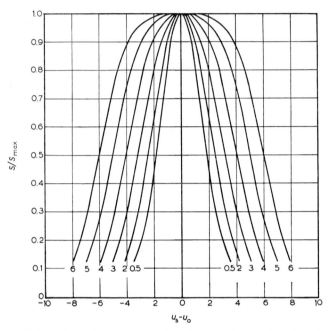

Fig. 5.6. Relative values of surface recombination velocity as a function of $u_s - u_0$ in the presence of single-charge centres situated at a discrete energy level E_t. The different curves are for various values of $|(E_t^f - E_i)/kT - u_0|$.

For $|(E_t^f - E_i)/kT - u_0| \gtrsim 3$, the unity in the numerator of (5.85) can be neglected, and we obtain that s reaches half its maximum value when the Fermi level at the surface coincides with either E_t^f or its reflection about u_0:

$$\frac{s}{s_{\max}} = \frac{1}{2} \quad \text{for} \quad |u_s - u_0| \approx |(E_t^f - E_i)/kT - u_0|. \tag{5.86}$$

We define the half-width Δ as the (positive) difference between the two values of u_s at which $s = \tfrac{1}{2} s_{\max}$. From (5.86) we obtain for Δ the approximate expression

$$\Delta \approx 2|(E_t^f - E_i)/kT - u_0|. \tag{5.87}$$

Experimentally, one can determine the parameters of the surface states involved in the recombination process from measurements of s as a function of u_s. The value of u_0 is obtained from the position of s_{max}, while $|(E_t^f - E_i)/kT - u_0|$ can be found either by fitting the experimental points to a theoretical curve of the type shown in Fig. 5.6 or from the approximate expressions given by (5.86) and (5.87). Such measurements at a single temperature, however, cannot determine $E_t^f - E_i$ uniquely because of the symmetrical character of the hyperbolic cosine functions in (5.83). To remove the ambiguity, additional information is required. This can be supplied by another s curve, taken at a different temperature. Assuming that the capture probabilities are temperature insensitive (constant u_0) we see from (5.87) that if $(E_t^f - E_i)/kT > u_0$, then a decrease in temperature would give rise to an increase in Δ, while if $(E_t^f - E_i)/kT < u_0$ the opposite is true. A second possibility of obtaining $E_t^f - E_i$ consists of measuring the temperature dependence of s_{max}. For $|(E_t^f - E_i)/kT - u_0| \gtrsim 3$, we can replace the hyperbolic cosine in (5.84) by an exponential function and, by substituting for n_i from (2.87), we obtain

$$s_{max} \approx K_p(N_t/N_c)(n_b + p_b)e^{(E_c - E_t^f)/kT} \quad \text{for} \quad (E_t^f - E_i)/kT \gtrsim u_0 + 3;$$
$$s_{max} \approx K_n(N_t/N_v)(n_b + p_b)e^{(E_t^f - E_v)/kT} \quad \text{for} \quad (E_t^f - E_i)/kT \lesssim u_0 - 3.$$
(5.88)

We see that s_{max} varies exponentially with $1/T$ and with the energy distance between E_t and either the conduction- or the valence-band edge. The measured slope, in conjunction with the value of Δ (determined at a single temperature), can thus yield the actual position of the recombination level.

In the derivation of (5.83)–(5.88) we have assumed that δp_b is small not only compared to $n_b + p_b$ but also compared to the *minority*-carrier density. While this latter restriction is usually adequate for germanium at room temperature, it may be impractical at low temperatures where the minority-carrier density is apt to be extremely small. (In the extrinsic range, the minority-carrier density decreases with temperature as $\exp(-E_g/kT)$.) In order to investigate the behaviour of s when the excess carrier density cannot be neglected with respect to the minority-carrier density (but is still small compared to that of the majority carriers), we again start with (5.81). We see that now $s = s_{max}$ at $u_s^* = u_0$ — that is, the maximum occurs at $u_s = u_0 + (E_F - E_F^*)/kT$ and not at $u_s = u_0$ as before (see definitions in Ch. 4, § 4). The value of s_{max} is given by

$$s_{max} = \frac{\sqrt{K_n K_p} N_t(n_b + p_b)}{2n_i^*\{(n_i/n_i^*)\cosh[(E_t^f - E_i)/kT - u_0] + 1\}},$$
(5.89)

and now depends on the size of the disturbance, as expressed by n_i^*. The surface recombination velocity is again a symmetrical function, this time about $u_s^* = u_0$:

$$\frac{s}{s_{max}} = \frac{(n_i/n_i^*)\cosh[(E_t^f - E_i)/kT - u_0] + 1}{(n_i/n_i^*)\cosh[(E_t^f - E_i)/kT - u_0] + \cosh(u_s^* - u_0)}. \quad (5.90)$$

In the event that unity can still be neglected in the denominator of (5.89) (and now this depends also on the size of the disturbance — that is, on n_i/n_i^*), the expression for s_{max} reduces to that for small disturbances (eq. (5.84)). We thus obtain

$$\frac{s}{s_{max}} = \frac{1}{2} \quad \text{for} \quad |(E_t^f - E_i)/kT - u_0| + \ln(n_i/n_i^*) \approx |u_s^* - u_0|, \quad (5.91)$$

and the half-width Δ is given by

$$\Delta \approx 2|(E_t^f - E_i)/kT - u_0| + 2\ln(n_i/n_i^*). \quad (5.92)$$

Since n_i^* is always greater than n_i, the value of Δ will decrease as the magni-

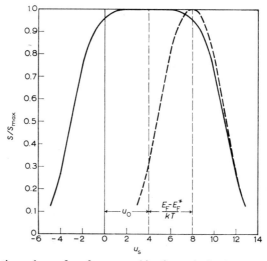

Fig. 5.7. Relative values of surface recombination velocity in an n-type sample as a function of surface potential for small disturbances (solid curve) and for large disturbances (dashed curve) using the following values for the parameters: $E_t^f - E_i = 11kT$, $u_0 = 4$, $n_i/n_i^* = 2 \times 10^{-2}$.

tude of the disturbance increases. This is illustrated in Fig. 5.7 where s/s_{max} is plotted as a function of u_s for an n-type sample. The solid curve corresponds to small disturbances (eq. (5.85)) and the dashed one to large disturbances (eq. (5.90)). In the construction of these curves, the following values were chosen for the various parameters entering into the calculations: $(E_t^f - E_i)/kT = 11$, $u_0 = 4$, $n_i/n_i^* = 2 \times 10^{-2}$. For germanium having an impurity concentration of 10^{14} cm^{-3}, this value of n_i/n_i^* applies at 220°K for a disturbance δp_b that constitutes about $\frac{1}{2}$% of the majority carrier density.

Throughout the above treatment, we have implicitly assumed that the velocity distribution of the carriers under non-equilibrium conditions remains Maxwellian. As has been pointed out by Berz [31], this may not be justified for steep barriers such that the potential drop across a mean free path in the space-charge region is comparable to kT/q. A modification of (5.83) then becomes necessary, especially if the surface scatters specularly. Since germanium and silicon surfaces have been found to be fairly diffuse scatterers (see Ch. 8, § 5), however, such a modification should be small.

8. More sophisticated models of surface recombination

In the treatment of the previous section we assumed that the recombination centres are single-charge centres having no excited levels in the forbidden gap. It may very well be that the surface levels are more complicated, resembling in this respect some of the localized bulk levels. In the present section we extend the surface recombination statistics to include two more complex models. The first considers the recombination centres as being multi-charge centres while the second treats single-charge centres having excited levels.

8.1. DOUBLE-CHARGE CENTRES

Recombination via multi-charge centres has been discussed by several authors [32-34]. Here we shall follow the general treatment of Sah and Shockley[33], but shall restrict ourselves to the case of double-charge centres.

As shown in § 5.2 above, the total number of electrons captured by double-charge centres is, to a good approximation, equal to that captured by two independent centres at energies E_{t1}^f and E_{t2}^f and having the same density (N_t). In order to calculate the surface recombination velocity, we once more write down the net electron and hole current densities flowing into the surface, but now bearing in mind that the states correspond to double-charge centres. Using the notation of § 5.2, we obtain (compare eqs. (5.71), (5.72))

$$\begin{aligned} U_n^{(1)} &= K_n^{(1)}[n_s^* M_0^* - n_1^{(1)} M_1^*]; \\ U_p^{(1)} &= K_p^{(1)}[p_s^* M_1^* - p_1^{(1)} M_0^*]; \\ U_n^{(2)} &= K_n^{(2)}[n_s^* M_1^* - n_1^{(2)} M_2^*]; \\ U_p^{(2)} &= K_p^{(2)}[p_s^* M_2^* - p_1^{(2)} M_1^*], \end{aligned} \quad (5.93)$$

where the stars again refer to non-equilibrium values. The recombination rates $U_n^{(2)}$ and $U_p^{(2)}$ are associated with the transition of the centres from being charged once to being charged twice and conversely. The emission constants $n_1^{(1)}$, $p_1^{(1)}$ and $n_1^{(2)}$, $p_1^{(2)}$ are obtained from the principle of detailed balance and are equal to the equilibrium carrier densities at the surface (n_s, p_s) when the Fermi level passes through E_{t1}^f and E_{t2}^f, respectively. Their values are given by (2.172) and (2.173) with E_t^f replaced by E_{t1}^f or E_{t2}^f. Under steady-state conditions, the electron and hole currents corresponding to each of the two types of transitions are equal. By equating the appropriate recombination rates in (5.93), we obtain

$$U^{(1)} = U_n^{(1)} = U_p^{(1)} = \frac{K_n^{(1)} K_p^{(1)} (n_s^* p_s^* - n_b p_b)}{K_n^{(1)} (n_s^* + n_1^{(1)}) + K_p^{(1)} (p_s^* + p_1^{(1)})} (M_0^* + M_1^*);$$

$$U^{(2)} = U_n^{(2)} = U_p^{(2)} = \frac{K_n^{(2)} K_p^{(2)} (n_s^* p_s^* - n_b p_b)}{K_n^{(2)} (n_s^* + n_1^{(2)}) + K_p^{(2)} (p_s^* + p_1^{(2)})} (M_1^* + M_2^*).$$

(5.94)

For small deviations from thermal equilibrium, the equilibrium densities may be used, and the surface recombination via the double-charge centres can be expressed as

$$s = \frac{U^{(1)} + U^{(2)}}{\delta p_b} = \frac{M_0 + M_1}{N_t} s_1 + \frac{M_1 + M_2}{N_t} s_2, \qquad (5.95)$$

where M_0, M_1, M_2 are given by (5.55) and

$$s_1 = \frac{(K_n^{(1)} K_p^{(1)})^{\frac{1}{2}} N_t (n_b + p_b)}{2 n_i \{\cosh[(E_{t1}^f - E_i)/kT - u_0^{(1)}] + \cosh(u_s - u_0^{(1)})\}}. \qquad (5.96)$$

The corresponding expression for s_2 is given by a similar equation with superscripts (1) replaced by superscripts (2) and E_{t1}^f by E_{t2}^f. The quantities $u_0^{(1)}$ and $u_0^{(2)}$ are defined by means of $K_n^{(1)}$, $K_p^{(1)}$, $K_n^{(2)}$, $K_p^{(2)}$ similarly to u_0 in (5.82). We see from (5.95) that recombination via the double-charge centres takes place as though they were two independent levels having effective densities $N_{t1} = M_0 + M_1$ and $N_{t2} = M_1 + M_2$. These densities are plotted in Fig. 5.8 as a function of $u_s - \frac{1}{2}[(E_{t1}^f - E_i) + (E_{t2}^f - E_i)]/kT$ for various values of $(E_{t2}^f - E_{t1}^f)/kT$. It is seen that throughout most of the range in which the Fermi level lies between E_{t1}^f and E_{t2}^f, both N_{t1} and N_{t2} are close to N_t. Outside this range, one of the densities is equal to N_t while the other approaches zero. Thus in general recombination via a double-charge centre will behave approximately as that via two independent levels if, in the range where N_{t1} (or N_{t2}) is less than N_t, s_1 (or s_2) is already small and so in any case does not affect the recombination. In Fig. 5.9 the function s is illustrated for two *independent* levels each of density N_t, where the following relevant parameters have been chosen: $u_0^{(1)} = -4$, $u_0^{(2)} = 4$, $E_{t1}^f - E_i = 0$, $E_{t2}^f - E_i = 7kT$. The ratio between the maximum values of s_1 and s_2 has

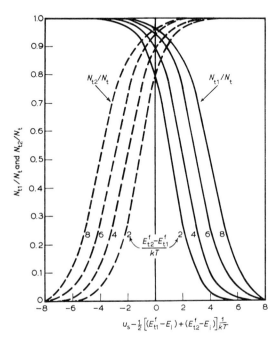

Fig. 5.8. Relative effective densities N_{t1} and N_{t2} for a double-charge centre as a function of the distance of the Fermi level at the surface from the mean value of the two effective energies E_{t1}^f and E_{t2}^f. The different curves correspond to various values of $(E_{t2}^f - E_{t1}^f)/kT$.

been taken as 0.4, which is equivalent to the choice $K_p^{(2)}/K_p^{(1)} \approx 3 \times 10^3$. The dashed curves show s_1 and s_2 separately, while the solid line corresponds to the composite curve $s_1 + s_2$. Values of s have also been calculated (using eq. (5.95)) for a double-charge centre with the same choice of parameters, and are indicated by the crosses in the figure. These values are seen to coincide with those of two independent centres except for small deviations in the region between the two maxima. We see that there is little hope of being able to distinguish experimentally between a double-charge centre and two independent single-charge centres: if there is no significant overlap between the s_1 and s_2 curves then the double-charge centre behaves essentially as two independent centres while, on the other hand, if the overlap is large then it is in any case difficult to resolve the experimental curve into its two components (s_1 and s_2).

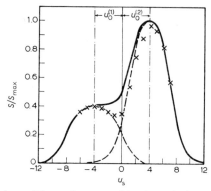

Fig. 5.9. Relative values of the surface recombination velocity arising from two independent single-charge centres as a function of surface potential. The crosses correspond to the case of a double-charge centre having the same parameters: $E_{t_1}^t - E_i = 0$, $E_{t_2}^t - E_i = 7kT$, $u_0^{(1)} = -4$, $u_0^{(2)} = 4$, $K_p^{(2)}/K_p^{(1)} \approx 3 \times 10^3$.

8.2. SINGLE-CHARGE CENTRE WITH EXCITED LEVELS

In all the models of surface recombination discussed so far, we have neglected the possibility of the existence of excited levels. In both single- and double-charge centres we assumed that for every condition of charge the surface states are at a single energy and can be at most degenerate. This assumption would seem to be reasonable considering that the lifetime of excited levels must be of the order of 10^{-8} sec or less and so cannot be important where the recombination lifetime is greater than 10^{-6} sec, as is usually the case (see Ch. 9, § 2). On the other hand, the theoretical work of Lax [35] shows that despite this fact excited states can be of paramount importance in the process of carrier capture by means of localized levels. In fact, there is almost no other way of understanding the mechanism of energy dissipation involved in such a process (see Ch. 2, § 7.2). Rzhanov [36] uses this model and concludes that when the excited level is near one of the energy-band edges, it is no longer permissible to neglect the probability that an electron (or hole) captured by the centre be re-emitted before the centre is able to relax to its ground state. An appropriate modification of the recombination statistics then becomes necessary, as is shown below.

Ch. 5, § 8.2   COMPLEX MODELS OF SURFACE RECOMBINATION   205

We shall assume [36] that the recombination centres are single-charge centres able to exist in four different quantum states: an unoccupied centre in the ground state, an unoccupied centre in the excited state, an occupied centre in the ground state, and an occupied centre in the excited state. The corresponding energies will be designated as in § 5.3 above by E_{00}, E_{01}, E_{10}, E_{11}, respectively. The recombination process is assumed to take place in the following manner. When the centre is unoccupied and in the *ground state* (E_{00}), it can capture an electron from the conduction band and as a result be converted into the excited state of an occupied centre (E_{11}). Subsequently, the centre relaxes to the ground state of an occupied centre (E_{10}). In this latter state, the centre is able to capture a hole from the valence band and thus be converted into the excited state of an unoccupied centre (E_{01}). Following this, the centre relaxes back to its ground state (E_{00}). The probability of capture by any other means will be neglected on the grounds that the energy dissipation involved is too large. We shall further assume that the energy differences between the excited and ground states, $E_{01}-E_{00}$ and $E_{11}-E_{10}$, are large compared to kT. The effective energy E_t^f, which in this case can be denoted simply by E_t, is then very nearly equal to $E_{10}-E_{00}$ (see §5.3 above). To the same approximation, the densities of the occupied and unoccupied centres in the ground states, n_t and p_t, are obtained from (5.56) as:

$$n_t/N_t = f_n(E_t); \qquad p_t/N_t = f_p(E_t). \tag{5.97}$$

This approximation means, in effect, that the excited states are neglected as far as occupation is concerned. (Their presence will assert itself, however, in the recombination processes.) The ionization energy of an occupied level in the excited state will be denoted by $E_c - E_n'$. In a similar way, the energy that must be given to a hole in order to transfer it to the valence-band edge from an unoccupied centre in the excited state will be designated by $E_p' - E_v$. Using this notation, we obtain from (5.56)

$$n_t' = n_t \exp\left[(E_t - E_n')/kT\right]; \qquad p_t' = p_t \exp\left[(E_p' - E_t)/kT\right], \tag{5.98}$$

and the energies E_n', E_p' are given by

$$E_n' = E_{11} - E_{00}; \qquad E_p' = E_{10} - E_{01}. \tag{5.99}$$

In the presence of external excitation, the rate of electron flow (per unit area) into the surface states can be expressed as

$$U_n = K_n[n_s^* p_t^* - n_1' n_t'^*] \tag{5.100}$$

where, as before, the stars represent non-equilibrium values. Here n'_1 is the emission constant of the electrons from the occupied centre in its excited state into the conduction band, and is obtained from (2.173) by replacing E^f_t by E'_n. In order to obtain a relationship between n^*_t and n'^*_t, we shall express the transition rate U'_n of the occupied centres from the excited to the ground state as

$$U'_n = r_n(n'^*_t - bn^*_t), \tag{5.101}$$

where r_n is the reciprocal of the lifetime of the excited state. The constant b is obtained from the principle of detailed balance which, for this case, requires that at thermal equilibrium the number of transitions in both directions be equal — that is, that $U'_n = 0$:

$$b = r_n n'_t/n_t = r_n \exp\left[(E_t - E'_n)/kT\right]. \tag{5.102}$$

The rate of change in the density of occupied centres in the excited state is given by

$$\frac{dn'^*_t}{dt} = U_n - U'_n = K_n[n^*_s p^*_t - n'_1 n'^*_t]$$
$$- r_n\{n'^*_t - n^*_t \exp\left[(E_t - E'_n)/kT\right]\}. \tag{5.103}$$

The left-hand side of this equation vanishes under steady-state conditions, and this can be used to calculate n'^*_t:

$$n'^*_t = \frac{K_n n^*_s p^*_t + r_n n^*_t \exp\left[(E_t - E'_n)/kT\right]}{K_n n'_1 + r_n}. \tag{5.104}$$

By substituting this value into (5.100), we obtain

$$U_n = \frac{r_n}{r_n + K_n n'_1} K_n[n^*_s p^*_t - n_1 n^*_t], \tag{5.105}$$

where n_1 is the emission constant for electrons from the occupied centre in the ground state (given by means of the energy E_t).

In a completely analogous manner, we obtain the expression for the rate of hole flow into the surface states:

$$U_p = \frac{r_p}{r_p + K_p p'_1} K_p[p^*_s n^*_t - p_1 p^*_t]. \tag{5.106}$$

The emission constants p_1 and p'_1 for holes from an unoccupied centre in the ground and the excited states are obtained from (2.172) by replacing E^f_t by E_t and E'_p, respectively.

Ch. 5, § 8.2   COMPLEX MODELS OF SURFACE RECOMBINATION   207

Since the lifetime of the excited states is short, we can assume (as in the thermal equilibrium case) that their density is negligible compared to that of the centres in the ground state. Thus $n_t^* + p_t^* \approx N_t$. We see that to this approximation (5.105) and (5.106) are identical to (5.71) and (5.72) with K_n and K_p replaced by K_n' and K_p', where the latter are given by

$$K_n' = \frac{r_n}{r_n + K_n n_1'} K_n; \qquad K_p' = \frac{r_p}{r_p + K_p p_1'} K_p. \qquad (5.107)$$

The physical significance of K_n' and K_p' can be understood from the following considerations. Take, for example, the case of electron capture by a vacant centre. The constant r_n represents the probability that an electron in an excited state drop to the ground state and thus be completely captured by the centre. The product $K_n n_1'$, on the other hand, represents the probability that an electron in an excited state be re-emitted into the conduction band before the centre is able to relax to its ground state. These two processes compete with each other, the former increasing and the latter decreasing the efficiency of electron capture. Thus the ratio $r_n/(r_n + K_n n_1')$ expresses the probability that an electron dropping into an excited state be permanently captured (by virtue of the centre relaxing to its ground state) and not be re-emitted back into the conduction band. The larger this ratio the more efficient is the electron capture process, and in the limit $K_n n_1' \ll r_n$ every electron dropping into an excited state reaches the ground state ($K_n' = K_n$). Since r_n and r_p are presumably of the order of 10^8 sec^{-1} or more, we see that the reemission process is important only for those excited levels close to the energy-band edges (large values of n_1' and p_1').

Using (5.107) we obtain, in a manner completely analogous to that in the preceding section, the expression for the surface recombination velocity

$$s = \frac{\sqrt{K_n' K_p'} N_t (n_b + p_b)}{2 n_i \{\cosh[(E_t - E_i)/kT - u_0'] + \cosh(u_s - u_0')\}}, \qquad (5.108)$$

where u_0' is given by

$$u_0' = \ln \sqrt{K_p'/K_n'} = u_0 + \ln \sqrt{\frac{r_p(r_n + K_n n_1')}{r_n(r_p + K_p p_1')}}. \qquad (5.109)$$

It is seen that (5.108) is identical to (5.83), except that here K_n, K_p, u_0 are replaced by their primed values. An important conclusion to be drawn from (5.108) and (5.109) is that the position of $s_{\max}$ may now vary with temperature. Moreover, the temperature dependence of $s_{\max}$ is no longer governed solely by $E_c - E_t$ (or $E_t - E_v$).

References

[1] I. E. Tamm, Z. Physik **76** (1932) 849; Physik. Z. Sowjetunion **1** (1932) 733.
[2] J. Koutecký, J. Phys. Chem. Solids **14** (1960) 233. (Review)
[3] W. Shockley, Phys. Rev. **56** (1939) 317.
[4] J. Koutecký, Z. Elektrochem. **60** (1956) 835; Collection Czechoslov. chem. Commun. **22** (1957) 669, 683; J. Koutecký and A. Fingerland, *ibid.* **25** (1960) 1; J. Koutecký, Phys. Status solidi **1** (1961) 554.
[5] T. B. Grimley, Proc. phys. Soc. London **72** (1958) 103; J. Phys. Chem. Solids **14** (1960) 227.
[6] A. B. Lipmann, Ann. Phys. **2** (1957) 16.
[7] A. W. Maue, Z. Physik **94** (1935) 717.
[8] E. T. Goodwin, Proc. Cambridge phil. Soc. **35** (1939) 205.
[9] H. Statz, Z. Naturforsch. **5a** (1950) 534.
[10] E. Antončík, *Proc. Int. Conf. Semiconductor Physics, Prague,* 1960 (Czechoslovak Academy of Sciences, Prague, 1961), p. 491; J. Phys. Chem. Solids **21** (1961) 137. (References)
[11] V. Heine, Proc. phys. Soc. London **81** (1963) 300.
[12] K. Artman, Z. Physik **131** (1952) 244.
[13] J. Koutecký and M. Tomášek, J. Phys. Chem. Solids **14** (1960) 241.
[14] J. Koutecký and M. Tomášek, *Proc. Int. Conf. Semiconductor Physics, Prague,* 1960 (Czechoslovak Academy of Sciences, Prague, 1961), p. 495.
[15] J. Koutecký, Czech. J. Phys. B **11** (1961) 565.
[16] J. Koutecký, Czech. J. Phys. B **12** (1962) 177.
[17] E. T. Goodwin, Proc. Cambridge phil. Soc. **35** (1939) 221, 232.
[18] J. Koutecký, Czech. J. Phys. B **12** (1962) 184.
[19] H. H. Woodbury and W. W. Tyler, Phys. Rev. **105** (1957) 84.
[20] G. K. Wertheim, Phys. Rev. **115** (1959) 37.
[21] S. G. Kalashnikov, *Proc. Int. Conf. Semiconductor Physics, Prague,* 1960 (Czechoslovak Academy of Sciences, Prague, 1961), p. 241. (References)
[22] H. J. Hrostowski in *Semiconductors* (Edited by N. B. Hannay), (Reinhold, New York, 1959), p. 437. (References)
[23] W. Shockley and J. T. Last, Phys. Rev. **107** (1957) 392. (References)
[24] G. G. E. Low, Proc. phys. Soc. London B **69** (1956) 1331.
[25] G. Rupprecht, Phys. Rev. **111** (1958) 75; Ann. N. Y. Acad. Sci. **101** (1963) 960.
[26] W. Shockley and W. T. Read, Phys. Rev. **87** (1952) 835.
[27] W. H. Brattain and J. Bardeen, Bell System tech. J. **32** (1953) 1.
[28] D. T. Stevenson and R. J. Keyes, Physica **20** (1954) 1041.
[29] A. Many, E. Harnik, and Y. Margoninski in *Semiconductor Surface Physics* (Edited by R. H. Kingston), (University of Pennsylvania Press, Philadelphia, 1957), p. 85.
[30] A. V. Rzhanov and T. A. Arkhipova, Fiz. tver. Tela **3** (1961) 1954 [translation: Soviet Phys. – Solid State **3** (1962) 1424].
[31] F. Berz, Proc. phys. Soc. London **71** (1958) 275.
[32] P. T. Landsberg, Proc. phys. Soc. London B **70** (1957) 282.
[33] C.-T. Sah and W. Shockley, Phys. Rev. **109** (1958) 1103.
[34] M. Bernard, J. Electron. Control **5** (1958) 15.
[35] M. Lax, Phys. Rev. **119** (1960) 1502.
[36] A. V. Rzhanov, Fiz. tver. Tela **3** (1961) 3691, 3698 [translations: Soviet Phys. – Solid State **3** (1962) 2680, 2684].

CHAPTER 6

EXPERIMENTAL METHODS I – THE FIELD EFFECT

1. Introduction

In chapters 4 and 5, we developed the phenomenological theories underlying the electronic behaviour of the space-charge region and the surface states. Particular attention was given to processes that can be directly correlated with measurable quantities so as to yield information on the surface parameters. In this and the following chapter we review the various experimental methods employed in studying such processes. We shall be concerned not so much with experimental detail as with the principles and theory of measurement in each method. In some cases the necessary theoretical background, to the extent that it has not been covered in earlier chapters, will also be included.

The electrical properties of real and clean surfaces are usually derived from measurements on regularly shaped homogeneous samples having as large a surface-to-volume ratio as is practical. Most of the electronic phenomena can be detected only for the sample as a whole. If the bulk properties are known, however, the contribution of the bulk can be accounted for and information obtained on the surface proper. Such a separation of bulk and surface effects is achieved most effectively by varying the barrier height at the free surface and following the resulting *change* in the electrical characteristics of the sample. At the same time this procedure enables one to determine the density and energy distribution of the surface states and the parameters characterizing their interaction with the free carriers in the space-charge region.

The equilibrium or quiescent value of the barrier height at the free surface of a given semiconductor sample is determined by the distribution of surface states. This, in turn, is a function of surface preparation as well as of the surrounding ambient. The effect of gaseous ambients on real surfaces was first utilized by Brattain and Bardeen[1] as a means of varying the barrier height in germanium. The so-called Brattain–Bardeen cycle consists of exposing the surface to wet and dry oxygen, wet and dry nitrogen, and ozone in a particular sequence. The wet gases give rise to n-type surfaces (V_s positive) while the ozone and the dry gases result in p-type surfaces. The cycling procedure produces continuous and gradual variation in surface potential over a range of about 0.4–0.5 eV. Many modifications of this gaseous cycle have since been employed, including the use of other gases[2]. For example, ammonia and alcohol vapours were found to produce n-type surfaces, while chlorine resulted in p-type surfaces.

Another method for varying the potential barrier, applicable to both real and clean surfaces, employs the effect of a capacitively applied field normal to the surface[3]. The charge induced by the field is distributed between the surface states and the space-charge region, and a change in the latter is accompanied by a corresponding change in barrier height. This method, in contrast to the gaseous ambient technique, does not affect the surface states, permitting the study of their characteristics as originally possessed by a given surface. The electric field can be produced in a number of ways. The semiconductor sample is usually shaped in the form of a thin rectangular filament with one cross sectional dimension large compared to the other, most of the surface area thus being confined to the two larger faces. A metal electrode, placed parallel to one of these faces and insulated from it by a suitable spacer, forms one plate of a plane parallel capacitor, the semiconductor sample constituting the other. Quite often two metal plates (in electrical contact), each parallel to one of the larger faces, serve as the field electrode. Thin sheets ($\approx 10\ \mu$) of Mylar or mica are used as spacers, and so are single crystals of barium or strontium titanate[4] because of their large dielectric constant. By applying a voltage across such a capacitor, fields as high as 2×10^6 V/cm can in practice be attained at the semiconductor surface. Comparable fields can also be produced by using a thin cylindrical filament surrounded by a large-diameter coaxial metal electrode. Higher applied voltages may be required in this case (due to the practical difficulties encountered in producing filaments of too small a diameter), but the dielectric spacers are dispensed with. This may be an important advantage whenever there is a possibility of spurious results arising from the intimate contact with the insulating medium. The fact that the cylindrical surface inherently consists of many crystallographic planes, however, is a serious limitation. Another method employs an electrolyte as the field electrode[5]. Such an arrangement is feasible only if the electrolyte–semiconductor interface can be made blocking to current flow. For clean surfaces the experimental procedure is more complicated. Here the use of dielectric spacers is inadvisable and, in addition, the field plate has to be manoeuvred into position (in the vacuum system) following the cleaning cycle. In the case of a cleaved surface, one of the severed parts of the sample may serve as the field electrode. In either configuration it is difficult to attain fields higher than about 10^5 V/cm.

The term "field effect" describes the change in sample conductance taking place as a result of a capacitively applied field normal to the surface. This effect has proved to be a powerful tool in studying surface phenomena and will be the subject of the present chapter. We begin by a general discussion of surface conductance and surface capacitance, both of which being intimately associated with the field-effect experiment. The measurement of surface conductance is particularly useful in determining the potential barrier height. Next we consider field-effect measurements under different experimental conditions. Section 4 treats the dc field effect, in which an electrostatic field is applied and the system allowed to reach steady-state conditions. Such measurements yield the overall charge trapped in the various surface states as a function of applied field. The following sections consider the transient behaviour under pulsed and alternating fields. For germanium and silicon, three time scales or frequency ranges have been found useful: (1) dc or low-frequency fields for studying the charge transfer mechanism between the slow states and the underlying space-charge region; (2) higher frequencies which eliminate the effect of slow states while, at the same time, permitting the fast states to maintain equilibrium with the space-charge region; and (3) short pulses and very high frequency fields, where the interaction of the fast states with the space-charge region can be investigated or in some cases even eliminated. The actual frequency at which one of these ranges ends and the other begins depends on the characteristics of the surface states involved. In the last section of the chapter we discuss a different field-effect configuration, one that employs p–n–p or n–p–n transistor structures.

2. Surface conductance

The presence of a potential barrier results in a surface layer — the space-charge region — whose conductance (parallel to the surface) is different from that of a parallel layer of comparable thickness in the underlying bulk. Thus a given semiconductor filament consists in effect of two conductors in parallel, one associated with the fixed bulk carrier densities and the other with the barrier-dependent surface densities. In most cases only *changes* in the conductance of the space-charge layer corresponding to different barrier heights can be measured directly. This experimental limitation suggests the feasibility of expressing the surface conductance in terms of these changes rather than in absolute terms. Conditions at flat bands ($V_s = 0$), where the excess surface-carrier densities are zero, are very convenient as a reference point. Thus, the surface conductance $\Delta\sigma$ at any barrier height V_s is defined as the change in filament conductance per square area of its surface resulting from a change in barrier height from 0 to V_s. For a homogeneous filament of uniform cross section, we have

$$\Delta\sigma = \frac{l^2}{A}\left(\frac{1}{R} - \frac{1}{R_0}\right). \tag{6.1}$$

Here A is the total surface area parallel to the direction of current flow, l is the filament length, R is the filament resistance, and R_0 is the value of R at $V_s = 0$. If we assume that the carrier mobilities in the space-charge layer are the same as those in the bulk then evidently

$$\Delta\sigma = q(\mu_n \Delta N + \mu_p \Delta P), \tag{6.2}$$

where ΔN and ΔP are the excess surface-carrier densities as defined in Ch. 4, § 3.1. Note that the dimensions of $\Delta\sigma$ are those of conductance (mhos) and do not involve length or area, the conductance associated with a square area of surface being independent of the size of the square.

The dependence of surface conductance on barrier height can be obtained from that of ΔN and ΔP calculated in Ch. 4, § 3. Conversely, if the surface conductance can be measured, it may be used to derive the corresponding value of barrier height[6-8]. It is apparent that both accumulation and inversion layers are characterized by high conductances. In accumulation layers this is due to the large number of majority carriers, in inversion layers to the large number of minority carriers. The surface conductance is less in depletion layers and passes through a minimum value $\Delta\sigma_{min}$ where very few mobile carriers are present in the space-charge region. The exact value V_{sm} of the barrier height at which the minimum occurs can be evaluated by differentiating (6.2). Substituting for $\Delta N, \Delta P$ from (4.37) and (4.38), we obtain

$$v_{sm} \equiv qV_{sm}/kT \approx -2u_b - \ln(\mu_n/\mu_p). \tag{6.3}$$

In Fig. 6.1, $\Delta\sigma$ is plotted as a function of the dimensionless barrier height v_s for an n- and a p-type germanium sample at 300°K. The impurity concentrations were chosen so as to yield bulk potentials of $u_b = +2$ and

$u_b = -2$, respectively. The dashed curves have been obtained from eq. (6.2) and Figs. 4.11, 4.12, using the normal (bulk) mobilities. The solid curves have been calculated employing surface mobility values, as explained further on. In the right-hand branches of the curves the surface conductance arises mainly from electrons, in the left-hand branches from holes. Only in the vicinity of the minimum does $\Delta\sigma$ consist of contributions from both electrons (whose density increases with v_s) and holes (whose density decreases with increasing v_s). The value $\Delta\sigma = 0$ is obtained twice: once for the case of flat bands ($v_s = 0$) where $\Delta\sigma = 0$ by definition, and again on the other side of the minimum where the minority carrier conductance just cancels the negative contribution of the depleted majority carriers.

Actually, the bulk mobilities μ_n, μ_p in (6.2) should be replaced by lower values because of the scattering by the surface of carriers moving in its vicinity. Theoretical estimates are available for the appropriate surface

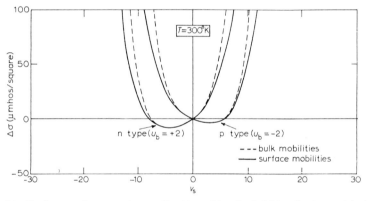

Fig. 6.1. Surface conductance $\Delta\sigma$ as a function of barrier height v_s for two extrinsic germanium samples ($N_D - N_A \approx \pm 2 \times 10^{14}$ cm^{-3}) at 300°K. The dashed curves correspond to bulk mobilities, the solid curves to surface mobilities.

mobilities to be used in this case, as discussed in detail in chapter 8. There is some uncertainty, however, regarding the exact nature of the surface scattering mechanism. The solid curves in Fig. 6.1 have been calculated on the basis of the diffuse scattering model (Figs. 8.2, 8.4–8.6 in Ch. 8, § 3), which assumes that each carrier scattered by the surface emerges with a completely random velocity. As such these curves represent the lowest limit to the surface conductance, the actual values lying somewhere in between the solid and dashed curves. It is seen that the surface mobilities depart significantly from the corresponding bulk values only in strong accumulation and inversion layers (large $|v_s|$), where the carriers in the space-charge region are constrained to move in deep potential wells. In these regions, however, the curves are steep and no appreciable error in the derivation of v_s from $\Delta\sigma$ is introduced by the uncertainty in surface mobility.

The situation is less satisfactory at lower temperatures, where the reduction in mobility by surface scattering is more pronounced. Figure 6.2 shows curves of $\Delta\sigma$ at 200°K for the two extrinsic samples of Fig. 6.1. It is seen that the values of v_s obtained from the solid curves (diffuse surface scattering) now depart more markedly from those derived from the dashed curves (no surface scattering). Another difficulty arises at low temperatures from the large extent of the depletion range ($0 \leq |v_s| \leq |v_{sm}|$). In this range $\Delta\sigma$ varies very slowly with v_s and it is inherently difficult to determine the

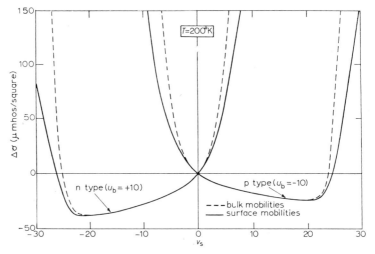

Fig. 6.2. Surface conductance $\Delta\sigma$ as a function of barrier height v_s for the two samples of Fig. 6.1, but now at 200°K.

barrier height by conductance measurements or, in fact, by any method involving ΔN and ΔP.

In order to determine the surface conductance, the filament is provided with two non-rectifying end contacts and its resistance is measured while the potential barrier is varied by external means. Obviously the maximum in filament resistance corresponds to the minimum in surface conductance $\Delta\sigma_{min}$. As can be seen from (6.3), the value v_{sm} for which this occurs is a unique (and known) function of temperature and impurity concentration.

The maximum in resistance can therefore be employed, in conjunction with the appropriate plot of $\Delta\sigma$ versus v_s, to correlate the filament resistance with the corresponding values of barrier height. By assuming that only the two large surfaces (of width w) contribute to $\Delta\sigma$, we obtain from (6.1) that

$$\Delta\sigma - \Delta\sigma_{\min} = \frac{l}{2w}\left(\frac{1}{R} - \frac{1}{R_M}\right), \qquad (6.4)$$

where R_M denotes the maximum value of the filament resistance R.

A convenient procedure for determining R_M consists in varying v_s *monotonically* (by an external field or by a gaseous ambient cycle) while the filament resistance is continuously monitored. The resistance first increases, as R_M is approached, and then decreases as v_s is changed further. It should be stressed that the usefulness of surface conductance as a tool for determining the barrier height hinges on the possibility of attaining $\Delta\sigma_{\min}$. In germanium the use of either an external field or an ambient cycle is generally sufficient by itself to provide the necessary range of variation in v_s. In some cases, particularly for silicon, the combined action of both gases and fields may be required, while in others even this procedure is of no avail. As we have seen in Ch. 5, § 4.2, the ease with which V_s can be varied depends on the density and distribution of surface states and these, in turn, are intimately related to surface treatment.

An interesting, if somewhat impractical method of determining the surface conductance directly has been described by Rubinstein and Fistul[9]. This consists of two-probe conductance measurements made on different sections of a wedge-shaped sample having a small subtended angle ($< 7°$). The conductance of the bulk is then eliminated by extrapolating the readings to zero wedge thickness, the limiting value representing the surface contribution only.

We have tacitly assumed above that the measurement of filament conductance represents the true parallel combination of bulk and surface conductances. For this to be the case, at least one of two requirements must be satisfied: either the metal contacts are non-rectifying towards the surface layer as well, or the conductance (perpendicular to the surface) between the surface and the underlying bulk is sufficiently large. While both requirements are evidently met when an accumulation layer exists at the surface, they may well be violated under inversion-layer conditions. Because the contacts were chosen so as to be non-rectifying to the bulk and thus capable of handling the *majority*-carrier flow, they will generally be unable to supply the large *minority* current flowing in the inversion layer. Under these conditions, the only way in which the inversion layer can contribute its share to the measured filament conductance is by maintaining good communication with the underlying bulk.

In order to get an idea as to the degree of communication necessary between surface and bulk and the conditions required for such communication, we shall represent the filament as being composed of two conductors, one corresponding to the (n-type) bulk section and the other to the inversion (p-type) surface layer, as shown in Fig. 6.3a. The end contacts are taken as non-rectifying to the bulk section. With the polarity shown, the contact to the p layer on the left will be in the forward direction, allowing

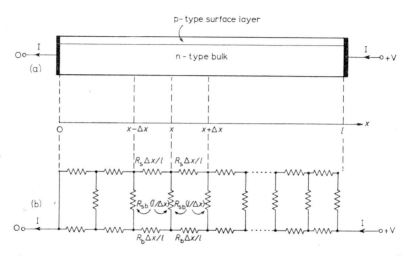

Fig. 6.3. Lumped-circuit representation of an inversion layer and its underlying bulk.

the flow of holes out of the filament. At this point ($x = 0$) the surface and bulk layers will therefore be at the same potential (zero). The contact at $x = l$ will be rectifying, blocking the flow of holes into the filament. For applied voltages small compared to kT/q, the filament can be represented to a good approximation by the equivalent lumped circuit shown in Fig. 6.3b (with $\Delta x \to 0$); R_b and R_s denote the bulk and surface resistances, respectively, while R_{sb} represents the resistance of the p–n junction existing between the bulk and the surface. The use of R_{sb} to describe the communication between the n and p regions is justified for the low applied voltages employed. Under these conditions, the current–voltage characteristics of the p–n junction are essentially linear, as can be seen from (2.209). Expanding the exponential function into a power series and neglecting second- and higher-order terms, we obtain for the current density (perpendicular to the surface) at any point x along the filament

$$J_{sb}(x) = (J_{ps}+J_{ns})q(Y_s-Y_b)/kT, \qquad (6.5)$$

where Y_s and Y_b are the potentials at the position x in the surface and the bulk layer, respectively. Thus

$$R_{sb} = \frac{kT}{qA(J_{ps}+J_{ns})}, \qquad (6.6)$$

where A is the surface area.

We now proceed to calculate the input impedance $R_i = V/I$ as measured by an external circuit. For any junction point x on the upper row of resistors we have, using Kirchhoff's law,

$$\frac{Y_s(x+\Delta x)-Y_s(x)}{R_s\Delta x/l} + \frac{Y_s(x-\Delta x)-Y_s(x)}{R_s\Delta x/l} + \frac{Y_b(x)-Y_s(x)}{R_{sb}l/\Delta x} = 0.$$

In the limit ($\Delta x \to 0$),

$$\frac{l}{R_s}\frac{d^2Y_s}{dx^2} = -\frac{1}{R_{sb}l}(Y_b-Y_s), \qquad (6.7)$$

and similarly for a junction point on the lower row of resistors

$$\frac{l}{R_b}\frac{d^2Y_b}{dx^2} = +\frac{1}{R_{sb}l}(Y_b-Y_s). \qquad (6.8)$$

Equations (6.7), (6.8) are two simultaneous differential equations. The appropriate boundary conditions are

$$Y_b(0) = Y_s(0) = 0, \qquad (6.9)$$

$$Y_b(l) = V, \qquad (6.10)$$

and

$$I = \frac{l}{R_b}\frac{dY_b}{dx}\bigg|_{x=l}. \qquad (6.11)$$

(The last condition follows from the fact that $(dY_s/dx$ at $x = l$ must vanish in the limit of $\Delta x \to 0$.)

The solution of (6.7) and (6.8) subject to these boundary conditions is

$$Y_b(x) = \frac{Rx}{R_i l} + \left(1+\frac{R}{R_i}\right)\frac{\sinh(\eta x/l)}{\sinh \eta} \qquad (6.12)$$

and

$$Y_s(x) = -\frac{R_s}{R_b} Y_b(x) + R_s I \frac{x}{l}, \tag{6.13}$$

where

$$\eta = \left(\frac{R_b + R_s}{R_{sb}}\right)^{\frac{1}{2}} \approx \left(\frac{R_s}{R_{sb}}\right)^{\frac{1}{2}}, \tag{6.14}$$

$$R = \frac{R_b R_s}{R_b + R_s}, \tag{6.15}$$

and

$$R_i = \frac{V}{I} = R\left(1 - \frac{1}{\eta}\tanh\eta\right) + R_b \frac{1}{\eta}\tanh\eta. \tag{6.16}$$

For $\eta \gg 1$, R_i reduces to R, the parallel combination of the surface and bulk resistances. In this case the measurement of the overall filament resistance yields the correct value of the surface conductance. For $\eta < 1$, however, R_i tends to R_b and the surface has little or no effect on the measured resistance.

We shall consider strong inversion layers (small R_s) where poor communication between bulk and surface (large R_{sb}) is most serious. Here R_0 in (6.1) may be replaced to a good approximation by the bulk resistance R_b, so that

$$\Delta\sigma \approx (l^2/A)(1/R_s). \tag{6.17}$$

The value of the surface conductance as deduced from the *measured* resistance, on the other hand, is given by

$$\Delta\sigma_i = (l^2/A)(1/R_i - 1/R_b). \tag{6.18}$$

Using (6.16), we obtain for the relative error in $\Delta\sigma_i$

$$\varepsilon \equiv \frac{\Delta\sigma - \Delta\sigma_i}{\Delta\sigma_i} \approx \frac{(1/\eta)\tanh\eta}{1 - (1/\eta)\tanh\eta}. \tag{6.19}$$

This function is plotted in Fig. 6.4 against $\eta^2 \equiv (R_s + R_b)/R_{sb}$. It is seen, for example, that for the error to be less than 10%, η^2 must be greater than 10^2.

For the strong inversion layers being considered, the electron surface density n_s is very small compared to the bulk hole density p_b, so that the surface-to-bulk electron current (J_{ns}) is usually negligible compared to the hole current (J_{ps}). Substituting for J_{ps} from (2.210) into (6.6) and using the Einstein relation (eq. (2.185)), we have

$$R_{sb} = \frac{1}{A} \frac{\sqrt{D_p \tau_p}}{q\mu_p p_b}. \tag{6.20}$$

The bulk-to-surface resistance is thus directly proportional to the diffusion length and inversely proportional to the bulk minority carrier density. By

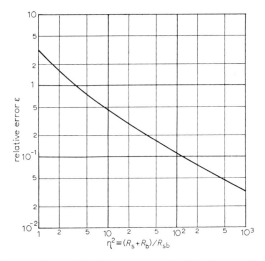

Fig. 6.4. Relative error incurred in the measurement of surface conductance in the presence of an inversion layer as a function of the degree of communication between surface and bulk (eq. (6.19)).

the use of (2.91) and (6.17), we can now express η^2 in the form

$$\eta^2 \approx \frac{R_s}{R_{sb}} \approx \frac{\mu_p \mu_n}{(\mu_p + \mu_n)^2} \frac{l^2}{\sqrt{D_p \tau_p}} \frac{\rho}{\varDelta\sigma\rho_i^2}, \tag{6.21}$$

where ρ is the resistivity of the n-type filament (assumed extrinsic) and $\rho_i \equiv 1/[q(\mu_p+\mu_n)n_i]$ is the intrinsic resistivity. The diffusion length is of the order of 0.1 cm or less, while the surface conductance is typically 100 μmho (see Figs. 6.1, 6.2). Substituting for ρ_i, μ_p, μ_n from Tables 2.2, 2.3 (pp. 57, 71), we obtain (taking $l = 1$cm) that for n-type germanium at room temperature, $\eta^2 > 10^2$ down to a resistivity of at least 10 ohm-cm, corresponding to an error of less than 10% in the measured surface conductance. The error in the deduced value of barrier height is considerably smaller, due to the steep variation of $\Delta\sigma$ with v_s in strong inversion layers. At lower temperatures, however, the error increases very rapidly since the intrinsic carrier density n_i decreases exponentially with decreasing temperature. In silicon n_i is very small even at room temperature, and the surface conductance can only be measured at temperatures higher than about 400°K.

Equation (6.21), combined with Fig. 6.4, may serve as a guide for estimating the accuracy in surface conductance measurements for inversion layers. The general conclusion to be drawn from the analysis presented here is that except for germanium at room temperature the method described above is usually inadequate for measuring surface conductance in inversion layers. Other, more suitable methods for this range employ the pulsed field effect and p–n–p or n–p–n structures, as will be described in §§ 6, 8 below.

3. Surface capacitance

The surface capacitance[10,11] is a convenient quantity by means of which the change in the overall surface charge can be correlated with the corresponding change in potential barrier height. Consider, for example, a semiconductor where no potential barrier exists at the surface. (Such a condition may arise from a special distribution of surface states or, more commonly, from an external biasing field of the proper magnitude.) If now a potential barrier is set up by varying the external field, a space-charge layer will form and, in addition, a change may take place in the charge stored in surface states. In Ch. 4, § 2.2 we defined the *space-charge capacitance* C_{sc} as the ratio of the space-charge density (per unit surface area) Q_{sc} to the barrier height V_s. In an analogous manner we can introduce the surface-state capacitance C_{ss} as

$$C_{ss} \equiv |\Delta Q_{ss}/V_s|, \qquad (6.22)$$

where ΔQ_{ss} represents the *change* in surface-state charge density brought about by changing the barrier height from 0 to V_s. The *surface capacitance* is defined as the ratio of the *total* change in surface charge density ΔQ_s to the barrier height V_s, and is given by

$$C_s \equiv |\Delta Q_s/V_s| = |(Q_{sc}+\Delta Q_{ss})/V_s| = C_{sc}+C_{ss}. \qquad (6.23)$$

The space-charge capacitance C_{sc} is a unique function of V_s for a given impurity concentration and temperature. It can be calculated from eqs.

(4.22), (4.23) or from Figs. 4.7, 4.9. The surface-state capacitance C_{ss}, on the other hand, depends on the particular distribution of surface states present on the surface.

In order to measure the surface capacitance C_s use can be made of the field-effect configuration described in § 1: a plane-parallel capacitor is formed between the semiconductor sample and a metal plate separated from the free surface by a thin insulating layer (Fig. 6.5a). The insulating medium

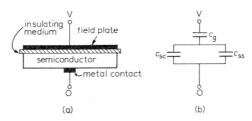

Fig. 6.5. Field plate–semiconductor capacitance.
(a) Schematic representation of experimental configuration.
(b) Equivalent capacitance circuit.

ensures that no conduction current flows between the metal and the semiconductor. Suppose that a change V_0 in the applied voltage across the capacitor is necessary in order to change the barrier height from 0 to V_s. In this process the overall surface charge is changed by ΔQ_s (and the charge on the metal plate by $-\Delta Q_s$). The change in potential drop across the insulating medium is evidently $V_0 - V_s$ and can be expressed by the relation

$$V_0 - V_s = -\Delta Q_s/C_g, \qquad (6.24)$$

where C_g is the geometric capacitance (per unit area) between the metal electrode and the surface proper — that is, the capacitance that would be measured were the semiconductor replaced by a metal. Combining (6.23) and (6.24), we obtain

$$\frac{1}{C_0} \equiv \left|\frac{V_0}{\Delta Q_s}\right| = \frac{1}{C_g} + \frac{1}{C_s} = \frac{1}{C_g} + \frac{1}{C_{sc} + C_{ss}}. \qquad (6.25)$$

Thus the effective surface capacitance C_0, as measured by an external circuit, consists of the geometric capacitance in series with the parallel combination of the space-charge and surface-state capacitances, as illustrated in Fig. 6.5b.

In practice, it is generally easier to measure the differential capacitance

$c_0 \equiv |dQ_s/dV_0|$. A small ac signal is superimposed on the dc bias voltage and the capacitance measured as a function of bias voltage. The frequency range is chosen so as to maintain equilibrium between the space-charge region and the particular surface states being studied (see §§ 4 and 5 below). Under these conditions one obtains, similarly to (6.25),

$$\frac{1}{c_0} = \frac{1}{C_g} + \frac{1}{c_s} = \frac{1}{C_g} + \frac{1}{c_{sc}+c_{ss}}, \qquad (6.26)$$

where C_g is the geometric capacitance as before, and c_{sc} and c_{ss} are the *differential* space-charge and surface-state capacitances, respectively:

$$\begin{aligned} c_{sc} &\equiv |dQ_{sc}/dV_s| = (q/kT)|dQ_{sc}/dv_s|; \\ c_{ss} &\equiv |dQ_{ss}/dV_s| = (q/kT)|dQ_{ss}/dv_s|. \end{aligned} \qquad (6.27)$$

For the case of fully ionized impurities and single-charge surface states (of density N_t and effective energy E_t^f), we obtain (see eq. (5.48))

$$c_{sc} = (\kappa\varepsilon_0/L)|dF_s/dv_s|; \qquad c_{ss} = (q^2 N_t/kT)|df(E_t^f)/dv_s|. \qquad (6.28)$$

A study of the characteristics of the functions F_s and $f(E_t^f)$ reveals that c_{sc} attains a minimum value in depletion layers and increases rapidly in both accumulation and inversion layers; c_{ss}, on the other hand, is maximum when the Fermi level at the surface crosses E_t^f and decreases rapidly on either side. This is illustrated in Fig. 6.6 for a typical n-type germanium sample ($u_b = 2$, $n_b + p_b \approx 2 \times 10^{14}$ cm^{-3}). The surface states are taken to be present with a density N_t of 6.5×10^{11} cm^{-2} and at an effective energy $4kT$ above the mid-gap.

If one can determine the differential surface capacitance $c_s \equiv c_{sc} + c_{ss}$ as a function of barrier height v_s, one can obtain substantial information concerning the surface-state structure. The experimental data are then compared with the known theoretical curve of c_{sc}, and the density and energy distribution of the surface states are derived from the deviations between the theoretical and experimental curves. This method is sensitive only in the regions where c_{sc} is not too large — that is, in depletion and weak accumulation and inversion layers. The principal difficulty in such measurements is the determination of c_s with sufficient accuracy. The surface capacitance is usually of the order of tenths of a microfarad per square centimetre, and in order to achieve the necessary sensitivity the geometric capacitance should not be too small by comparison (see eq. (6.26)). The ordinary field-effect configuration, which employs a Mylar spacer or (in the case of clean surfaces) a vacuum gap, is inadequate since the geometric capacitance in such structures cannot in practice exceed a few hundred picofarads per square centimetre, a value that is much smaller than c_s. In the case of real surfaces it is sometimes possible to increase the geometric capacitance considerably by evaporating a very thin insulating layer, or by growing an oxide film, on the surface and then evaporating a metal electrode on top. This procedure is not easy, however, and unless

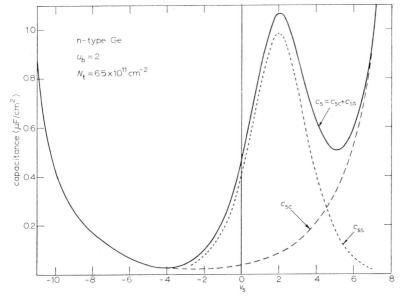

Fig. 6.6. Space-charge (c_{sc}), surface-state (c_{ss}), and surface (c_s) differential capacitances as a function of potential barrier for a typical n-type germanium sample ($u_b = 2$, $n_b + p_b \approx 2 \times 10^{14}$ cm^{-3}, $L \approx 3.4 \times 10^{-5}$ cm) at room temperature. The surface states are present with a density of 6.5×10^{11} cm^{-2} and at an effective energy $4kT$ above the mid-gap.

special precautions are taken the layers contain pinholes which form conducting channels between the metal and the semiconductor. Evaporated layers have been successfully applied on cadmium sulphide films[12], while on silicon a thermally grown oxide has proved very effective[13]. In both cases insulating layers as thin as $0.1 - 1$ μ having high dielectric strength were obtained.

The semiconductor–electrolyte system[5] is perhaps the most suitable one for surface capacitance measurements. The electronic processes taking place at the interface of such a system have been reviewed by Dewald[14] and more recently by Harten[15]. In certain cases a potential barrier at the electrolyte side of the interface limits the flow of majority carriers (and sometimes of minority carriers as well) into the semiconductor (see Ch. 9, § 2), such that under the proper polarity the semiconductor–electrolyte contact is blocking to a good approximation. At the same time, for not too dilute electrolytes the "geometric capacitance" associated with the blocking layer is very large (10–100 $\mu F/cm^2$), so that the measured capacitance is in effect equal to the semiconductor surface capacitance $c_{sc}+c_{ss}$. This in turn results in almost the entire voltage drop between the semiconductor and the electrolyte occurring across the space-charge layer of the former. Thus the changes in barrier height are given directly by the values of the applied voltage. In practice, one measures the potential difference between the semiconductor and a standard electrode immersed in the electrolyte while the barrier height is being varied by passing a small current between the semiconductor

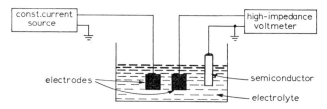

Fig. 6.7. Field-effect configuration employing a semiconductor–electrolyte system.

and another electrode. The experimental arrangement used for such measurements is illustrated in Fig. 6.7. A constant-current source is connected to the metal electrode on the left-hand side of the diagram and the voltage drop between the electrolyte and the semiconductor sample is measured by means of a high-impedance voltmeter attached to the metal electrode on the right. The voltage changes brought about by varying the current yield in most cases the changes in barrier height V_s, except for the small ohmic drop across the electrolyte and the semiconductor bulk. The better the blocking characteristics of the interface, the lower the current needed to produce a given change in barrier height. The differential surface capacitance c_s can be conveniently measured by using a third electrode (not shown in the figure) connected to a small ac or pulsed voltage source[16,17]. In the former case, a conventional ac bridge is used while in the latter, the capacitance is derived from the transient current response (see also § 6.1 below). In order to determine the absolute magnitude of the barrier height, a reference point is necessary. This is provided by the minimum value of the measured capacitance (neglecting the surface-state contribution to c_s), by surface-photovoltage measurements (see Ch. 7, § 2), or by surface conductance data[18]. Even though the last procedure may seem the most direct, some uncertainty is present because of possible electrolytic conduction parallel to the surface[19].

4. Dc field effect

The dc field effect consists of the steady-state changes in surface conductance induced by electrostatic fields. For each applied field the system is allowed to reach equilibrium and the filament resistance is measured when no further change in its value is detectable. The measuring voltage across the filament is kept small compared to that applied at the field plate so as to maintain the entire surface at effectively the same potential. As has been pointed out in the preceding section, the geometric capacitance C_g is usually much smaller than the surface capacitance C_s. The total charge density Q_s induced at the semiconductor surface is then practically independent of the barrier height and is given by $Q_s = C_g V_p$, where V_p is the voltage applied at the field plate. The induced charge is distributed between the space-charge region and the surface states, giving rise to changes δQ_{sc} and δQ_{ss} in the free and trapped charge densities, respectively:

$$Q_s = \delta Q_{sc} + \delta Q_{ss}. \tag{6.29}$$

The change in resistance that one measures results almost entirely from δQ_{sc}, the mobility of the carriers in the surface states being normally orders of magnitude lower than that of the free carriers in the space-charge region. If the range of variation in filament resistance can be made to include the maximum value R_M, then the barrier height V_s can be determined for each value of applied field. Since for a given semiconductor (and temperature) δQ_{sc} is a known function of V_s (eqs. (4.16), (4.19)), the trapped charge δQ_{ss} can be evaluated from (6.29) as a function of V_s. By comparing these results with the theoretical expression for the occupation statistics (see Ch. 5, §§ 4, 5), one can determine, at least in principle, the density and energy distribution of the various surface states[20]. This procedure is straightforward only if the distribution of the surface states is discrete and if the dominant sets of states are widely separated in energy. Otherwise the comparison does not yield the distribution in a unique manner and additional information (derived from measurements discussed in chapter 7) becomes necessary.

It should be noted that in the case being considered — dc fields and steady-state conditions — all the surface states participate, whatever their time constants. Thus on real surfaces of germanium (and often of silicon as well), charge is trapped in both the fast and the slow states. The slow states, however, are usually present with a much higher density than the fast states, so that they will completely dominate the trapping process. Typically, the slow-state density is at least 10^{13} cm^{-2}, a value that is sufficient to anchor the barrier height against any significant variation, even under the highest fields feasible. As a result, the steady-state field effect is very small and in practice it is difficult to determine the slow-state density apart from setting a lower limit to its value. The charge transfer mechanism between the bulk and the slow states can be conveniently studied by observing the *transient* response of the surface conductance[21]. The field is applied in the form of a step function, and the filament conductance is recorded as it relaxes towards its steady-state value (which, as has just been pointed out, is very close to the initial, field-free value).

Care should be taken in these measurements to maintain the filament at a constant temperature. The changes in surface conductance are typically of the order of 1 % of the overall filament conductance. In extrinsic germanium and silicon, a comparable change in bulk conductance results from a drift in temperature of 1°C (see Table 2.3, p. 71). This problem is particularly serious in dc field-effect measurements since these take a considerable time to carry out. Another source of error is introduced[22] by the application of the field itself if the ambient temperature is different from that of the sample (as is often the case in measurements above or below room temperature). The electrostatic forces brought into play by the action of the field press the field plate more tightly against the filament, thereby enhancing the thermal contact between the latter and the surrounding ambient. The resulting change in sample temperature can give rise to a relaxation in bulk conductance which may well be comparable (both in magnitude and in time constant) to that in the surface conductance.

5. Low-frequency field effect

It is evident from the above considerations that the use of dc fields under steady-state conditions is not suitable as a means of varying the potential barrier in germanium and silicon when a large density of slow states is present on the surface. Fortunately, the long time constants associated with the slow states make it easy to eliminate their effect, leaving only the fast states as active participants in the charge trapping process. One way of achieving such a separation consists in measuring the surface conductance (as well as other surface characteristics) a short time following the onset of an applied dc field, before any appreciable charge can be trapped in or released from the slow states [23,24]. This procedure is very useful at low temperatures where the time constant of the slow states is long (hours). At room temperature, alternating fields are more convenient [4,20,25]. Since the time constants of the slow and fast states are usually orders of magnitude apart, a considerable frequency range is available in which the slow states are inoperative while the fast states are in complete equilibrium with the space-charge region. In germanium at room temperature, for example, this range extends from a fraction of a cycle to several kilocycles per second, and it is possible to swing the potential barrier over practically the entire energy gap. The filament resistance is then measured as a function of the ac voltage applied to the field plate, and the density and energy distribution of the fast states are derived as before from a comparison between the experimental $\Delta\sigma$ versus v_s curve and the theoretical expression for the occupation statistics.

Several methods have been used to measure the ac field effect. In one arrangement [4,20] a dc voltage is applied across the filament and a series resistor. The signal across the resistor due to the modulation of the filament conductance is amplified and displayed directly on a cathode ray oscilloscope (CRO) whose sweep is driven by a portion of the ac voltage applied to the field plate. This method is very convenient since the whole field effect curve is displayed at once. Provisions should be made, however, especially at higher frequencies, to balance out the displacement current through the plate–semiconductor capacitor. The spurious signal produced in this manner across the series resistor may not be negligible compared to the field-effect signal, even though the measurement is carried out at frequencies well below the reciprocal of the charge relaxation time (see § 6.1 below). The reason for this is that a relatively large ac plate voltage produces only a small field-effect signal. A simple and effective circuit [26] for minimizing the contribution of the displacement current is shown in Fig. 6.8. With the dc measuring voltage shorted out, the resistor R_1 is adjusted until the signal on the CRO is at a minimum. Under these conditions the displacement current is distributed evenly across the two input terminals of the differential amplifier.

When the measuring voltage is now switched on, the trace on the CRO represents very nearly the true field-effect signal. This circuit is also very useful in pulsed field-effect measurements (§ 6 below). Conditions can be improved further by increasing the dc current through the filament so as to reduce the relative contribution of the displacement current. This is limited, however, by contact injection which may take place at higher currents, and by joule heating of the sample.

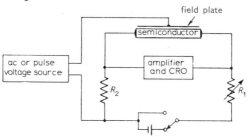

Fig. 6.8. Circuit for minimizing the effect of the displacement current on the ac or pulsed field-effect measurement. (After Low, reference 26.)

A more accurate procedure consists of measuring the filament resistance by a pulse-activated Wheatstone bridge[27]. The use of pulses permits a much higher current through the filament without excessive heating. The main

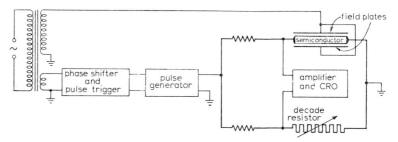

Fig. 6.9. Bridge circuit used for ac field-effect measurements. The pulse activating the bridge is synchronized with either the positive or the negative crest of the ac plate voltage. (After Many and Gerlich, reference 27.)

advantage, however, is that the measurement of filament resistance (taken point by point) can be made at the *onset* of the pulse, before any contact-injection effects (if present) can interfere with the measurement. A schematic diagram of the bridge is shown in Fig. 6.9. The ac voltage applied to the field plates and the triggering signal for the pulse generator are supplied by two secondary windings of the same transformer, the triggering signal passing through a phase shifting network first. The phase shift is adjusted so that the pulse applied to the bridge coincides (in time) with either the positive or the negative crest of the ac voltage. The period of the ac voltage is made sufficiently long so as to ensure a constant field at the surface throughout the duration of the pulse. (This is particularly important in connection with pulsed field-effect and surface recombination velocity measurements, as will be discussed later.) The off-balance pulse signal from the bridge is displayed on a CRO, and the decade resistor in the opposite arm is adjusted until balance is attained. This procedure is repeated at different amplitudes (positive and negative) of the ac voltage, and the filament resistance, as read from the decade resistor, is determined as a function of the field at the surface. The measurement is thus similar to that of the dc field effect, but now the slow states are inactive under the quasi steady-state conditions prevailing at the crest of the ac field.

6. Pulsed field effect

In the low-frequency field effect, the variation of the field occurs on a time scale that is large compared to the time constants associated with the fast states, so that the latter are at all times in equilibrium with the underlying bulk. The primary objective of the pulsed field effect, on the other hand, is to study the transient behaviour of the induced charge and the interaction processes leading up to steady-state conditions[26,28-32]. A fast rise-time pulse is applied to the field plate, and the resulting change in surface conductance is observed on a CRO starting from a point as close to the pulse onset as is permitted by the experimental resolution (normally 1–10 μsec). Some of the processes taking place under these conditions have been analysed in Ch. 5, § 6. Here we shall consider the various experimental techniques based on this analysis and used to obtain the characteristics of the fast surface states and the underlying space-charge region.

6.1. CHARGE RELAXATION TIME

In the analysis of Ch. 5, § 6 we assumed that the charge induced by the field appears in the space-charge region in a time that is short compared to the interaction time between the free carriers and the surface states. Such an assumption leads to a simple interpretation of pulsed field-effect data, and in this subsection we examine the conditions under which it is valid in practice. The charging process involves carrier flow through the bulk, between the end contacts and the space-charge region adjacent to the field electrode. On the basis of simple physical arguments it is expected that the decay time characterizing this displacement current should be of the order of RC_0 where R is the resistance of the bulk region between the contacts and the surface and C_0 the effective capacitance between the field plate and the semiconductor bulk (see § 3). To show this on a more quantitative basis we consider a uniform, extrinsic semiconductor sample having planar geometry. The plane $z = 0$ will represent the free surface adjacent to the field plate, and the plane $z = z_1$ the position of the (non-rectifying) metal contact. To simplify the calculation, we neglect surface states and assume that the quiescent barrier height is zero. The polarity of the pulsed field is taken such as to repel majority carriers from the surface and, in order to eliminate minority-carrier effects, its magnitude is restricted to values corresponding to depletion layers.

The total current density $J_z(t)$ (perpendicular to the surface) at any time t following the onset of the pulse is given by the sum of the conduction and displacement current densities. Neglecting diffusion currents, we may write (for an n-type sample)

$$J_z(t) = q\mu_n n(z, t)\mathscr{E}_z(z, t) + \kappa\varepsilon_0 \frac{\partial \mathscr{E}_z(z, t)}{\partial t}, \qquad (6.30)$$

where $\mathscr{E}_z(z, t)$ and $n(z, t)$ are the position- and time-dependent electric field and electron density, respectively, μ_n is the electron mobility, and κ is the dielectric constant of the semiconductor. In the absence of un-ionized impurities, Poisson's equation becomes

$$\frac{\partial \mathscr{E}_z(z, t)}{\partial z} = \frac{q}{\kappa\varepsilon_0}[n_b - n(z, t)], \qquad (6.31)$$

where n_b is the equilibrium electron density. Elimination of n from (6.31) and substitution into (6.30) gives

$$J_z(t) = q\mu_n n_b \mathscr{E}_z - \tfrac{1}{2}\mu_n \kappa\varepsilon_0 \frac{\partial \mathscr{E}_z^2}{\partial z} + \kappa\varepsilon_0 \frac{\partial \mathscr{E}_z}{\partial t}. \qquad (6.32)$$

Obviously, the total current density $J_z(t)$ is position independent. Hence, integration of (6.32) with respect to z yields

$$J_z(t) = \frac{q\mu_n n_b}{z_1} V_s + \frac{\mu_n \kappa\varepsilon_0}{2z_1}(\mathscr{E}_0^2 - \mathscr{E}_1^2) + \frac{\kappa\varepsilon_0}{z_1}\frac{\partial V_s}{\partial t}, \qquad (6.33)$$

where $\mathscr{E}_0(t)$ and $\mathscr{E}_1(t)$ are the values of the electric field at $z = 0$ and $z = z_1$, respectively, and

$$V_s = \int_0^{z_1} \mathscr{E}_z(z, t) dz. \tag{6.34}$$

Here V_s represents the total voltage drop between the surface and the end contact and is negative under the present conditions (depletion layers).

Consider first the case in which $C_g \gg C_{sc}$. Here the voltage drop across the blocking layer is negligible and V_s is a constant equal to the pulse amplitude. The last term in (6.33) then vanishes, a condition that simplifies the analysis considerably. At the surface ($z = 0$) the conduction current is zero, so that we may also write (see eq. (6.30))

$$J_z(t) = \kappa\varepsilon_0 (d\mathscr{E}_0/dt).$$

Combining this equation with (6.33) we obtain

$$q\mu_n n_b V_s + \tfrac{1}{2}\mu_n \kappa\varepsilon_0(\mathscr{E}_0^2 - \mathscr{E}_1^2) = \kappa\varepsilon_0 z_1 (d\mathscr{E}_0/dt). \tag{6.35}$$

Another differential equation relating $\mathscr{E}_0(t)$ and $\mathscr{E}_1(t)$ may be obtained by equating the current densities at $z = 0$ and $z = z_1$. This, however, is not necessary since, as we shall see shortly, $\mathscr{E}_1^2(t)$ in (6.35) can be neglected to a good approximation with respect to $\mathscr{E}_0^2(t)$.

At the onset of the pulse ($t = 0$), prior to any charge displacement, the field is uniform throughout the sample. Thus

$$\mathscr{E}_0(0) = \mathscr{E}_1(0) = V_s/z_1. \tag{6.36}$$

Thereafter, as more and more (negative) charge leaves the sample through the contact, the voltage drop concentrates in the depletion layer at the surface. At the same time $\mathscr{E}_1(t)$ tends to zero. The steady-state value of $\mathscr{E}_0$ is obtained from (6.35) by equating $d\mathscr{E}_0/dt$ and $\mathscr{E}_1$ to zero:

$$\mathscr{E}_0(\infty) = V_s/L_c, \tag{6.37}$$

where

$$L_c = \sqrt{\frac{\kappa\varepsilon_0 |V_s|}{2qn_b}} = \sqrt{\frac{\kappa\varepsilon_0 kT|v_s|}{2q^2 n_b}}. \tag{6.38}$$

It will be recognized that L_c is just the effective charge distance in strong depletion layers (see eqs. (4.22), (4.26)). For any case of practical interest L_c is much smaller than z_1, so that the absolute magnitude of the final steady-state field at the surface is very large compared to its initial magnitude.

It is also evident that a short time following the onset of the pulse, $|\mathscr{E}_1(t)|$ (which is smaller than $|V_s/z_1|$) may already be neglected compared to $|\mathscr{E}_0(t)|$. Using this approximation and introducing dimensionless variables, we obtain from (6.35) that

$$1 - \alpha^2(t) = 2\tau_c \frac{d\alpha(t)}{dt}, \qquad (6.39)$$

where

$$\alpha(t) = \frac{\mathscr{E}_0(t)}{V_s/L_c} \qquad (6.40)$$

and

$$\tau_c \equiv \tfrac{1}{2} R C_{sc} = \tfrac{1}{2} \frac{z_1}{q\mu_n n_b} \frac{\kappa\varepsilon_0}{L_c}. \qquad (6.41)$$

Here R is the resistance between the metal contact and the surface and C_{sc} is the space-charge capacitance as defined in (4.23), both per unit surface area.

The solution of (6.39), subject to the boundary conditions (6.37), is

$$\alpha(t) = \frac{1 - (1 - 2L_c/z_1)e^{-t/\tau_c}}{1 + (1 - 2L_c/z_1)e^{-t/\tau_c}}. \qquad (6.42)$$

The current density is obtained by differentiation. Neglecting L_c/z_1 with respect to unity, we have

$$J_z(t) = \frac{V_s}{R} \frac{4e^{-t/\tau_c}}{(1 + e^{-t/\tau_c})^2}. \qquad (6.43)$$

It is seen that the displacement-current decay is approximately exponential, the time constant associated with the charging of the space-charge region being given by $\tfrac{1}{2} R C_{sc}$. Physically, the reason for the factor $\tfrac{1}{2}$ follows from the fact that the effective capacitance of the system is a function of time. For small t it equals the capacitance between the field plate and the metal contact ($\kappa\varepsilon_0/z_1$), which is very small in the case under consideration ($z_1 \gg L_c$). Subsequently, as the space-charge region builds up, the capacitance increases towards its steady-state value of $C_{sc}(= \kappa\varepsilon_0/L_c)$. The mean value is thus close to $\tfrac{1}{2} C_{sc}$. If a series resistance is present in the circuit, its value should of course be added to that of the filament resistance.

The field-effect configuration considered thus far ($C_g \gg C_{sc}$) can be realized in practice only in an electrolyte–semiconductor system. Under these conditions the charge relaxation time τ_c may be quite large. For a typical germanium sample, $C_{sc} \approx 0.1 - 1\ \mu\mathrm{F/cm}^2$ and $R \approx 10^2 - 10^3\ \mathrm{ohm \cdot cm}^2$, so that τ_c is of the order of 100 μsec. Such long time constants (and the large current associated with the charging process) permit an easy measurement of the transient response of the system[16,17]. The surface capacitance is derived by analysing this response in terms of the equivalent circuit shown in Fig. 6.10.

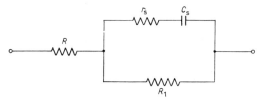

Fig. 6.10. Equivalent circuit representing the electrical behaviour of the semiconductor–electrolyte system.

Here R represents the filament resistance (together with the electrolyte resistance) and R_1 the (much larger) "leakage" resistance of the interface. The surface capacitance C_s now includes the contribution of those surface states that can participate in the charging process within the measurement time. The introduction of the small series resistance r_s is prescribed by the experimental data[17] and probably represents the charge exchange between the surface states and the space-charge region.

In the general case where the geometric capacitance C_g is not large compared to the space-charge capacitance C_{sc}, a similar calculation to that presented above yields for the charge relaxation time

$$\tau_c = RC_0', \tag{6.44}$$

where

$$C_0' = [(1/C_g)^2 + 4(1/C_{sc})^2]^{-\frac{1}{2}}. \tag{6.45}$$

(The effective capacitance C_{eff}' characterizing the relaxation process is nearly but not quite equal to the series combination of C_g and C_{sc}.) It is readily seen that for $C_g \gg C_{sc}$, τ_c reduces to $\tfrac{1}{2}RC_{sc}$ as it should, while if $C_g \ll C_{sc}$, then $\tau_c = RC_g$.

The above calculations are based on planar geometry in which the field plate and the metal contact are parallel and on opposite sides of a semiconductor slab. In most field-effect measurements, on the other hand, the pulse is applied between the field plate and the two end contacts of a thin filament (see, for example, Fig. 6.9). Here the effective resistance through which the charging current flows to the surface (or away from it) is approximately that of the two halves of the filament connected in parallel. Thus the charge relaxation time would be approximately $\frac{1}{4}R_f C_f$, where R_f denotes the (longitudinal) filament resistance and C_f the overall effective capacitance (equal to C_0' multiplied by the surface area of the field plate). Clearly, the product $R_f C_f$ is directly proportional to the square of the filament length, so that a considerable reduction in the relaxation time can be effected by keeping the filament short. As we shall see in Ch. 9, § 2, use is made of this characteristic to attain very high frequency performance in the field-effect transistor.

The charge relaxation time characterizes the decay of the displacement current following the onset of the pulse. As such it represents the time taken by the space-charge layer to charge up. Only after the displacement current has died out does the induced mobile charge attain an equilibrium distribution and the interpretation of the field-effect data in terms of the surface-state parameters become feasible. The charge relaxation time should therefore be kept small compared to the surface-state time constants. Actually, it is sufficient that it be small compared to the experimental resolution time available, since any surface states characterized by shorter time constants are in any case inaccessible to measurement. This condition is readily satisfied in the ordinary field-effect configuration where the Mylar or mica spacer is inherently associated with a relatively small capacitance. Typically, $C_g \approx 100$ pF cm^{-2}, giving rise to a charge relaxation time (in germanium and silicon) of 0.01–0.1 μsec, whereas the experimental resolution is usually 1–10 μsec. When necessary, τ_c can be reduced still further by using short filaments, as has been pointed out above. Decreasing the geometric capacitance and the measuring series resistor is also helpful, but then the field-effect signal is reduced proportionally.

The displacement current poses more of a problem to the actual measurement, since quite a few time constants must elapse before the current decays to the much lower level of the field-effect signal. Compensating circuits such as the one shown in Fig. 6.8 are very useful in minimizing the spurious signal due to the displacement current, but this becomes increasingly more difficult the closer one approaches the onset of the pulse.

6.2. SURFACE-STATE TIME CONSTANTS

In this and the next subsection we consider pulsed field-effect measurements involving majority carriers only. This corresponds to an extrinsic sample in which the potential barrier at the surface is restricted to depletion and accumulation layers. For an n-type sample, a positive potential applied at the field plate increases the electron surface density, and electrons tend to drop into unoccupied surface states. In the reverse polarity, free electrons are rapidly depleted from the space-charge region and are subsequently replaced by emission from occupied surface states. The latter process, being a thermal one, is generally the slower of the two and thus more amenable to experimental study. For a large negative pulse such that the surface free electron density n_{s0} just following the onset of the pulse is vanishingly small, the initial emission process is essentially exponential (see eqs. (5.67), (5.68)).

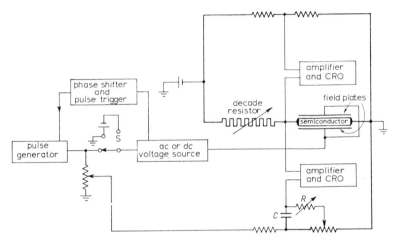

Fig. 6.11. Bridge circuit for determining the relaxation time of the pulsed field effect. (After Harnik et al., reference 31.)

The time constant increases rapidly (exponentially) with decreasing temperature and increasing separation between the conduction-band edge and the surface states. Its measurement can therefore be used to determine the energy position of the surface states. In practice one can measure only the relaxation in filament conductance, but since any decrease in the occupation density n_t of the surface states must be accompanied by an equal increase in the excess surface electron density ΔN (see eq. (5.59)), the observed relaxation in conductance is a direct reflection of the trapped-charge emission process.

A convenient experimental arrangement[31] for measuring the pulsed field effect is shown in Fig. 6.11. The bridge in the upper part of the diagram is used to determine the filament resistance at the onset of the pulse, the

CRO serving as null indicator and also to display the subsequent relaxation in filament conductance. The bridge can be activated by a dc voltage source (as shown) or, when higher sensitivity is required, by a properly synchronized pulse. (The larger signal attainable in the latter case usually eliminates the need for balancing out the displacement current by a circuit of the type shown in Fig. 6.8.) The field plate is connected to a series combination of an ac or dc voltage source and a pulse generator. In the former case the pulse is synchronized with either the positive or the negative crest of the ac voltage, as described previously (Fig. 6.9). At low temperatures, where the time constant of the slow states is very large, dc voltages are more convenient as a means of varying the barrier height. The same pulse is also used (via an attenuator) to activate the RC network shown in the bottom

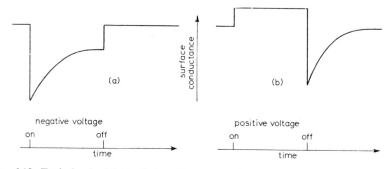

Fig. 6.12. Typical pulsed field-effect oscillograms under accumulation- or depletion-layer conditions at the surface of an n-type sample. The displacement-current spikes at the onset and the termination of the pulse (as well as the fast, experimentally undetectable relaxation associated with the charging of the surface states) have been omitted.

part of the diagram. The exponential signal derived from this network is balanced against the exponential portion of the filament relaxation curve, a CRO again serving as null indicator. Such a detection method is very sensitive to the degree of exponentiality of the relaxation process, and in addition it enables the time constant to be read directly off the variable resistor R. The method is capable of measuring time constants ranging from microseconds to milliseconds. Longer time constants are more easily obtained by analysing the relaxation trace on the CRO or, when in the seconds range, by the use of a mechanical recorder.

Typical pulsed field-effect oscillograms obtained for an n-type sample are illustrated schematically in Fig. 6.12. (The displacement-current spikes at the onset and termination of the pulse have been omitted for simplicity.) For negative pulses (Fig. 6.12a), electrons are expelled from the filament and a depletion layer is formed at the surface. The initial rapid decrease in surface conductance corresponds to the entire induced charge appearing in the form of uncompensated positive donors. Thereafter, electrons are thermally emitted from the surface states into the conduction band and the surface conductance relaxes towards its new steady-state value. When only one set of states is involved, the initial portion of the curve (where n_s is small compared to its final value) is exponential and can be used to determine the emission time constant [29].

Alternatively, the emission time constant can be derived from the relaxation process following the termination of a *positive* pulse [31]. At the onset of the pulse, an accumulation layer is produced and electrons tend to drop into unoccupied surface states. Due to the large value of n_s established in this manner, the charging of the states is very fast and cannot usually be resolved experimentally. The observed trace on the CRO (Fig. 6.12b) is accordingly a horizontal line throughout the pulse duration, corresponding to equilibrium conditions between the trapped and mobile electrons. At the termination of the pulse, the excess trapped electrons initially repel from the surface an equal number of free electrons and the surface conductance drops abruptly. The situation here is equivalent to that following the onset of a negative pulse. The subsequent thermal release of carriers from the surface states gives rise to a relaxation of the surface conductance back to its value prior to the application of the pulse. This procedure has the important advantage that the measurement of the decay constant is carried out under field-free conditions. Moreover, the final equilibrium value of the surface conductance is now known *ab initio*, which is particularly useful when the emission time constant is very long.

Usually, several sets of fast states are involved in the charge emission process. Unless the sets are very close in energy, however, each will dominate the relaxation process in a different temperature range and can be singled out accordingly. In addition, some separation of the sets can be effected at the outset by proper control of their steady-state occupation prior to the application of the pulse. By varying the ac or dc bias voltage in series with the pulse generator, different states can be emptied (or filled) and their participation in the field-effect measurement eliminated.

The amplitude of the applied pulse should be chosen very carefully. In principle the essential characteristics of the relaxation process are independent of the magnitude of the field, provided a certain threshold is exceeded to satisfy the assumption of small n_{s0} underlying (5.67). In reality, however, it is found[31,33] that field-enhanced emission from surface states sometimes takes place at large fields, a process that may completely mask the thermal release (see Ch. 7, § 4 and Ch. 9, § 2). At too low amplitudes, on the other hand, the experimental sensitivity may be insufficient. A procedure useful in eliminating high-field effects is to observe the relaxation in surface conductance at different amplitudes, in each case checking that the decay is exponential (over most of its range) and characterized by the same time constant.

6.3. MEASUREMENT OF MAJORITY-CARRIER SURFACE MOBILITY

As has been pointed out in § 2, surface scattering will generally reduce the mobility of the carriers moving close to the surface. The measurement of this effect is rendered difficult by the presence of fast surface states. Were such states absent, the surface mobility at each value of barrier height would have been easily deducible from low-frequency field-effect data. In this case, ΔN (for accumulation layers in an extrinsic n-type sample) is given by the known induced charge, and the change in surface conductance is a direct measure of the majority-carrier surface mobility. The presence of surface states prevents such a simple analysis because the fraction of the induced charge that is mobile is not known. The effect of the surface states can be considerably reduced, however, by the use of the pulsed field effect[30]. Here advantage is taken of the relatively slow thermal release of charge from surface states at low temperatures. The polarity of the applied pulse is again chosen so as to repel majority carriers (electrons) from the surface (see Fig. 6.12a). The change in surface conductance is then measured at the *onset* of the pulse, before the free carrier deficit induced by the field can be redistributed between the surface states and the majority-carrier band. In this way one eliminates the effect of all states except those characterized by release times shorter than the experimental resolution time (≈ 1 μsec). Referring to (5.68) and (2.173), we see that if the surface states have atomic cross sections ($\approx 10^{-15}$ cm^2) then only those states (if present) separated by less than about $12kT$ from the majority-carrier band ec ge will release charge in a time shorter than 1 μsec. At liquid nitrogen temperature, for example, this energy interval is 0.08 eV and constitutes only a small fraction of the energy gap. The measurement thus yields a fairly good lower limit to the true surface mobility.

The bridge shown in the upper part of Fig. 6.11 is well suited to surface mobility measurements. The ac or dc biasing field is used to produce a strong accumulation layer at the surface, and the change in filament resistance is measured as a function of the amplitude of the superimposed pulse. The measuring procedure is as follows. The bridge, now activated by the pulse, is balanced and the filament resistance R_f read on the decade resistor. A negative pulse of amplitude δV is then switched on at the field plate. The decade resistor is readjusted to balance the bridge as close to the onset of the pulse as the circuitry resolution permits. The pulse is then switched off the plate and the ac or dc biasing field is adjusted to restore bridge balance. The pulse is now switched on again and the process repeated. Figure 6.13 illustrates this sequence of operations. The negative pulse is synchronized with

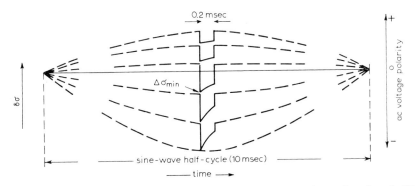

Fig. 6.13. Illustration of the measuring sequence used for deriving $\Delta\sigma$ as a function of $n\delta V$. The solid lines (pulses) correspond to the portions displayed on the CRO. (After Many et al., reference 30.)

the positive (or negative) crest of a 50 cps biasing voltage, as explained above. Each curve represents the variation of surface conductance with time under the combined action of the ac bias and the superimposed pulse, starting from an accumulation layer (upper curve) and down to depletion layers. In this manner a series of values of filament resistance $(R_f)_n$ is obtained, each corresponding to the number of times n that the pulse has been switched on (and off). This is equivalent to a measurement of filament resistance as a function of $n\delta V$, the voltage of a pulse whose amplitude is varied in steps of δV (with no biasing field). The difficulties incurred by the use of such high-voltage pulses, however, are avoided in the present technique. It should be noted that the measurement is carried out under non-equilibrium conditions as far as the surface states and *minority*-carrier band are concerned. The conduction band, however, is in complete equilibrium with the induced (mobile) charge, the majority-carrier Fermi level being well defined both in the bulk and at the surface.

The minimum in surface conductance $\Delta\sigma_{\min}$ is derived from the usual field-effect measurement, in which sufficient time is allowed for steady-state conditions to be established between both bands and the fast surface states. The various $(\Delta\sigma)_n$ values corresponding to $(R_f)_n$ are computed by reference to $\Delta\sigma_{\min}$. If one assumes that no surface states are involved in this process, then the mobile charge induced is proportional to $n\delta V$. This enables the evaluation of the barrier heights $(V_s)_n$ for each reading, and $(\Delta\sigma)_n$ can be plotted against $(V_s)_n$. The surface mobility can then be evaluated as a function of V_s, as will be discussed in greater detail in Ch. 8, § 5.1.

A good estimate of surface mobility is particularly necessary at low temperatures, where the use of bulk mobilities may introduce appreciable errors in the derivation of V_s from $\Delta\sigma$ data. The use of the method described above can fulfil such a need in accumulation and depletion layers. This method can also be extended to yield directly the density and energy distribution of the fast states in the upper half of the energy gap. All that is required is to repeat the sequence of measurements in the *reverse* order. One starts with a depletion layer, and by switching on (and off) a positive pulse, so as to charge the surface states, one gradually progresses towards accumulation layers. Because of the fast trapping under these conditions, the surface states are in equilibrium with the conduction band throughout the measurement. Thus the measured $(R_f)_n$ values now represent only the mobile fraction of the induced charge. The barrier height corresponding to each value of $(R_f)_n$ is determined from the non-equilibrium measurements, and the curve of trapped charge density *versus* barrier height is derived from a comparison of the results obtained by the two measurement procedures [31].

6.4. RELAXATION PROCESSES INVOLVING BOTH CARRIER TYPES

The discussion so far has been confined to accumulation and depletion layers of an extrinsic sample, in which minority carrier effects are negligible. The situation is quite different when an inversion layer is either present or induced by the pulsed field. At the onset of the pulse the initial charging current through the sample is carried almost entirely by majority carriers as before, whereas under steady-state conditions the induced charge appears, at least partially, as a change in *minority*-carrier density in the inversion layer. Thus, in addition to the previously discussed charge exchange between the bands and the surface states, the relaxation process must now involve hole–electron recombination (or generation), whether via the surface states or by some other mechanism in the bulk. By proper choice of the experimental conditions, the recombination process can be isolated from all other processes and employed to determine the excess surface density of minority carriers (ΔP or ΔN)[32]. This technique is particularly useful for determining the barrier height of inversion layers at low temperatures (or in large-gap semiconductors), where lack of communication between bulk and surface may invalidate ordinary surface-conductance measurements (see § 2).

Consider an inversion layer on an extrinsic n-type sample with a positive pulse applied to the field plate. At the onset of the pulse, electrons (the majority carriers) flow towards the surface from the end contacts while at the same time holes are injected from the surface into the bulk. The charge relaxation time associated with this process will again be assumed small compared to the time constants of all other processes in the sample. Suppose first that no surface states are present and that the induced charge density Q_s (per unit area) is small compared to the mobile charge density $q\Delta P$ in the inversion layer. Under these conditions very few excess electrons will reach the space-charge region, most of them remaining in the adjacent bulk to neutralize the charge of the injected holes. Thus the initial rise in surface conductance will be approximately given by $\mu_n|Q_s| + (\mu_p - \mu_{ps})|Q_s|$. The first term represents the conductance of the excess induced electrons while the second term results from the enhanced mobility of the holes injected into the bulk (mobility μ_p) compared to that in the inversion layer (surface mobility μ_{ps}). The subsequent decay in surface conductance (see Fig. 6.14) arises from the recombination of the excess hole–electron pairs. The time constant of this process is evidently of the order of the minority-carrier lifetime (usually tens to hundreds of microseconds in germanium and silicon) and can easily be resolved by conventional techniques. The surface conductance eventually reaches its steady-state value which, in the case being considered, is less than what it was prior to the application of the pulse by the amount $\mu_{ps}|Q_s|$. Thus the overall amplitude $\delta\sigma$ of the decay — that is, the conductance change from the initial value to the final, steady-state value — is given by $(\mu_n + \mu_p)|Q_s|$. A similar process takes place following the termination of the pulse, the relaxation towards equilibrium now involving hole–electron pair generation.

For high pulse amplitudes the expressions derived above for the initial and final values of surface conductance should be modified to include the contribution of the unpaired electrons appearing in the space-charge region. The decay amplitude $\delta\sigma$, however, is still a measure of the total density of injected holes δP:

$$\delta\sigma = q(\mu_n + \mu_p)\delta P, \tag{6.46}$$

but now $q\delta P$ may be smaller than $|Q_s|$. As the pulse amplitude is raised, more and more holes are injected and $\delta\sigma$ increases accordingly. When the induced charge $|Q_s|$ exceeds a certain threshold, then $\delta\sigma$ will saturate, corresponding

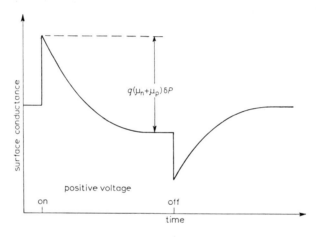

Fig. 6.14. Typical pulsed field-effect oscillogram under inversion-layer conditions at the surface of an n-type sample. The decay amplitude $q(\mu_n+\mu_p)\delta P$ corresponds to hole–electron pair recombination at and near the surface.

to all the available minority carriers in the inversion layer (of excess surface density ΔP) being injected into the bulk:

$$(\delta\sigma)_{\text{sat}} = q(\mu_n + \mu_p)\Delta P. \tag{6.47}$$

It should be noted that (6.46) and (6.47) involve the bulk and not the surface mobilities, since the recombining hole–electron pairs move in the bulk, outside the space-charge region. It is for this reason, also, that the excess surface density ΔP contributes its full share to the measured decay amplitude, thus eliminating the difficulty arising in ordinary surface-conductance measurements from lack of communication between surface and bulk.

The presence of surface states will clearly complicate the relaxation process considerably. In addition to the recombination decay just considered, electrons in the conduction band could be trapped at the surface and, at the same time, trapped holes may be emitted into the valence band, the surface states tending to increase their net electron occupation. By the use of sufficiently large pulse amplitudes such that the inversion layer converts into an accumulation layer, however, the observed decay amplitude $(\delta\sigma)_{sat}$ will still be given to a good approximation by (6.47), irrespective of whether surface states are present or not. Two objectives are achieved in this manner. First, the free electron density n_s is made so large that any trapping by surface states will take place in a very short time ($10^{-8}-10^{-7}$ sec in germanium). The trapping time constant thus becomes much smaller than the minority-carrier lifetime, and in fact is undetectable in a CRO trace of the type shown in Fig. 6.14. Second, the large value of n_s inhibits the emission of holes from surface states. The electron capture rate is given by terms of the type $K_n n_s p_t$ and the hole emission rate by $K_p p_1 p_t$, and as long as $K_n n_s \gg K_p p_1$ the states will trap electrons more readily than they will emit holes. The higher n_s the less trapped holes will be injected into the bulk and the more nearly will the observed $(\delta\sigma)_{sat}$ represent the *free* excess hole density ΔP in the inversion layer.

In practice, one measures the decay amplitude $\delta\sigma$ as a function of pulse amplitude and takes the (saturated) value at the highest available pulse. Once ΔP has been obtained in this manner, it can be used to evaluate the barrier height prior to the application of the pulse (see Ch. 4, §§ 2, 3). By combining this measurement with dc or low-frequency ac field-effect measurements, the trapped charge density in the lower half of the gap can be obtained as a function of barrier height. Moreover, if the communication between surface and bulk is good (germanium at room temperature), this technique can be used to estimate the minority carrier surface mobility (see Ch. 8, § 5.1).

An elegant demonstration of such surface injection effects has been given by Harrick[34] using a modified Haynes–Shockley drift experiment[35]. A pulse applied to a small-area field plate is used to produce a local excess (or deficit) of minority carriers in the adjacent bulk region. The disturbance is then swept down the filament and detected at a later time by a collector probe. This method can serve as a quick test for the conductivity type at the surface. It can also be used to estimate the minority-carrier excess surface density (ΔP or ΔN) in a manner similar to that described above.

7. High-frequency field effect

The resolution time of the pulsed field-effect technique is limited, on the one hand, by the difficulty of eliminating completely the displacement current and, on the other, by the rise time of the amplifiers. In order to reduce the overall resolution time below 1–10 μsec, considerable modifications in the experimental apparatus would be required. An alternative approach is provided by the use of high-frequency fields. In this case the displacement current is 90° out of phase with the field-effect signal over a wide frequency range and can more easily be separated out. Moreover, by restricting the ac field to small amplitudes, the field-effect is very nearly sinusoidal, permitting highly sensitive detection. This method was first used by Aigrain[36] and was later developed by Montgomery[37], who was able to extend the measurements up to 50Mc. The higher frequency range corresponds to a resolution time of $10^{-8}-10^{-7}$ sec and is therefore of particular interest. Similar measurements, but not extending to such high frequencies, have been described by Yunovich[38].

A quantitative analysis of the high-frequency field effect was carried out by Garrett[39] and by others[40–42]. Here we shall discuss only the main features of this effect and describe the experimental apparatus used for its measurement.

In general the change in surface conductance will be out of phase with the applied ac voltage at the field plate. If we express the charge on the field plate by

$$Q_s = Q_{s0} \cos \omega t, \qquad (6.48)$$

then for the small-signal case being considered we may write for the change in surface conductance

$$\delta\sigma = \delta\sigma_0 \cos(\omega t - \theta), \qquad (6.49)$$

where θ is the phase angle between Q_s and $\delta\sigma$. It will be useful at this point to introduce the so-called field-effect mobility[6, 43] μ_{fe}, defined as a complex vector whose magnitude is equal to the ratio of the amplitudes of the conductance $\delta\sigma_0$ and the plate charge Q_{s0} and whose argument is given by the phase angle θ. The real part μ'_{fe} of μ_{fe} represents the in-phase component (positive or negative) of the field-effect mobility with respect to the plate charge (or the ac field),

$$\mu'_{fe} = \frac{\delta\sigma_0}{Q_{s0}} \cos\theta. \qquad (6.50)$$

As we shall see below, μ'_{fe} can be measured very conveniently and is therefore the quantity of interest.

At low frequencies the induced charge is in equilibrium with the two bands and the surface states, and conditions are the same as those considered in § 5. The surface conductance is then in phase with the applied field ($\theta = 0$ or $\theta = 180°$, depending on the bulk conductivity type and on the quiescent value of the barrier height). In the absence of surface states the field-effect mobility will be equal or opposite to the majority-carrier surface mobility in accumulation and depletion layers and to the minority-carrier surface mobility in inversion layers. The presence of surface states will of course reduce the value of μ_{fe}.

As the frequency is raised, dispersion phenomena set in, the field-effect mobility being no longer in phase with the applied voltage. The nature of these phenomena depends on the characteristics of the fast surface states and on the type of space-charge layer at the surface. Here again the simplest case to analyse corresponds to accumulation and depletion layers, where only majority-carriers contribute to the field-effect mobility. We consider an n-type extrinsic sample and assume that only one discrete set of surface states is involved. The frequency dependence of the field-effect mobility can be obtained in this case by a modification of the treatment of the small-signal pulsed field effect given in Ch. 5, § 6. The same notation is used, except that $n_{s\infty}$, $n_{t\infty}$, ΔN_∞ are now replaced by the corresponding quiescent values n_{s0}, n_{t0}, ΔN_0. In addition, (5.59) must be replaced by

$$\delta(\Delta N) + \delta n_t = Q_s/q = (Q_{s0}/q)e^{i\omega t}, \qquad (6.51)$$

where all quantities are represented for convenience in complex notation. This equation expresses the equality between the total induced charge density and the overall change in the trapped and mobile charge densities. Analogously to the previous section, we assume that the charge relaxation time is small compared to $1/\omega$, so that the electrons in the conduction band have an equilibrium distribution. In particular, we assume the validity of (5.61) at all times. Using (5.57) we now obtain the differential equation (compare eq. (5.62)):

$$\frac{d(\Delta N)}{dt} + \frac{\delta(\Delta N)}{\tau_{ss}} = \left(\frac{1}{\tau_{ss}} - \frac{1}{\tau_1}\right)\frac{Q_s}{q} + \frac{1}{q}\frac{dQ_s}{dt}. \qquad (6.52)$$

Here τ_{ss} is the relaxation time of the surface states, defined similarly to (5.64):

$$\tau_{ss} = \cfrac{1}{K_n\left[(N_t-n_{t0})\left.\cfrac{dn_s}{d(\Delta N)}\right|_0 + n_{s0}+n_1\right]}, \tag{6.53}$$

while

$$\tau_1 = \cfrac{1}{K_n(N_t-n_{t0})\left.\cfrac{dn_s}{d(\Delta N)}\right|_0}. \tag{6.54}$$

By substituting for Q_s from (6.51), one readily finds that the stationary solution of (6.52) is

$$\delta(\Delta N) = \delta(\Delta N)_0 e^{i\omega t}, \tag{6.55}$$

where $\delta(\Delta N)_0$ is the complex amplitude of the excess surface-electron density:

$$\delta(\Delta N)_0 = \left(1 - \frac{\tau_{ss}}{\tau_1}\frac{1-i\omega\tau_{ss}}{1+\omega^2\tau_{ss}^2}\right)\frac{Q_{s0}}{q}. \tag{6.56}$$

Thus, the field-effect mobility is given by

$$\mu_{fe} = \frac{q\mu_{ns}\delta(\Delta N)_0}{Q_{s0}} = \mu_{ns}\left(1 - \frac{\tau_{ss}}{\tau_1}\frac{1-i\omega\tau_{ss}}{1+\omega^2\tau_{ss}^2}\right), \tag{6.57}$$

and the in-phase (real) component by

$$\mu'_{fe} = \mu_{ns}\left(1 - \frac{\tau_{ss}}{\tau_1}\frac{1}{1+\omega^2\tau_{ss}^2}\right). \tag{6.58}$$

The frequency dependence of these quantities is illustrated in Fig. 6.15. It is

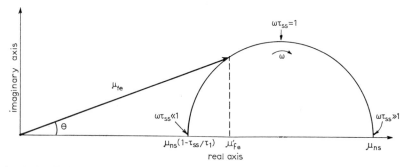

Fig. 6.15. The frequency dependence of the field-effect mobility μ_{fe} in the complex plane for accumulation or depletion layers on an n-type sample. The semicircle represents the locus of the end points of the complex vector μ_{fe} at different frequencies.

seen that the locus of the end points of the complex vector representing μ_{fe} traces out a semicircle whose centre is on the real axis. At low frequencies ($\omega\tau_{ss} \ll 1$), μ_{fe} is real, and because of surface trapping its magnitude is smaller than μ_{ns}. At higher frequencies μ_{fe} becomes complex, the phase angle θ first increasing and then decreasing. At very high frequencies ($\omega\tau_{ss} \gg 1$) hardly any surface trapping can take place, and both μ_{fe} and μ'_{fe} approach the limiting value of μ_{ns}. If the latter frequency range ($\approx$ 100 Mc in germanium at room temperature) were within experimental reach, this method could be used to determine the majority-carrier surface mobility. The frequency dependence of μ'_{fe} is shown also by the solid curve of Fig. 6.16.

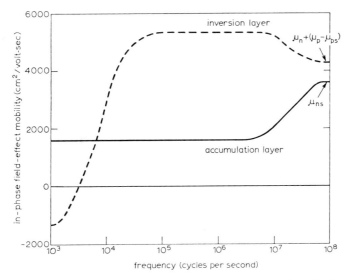

Fig. 6.16. Schematic representation of the frequency dependence of the in-phase field-effect mobility μ'_{fe} under accumulation-layer (solid curve) and inversion-layer (dashed curve) conditions for an n-type germanium sample.

In the case of inversion layers, dispersion in the field-effect mobility may also occur in a much lower frequency range, comparable to the reciprocal of the minority-carrier lifetime τ_p. (If the surface recombination velocity is not too high, τ_p represents the fundamental mode lifetime[39] — see Ch. 7, § 1.) The situation here is similar to that considered in the preceding subsection. As long as $\omega \ll 1/\tau_p$, the field-effect mobility is equal (and opposite in sign) to the surface hole mobility, except in so far as it is reduced by surface trapping. When the frequency approaches $1/\tau_p$, there is not sufficient time for the holes injected from (or extracted by) the surface to be taken up by recombination (or to be supplied by thermal generation), and dispersion sets in — see dashed curve in Fig. 6.16. For $\omega \approx 1/\tau_p$, the field effect is 90° out of phase with the applied voltage and μ'_{fe} goes to zero. This occurs when the in-phase contribution of the excess hole–electron pairs injected into the bulk just balances the "true" (low-frequency) minority-carrier contribution. At higher frequencies μ'_{fe} changes sign and then approaches a (positive) saturation value where all holes injected into the bulk contribute fully (and in phase) to the surface conductance. The saturation value lies somewhere between the limits $\mu_n + (\mu_p - \mu_{ps})$ and $\mu_n + \mu_p$, depending on the density of surface states. The upper limit obtains when most of the holes injected into the bulk originate from surface states. The lower limit corresponds to no hole emission from surface states, the contribution of the hole members of the

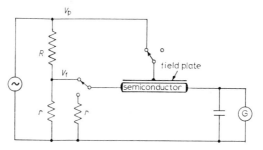

Fig. 6.17. Experimental arrangement for measuring the in-phase field-effect mobility μ'_{fe} at different frequencies. (After Montgomery, reference 37.)

hole–electron pairs arising solely from their enhanced mobility in the bulk over that in the space-charge layer. Whatever the actual saturation value in this intermediate frequency range, the lower limit will be approached at very high frequencies, where insufficient time is allowed for hole emission to take place. This second dispersion phenomenon is similar to that in accumulation layers and sets in when $1/\omega$ is comparable to the surface-state relaxation time. In the present case, however, the final saturation value is approached from above (see Fig. 6.16) and is a measure of the *minority*-carrier surface mobility.

The experimental apparatus used by Montgomery[37] is outlined in Fig. 6.17. An ac voltage $V_p = V_{p0} \cos \omega t$ is applied to the field plate and a smaller voltage $V_f = V_{f0} \cos \omega t$, of the same frequency and phase, is applied across the filament. Product modulation of the latter voltage by the conductivity changes due to the field effect produces a dc current in the galvanometer proportional to the in-phase field-effect mobility μ'_{fe}. Under small-signal conditions the filament conductance may be expressed as $G = G_0 + \delta G \cos(\omega t - \theta)$, where θ is the phase angle between the field effect and the applied field. The current through the sample is thus given by

$$I = V_f G = V_{f0} G_0 \cos \omega t + \tfrac{1}{2} V_{f0} \delta G [\cos \theta + \cos(2\omega t - \theta)], \quad (6.59)$$

and the measured dc current by

$$I_{dc} = \tfrac{1}{2} V_{f0} \delta G \cos \theta. \quad (6.60)$$

The induced charge density is given by $Q_s = C_g V_p$, where C_g is the geometric capacitance (per unit area). Thus, using (6.1), (6.50), and (6.60), we obtain for the in-phase field-effect mobility

$$\mu'_{fe} = \frac{2 I_{dc} l^2}{C_g A V_{f0} V_{p0}}, \quad (6.61)$$

where l is the filament length and A is the area under the field plate.

It is evident that as long as the frequency is lower than the reciprocal of the charge relaxation time ($\approx 4G_0/C_g A$), the displacement current is 90° out of phase with the field effect and does not contribute to I_{dc}. For $\omega C_g A \approx 4G_0$, the assumptions of the underlying theory break down. Experimentally this will manifest itself in a substantial contribution of the displacement current to the measured current. As has been pointed out in § 6.1, the upper frequency limit can be extended by reducing the filament length and by choosing as small a value of the geometric capacitance C_g as is compatible with the available experimental sensitivity. The highest frequency attained by Montgomery[37] was 50 Mc. For germanium at room temperature this is not quite high enough to enter the range where the surface states are completely inoperative, but a decrease in temperature should make this possible (see Ch. 5, § 6).

8. Channel conductance

In the techniques discussed so far, the surface conductance constituted a small fraction of the measured filament conductance. Another approach to the study of the field effect is to isolate the underlying bulk from the measurement and to observe directly the characteristics of the space-charge layer.

This can be done in the case of strong inversion layers if the end contacts are made blocking to the flow of majority carriers and non-rectifying towards the inversion layer. The p–n–p (or n–p–n) junction-transistor structure with an inversion layer on the n (or p) base region is ideally suited to this purpose[44-46]. The shorting action of the surface under these conditions[47-49] poses a serious technological problem in device fabrication, and it was the need for eliminating such undesirable effects that initially prompted a systematic study of channel conductance.

The experimental arrangement employed by Kingston[45] and by Statz and co-workers[50] is outlined schematically in Fig. 6.18. The inversion layer at the surface of the n base region is shown as part of the p-type end regions.

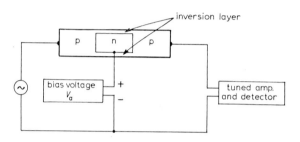

Fig. 6.18. Circuit used to measure inversion-layer channel conductance as a function of bias voltage. (After Statz et al., reference 50.)

The base is biased reversely with respect to the p regions, and the conductance of the structure is measured by applying a small ac voltage (≈ 0.1 V at 1000 cps) across the (non-rectifying) end contacts. It is evident that the ac current is confined mostly to the narrow channel comprising the p inversion layer. This will be particularly so for not too small biases ($\gg kT/q$), since the saturation current through the junction is then practically independent of bias and the superimposed ac voltage will draw negligible current through the bulk. The situation here is similar to that considered in § 2, with the roles of the inversion layer and the underlying bulk region interchanged. In order for the bulk contribution to the overall measured conductance to be negligible, we now require that the channel resistance R_s be small compared to the differential junction resistance at any given reverse bias. This condition is easily satisfied in strong inversion layers. The general problem of surface leakage in p–n junctions in the presence of an inversion layer has been analysed in detail by Cutler and Bath[51] and by Groschwitz and Ebhardt[52].

The reverse bias V_a is used to vary the barrier height in the inversion layer in much the same way as an external field is used in the ordinary field-effect configuration considered above. The resulting change in channel conductance can then be employed to derive the characteristics of the various surface states (slow and fast), below the mid-gap on an n-type base and above on a p-type base. This technique has the special advantage that the measured conductance is very nearly equal to the surface conductance, so that the barrier height can be deduced directly from the data without recourse to the minimum as a reference point. In addition, the effect of temperature variation, which may be quite serious in the ordinary dc or

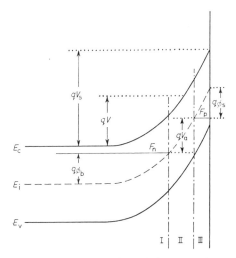

Fig. 6.19. Energy-band diagram at the surface of the n-type base of a p–n–p transistor structure.

low-frequency field effect, is of little concern in the present case. The disadvantage of the technique lies in the rather involved computational procedure required to interpret the data, even when a number of simplifying assumptions are introduced. The method is also inherently limited to inversion layers, which must be produced by special surface treatments. These consist of oxidation of the etched structure and exposure to suitable gaseous ambients[50].

Figure 6.19 shows the energy-band diagram at the surface of an n-type base. Because of its relatively large conductance, the inversion layer assumes the negative potential of the end p regions. Thus, the entire bias V_a appears across the insulating region between the inversion layer and the bulk (region II), which has been depleted of both electrons and holes. This is indicated in the figure by a separation of qV_a between the majority-carrier *quasi*-Fermi levels F_n and F_p in the bulk and at the surface, respectively. In each region the pertinent quasi-Fermi level (for electrons in I, for holes in III) is essentially horizontal, as discussed in Ch. 2, § 7.4.

In order to interpret the experimental data, one must derive a number of expressions relating the channel conductance, the applied bias, the space-charge density, and the barrier height[50]. It will be convenient to consider each of the regions in Fig. 6.19 separately. We assume that the base material is extrinsic and that the channel consists of a strong inversion layer — in other words, the hole density in the bulk and the electron density at the surface are both negligibly small.

Region I: the n-type bulk. This region extends from the bulk up to the point where F_n intersects E_i (that is, from $V = 0$ to $V = -\phi_b$). The space-charge density here is given approximately by (see Ch. 4, § 2.2)

$$\rho = qn_b(1-e^v) \quad \text{for} \quad 0 \geq v \geq -u_b, \tag{6.62}$$

where v and u_b are the dimensionless potentials defined in Ch. 4, § 2.1.

Region II: the intrinsic region. This region extends from $V = -\phi_b$ to $V = -\phi_b - |V_a|$. Here the free-carrier densities may be neglected, so that the space-charge density is given to a good approximation by the static ionized impurity charge:

$$\rho = qn_b \quad \text{for} \quad -u_b \geq v \geq -(u_b + q|V_a|/kT). \tag{6.63}$$

Region III: the inversion layer. The space charge in this region consists of both the static charge and the free-hole charge. Thus

$$\rho = q(n_b + n_i e^{-u}) \quad \text{for} \quad 0 \geq u \geq u_s, \tag{6.64}$$

where

$$u \equiv (F_p - E_i)/kT = u_b + q|V_a|/kT + v \tag{6.65}$$

and u_s is the value of u at the surface. The quantity u_s is analogous to the surface potential defined in Ch. 4, § 2.1.

Physically, the effect of the bias voltage on the charge distribution in the three regions and in the surface states can be seen from the following considerations. As $|V_a|$ is increased, more electrons are removed from the base, and the intrinsic region (II) widens. The resulting increase in positive (immobile) space charge is accompanied by an equal increase in negative charge in the inversion layer and in the surface states. Free holes initially flow out of the inversion layer (via the end p regions), and the surface potential $|u_s|$ decreases (eq. (6.67) below). The equilibrium between the surface states and the valence band is now upset, and the former tend to lose holes (by capturing electrons from the latter). As a result the barrier subsequently relaxes towards its new steady-state value, with the hole deficit distributed between the surface states and the inversion layer. If the surface states are present with a large enough density, they may effectively anchor the surface potential against any change. Even in this case, however, the excess surface-hole density ΔP decreases since, due to the higher field at the surface, the inversion layer becomes narrower and can accomodate less free holes (see eq. (6.69)).

Poisson's equation (4.9) can readily be integrated once using the charge distribution given by (6.62)–(6.64). In particular, for region III (the inversion layer) we obtain

$$\frac{du}{dz} = \sqrt{\frac{2q^2}{\kappa\varepsilon_0 kT}} \sqrt{n_b(u_b+q|V_a|/kT-u-1)+n_i e^{-u}}. \quad (6.66)$$

The field at the surface is given by $\mathscr{E}_s = (kT/q)(du/dz)_{u=u_s}$, and from this value one can derive the total space-charge density Q_{sc} using Gauss's law (4.18):

$$Q_{sc} = \sqrt{2\kappa\varepsilon_0 kT} \sqrt{n_b(u_b+q|V_a|/kT+|u_s|-1)+n_i e^{|u_s|}}. \quad (6.67)$$

The space-charge density must also be equal in magnitude and opposite in sign to the total charge density in the surface states:

$$Q_{sc} = -Q_{ss}. \quad (6.68)$$

This follows from the requirement of overall neutrality at the surface, the field outside the semiconductor being zero.

In order to obtain the free-hole excess surface density ΔP we proceed as in Ch. 4, § 3.1, but now using (6.66). The integral $\int p\,dz = \int [p/(du/dz)]du$ can be evaluated in closed form provided we neglect in (6.66) the linear term in u. Thus

$$\Delta P \approx \sqrt{2\kappa\varepsilon_0 kT/q^2} \left\{ \sqrt{n_b(q|V_a|/kT+|u_b|-1)+n_i e^{|u_s|}} - \sqrt{n_b(q|V_a|/kT+|u_b|-1)+n_i} \right\}. \quad (6.69)$$

This equation is a good approximation[50] for bias voltages exceeding 2V.

The measured channel conductance $\Delta\sigma$ is very nearly equal to $q\mu_{ps}\Delta P$ and can thus be used (provided a good estimate of μ_{ps} is available) to obtain ΔP. Equation (6.69) is then solved (graphically) to derive the surface potential u_s for each pair of values $\Delta\sigma$, V_a. The calculated value of u_s is inserted in (6.67) and Q_{sc} (and hence Q_{ss}) determined. In this manner one obtains the trapped charge in surface states as a function of surface potential. All the experimental procedures discussed in §§ 4–7 apply equally well to channel conductance measurements. Dc biases can be used to study the slow states and the trapped charge distribution in fast states[50], while ac[53] and pulsed biases can serve to investigate the interaction processes between the fast states and the space-charge layer. Measurements of surface capacitance would be expected to prove particularly useful in this respect.

It should be noted that in the ordinary field-effect technique (§§ 4, 5), the measured curve of Q_{ss} *versus* u_s (or v_s) corresponds to true equilibrium conditions between the surface states under study and *both* bands. This is not quite the situation in the present case, since current (the reverse saturation current) flows continuously between the bulk and the surface, and the minority carrier densities in both regions are displaced from their equilibrium values. For the strong inversion layers being considered, however, the electron surface density is in any event very small and plays a negligible role in controlling the occupation of the surface states at the bottom of the gap. (Referring to (5.73), we see that both n_s^* and n_1 are small compared to p_s^* and p_1, so that n_t^* (or p_t^*) is essentially determined by the large surface-hole density p_s^*.) In other words, the surface states are in "intimate contact" with the valence band, and the same quasi-Fermi level F_p governs the density of both free and trapped holes. Thus to a good approximation the experimental data yield the density and energy distribution of the surface states, just as in the case of true equilibrium conditions.

References

[1] W. H. Brattain and J. Bardeen, Bell System tech. J. **32** (1953) 1.
[2] R. H. Kingston, J. appl. Phys. **27** (1956) 101. (References)
[3] W. Shockley and G. L. Pearson, Phys. Rev. **74** (1948) 232.
[4] H. C. Montgomery and W. L. Brown, Phys. Rev. **103** (1956) 865.
[5] W. H. Brattain and C. G. B. Garrett, Bell System tech. J. **34** (1955) 129.
[6] J. R. Schrieffer, Phys. Rev. **97** (1955) 641.
[7] R. H. Kingston and S. F. Neustadter, J. appl. Phys. **26** (1955) 718.
[8] C. G. B. Garrett and W. H. Brattain, Phys. Rev. **99** (1955) 376.
[9] R. N. Rubinstein and V. I. Fistul, Doklady Akad. Nauk U.S.S.R. **125** (1959) 542 [translation: Soviet Phys. — Doklady **4** (1959) 431].

REFERENCES

[10] W. L. Brown, W. H. Brattain, C. G. B. Garrett, and H. C. Montgomery in *Semiconductor Surface Physics* (Edited by R. H. Kingston), (University of Pennsylvania Press, Philadelphia, 1957), p. 117.
[11] H. M. Bath and M. Cutler, Bull. Am. phys. Soc. Ser. II **3** (1958) 378.
[12] P. K. Weimer, Proc. Inst. Radio Engr. **50** (1962) 1961.
[13] S. R. Hofstein and F. P. Heiman, Proc. Inst. Elect. Electronics Engrs. **51** (1963) 1190. (References)
[14] J. F. Dewald in *Semiconductors* (Edited by N. B. Hannay), (Reinhold, New York, 1959), p. 727. (References)
[15] H. U. Harten in *Festkörperprobleme* vol. III (Edited by F. Sauter), (Friedr. Vieweg, Braunschweig). To be published. (References)
[16] K. Bohnenkamp and H. J. Engell, Z. Electrochem. **61** (1957) 1184.
[17] W. H. Brattain and P. J. Boddy, J. electrochem. Soc. **109** (1962) 572.
[18] P. J. Boddy and W. H. Brattain, Ann. N. Y. Acad. Sci. **101** (1963) 683.
[19] W. W. Harvey, Ann. N. Y. Acad. Sci. **101** (1963) 904.
[20] W. L. Brown, Phys. Rev. **100** (1955) 590.
[21] S. R. Morrison in *Semiconductor Surface Physics* (Edited by R. H. Kingston), (University of Pennsylvania Press, Philadelphia, 1957), p. 169.
[22] N. B. Grover and Y. Goldstein, J. Phys. Chem. Solids **17** (1961) 338.
[23] A. Many, Y. Margoninski, E. Harnik, and E. Alexander, Phys. Rev. **101** (1955) 1433, 1434.
[24] A. Many, E. Harnik, and Y. Margoninski in *Semiconductor Surface Physics* (Edited by R. H. Kingston), (University of Pennsylvania Press, Philadelphia, 1957), p. 85.
[25] J. Bardeen and S. R. Morrison, Physica **20** (1954) 873.
[26] G. G. E. Low, Proc. phys. Soc. (London) B **68** (1955) 10, 1154.
[27] A. Many and D. Gerlich, Phys. Rev. **107** (1957) 404.
[28] P. C. Banbury, G. G. E. Low, and J. D. Nixon in *Semiconductor Surface Physics* (Edited by R. H. Kingston), (University of Pennsylvania Press, Philadelphia, 1957), p. 70.
[29] G. Rupprecht, Phys. Rev. **111** (1958) 75; Ann. N. Y. Acad. Sci. **101** (1963) 960.
[30] A. Many, N. B. Grover, Y. Goldstein, and E. Harnik, J. Phys. Chem. Solids **14** (1960) 186.
[31] E. Harnik, Y. Goldstein, N. B. Grover, and A. Many, J. Phys. Chem. Solids **14** (1960) 193.
[32] A. Many, Y. Goldstein, N. B. Grover, and E. Harnik, *Proc. Int. Conf. Semiconductor Physics, Prague*, 1960 (Czechoslovak Academy of Sciences, Prague, 1961), p. 498.
[33] A. Many and Y. Goldstein, *Proc. Int. Conf. Physics and Chemistry of Solid Surfaces, Providence*, 1964 (Surface Science **2** (1964) 114).
[34] N. J. Harrick, Phys. Rev. Letters **2** (1959) 199.
[35] J. R. Haynes and W. Shockley, Phys. Rev. **81** (1951) 835.
[36] P. Aigrain, J. Lagrenaudi, and G. Liandrat, J. phys. radium **13** (1952) 587.
[37] H. C. Montgomery, Phys. Rev. **106** (1957) 441.
[38] A. E. Yunovich, Zhur. tekh. Fiz. **27** (1957) 1707 [translation: Soviet Phys. — Tech. Phys. **2** (1957) 1587].
[39] C. G. B. Garrett, Phys. Rev. **107** (1957) 478.
[40] A. E. Yunovich, Zhur. tekh. Fiz. **28** (1958) 689 [translation: Soviet Phys. — Tech. Phys. **3** (1958) 646]; Fiz. tver. Tela **1** (1960) 1092 [translation: Soviet Phys. — Solid State **1** (1960) 998].
[41] V. L. Bonch-Bruevich, Zhur. tekh. Fiz. **28** (1958) 70 [translation: Soviet Phys. — Tech. Phys. **3** (1959) 63].
[42] F. Berz, J. Electron. Control **6** (1959) 97; J. Phys. Chem. Solids **23** (1962) 1795.

[43] J. R. Schrieffer in *Semiconductor Surface Physics* (Edited by R. H. Kingston), (University of Pennsylvania Press, Philadelphia, 1957), p. 55.
[44] W. L. Brown, Phys. Rev. **91** (1953) 518.
[45] R. H. Kingston, Phys. Rev. **93** (1954) 346; **98** (1955) 1766.
[46] G. A. deMars, H. Statz, and L. Davis, Jr., Phys. Rev. **98** (1955) 539.
[47] H. Christensen, Proc. Inst. Radio Engrs. **42** (1954) 1371.
[48] A. L. McWhorter and R. H. Kingston, Proc. Inst. Radio Engrs. **42** (1954) 1376.
[49] M. Cutler and H. M. Bath, J. appl. Phys. **25** (1954) 1440.
[50] H. Statz, G. A. deMars, L. Davis, Jr., and A. Adams, Jr. in *Semiconductor Surface Physics* (Edited by R. H. Kingston), (University of Pennsylvania Press, Philadelphia, 1957), p. 139; Phys. Rev **101** (1956) 1272.
[51] M. Cutler and H. M. Bath, Proc. Inst. Radio Engrs. **45** (1957) 39.
[52] E. Groschwitz and R. Ebhardt, Z. angew. Physik **11** (1959) 296.
[53] L. M. Terman, Solid-State Electron. **5** (1962) 285.

CHAPTER 7

EXPERIMENTAL METHODS II – OTHER MEASUREMENTS

The field-effect techniques discussed in the preceding chapter all employ the surface conductance as a means of studying the various electronic processes at the surface. The present chapter reviews a number of other methods, based on different phenomena, which can be used as well for this purpose, either separately or in conjunction with the field effect. Particular attention is given to surface recombination velocity measurements, which have proved extremely valuable in determining the characteristics of the surface states on germanium and silicon. The other measurements will be discussed in greater or lesser detail, depending on the amount of information they can yield on the surface parameters of interest. In almost all cases the measurements are carried out as a function of barrier height, which is varied either by external fields or by gaseous ambients.

1. Surface recombination velocity

In general the value of the surface recombination velocity s is derived indirectly from lifetime measurements. The presence of surface recombination in addition to bulk recombination gives rise to diffusion currents towards the surface and results in the measured sample lifetime being smaller than the bulk lifetime. It is this effect that is utilized in determining s. In the present section we first derive a number of expressions (corresponding to different experimental conditions) for the effective sample lifetime in terms of the bulk lifetime, the recombination velocity at the surface, the geometry of the sample, and the electric field used in the measurement. Next we review briefly some of the important experimental techniques used for measuring the effective sample lifetime. The discussions will be confined to the simplest and most common case (in germanium and silicon) — low-level excitation and no trapping. In other words, we assume that $\delta p_b = \delta n_b \ll p_b + n_b$ and that the hole and electron recombination processes in the bulk are characterized by a common lifetime τ. An extrinsic n-type sample will be considered throughout, the case of a p-type sample being completely analogous. For non-extrinsic material, the minority-carrier mobility and diffusion constant appearing in all expressions must be replaced by the ambipolar mobility and diffusivity (see Ch. 2, § 7.3).

1.1. SAMPLE EFFECTIVE LIFETIME

The effect of diffusion and drift on the measured lifetime will be analysed for a homogeneous, rectangular filament along the x axis, with the surfaces of interest perpendicular to the y and z directions. As a first step we treat the case of an infinitely long filament in the absence of external fields. This will take care of diffusion in the y and z directions and enable us to derive the sample effective lifetime as determined by recombination in the bulk and on the cross-sectional surfaces. Thereafter (§ 1.2), it will be sufficient to consider one-dimensional motion in the x direction of a finite filament in which both diffusion and field currents are present along this direction only.

For the case of steady excitation we consider a filament that is infinite in the y direction as well, the calculations otherwise becoming too cumbersome. We thus have an infinite slab of material bounded by the planes $z = \pm d$. The external excitation is assumed uniform over both the x and y directions, producing $\mathscr{L}$ pairs of carriers per unit volume per unit time. Under these conditions $\mathscr{L}$ and δp_b will be functions of z only, and (2.194) reduces to (for an extrinsic n-type semiconductor):

$$\mathscr{L} - \frac{\delta p_b}{\tau} + D_p \frac{d^2(\delta p_b)}{dz^2} = 0. \tag{7.1}$$

The reader should note that here (as well as in Ch. 5, § 7), we have used the subscript b in the symbols for the excess carrier densities even though in the presence of surface recombination these densities will generally be functions of position z, contrary to our practice in Ch. 2, § 7.3. The purpose of this notation is to stress the point that δp_b and δn_b refer to the bulk region, where the variation of the excess density with z is usually very slow compared to that in the space-charge region. In fact, (7.1) need be solved only in the bulk, the detailed processes at the surface and in the adjacent space-charge layer (whose width is very small compared to the sample thickness) being introduced by appropriate boundary conditions.

Two forms of excitation are of interest in practice: (a) uniform excitation and (b) surface excitation. The first can be realized by using light in the absorption edge, X rays, or high-energy particle bombardment. If the absorption coefficient is small compared to the reciprocal of the slab thickness, $\mathscr{L}$ is essentially constant, independent of z. Case (b) can be achieved by employing light in the fundamental absorption band. Usually the light penetration depth is much less than both the sample thickness and the diffusion length, so that $\mathscr{L}$ in (7.1) can be taken as zero everywhere in the sample except at the illuminated surface, where excitation proceeds at a rate of, say, $\mathscr{L}_s$ per unit area. In both cases the net rate of surface recombination will be taken to be directly proportional to the excess carrier density δp_b at the surface just outside the space-charge layer, the proportionality constant[1] being the surface recombination velocity s. The boundary conditions, however, will be different in the two cases. For uniform excitation (a), the diffusion current towards each surface (assumed to be characterized by the same value of s) satisfies the relation

$$-D_p \frac{d(\delta p_b)}{dz}\bigg|_{z=\pm d} = \pm s \delta p_b(\pm d), \tag{7.2}$$

while for surface excitation (b)

$$-D_p \frac{d(\delta p_b)}{dz}\bigg|_{z=+d} + \mathscr{L}_s = s \delta p_b(d);$$

$$D_p \frac{d(\delta p_b)}{dz}\bigg|_{z=-d} = s \delta p_b(-d). \tag{7.3}$$

Considering (7.2) for example, one may look upon each surface as a "sink" for the excess holes (and electrons) present with a density $\delta p_b(\pm d)$ and flowing into it with a velocity s.

For the case of uniform excitation (a), the solution of (7.1) is [2]

$$\delta p_b = \mathscr{L}\tau \left[1 - \frac{(s/D_p)\cosh(z/L_p)}{(1/L_p)\sinh(d/L_p)+(s/D_p)\cosh(d/L_p)}\right], \tag{7.4}$$

where $L_p = (D_p \tau)^{\frac{1}{2}}$ is the hole diffusion length (see Ch. 2, § 7.3). The average value of δp_b, of interest in calculating the integrated excess conductivity (see below), is given by

$$\overline{\delta p_b} = \frac{1}{2d}\int_{-d}^{+d} \delta p_b \, dz = \mathscr{L}\tau_{\text{eff}}, \tag{7.5}$$

where

$$\tau_{\text{eff}} = \tau \left[1 - \frac{(sL_p/D_p d)\sinh(d/L_p)}{(1/L_p)\sinh(d/L_p)+(s/D_p)\cosh(d/L_p)}\right]. \tag{7.6}$$

Note that (7.5) has the same form as the steady-state solution (eq. (2.167)) corresponding to the same slab but without surface recombination ($s = 0$) or, more generally, to an infinite sample. The bulk lifetime is replaced in the present case by an effective sample lifetime τ_{eff} while δp_b is replaced by the average value $\overline{\delta p_b}$. In order that it be useful in practical cases, however, one would want the expression for τ_{eff} to be simple in form. A few such cases, applicable under special limiting conditions, will be cited. For s very large

$$\tau_{\text{eff}} = \tau[1 - (L_p/d)\tanh(d/L_p)], \tag{7.7}$$

the effective lifetime being independent of s. For $sL_p/D_p \ll \tanh(d/L_p)$

$$\tau_{\text{eff}} = \tau[1 - (sL_p/D_p)(L_p/d)], \tag{7.8}$$

while for $d/L_p \ll 1$

$$1/\tau_{\text{eff}} = 1/\tau + s/d. \tag{7.9}$$

This last expression is of particular interest both because of its simple form and because it usually applies to germanium and silicon, where the bulk lifetime can be made sufficiently large.

For the case of surface excitation (b), the solution of (7.1) (with $\mathscr{L} = 0$) subject to the boundary conditions (7.3) yields [2,3]

$$\delta p_b = \frac{\mathscr{L}_s}{2d}\tau\left(\frac{d}{L_p}\right)^2 \times$$

$$\times \frac{2(d/L_p)\cosh[(d+z)/L_p] + 2(sd/D_p)\sinh[(d+z)/L_p]}{[(sd/D_p)^2 + (d/L_p)^2]\sinh(2d/L_p) + 2(sd/D_p)(d/L_p)\cosh(2d/L_p)}. \tag{7.10}$$

The average value $\overline{\delta p_b}$ is given by

$$\overline{\delta p_b} = (\mathscr{L}_s/2d)\tau_{\text{eff}}, \tag{7.11}$$

where

$$\tau_{\text{eff}} = \frac{L_p\{(D_p/L_p)\sinh(2d/L_p) + s[\cosh(2d/L_p) - 1]\}}{[s^2 + (D_p/L_p)^2]\sinh(2d/L_p) + (2sD_p/L_p)\cosh(2d/L_p)}. \tag{7.12}$$

Here again the situation is equivalent to that in a sample with no surface recombination, in which the volume excitation rate is $\mathscr{L}_s/2d$ and the bulk lifetime is τ_{eff}. Equation (7.12) reduces to (7.9) under more stringent requirements than in case (a), namely when $d/L_p \ll 1$ and, at the same time, $s \ll (D_p/\tau)^{\frac{1}{2}}$.

The excess carrier distribution in the slab for the two forms of excitation is illustrated in Fig. 7.1 for typical conditions in germanium ($L_p = d$, $sd/D_p = 3$). As expected, for uniform excitation δp_b is maximum in the middle ($z = 0$) and decreases symmetrically on either side. For surface excitation, on the other hand, δp_b is peaked at the illuminated surface and decays as the other surface is approached.

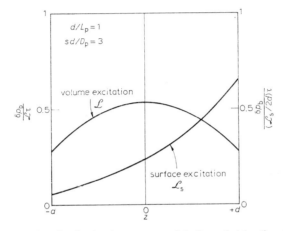

Fig. 7.1. Excess-carrier distribution in an n-type slab (bounded by the planes $z = \pm d$) for volume and surface excitation.

We turn now to the transient response, considering here the more general case of a rectangular filament bounded by the planes $y = \pm w$ and $z = \pm d$. The excitation will be assumed as before to be uniform along x but arbitrary along the y and z directions. The continuity equation (2.194) now becomes

$$\frac{\partial(\delta p_b)}{\partial t} = \mathscr{L}(y, z) - \frac{\delta p_b}{\tau} + D_p \left[\frac{\partial^2(\delta p_b)}{\partial y^2} + \frac{\partial^2(\delta p_b)}{\partial z^2} \right], \tag{7.13}$$

with the boundary conditions

$$-D_p \frac{\partial(\delta p_b)}{\partial y}\bigg|_{y=\pm w} = \pm s \delta p_b(\pm w); \tag{7.14}$$

$$-D_p \frac{\partial(\delta p_b)}{\partial z}\bigg|_{z=\pm d} = \pm s \delta p_b(\pm d). \tag{7.15}$$

Equation (7.13) is a linear partial differential equation. For practical purposes it is sufficient to solve the homogeneous equation (with $\mathscr{L} = 0$), corresponding to the decay of δp_b following a sudden interruption of the external excitation. For the case in which the excitation (prior to its interruption) is uniform, the solution[1] of (7.13) is given by a sum of terms of the form

$$\delta p_b(y, z) = \sum_i a_i e^{-t/\tau_i} \cos(b_i y) \cos(c_i z), \tag{7.16}$$

where the constants τ_i, b_i, c_i must satisfy the relationship

$$1/\tau_i = 1/\tau + D_p(b_i^2 + c_i^2). \tag{7.17}$$

The boundary conditions (7.14) and (7.15) then reduce to

$$sw/D_p = b_i w \tan(b_i w) = \eta_i \tan \eta_i; \tag{7.18}$$

$$sd/D_p = c_i d \tan(c_i d) = \zeta_i \tan \zeta_i. \tag{7.19}$$

For any given values of sw/D_p and sd/D_p, there will be an infinite number of pairs of solutions η_i and ζ_i, each lying in the interval $i\pi$ to $(i+\tfrac{1}{2})\pi$, where $i = 0, 1, 2, \ldots$. In term of these solutions, (7.17) can be written as

$$\frac{1}{\tau_i} = \frac{1}{\tau} + D_p \left[\frac{\eta_i^2}{w^2} + \frac{\zeta_i^2}{d^2} \right]. \tag{7.20}$$

From (7.16) and (7.20) it is evident that, strictly speaking, the decay of δp_b will not be represented by a simple exponential function as in the case of no surface recombination, but by a sum of terms each having its own time constant. However, the first term in the series, corresponding to the smallest solutions η_0 and ζ_0 (both lying in the range 0 to $\tfrac{1}{2}\pi$) and thus to the *largest* time constant τ_0, will dominate the decay characteristics. The time constant τ_0 is accordingly referred to as the sample effective lifetime τ_{eff} for transient response:

$$\frac{1}{\tau_{\text{eff}}} = \frac{1}{\tau} + D_p \left[\frac{\eta_0^2}{w^2} + \frac{\zeta_0^2}{d^2} \right], \tag{7.21}$$

and the excess carrier distribution represented by the dominant term, as the fundamental mode distribution. The other, higher-order terms can be regarded as initial transients which die away rapidly. Once the fundamental mode has been established, the excess carriers at *every region* of the sample, including the surfaces, will decay at the *same* rate. The distribution of δp_b under these conditions is peaked at $y = z = 0$ and decreases symmetrically as the surfaces are approached. In one of the z planes, for example, δp_b is represented by a half-wave cosine function for $s = \infty$ ($\delta p_b(\pm w) = 0$), by part of a half wave for intermediate values of s, and by a straight line for $s = 0$ ($\tau_{\text{eff}} = \tau$).

So far it has been assumed that the excitation prior to its interruption is uniform over the sample. For other modes of excitation, corresponding to an arbitrary spatial distribution of δp_b at $t = 0$, the general solution of (7.13) will include sine as well as cosine terms. The fundamental mode, however, will always be established eventually, when the higher-order terms die away. It is only the higher-order terms, which rapidly transform the initial spatial distribution into the fundamental-mode distribution, that depend on the form of excitation. For example, it is possible to suppress such terms completely by making the initial distribution identical to that of the fundamental mode. This can best be approximated in practice by employing uniform excitation, as assumed previously. Surface excitation, on the other hand, may give rise to higher-order terms of large amplitude, since the initial distribution, which is peaked at the illuminated surface, must change appreciably in the process of establishing the symmetrical fundamental mode. The amplitude of the higher-order terms will be lower, the lower the value of s and the smaller the filament cross section.

To summarize, the problem of carrier motion and decay has been reduced to a one-dimensional problem. For this purpose, an effective sample lifetime was introduced appropriate to the conditions at hand, whether for steady-state or transient response. The effective lifetime takes into account the cross-sectional diffusion current and is related to the bulk lifetime, surface recombination velocity, and cross-sectional dimensions by any of the expressions derived above. It is evident that the separate surface and bulk parameters can be obtained from a measurement of effective lifetime τ_{eff}, provided the dimensions of the sample are properly controlled. In filaments of large cross sections and/or small bulk lifetimes, $\tau_{\text{eff}} \to \tau$. For thin samples, on the other hand, surface recombination may become dominant, and s can be determined directly. For intermediate conditions, s can be deduced only if τ is known or if more elaborate procedures that provide additional data are used.

1.2. MEASUREMENT OF EFFECTIVE LIFETIME

Numerous methods have been devised for measuring minority-carrier lifetime in semiconductors, each with its specific requirements and objectives. No single technique is suitable under all conditions, and in choosing the most appropriate means one is guided by such considerations as structural and geometrical limitations on the sample, the range and type of lifetime to be measured (local or average), the likelihood of the occurrence of trapping effects, and the possibility that the lifetime depends on the excitation level. No attempt will be made here to list all these methods, much less to describe them, and the interested reader is referred to the literature[4]. The discussion will be confined mainly to those methods that have proved especially useful in surface recombination velocity measurements. In order to be classified as useful, such measurements must permit an accurate and rapid determination of the variation of s with barrier height, and it is the extent to which this is achieved in any particular instance that governs the choice of method.

Most commonly, the effective lifetime is deduced from conductance measurements under non-equilibrium conditions. In view of the discussions in the previous subsection, we may consider a rectangular filament in which the excess carrier density over any cross section is uniform and given by $\overline{\delta p_b}$. In general $\overline{\delta p_b}$ will vary along the length of the filament, either because of non-uniform excitation or due to contact effects, as will be discussed in a moment. It will therefore be useful to consider the total number of excess carriers δP_{tot}, obtained by integrating $\overline{\delta p_b}$ over the filament length l,

$$\delta P_{tot} = 4wd \int_0^l \overline{\delta p_b}\, dx. \tag{7.22}$$

The change in filament conductance δG is proportional to δP_{tot} irrespective of the excess-carrier distribution along the filament, provided that the following conditions are satisfied: (a) δp_b is everywhere small compared to the equilibrium density $p_b + n_b$, (b) $\overline{\delta p_b}$ is equal at the two ends ($x = 0$, $x = l$), and (c) the end contacts are non-rectifying. To show this we write the current along x in the form[2,5]

$$I = I_0 + \delta I = 4qwd\{(\mathscr{E}_0 + \delta\mathscr{E})(\mu_n n_b + \mu_p p_b) \\ + (\mathscr{E}_0 + \delta\mathscr{E})(\mu_n + \mu_p)\overline{\delta p_b} + (D_n - D_p)[d(\overline{\delta p_b})/dx]\}, \tag{7.23}$$

where the subscript o refers to conditions in the absence of excitation ($\mathscr{L} = 0$). We now integrate (7.23) with respect to x. Condition (a) permits us to neglect second order terms of the type $\delta\mathscr{E}\overline{\delta p_b}$, since the variation of the field $\delta\mathscr{E}$ along x is small. From (b) it follows that the contribution of the diffusion term to the integral vanishes, while (c) ensures that no voltage drop occurs across the contacts. Condition (a) holds for sufficiently low excitation levels, while (b) and (c) can be easily satisfied (in germanium and silicon) by proper alloying or soldering techniques. The contacts obtained in this manner are associated with high recombination rates in their immediate vicinity and thus are to be regarded as end surfaces with essentially infinite recombination velocity. Hence δp_b vanishes at both ends, and condition (b) is automatically satisfied. Recalling that the current is position-independent, we finally obtain after integration of (7.23) that

$$\delta I = V_0 \delta G + G_0 \delta V = q(\mu_n + \mu_p) V_0 \delta P_{tot}/l^2 + G_0 \delta V, \quad (7.24)$$

where G_0 is the conductance and V_0 is the voltage drop across the filament in the absence of excitation; $\delta V \equiv \int_0^l \delta \mathscr{E} dx$ is the change in voltage as a result of excitation. The deviation δV will be zero if a constant-voltage source is employed and non-vanishing if a series resistor is interposed between the source and the filament. In either case it follows from (7.24) that the change in filament conductance δG is given by

$$\delta G = q(\mu_n + \mu_p) \delta P_{tot}/l^2, \quad (7.25)$$

which is just what we set out to show.

The question now arises as to how well δP_{tot}, and through it δG, is a measure of the effective lifetime τ_{eff}. As was pointed out above, $\overline{\delta p_b}$ vanishes at the contacts. More important still, in the presence of an applied field excess carriers may be injected into and/or swept out of the filament through the contacts. Such end effects must be minimized, or controlled, in order for a simple and unambiguous relationship to hold between δP_{tot}, $\mathscr{L}$, and τ_{eff}. Consider first steady-state measurements. In the absence of end effects we have

$$\delta P_{tot} = 4lwd \overline{\mathscr{L}} \tau_{eff}, \quad (7.26)$$

where $\overline{\mathscr{L}}$ is the mean excitation rate along the filament. The excess conductance δG can be measured by a variety of methods, the most effective of which employs a Wheatstone bridge. If the carrier mobilities are known, as well as the excitation rate $\overline{\mathscr{L}}$, then τ_{eff} can be directly evaluated from (7.25) and (7.26). It is usually difficult to determine $\overline{\mathscr{L}}$ directly, and it may be easier to calibrate the external source with a sample of known lifetime. The main advantage of this method lies in its simplicity. It is particularly useful when relative values of effective lifetime are of interest, such as in measurements of surface recombination velocity under varying conditions at the surface.

In order to determine the limitations imposed by end effects on such conductance measurements, it is necessary to solve the continuity equation applicable to the present case. Only the effect of carrier loss by sweep-out through the end contacts will be considered, contact injection normally being easier to suppress. Thus the excitation rate $\mathscr{L}$ will be assumed to originate entirely from an external source (such as light). Under steady-state conditions, the continuity equation reduces to (see eq. (2.197))

$$\mathscr{L} - \frac{\overline{\delta p_b}}{\tau_{\text{eff}}} + D_p \frac{d^2(\overline{\delta p_b})}{dx^2} - \mu_p \mathscr{E} \frac{d(\overline{\delta p_b})}{dx} = 0. \qquad (7.27)$$

For constant and uniform excitation rate, the solution of (7.27) subject to the boundary conditions $\overline{\delta p_b} = 0$ at the two ends yields for the total number of excess carriers [2]

$$\delta P_{\text{tot}} = 4lwd\mathscr{L}\tau_{\text{eff}} \left[1 - \frac{2(\alpha_1 - \alpha_2) \sinh\left(\frac{1}{2}l\alpha_1\right) \sinh\left(\frac{1}{2}l\alpha_2\right)}{l\alpha_1 \alpha_2 \sinh\left[\frac{1}{2}l(\alpha_1 - \alpha_2)\right]} \right], \qquad (7.28)$$

where

$$\alpha_{1,2} \equiv \frac{\mu_p \mathscr{E}_0}{2D_p} \pm \left[\left(\frac{\mu_p \mathscr{E}_0}{2D_p}\right)^2 + \frac{1}{D_p \tau_{\text{eff}}} \right]^{\frac{1}{2}}. \qquad (7.29)$$

The second term in the brackets of (7.28) represents a correction to the ideal case (eq. (7.26)) arising from the diffusion currents towards the end contacts (where $\overline{\delta p_b}$ is zero) and from the sweep-out of excess carriers by the applied field. The correction is small only if both the diffusion and the drift lengths (see Ch. 2, § 7.3) are small compared to the filament length:

$$\sqrt{D_p \tau_{\text{eff}}} \ll l; \\ \mu_p \mathscr{E}_0 \tau_{\text{eff}} \ll l. \qquad (7.30)$$

This is intuitively obvious, since the diffusion and drift lengths represent the distances from either contact over which end effects extend. For the case of uniform excitation being considered, these distances must be kept small compared to the filament length in order for the ideal situation to be well approximated. To see how stringent such conditions are, consider an n-type germanium filament with an effective lifetime of 200 μsec, a value typical in surface studies. For the correction due to diffusion to be less than 10%, the sample should be 2 cm long. A field of 1 V/cm applied to the filament will increase the correction term to 20%.

Conditions can be improved considerably by confining the carrier excitation to a small region of the filament well removed from the contacts, rather than using excitation over the entire length. This enables the use of much higher fields without serious end effects and thus leads to a greatly increased sensitivity. For the example just cited, the correction term amounts to only 0.4 % when the excitation is confined to the middle of the filament.

Conductivity methods are more widely used to determine the effective lifetime under transient conditions. In contrast to steady-state measurements, a knowledge of the carrier mobility or the excitation rate is not required, the lifetime being derived directly from the rate of change of conductance. Here again loss of excess carriers by sweep-out is minimized by keeping the diffusion and drift lengths small compared to the filament length or, still better, by employing localized rather than uniform excitation. Measurements of the rise or decay of conductance for pulsed excitation is especially simple and popular. Consider excitation in the form of a square pulse with flat top and extended duration ($\gg \tau_{\text{eff}}$). After the fundamental mode has been established, the rise of the excess conductance is given by (see eqs. (7.16), (7.21))

$$\delta G = \delta G_\infty (1 - e^{-t/\tau_{\text{eff}}}), \tag{7.31}$$

and the decay by

$$\delta G = \delta G_\infty e^{-t/\tau_{\text{eff}}}, \tag{7.32}$$

where δG_∞ is the steady-state excess conductance, given by (7.25) and (7.26). Frequently the excitation pulse is a short spike. Under such conditions insufficient time is allowed for the excess conductance to attain its steady-state value δG_∞. The decay, however, has the same form as (7.32) (but with a smaller amplitude) and can be used equally well to derive τ_{eff}.

The detection system, in its simplest form, consists of the filament, a series resistor, and a dc voltage source. The transient voltage change across the series resistor is amplified and displayed on a CRO, and the effective lifetime is obtained from an analysis of the trace. For greater accuracy and convenience, it is preferable to use a bridge circuit of the type shown in Fig. 6.11, and to balance the transient against an exponential signal derived from an *RC* network[6]. The effective lifetime can be read directly off the calibrated resistor *R*, similarly to the surface-state time constant in the pulsed field-effect measurement (Ch. 6, § 6.2). Here, as well, the procedure permits an immediate check on the exponentiality of the transient signal. Either the rise or the decay in conductance can be measured in this manner, the voltage pulse that activates the *RC* network (and sometimes the rest of the bridge as well) being appropriately synchronized with the excitation pulse.

The various pulse methods differ mainly in the source of excitation: light, particle bombardment, or contact injection. Short light spikes from a spark gap[7] or flash tube[8] are frequently used. They give adequate intensity, but the spectral range is limited mostly to the visible and ultraviolet. In the case of spike excitation one naturally measures the decay characteristics following the spike. A resolution time of the order of a microsecond may be achieved by these methods. Much shorter pulses can be obtained by elaborate rotating mirror systems[9] or by high-energy electron beams from a pulsed Van de Graaff accelerator[10], enabling determination of lifetimes as low as 10^{-9} sec. Such short lifetimes, however, are rarely of interest in surface recombination velocity measurements.

Electrical injection of carriers from contacts[6,11,12] can provide very fast excitation pulses (both rise and decay) of variable duration over an extremely wide range of excess carrier densities. Injection excitation is very useful in transient measurements, as opposed to steady-state measurements. In the latter, the lifetime can be derived only if the excitation level is well known, a requirement that is hard to attain by injection. In transient measurements, on the other hand, considerably more information is embodied in the rise or decay characteristics, so that both the lifetime and the excitation level can be determined simultaneously[6]. Another attractive feature is that uniform distribution of excess carriers in cross section is automatically achieved, so that higher-order decay terms are largely suppressed. Non-rectifying, injecting contacts are easy to produce (by soldering techniques) in germanium, but not so readily in silicon. In this latter case, it is sometimes expedient to apply the injecting pulse to an auxiliary contact (usually a point contact) near one end of the filament while measuring conductance changes with non-rectifying end contacts. It is worth considering at this point the relative merits of measuring either the rise of the pulse or its decay. The change in conductance during the rise can be measured most simply and sensitively from the change in injection current itself[6]. Sweep-out effects cannot occur during the rise until the front of injected carriers reaches the other end contact. Even if the transit time is no longer than the effective lifetime (or even about $\frac{1}{3}\tau_{\text{eff}}$), the lifetime can still be determined by the null method using the balancing RC network. For very long lifetimes it is preferable to measure the decay characteristics, provided the applied field can be kept sufficiently low to minimize sweep-out effects.

In addition, conditions corresponding to the fundamental-mode distribution are more nearly realized. While the level of excitation may not be easily controllable during the rise, it is always possible to measure the low-level lifetime in the decay by waiting for the signal to fall to the appropriate level. Sweep-out effects, however, may be a limiting factor in this case.

The lifetime measurements discussed so far are all based on conductance changes as a means of determining the excess-carrier densities. Quite a few methods employ other detection techniques. One of the earliest and most widely used methods of this sort is based on the photoelectromagnetic (PEM) effect. This may be described as the Hall effect associated with the diffusion of photoexcited carriers. We shall now outline the theory[3] of the PEM effect and comment briefly on the technique of measurement. Consider a rectangular slab of material bounded by the planes $z = \pm d$ and illuminated by uniform light which is highly absorbed at the $z = +d$ surface. Carriers are continuously diffusing towards the $z = -d$ surface, resulting in a steady-state distribution described by (7.10). The hole and electron current densities are given by

$$J_{pz} = -J_{nz} = -qD_p \frac{d(\delta p_b)}{dz}. \tag{7.33}$$

If now a magnetic field is applied along the y direction, the excess holes and electrons will be deflected to the right ($x > 0$) and to left ($x < 0$), respectively, establishing in this manner a net current in the $+x$ direction (Fig. 7.2).

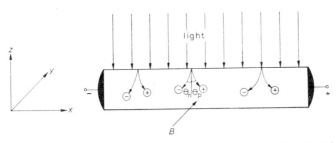

Fig. 7.2. A schematic diagram illustrating the magnetic deflection of holes and electrons diffusing towards the bottom surface.

For sufficiently weak magnetic fields, the hole and electron components of this current are to a good approximation $-\theta_p J_{pz}$ and $-\theta_n J_{nz}$, where θ_p and θ_n are the respective Hall angles (see Ch. 2, § 6.3). Thus, in the general case where, in addition, an electric field ($\mathscr{E}_x$) is applied along x, we have

$$J_{px} = q\mu_p(p_b+\delta p_b)\mathscr{E}_x + q\theta_p D_p \frac{d(\delta p_b)}{dz}; \tag{7.34}$$

$$J_{nx} = q\mu_n(n_b+\delta p_b)\mathscr{E}_x + q|\theta_n|D_p \frac{d(\delta p_b)}{dz}. \tag{7.35}$$

Adding the two equations, we obtain

$$J_x = (\sigma_0+\delta\sigma)\mathscr{E}_x + q\theta D_p \frac{d(\delta p_b)}{dz}, \tag{7.36}$$

where $\theta = \theta_p + |\theta_n|$, and σ_0 and $\delta\sigma$ are the equilibrium and excess conductivity, respectively. It should be noted that for the case of uniform and low-level excitation, and in the absence of end effects, the values of $\delta\sigma$, δp_b, and $\mathscr{E}_x$ are independent of x (and y).

Two conditions of measurement will be considered, one when the filament is shorted, the other when it is open. In the former case, $\mathscr{E}_x = 0$ and the short-circuit current $I_x^{(sc)}$ (per unit width along y) is given from (7.36) as

$$I_x^{(sc)} = q\theta D_p \int_{-d}^{d} \frac{d(\delta p_b)}{dz} dz = q\theta D_p[\delta p_b(d) - \delta p_b(-d)], \tag{7.37}$$

where $\delta p_b(d)$ and $\delta p_b(-d)$ are the excess carrier densities at the illuminated and dark surfaces, respectively. Under open-circuit conditions, on the other hand, $J_x = 0$ and we obtain from (7.36) and (7.37) for the open-circuit field $\mathscr{E}_x^{(oc)}$

$$2d(\sigma_0+\overline{\delta\sigma})\mathscr{E}_x^{(oc)} = -q\theta D_p[\delta p_b(d) - \delta p_b(-d)], \tag{7.38}$$

where $\overline{\delta\sigma} \equiv q(\mu_p+\mu_n)\overline{\delta p_b}$ is the mean excess conductivity. Combining (7.37) and (7.38) we obtain

$$I_x^{(sc)} = -2d(\sigma_0+\overline{\delta\sigma})\mathscr{E}_x^{(oc)}. \tag{7.39}$$

The consistency of (7.37) and (7.38) can be used in practice to test the validity of the assumption of small Hall angles under the actual conditions of measurement.

The difference $\delta p_b(d) - \delta p_b(-d)$ in (7.37) can be expressed in terms of the surface excitation rate $\mathscr{L}_s$ and the surface recombination velocity s by the use of (7.10). Equation (7.11) relates $\mathscr{L}_s$ to $\overline{\delta p_b}$ and thus to $\overline{\delta\sigma}$. Introducing these values into (7.37) we obtain, after some manipulation[3],

$$\frac{I_x^{(sc)}}{2d} = \theta \frac{D_p}{L_p} \frac{s + (D_p/L_p)\tanh(d/L_p)}{s \tanh(d/L_p) + D_p/L_p} \frac{\overline{\delta\sigma}}{\mu_p + \mu_n}. \quad (7.40)$$

It is seen that $I_x^{(sc)}/\overline{\delta\sigma}$ is independent of light intensity. Actually, it is also independent of the surface recombination velocity at the *illuminated* surface, as can be shown[3] by a somewhat more involved calculation, the value of s that appears in (7.40) referring to the *dark* surface. Several limiting cases may be considered. For thick slabs such that $d/L_p \gg 1$, the above equation reduces to

$$\frac{I_x^{(sc)}}{2d} = \theta \frac{D_p \overline{\delta\sigma}}{L_p(\mu_p + \mu_n)}, \quad (7.41)$$

a useful expression for determining the bulk lifetime (through the relation $L_p^2 = D_p \tau_{\text{eff}}$). For negligible bulk recombination and/or very thin slabs such that $d/L_p \ll 1$,

$$\frac{I_x^{(sc)}}{2d} = \theta s \frac{\overline{\delta\sigma}}{\mu_p + \mu_n}, \quad (7.42)$$

which is a direct measure of the surface recombination velocity. Finally, for s very large (on the dark surface), as obtained for example by sand blasting, we have

$$\frac{I_x^{(sc)}}{2d} = \theta \frac{D_p}{L_p} \coth(d/L_p) \frac{\overline{\delta\sigma}}{\mu_p + \mu_n}. \quad (7.43)$$

This expression is valid even for thick slabs ($d/L_p \gg 1$) and is particularly useful for measuring large values of bulk lifetime.

In practice the surface under study is kept in the dark, the illuminated surface having been prepared with as low a value of surface recombination velocity as possible, in order to obtain maximum sensitivity (maximum $\overline{\delta p_b}$). The mean excess conductivity $\overline{\delta\sigma}$ can be measured by any of the methods discussed above, while the Hall angle $\theta (\equiv \theta_p + |\theta_n|)$ can be derived from Hall effect measurements using the same experimental configuration.

Another method employing magnetic fields is that based on the Suhl effect[13]. Here injected carriers are deflected towards the surface, where their recombination rate is enhanced compared to the case of no magnetic field. This effect can be used to derive the surface recombination velocity. The measurement, however, is complicated and time-consuming and requires tedious numerical calculations. It is now chiefly of historical importance, being one of the earliest methods used for measuring surface recombination.

To conclude this discussion of effective-lifetime measurements, we shall consider briefly several methods based on the detection of *local* rather than integrated excess-carrier distributions. A local detector probes the *spatial* distribution, from which the recombination parameters can be determined. The most sensitive detector of this sort is a collector probe[1], which consists of a point-contact rectifier or a p-n junction operated at reverse bias. The saturation current is proportional to the fractional change in *minority*-carrier density — in contrast to the filament conductance, which responds to the total number of carriers. The collector is ordinarily used to detect relative rather than absolute values of excess densities. For quantitative measurements its response can be calibrated against conductance changes using voltage probes on either side of the collector[14]. Under ideal conditions only those minority carriers within a diffusion length will contribute to the saturation current, the bias field being almost entirely confined to the junction region (see Ch. 2, § 7.4). In the presence of a highly conducting inversion layer at the surface beneath the collector, however, this may no longer be the case. The collector will then respond to excess carriers located as far out as the edges of the conducting channel, the effective area of the collector thus extending well beyond its geometrical limits[15]. Serious errors may be introduced in this manner leading, in measurements based on collector detection, to gross overestimation of lifetime[16].

A collector probe may also be operated under open-circuit conditions. In the presence of excess carriers, a "floating" voltage is generated across the collector junction which, for low-level excitation, is directly proportional to the excess minority-carrier density. (This is analogous to the photovoltaic effect.)

The use of a collector probe in the Haynes–Shockley[17] technique has already been mentioned in Ch. 2, § 7.3 and in Ch. 6, § 6.4. Minority carriers are injected at one point of a filament (electrically or optically), are swept down the filament, and are detected by a collector. The minority-carrier drift mobility and effective lifetime are derived from the time of flight and the attenuation of the injected pulse. The diffusion length $(D_p \tau_{eff})^{\frac{1}{2}}$ rather than the drift length $\mu \mathscr{E} \tau_{eff}$ can be measured by omitting the sweeping field in the Haynes–Shockley arrangement[18-20]. A thin line of light provides a constant source of excitation, and the collector response is measured as a function of its distance from the source. The diffusion length is derived from the exponential attenuation (eq. (2.198)). This method is most valuable for probing the lifetime of local regions[21,22]. As pointed out previously, however, care must be taken to avoid conducting channels. In order to minimize any influence of the collector field on the carrier motion, it is preferable to operate the collector under open-circuit conditions.

Localized detection has also been employed in transient measurements. One such measurement, the pulse reverse method[23], consists of observing the transient recovery of a p–n junction when it is suddenly switched from forward to reverse operation. In the forward stage, excess minority carriers are injected into the base region, while in the reverse stage the junction is employed to detect their rate of recombination. For a simple rectangular geometry, the lifetime determined in this manner can be correlated with the pertinent bulk and surface parameters, as before.

A few methods are available for measuring effective lifetime without the use of contacts. One employs infrared absorption[24] to determine the spatial distribution of excess carriers, as will be described in § 5 below. Another method consists of detection by surface photovoltage[25] (see § 2 below). Excess carriers are generated by a flash of light, giving rise to a surface photovoltage which, for sufficiently low light intensity, is proportional to the excess-carrier density (see Ch. 4, § 4). A field-effect electrode, capacitively coupled to the surface, is used to detect the subsequent decay of carriers. Here again the fundamental-mode decay constant is the effective sample lifetime given by (7.21).

Finally, let us say just a word about trapping effects. Such effects are sometimes encountered in germanium at reduced temperatures and in silicon even at room temperature. They are much more common in large-gap materials. In the presence of trapping, $\overline{\delta p_b} \neq \overline{\delta n_b}$ and the hole and electron lifetimes are different. The derivation of the surface recombination velocity under these conditions becomes so complicated as to be impractical. Usually, trapping can be readily detected in transient measurements from the difference in the rise and decay characteristics (which are identical only in the absence of trapping), and by the appearance of a long tail in the decay. In germanium and silicon, a weak background illumination can often saturate the traps and thus eliminate their detrimental effect.

2. Contact potential and surface photovoltage

Measurements of contact potential are widely used to determine the electron affinity and other surface properties of semiconductors. As was discussed in Ch. 4, § 1, the contact potential between two conducting solids is given by the difference in their work functions:

$$V_{cp} = (W_\phi - W'_\phi)/q. \tag{7.44}$$

Considering for example an n-type semiconductor, we may express the work function as

$$W_\phi = \chi + W_{bn} - qV_s, \tag{7.45}$$

where χ is the electron affinity, W_{bn} the energy separation between the Fermi level and the conduction-band edge in the bulk, and V_s the potential barrier height. If both χ and W_{bn} are known it is possible, at least in principle, to determine V_s from the measurement of W_ϕ. In practice, however, $q|V_s|$ is usually small compared to W_ϕ so that a small relative error in the absolute magnitude of the latter may introduce a large error in the former. Accordingly, it is more feasible to derive the surface characteristics from the observed *changes* in work function under varying conditions at the surface. Such changes can occur during the different stages of producing a clean surface (Ch. 3, § 2.2), by varying the ambient gas (Ch. 6, § 1), or by illumination (Ch. 4, § 4). While in the first two cases the changes in work function may be due to a variation in both χ and V_s, illumination can alter only V_s.

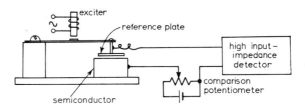

Fig. 7.3. Schematic diagram of an apparatus used to measure contact potential by the Kelvin method. (After Brattain and Garrett, reference 28.)

In practice one measures the contact potential between the semiconductor and some reference electrode whose work function is not expected to alter in the course of varying the conditions at the semiconductor surface. The use of molybdenum seems adequate for this purpose, but it should be borne in mind that possible variations in the work function of any reference electrode cannot be completely ruled out.

Several techniques have been developed for measuring contact potential[26], the most common of which is still the classical Kelvin method[27]. Figure 7.3 shows a schematic diagram of the measuring apparatus used[28]. The reference electrode is placed parallel to the semiconductor surface, about 1mm away. It is mounted on a vibrating reed which is driven by the electromagnet at the left. This varies the plate–semiconductor capacitance C_g sinusoidally and gives rise to an electric signal (of the same frequency) across the detector when any potential drop (contact or other) is present between the two media. The circuit includes a series dc voltage source which can be adjusted to balance out the contact potential, a high-input-impedance amplifier serving as a null indicator. At balance, the voltage of the external source is clearly equal and opposite to the contact potential difference. Brattain and Bardeen[29], applying this method to germanium, were able to achieve an accuracy of about one millivolt. Evidently, the less the separation between the two media (the larger C_g) the higher the contact-potential signal. By using high-gain amplifiers, however, sufficient detection sensitivity can be achieved even for extremely low values of C_g. This is the case, for example, in measurements employing thin wire probes as reference electrodes, a configuration used to explore the work function of very small regions of the surface[30].

Of the various measurements of contact potential, that of its change δV_{cp} with light can be most readily correlated with the characteristics of the space-charge region and the surface states. This measurement and its interpretation in terms of the pertinent surface parameters will accordingly be discussed in some detail. In the Kelvin method, δV_{cp} is simply given by the change in the compensating voltage required to restore balance under illumination. The periodic variation of the plate–semiconductor spacing can be dispensed with, however, by the use of chopped rather than steady light[29,31,32]. The signal due to the modulation in contact potential is amplified and displayed on a CRO. The magnitude of δV_{cp} can be derived from the constants of the circuit or, more accurately, by comparison with the signal obtained from an external oscillator placed in series with the plate–semiconductor capacitor and operated at the chopped-light frequency.

Several processes may contribute to the measured change in V_{cp}. First and foremost is the surface photovoltage discussed in Ch. 4, § 4. The change δV_s in barrier height brought about by the light is picked up by the reference electrode, and if no other voltage drops occur in the semiconductor bulk, δV_{cp} must be equal to δV_s. Moreover, because of the large separation between the metal plate and the semiconductor as compared to the width of the space-charge layer ($C_g \ll C_{sc}$), the change in surface charge is negligible and the value of δV_s is essentially unaffected by the proximity of the reference electrode.

Other contributions to the measured change in V_{cp} may originate from a photovoltage generated at the metal–semiconductor contact and from the Dember potential[33]. The former can be eliminated by careful preparation of the contact and, more important, by keeping the light away from the contact region. (Inhomogeneities in the bulk resistivity may give rise to similar photovoltages and must therefore be avoided.) The Dember potential arises from the electric field set up internally to equalize the otherwise different hole and electron diffusion currents in a direction normal to the surface. Such diffusion always takes place in the presence of surface recombination or when the excitation in depth is not uniform (see § 1). To calculate the magnitude of the Dember potential, we consider one dimensional flow (along z) in a homogeneous slab under external excitation. For the case of no trapping, we have from the neutrality condition (see Ch. 2, § 7.3) that $\delta p_b = \delta n_b$ everywhere in the sample. In the absence of an externally applied voltage across the end planes of the slab ($z = \pm d$), the total current must vanish and (2.179) can be written as

$$J = J_p + J_n = [\sigma_0 + q(\mu_p + \mu_n)\delta p_b]\mathscr{E}_D - q(D_p - D_n)\frac{d(\delta p_b)}{dz} = 0, \quad (7.46)$$

where $\mathscr{E}_D$ is the Dember field established by the hole and electron diffusion currents. Integration of (7.46) over the slab thickness and use of the Einstein relation (eqs. (2.184), (2.185)) immediately yield for the Dember potential difference

$$V_D = -\int_{-d}^{d} \mathscr{E}_D \, dz = \frac{kT}{q}\frac{b-1}{b+1} \ln \frac{\sigma_0 + \delta\sigma(d)}{\sigma_0 + \delta\sigma(-d)}, \quad (7.47)$$

where

$b \equiv \mu_n/\mu_p$, $\delta\sigma(d) \equiv q(\mu_p + \mu_n)\delta p_b(d)$, $\delta\sigma(-d) \equiv q(\mu_p + \mu_n)\delta p_b(-d)$,

the last two expressions representing the excess conductivities at the two end planes of the slab. It is evident that in semiconductors with no trapping, V_D rarely exceeds kT/q. In germanium at room temperature, for example, $b \approx 2$, so that even if $\delta\sigma(d) = \sigma_0$ (a large excess density indeed) and $\delta\sigma(-d) = 0$, the Demper potential amounts to only about 0.01 V.

Thus, apart from a small correction term, the change in contact potential with light yields, under the proper experimental conditions, the surface photovoltage δV_s. In Ch. 4, § 4, δV_s was calculated for a semiconductor devoid of surface states. From the condition of overall neutrality at the surface ($Q_{sc} = $ const), we obtained that the sign of δV_s is always opposite to that of V_s — that is, the light tends to flatten the bands. This effect can readily be used to determine the sign of V_s. Moreover, if the barrier height is varied during the photovoltage measurement, the reversal in sign of δV_s as the bands go through flat-band conditions can serve as a reference point in place of the surface conductance minimum. In principle it is possible to determine the absolute magnitude of the barrier height as well, since in the limit of very strong excitation δV_s tends to $-V_s$. In depletion layers δp_b must be comparable to n_b for saturation in δV_s to occur, while in accumulation or inversion layers a much higher excitation level may be required[32]. Such large excess densities are hard to attain by light, but they can be produced by injection from a p–n junction placed close to the surface[34].

In the presence of fast surface states, charge exchange can take place (during the measurement time) between these states and the space-charge region, and it is only the sum of free and trapped charge that will remain constant with light. This constant is given by the charge in slow states and/or the charge induced capacitively by an external field, both assumed not to alter upon illumination. The calculation of the surface photovoltage as a function of the steady-state excess density δp_b just below the space-charge layer is a straightforward but tedious task. One begins by constructing a family of curves of Q_{sc} versus v_s^*, using δp_b (or u_b^*) as the variable parameter. This can readily be done on the basis of the analysis in Ch. 4, § 4, and the use of eqs. (4.16), (4.19) or Figs. 4.7, 4.8. Next, one considers a discrete set of fast states and assumes tentative values for their energy position E_t^f, density N_t, and capture cross-section ratio K_p/K_n. A second family of curves, this time of the trapped charge density Q_{ss} versus v_s^*, where $Q_{ss} = \text{const} - qn_t^*$, is then constructed with the help of (5.73), δp_b again being used as the variable parameter. This procedure is illustrated in Fig. 7.4 for an n-type germanium sample at room temperature, with $u_b = 2$, $v_s = -6$, $(E_t^f - E_i)/kT = -3$, $N_t = 2 \times 10^{11}$ cm^{-2}, and $K_p/K_n = 10^{-3}$. To simplify the presentation, the constant charge $Q_{sc} + Q_{ss}$ has been taken equal to zero. In this case the point of intersection of any two curves (one from each family) having the same δp_b yields the corresponding value of v_s^*. Such points are joined by the dashed curve, which thus represents the variation of v_s^* with δp_b. The dotted horizontal line indicates the path the system would have taken in the absence of surface states. In both cases the light tends to flatten the bands ($v_s^* \to 0$), but the rate of change of $|v_s^*|$ with δp_b is slower in the presence of surface states. While this behaviour is generally followed, special surface-state parameters can be specified for which $|v_s^*|$ first *increases* with δp_b before it reverses direction and finally tends to zero at very high light intensities.

In principle, one can derive the surface-state parameters from a comparison of experimental data of δV_s versus δp_b, taken at different values of the quiescent barrier height, with theoretical curves of the type just discussed, using trial and error procedures. Since many parameters are involved, however, particularly if more than one set of surface states are present, the number of different possibilities is huge and the analysis becomes prohibitive. A more practical approach is to combine measurements of the surface photovoltage with those of the field effect[31]. The sample is illuminated with chopped light, and both the photovoltage signal and the

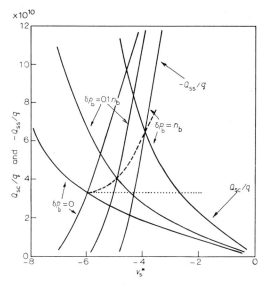

Fig. 7.4. Curves of Q_{sc} and Q_{ss} versus v_s^* at different values of δp_b for an n-type germanium sample ($u_b = 2$) at room temperature. The equilibrium barrier height v_s (for $\delta p_b = 0$) was chosen as -6. The surface-state parameters used in the calculations are $(E_t^t - E_1)/kT = -3$, $N_t = 2 \times 10^{11}$ cm^{-2}, and $K_p/K_n = 10^{-3}$. The dashed curve represents the path the system takes with increasing excitation level. The dotted curve corresponds to the absence of surface states.

change in dark conductance are monitored as the barrier height is continuously varied. The field-effect measurement yields the density and energy distribution of the surface states, as described in Ch. 6, § 5. With this information available, the interpretation of the surface photovoltage data becomes much less ambiguous. If one assumes that all surface states are characterized by the same ratio of hole to electron capture cross sections

(K_p/K_n), the value of this ratio can readily be deduced from the rate of change of δV_s with δp_b.

A serious disadvantage of the surface photovoltage technique as a tool for deriving surface-state parameters lies in its insensitivity to a large class of surface states[35]. Extensive calculations[32] of the type illustrated in Fig. 7.4 show that in a large variety of cases, the δV_s *versus* δp_b curves are either not affected at all or else their general shape is essentially unaltered by surface states. Evidently, the surface states that are most likely to alter their occupation with light and thus affect the surface photovoltage to an appreciable extent are those lying at or close to the Fermi level. For strong accumulation or inversion layers and not too extreme values of K_p/K_n, the occupation of such states cannot change significantly since it is controlled by the *dominant* carriers at the surface, whose density is relatively insensitive to light (see discussion of a similar case at the end of Ch. 6, §8). Only in depletion layers, where $K_p p_s^*$ and $K_n n_s^*$ may be comparable, can the surface states play any important role in the charge redistribution at the surface. The insensitivity of the photovoltage to surface states, however, makes it a most valuable tool for determining the barrier height; both large and small signal conditions can be used to advantage, as explained earlier.

3. Photoelectric emission

The photoelectric effect consists of the emission of electrons from the surface of a solid as a result of illumination in the proper wavelength range. The threshold wavelength for photoemission corresponds to the energy separation between the electron energy in vacuum, just outside the surface, and the highest level in the crystal that is sufficiently populated with electrons. In metals the energy levels are occupied essentially up to the Fermi level, so that the threshold energy is just the work function. Such is also the case in degenerate semiconductors, since here too the highest densely occupied states lie at the Fermi level (above the conduction-band edge in n-type material, below the valence-band edge in p-type). In non-degenerate semiconductors, on the other hand, the densely occupied levels are in the valence band, and the threshold for the corresponding photoemission process is given by the sum of the electron affinity χ and the energy gap E_g. Other, lower thresholds may also exist, those associated with transitions from the conduction band, from localized bulk levels, and from surface states[30,36-39]. Photoemission due to such transitions will be much weaker, however, and will be experimentally detectable only under special conditions (a degenerate n-type surface layer in the first instance, a very large density of states in the other two cases). Accordingly, unless otherwise stated, we shall use the term threshold to refer to transitions from the valence band.

The photoelectric yield, defined as the number of electrons emitted per incident photon, is a complicated function of the energy-band structure, the type of optical transition involved, and the nature of the dominant scattering processes in the crystal. Photoemission arises chiefly from volume excitation of electrons, but its detailed characteristics are controlled by both bulk and surface effects. In order to be able to derive the surface properties from the experimental data, one has to be well acquainted with the main physical processes taking place in photoemission. These will now be reviewed, followed by a brief discussion of experimental technique.

Photoemission involves basically the following steps: the excitation by light of electrons into higher energy states, the diffusion of the excited electrons towards the surface, and (if they are sufficiently energetic) their emission into the vacuum outside. These processes determine the emission yield and are responsible for a given material being a good or a poor emitter. Metals whose threshold lies in the visible range, for example, belong in the second category because of their high reflectivity and the strong electron–electron scattering at the energies involved. Due to the high reflectivity, only a small fraction of the incident light is absorbed, while the strong scattering reduces the mean free path and thus the escape depth — the effective thickness of the layer below the surface from which photoemission can take place. Electrons excited deeper in the bulk lose a large fraction of their energy (to other electrons) and are unable to reach the surface with sufficient energy to surmount the barrier confining them to the crystal (in this case the work function). The yield is higher for semiconductors, where the reflectivity is smaller and electron–electron scattering is negligible. The dominant energy loss here is usually impact ionization — the production by the excited electron of a hole–electron pair across the gap[40]. The mean free path determined by this process is typically of the order of 20–30 Å, which is a sizable fraction of the light penetration depth (100–1000 Å, corresponding to an absorption coefficient of 10^5–10^6 cm^{-1}). Energy loss by impact ionization becomes unimportant when the electron affinity of the semiconductor is smaller than a certain minimum value (usually about 2–4 times the energy gap). In this case, excited electrons corresponding to the photoemission threshold, and thus able to escape the crystal, have insufficient energy for pair production across the gap. The only way in which such electrons can lose energy is by the relatively inefficient process of lattice scattering. As a result the electron escape depth and hence the photoelectric yield are increased considerably[41]. The electron affinity of a given semiconductor can often be lowered by suitable surface treatments. Adsorption of cesium ions on germanium and silicon may be cited as an important example of such treatments[38].

Excitation of electrons from the valence band by light may occur by either direct (vertical) or indirect (phonon-aided) optical transitions (see Ch. 2, § 3.4). The two processes are illustrated[37] in Fig. 7.5. Figure 7.5a shows the

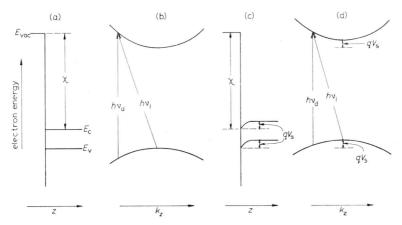

Fig. 7.5. Schematic representation of the direct and indirect transitions involved in photoemission under flat-band (a and b) and accumulation-layer (c and d) conditions.

usual diagram of the valence- and conduction-band edges as a function of distance (z) from the surface, in the absence of a potential barrier (flat bands). Figure 7.5b is a plot of electron energy *versus* the z component of the wave vector k for the valence band and for some higher conduction band whose curve intersects the vacuum level (E_{vac}). (For simplicity, the extrema of both bands are assumed to lie at $k = 0$.) For indirect transitions, the threshold E_{Ti} for photoemission corresponds to a photon energy hv_i equal to the energetic separation between the top of the valence band E_v and the vacuum level E_{vac}, the energy of the phonon involved in the necessary momentum transfer being negligible. Thus $E_{Ti} = \chi + E_g$. For direct transitions, on the other hand, the wave vector k is conserved and, as a consequence, a higher-energy photon (hv_d in the figure) is generally needed to induce photoemission. The direct threshold E_{Td} may be influenced by whether or not the excited electrons suffer appreciable phonon scattering prior to leaving the crystal. While the energy losses incurred by such scattering ($\approx kT$) are much too small to affect the electron energy (several electron volts), the momentum exchange can be quite large (see Ch. 2, § 6.1). In the absence of scattering, E_{Td} will be larger than E_{Ti} by the energy of the hole left behind (which is now produced below the valence-band maximum) plus the kinetic energy of the emitted electron in the tangential direction, which is not affected by the emission. Scattering may reduce the direct threshold, since there is no requirement of transverse momentum conservation. Some, lower-energy electrons, which would otherwise be unable to leave the crystal, may be reoriented by scattering so as to have the necessary energy in the z direction.

TABLE 7.1

Threshold energies and energy dependence of the photoelectric yield for different excitation and scattering processes. (After Kane, reference 42.)

Excitation from	Transition	Scattering process	Threshold E_T	r
BULK PROCESSES				
Valence band	indirect	unscattered / scattered	$E_T = \chi + E_c - E_v$	5/2
	direct	unscattered	$E_T \geq \chi + E_c - E_v$	1
		scattered	$E_T \geq \chi + E_c - E_v$	2
SURFACE PROCESSES				
Valence band		diffuse surface scattering	$E_T = \chi + E_c - E_v$	5/2
		specular surface scattering		3/2
Discrete localized surface states below E_F			$E_T = \chi + E_c - E_t$	1
Continuous distribution of surface states at E_F			$E_T = \chi + E_c - E_F$	2
Surface-state band below E_F	indirect		$E_T > \chi + E_c - E_F$	2
	direct		$E_T > \chi + E_c - E_F$	1
Surface-state band at E_F	indirect		$E_T = \chi + E_c - E_F$	5/2
	direct			3/2

The spectral photoelectric yield due to the two types of transitions would be expected to rise slowly as the photon energy is raised above the indirect threshold and then to show an abrupt increase as the direct threshold is exceeded and the more efficient excitation process becomes dominant. Kane [42] has derived theoretically the form of the yield *versus* energy curve near the threshold points for a general shape of the band structure and for a variety of excitation and scattering processes. (Actually, the quantity considered in all such calculations is the yield per *absorbed* photon. The yield per *incident* photon defined above can readily be evaluated if the spectral reflectivity of the semiconductor is known.) The theory is based on density-of-states considerations, the optical absorption and transition probabilities being assumed not to vary rapidly near threshold. The analysis shows that each of the processes involved should give for the yield Y above the corresponding threshold E_T a power law of the form

$$Y \propto (hv - E_T)^r, \qquad (7.48)$$

where r assumes integral or half-integral values lying between 1 and 5/2, depending on the type of excitation and scattering. These values are listed in Table 7.1, taken from Kane[42]. Also included are the corresponding threshold energies for the case of flat bands at the emitting surface. The upper part of the Table refers to the purely bulk processes discussed above, the lower part to processes in which the surface is involved in one way or another. The latter will be considered a little further on.

So far the energy bands were assumed flat. The threshold energies will evidently be different when a potential barrier exists at the surface, lower for bands bending downward ($V_s > 0$) and higher for bands bending upward ($V_s < 0$). The former case is illustrated in Figs. 7.5c and 7.5d, where transitions are considered from a point just below the space-charge layer. (The light penetration and the electron escape depths are both assumed sufficiently large for such transitions to be effective in photoemission.) For indirect transitions, E_{Ti} will clearly be reduced by qV_s. For direct transitions, on the other hand, one has to take into account the fact that the threshold excitation now involves a smaller value of $|k_z|$, corresponding to a lower energy of the hole left behind. The reduction in E_{Td} will accordingly be greater than qV_s, by an amount depending on the detailed structure of the two bands. The direct threshold may thus be expressed approximately as $E_{Td} - bqV_s$, where the numerical factor b is larger than unity and is usually determined experimentally[36].

In order to calculate the overall yield[36,43] one has to integrate over the distance below the surface, considering both the effect of the bending of the bands and also the attenuation of light and of the emitted electrons with depth. The former is introduced by taking the indirect and direct thresholds at any point z (and potential V) as $E_{\text{Ti}} + q(V - V_s)$ and $E_{\text{Td}} + bq(V - V_s)$, respectively. (The factor b is assumed independent of z.) The attenuation with depth is introduced by means of a parameter l defined as $(1/l_v + 1/l_e)^{-1}$, where l_v and l_e are the optical absorption and the electron escape depths, respectively, both assumed constant over the range of wavelengths treated. Thus, if we consider both direct transitions with no scattering and indirect transitions, the yield can be expressed in the form (see Table 7.1)

$$Y(v) = a_i \int_0^\infty [hv - E_{\text{Ti}} - q(V - V_s)]^{\frac{5}{2}} e^{-z/l} dz$$
$$+ a_d \int_0^\infty [hv - E_{\text{Td}} - bq(V - V_s)] e^{-z/l} dz. \quad (7.49)$$

The integration is carried out only over those values of z which, for a given v, yield positive values for the integrands. The coefficients a_i and a_d are the appropriate proportionality factors for the indirect and direct processes, respectively. These coefficients, as well as the factor b, can be determined by a numerical evaluation of the integrals and curve fitting to the experimental data. Clearly, only for photon energies lower than the direct threshold will the much weaker indirect process be resolvable experimentally.

In addition to the bulk excitation and scattering processes discussed above, one has to consider similar processes involving the surface (lower part of Table 7.1). Two main categories may be distinguished. The first consists of bulk excitation, as before, but now the surface rather than the bulk participates in the momentum exchange. The threshold for this type of transition will be the same as the indirect bulk threshold, namely $\chi + E_g$. The values of the exponent r in (7.48), however, will depend on whether the surface is a diffuse or a specular electron scatterer (see definitions in Ch. 8, § 1), as indicated in the Table.

The second category of surface processes consists of photoemission from occupied surface states. Due to the relatively low density of such states, the photoelectric yield in this case will obviously be much smaller than for bulk excitation. Since the threshold for surface-state emission lies below $\chi + E_g$, however, such emission may be detectable as a small tail in the Y versus $h\nu$ plot. For photoemission from a single discrete set of localized surface states at energy E_t, the threshold will be just $\chi + E_c - E_t$. In the case of a continuous distribution of localized states, the threshold will correspond to emission from the topmost levels if all states lie below E_F (and are thus fully occupied) or to emission from E_F if the Fermi level passes through the states. The functional dependence of the yield will be different in the two cases (Table 7.1). Kane[42] also considers transitions from surface-state bands which, as we have seen in Ch. 5, §§ 2, 3, may be present on an ideal surface. The photoemission characteristics depend in this case on whether the momentum tangential to the surface is or is not conserved ("direct" or "indirect" transitions) and whether the threshold occurs at or away from the Fermi level. These four processes are indicated in the last rows of the Table.

Experimentally, one measures the photoelectric yield as a function of the wavelength of the exciting light[44]. The sample is provided with an electric contact and is inserted into a high-vacuum chamber. In the case of real surfaces, the sample is etched or otherwise treated prior to mounting. Clean or cleaved surfaces are produced in the same high-vacuum chamber. Several cleaning cycles or cleavage operations can be performed successively by manipulating all elements in the high-vacuum chamber from the outside through bellows[30,36-39]. Quite often, photosensitive materials are produced by evaporation techniques[45].

The optical system consists of a light source and a monochromator. The former is typically a mercury arc, the wavelengths of interest (for germanium and silicon) usually lying in the ultraviolet range. A well-defined spot of monochromatic light is focused through a quartz window on the desired region of the surface. To minimize photoemission from other parts of the chamber, the beam is oriented so as to strike the surface at normal incidence. In this manner the specularly reflected light can leave the chamber by the same window through which it entered. A portion of the beam ($\approx 10\%$) is sometimes split off and focused on a photosensitive detector for monitoring the main-beam intensity. To determine the flux of photons actually absorbed by the sample, one also needs to know the reflectivity of the surface at each wavelength. This is available for many semiconductors; in the other cases, provisions can be made to measure the reflectivity while the sample is in the chamber[36].

The emission current is often quite low ($\approx 10^{-12}$A), necessitating the use of sensitive measuring techniques such as an electrometer or chopped light with ac amplification. (The latter procedure eliminates the effect of the dark current.) To suppress spurious currents due to emission from surfaces other than that of the sample, the sample is mounted on an insulated support and only the current leaving it is registered — all other parts of the system are at ground potential and serve as the collector. This precaution is not necessary when the threshold lies in the visible range, the emission from the other surfaces of the system then being negligible. The voltage applied between the sample (cathode) and the collector (anode) is made sufficiently high to be well within the range of current saturation (negligible space charge), but excessive voltage should be avoided[46]. For single-crystal samples a voltage of the order of 10 V is usually sufficient for current saturation, but larger voltages may be necessary in the case of thin evaporated layers because of their high resistance. The energy distribution of the emitted electrons can be determined by retarding potential measurements or by magnetic deflection techniques.

4. High-field effects

A variety of electronic processes in solids may set in or undergo a drastic change in the presence of high electric fields. In the bulk, carriers may acquire energies well above thermal due to the inability of the scattering process to dissipate completely the energy gained from the field. The transport of such "hot carriers" is generally no longer characterized by a constant, field-independent mobility, and the simple proportionality relation between current and applied voltage (Ohm's law) breaks down[47]. At still higher fields, the energy gain may be sufficient to produce excess carriers by impact ionization[48,49], either from localized levels in the forbidden gap or directly across the gap. Such processes can become multiplicative, giving rise to very large currents — avalanche breakdown. A second category of high-field effects, in which the surface is often involved, consists of quantum-mechanical tunnelling through a potential barrier between two allowed states. This can take place either from the semiconductor into the space outside — external field emission[50], or inside the semiconductor — internal field emission[51]. The latter may proceed between the valence and conduction bands and also between localized bulk or surface states and either of the bands.

High-field effects are sometimes inadvertently present in surface measurements, or else they may be specifically invoked to study surface properties. It would therefore be useful to review briefly some of the more important effects of this sort and to note the manner in which they manifest themselves experimentally.

External field emission from semiconductors has been treated [52,53] along lines similar to those employed by Fowler and Nordheim [54,55] for metals. In semiconductors, however, there are special effects associated with the state of the surface and with the not insignificant penetration of the external field into the interior. Figure 7.6 shows a potential-energy diagram inside and outside of an n-type semiconductor in the presence of an external field of the relevant polarity. The potential barrier through which electrons are required to tunnel consists, in its simplest form, of an abrupt rise corresponding to the

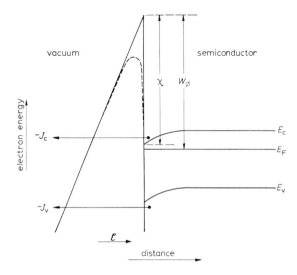

Fig. 7.6. Schematic representation of external field emission from the conduction and valence bands. The dashed curve illustrates the modification in the potential barrier introduced by the image force.

electron affinity χ, followed by a sloping side whose gradient is the external field $|\mathscr{E}|$ (solid line in the figure). The tunnelling currents from the conduction and valence bands (J_c and J_v) are represented by horizontal arrows to signify that the total energy must be conserved in the transition (phonon interactions being neglected). For conduction-band electrons, the emission current density is approximately given by [53]

$$J_c \propto n_s \exp\left[-\frac{4\sqrt{2m^*}}{3\hbar q} \frac{\chi^{\frac{3}{2}}}{|\mathscr{E}|}\right] \quad (7.50)$$

under non-degenerate conditions, and by

$$J_c \propto \exp\left\{ - \frac{4\sqrt{2m^*}}{3\hbar q} \frac{[\chi-(E_F-E_c)]^{\frac{3}{2}}}{|\mathcal{E}|} \right\} \qquad (7.51)$$

when the Fermi level is above the conduction-band edge at the surface. The proportionality factors in (7.50) and (7.51) vary relatively slowly with the applied field (and temperature), and may be regarded as constants. The exponential terms express the rapid attenuation of the electron wave function, and thus of the tunnelling probability, with tunnelling distance. The field is assumed to be sufficiently strong and the temperature not too high so that most of the emitted electrons originate from a small energy interval, situated above E_c in the first case (eq. (7.50)) and around E_F in the second (eq. (7.51)). At the other extreme, low fields and high temperatures, emission *over* the potential barrier, rather than through it, will predominate. This of course corresponds to thermionic emission. There is also an intermediate range, where electrons tunnel through the barrier but come mainly from energy levels well above E_c or E_F. This, the so-called T−F emission[56], will depend strongly on both temperature and field.

In reality, the sharp peak of the surface barrier in Fig. 7.6 is rounded off due to the image force between the electron and its electrostatic image in the semiconductor[55]. The actual barrier looks more like that illustrated by the dashed curve in the figure. This introduces a correction[53] into (7.50) and (7.51).

Equation (7.50) is expected to hold at low fields, while at high fields (7.51) will be applicable. The transition from the one to the other depends largely on the density and energy distribution of the surface states. A large density may anchor the position of the Fermi level at the surface and maintain non-degenerate conditions over a considerable range of applied field. Only after all surface states have been filled can the external field penetrate into the interior and produce, eventually, a degenerate accumulation layer. At this point a pronounced rise in emission current will take place, with J_c being given by (7.51) rather than by (7.50). If the rise in emission can be observed experimentally, it may serve as an indirect indication of the presence of surface states. Although emission from surface states may also occur, it will probably be masked by the much larger emission from the degenerate conduction band.

Due to its high electron density, the contribution of the valence band to the emission current may not be negligible even though the electrons must tunnel through a barrier higher by E_g, the forbidden gap. The results of the calculations[53] in this case are similar to those for tunnelling from the conduction band, except for the difference in the proportionality factors and in the exponents (where χ is replaced by $\chi + E_g$). In germanium at room temperature, for example, the emission current from the two bands may be comparable.

The experimental arrangement for studying external field emission[57] is very similar to that employed in the field emission microscope (Ch. 3, §2.3). Fields in the range 10^7 to 10^8 V/cm, required for a measurable emission current, are produced as before by shaping the emitter in the form of a fine tip (radius of curvature 10^{-5}–10^{-4} cm). The emitter is placed inside an evacuated chamber lined with a conducting layer to collect the electrons. The measured quantities are the applied voltage and the emitted current. The field itself, which depends on the emitter tip geometry and is not known accurately, is simply taken to be proportional to the voltage. Actually it varies along the surface, and the emission current is an average over the entire emitter area. Unfortunately, such an average is not likely to exhibit the sharp transition in emission characteristics due to surface states.

Similar tunnelling processes are involved in *internal* field emission. One usually considers the case of a uniform field acting along one of the directions in the crystal. For band-to-band transitions (Zener effect), the barrier for tunnelling is of height E_g and of width $E_g/q|\mathscr{E}|$. The emission current density is given by [58,59]

$$J_{vc} \propto \exp\left[-\frac{\pi\sqrt{2m^*}}{4\hbar q}\frac{E_g^{\frac{3}{2}}}{|\mathscr{E}|}\right]. \qquad (7.52)$$

For emission from localized bulk or surface states into the conduction band (field ionization), on the other hand, the barrier height is $E_c - E_t$ and its width is $(E_c - E_t)/q|\mathscr{E}|$. The emission current density is then [60]

$$J_{tc} \propto \exp\left[-\frac{4\sqrt{2m^*}}{3\hbar q}\frac{(E_c - E_t)^{\frac{3}{2}}}{|\mathscr{E}|}\right]. \qquad (7.53)$$

These transitions are illustrated by the horizontal arrows in Fig. 7.7. Here again, the proportionality coefficients vary relatively slowly with applied field and with temperature. Regarding the effective masses m^* appearing in (7.52) and (7.53) (as well as in (7.50) and (7.51)), there is no obvious relation between them and the usual conductivity masses. They should be considered as "tunnelling effective masses" whose values are to be determined by experiment [61].

Impact ionization often sets in at the same range of fields as for internal emission. The two processes, while completely different, are not always easily distinguishable experimentally. In general, the current due to impact ionization increases more steeply with field than the tunnelling current, a criterion that is sometimes employed in separating the two. A more reliable test, however, is provided by photoconductivity measurements, as will be

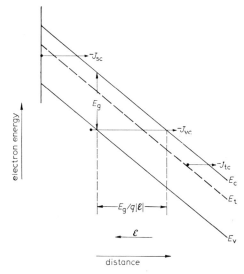

Fig. 7.7. Illustration of internal field emission into the conduction band from surface states (J_{sc}), from localized bulk levels (J_{tc}), and from the valence band (J_{vc}).

discussed below. Impact ionization is governed largely by the particular scattering processes in each crystal, and there is no simple universal expression relating the multiplication rate to the applied field. This problem has received considerable attention, and the interested reader is referred to the literature on the subject [48,49].

In insulators, the high fields (in the range 10^5 to 10^7 V/cm) required for the study of field emission and impact ionization can be produced by applying high voltages across a thin slab of the material [62]. Due care must be taken, however, to eliminate carrier injection from the end contacts (by field emission or otherwise), a condition that is hard to satisfy in most cases. Such an arrangement is not very feasible in semiconductors, where the high conductance makes the current prohibitively large. Instead, advantage may be taken of the high field present at the depletion layer of a reverse-biased p–n junction. Detailed investigations of the current–voltage characteristics in such structures have shed considerable light on breakdown processes in semiconductors[48]. For not too low resistivities of the p and n sections, impact ionization is the dominant mechanism in breakdown. In degenerate materials, on the other hand, the junction width can be as thin as 100 Å, and tunnelling from the valence into the conduction band predominates. This is the basis of operation of the well-known tunnel diode [63, 64].

Another technique used to produce high fields employs a Schottky barrier at the free surface. Such a depletion layer is formed by a capacitively applied field, using either an insulated field-effect electrode or an electrolyte as the blocking contact. In the former case, the measuring procedure[65, 66] is similar to that used in determining the time constant associated with the thermal emission of majority carriers from surface states (see Ch. 6, § 6.2). Now, however, the measurements are carried out at much higher pulse amplitudes, corresponding to fields of 10^4 to 10^5 V/cm at the semiconductor surface. Any impact ionization or field emission that may take place under these conditions will immediately be detectable by an enhanced rate of decay of the field-effect signal due to the faster redistribution of the induced charge among the surface states and the two bands. Rather than be emitted thermally, carriers in surface states will recombine with the minority carriers produced by impact in the first case, or tunnel directly into the majority-carrier band in the second case. The measurement is taken sufficiently close to the onset of the pulse, before any significant distortion of the initially induced Schottky barrier can take place due to screening by minority carriers (produced either thermally or by impact ionization). The initial field at the surface is thus well defined and can readily be calculated from the observed change in surface conductance. An attractive feature of this method is that the relaxation process acts as an automatic safety valve in limiting the total charge transported and thus in guarding against destructive breakdown of the sample.

In the case of conducting cadmium sulphide crystals, strong depletion layers corresponding to fields as high as 2×10^6 V/cm can be produced[67] by using an electrolyte as a blocking contact. A non-rectifying metal electrode is applied at one end of the sample and the other end is dipped into an aqueous solution of sodium or potassium chloride negatively biased with respect to the metal electrode. Due to the fact that the electrolyte blocks the flow of majority carriers (electrons) *into* but not the flow of minority carriers *out* of the crystal, no inversion layer can form at the surface and the Schottky barrier produced by the bias is maintained indefinitely. It is therefore sufficient to carry out all measurements under dc conditions. Here again, any breakdown process taking place in the space-charge layer will manifest itself by a sharp rise in current with increasing bias voltage. In order to distinguish between tunnelling and impact ionization[67], the same procedure used in reverse-biased p–n junctions[68] can be employed. This consists in observing the change in the current–voltage characteristics (as displayed, for example, on a CRO) upon illumination by strongly absorbed light. If tunnelling is the dominant mechanism (either from surface states or across the gap), no significant change will occur, except for a relatively small *constant* contribution representing the saturated photocurrent. For impact ionization, on the other hand, the pre-breakdown current should increase considerably because of the increased number of primary electrons made available for multiplication.

The dc measurements just described are not likely to reveal tunnelling from surface states which, because of their limited number, will be rapidly exhausted before their effect can be detected. Pulse techniques, in which the current–voltage characteristics are taken at the onset of the pulse, are much more suitable for this purpose[69]. The method is similar to that of the pulsed field effect, except that now (as in the dc case) one measures the current normal to the surface rather than parallel to it.

5. Optical measurements

Light absorption in a non-polar semiconductor at wavelengths longer than the fundamental absorption edge arises chiefly from transitions of free carriers within an energy band or between localized levels and either band. Free-carrier absorption usually consists of a broad band in the infrared and increases with wavelength[70] roughly as λ^2. Localized-state absorption, on the other hand, is expected to exhibit absorption edges characteristic of the energy position of the states relative to the bands. These effects have been very successfully utilized by Harrick[71] for measuring the excess surface-carrier densities in the space-charge layer as well as for studying the characteristics of surface states and the structure of the adsorbed layer at the surface. The present section will outline the principles involved in these various measuring techniques.

The spatial distribution of excess carriers in the semiconductor bulk can be determined by narrow-beam light probes using conventional infrared spectroscopy[24], but the study of absorption due to the much fewer carriers in the space-charge layer or in surface states requires a considerable increase in sensitivity[72,73]. This is achieved in two ways, as illustrated[71] in Fig. 7.8. First, multiple internal reflections are used to enhance the signal strength. The semiconductor sample is prepared in the form of a thin slab with carefully polished surfaces. The two end surfaces are cut at an angle such that the monochromatic beam entering the sample at normal incidence strikes one of the large surfaces at an angle exceeding the critical angle. Under these conditions, the light penetrates only slightly into the rarer medium outside and is then totally reflected towards the other, parallel surface. From thereon

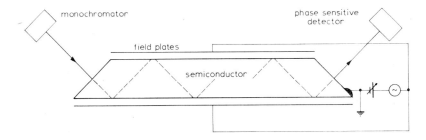

Fig. 7.8. Experimental arrangement for measuring infrared absorption at the surface, utilizing multiple reflections. (After Harrick, reference 71.)

the beam is reflected back and forth inside the slab, emerging finally through the other end (again at normal incidence), where its intensity is measured by means of the detector. With a sample 5 cm long and 0.5 mm thick, one hundred internal reflections can be attained for an angle of incidence of 45°, corresponding to a hundredfold increase in the sensitivity of the absorption measurement. An even larger number of reflections is obtainable with thinner samples and smaller angles.

Since no power is lost by transmission out of the sample, the attenuation of light is due entirely to absorption in the semiconductor and in the thin layer outside (within a penetration depth of the totally reflected light). In order to determine the contribution of the carriers at the *surface* (both free and trapped) to the overall absorption, their density is modulated by means of a small capacitively applied ac field, as shown in the figure. The resulting modulation in the beam intensity is then a direct measure of the surface-carrier absorption. Through the use of a phase-sensitive detector, very high gains can be achieved. Excess surface-carrier densities as low as 10^{10} cm^{-2} were detected by Harrick[73], corresponding to a gain of about 10^4 in sensitivity over that obtained by conventional infrared measurements.

Having determined the absorption spectrum due to the surface-carriers, one's next problem is to isolate the various transitions involved. Evidently, free carrier absorption in the space-charge region will be the dominant process at long wavelengths because of its λ^2 dependence. Accordingly, measurements in this range can serve to determine the excess surface-carrier densities if the absorption coefficient of holes and electrons is known[74]. Moreover, by extrapolating the results to shorter wavelengths (using the λ^2 law) and subtracting the calculated values from the measured absorption spectrum, one can derive the absorption due exclusively to the surface states[73]. Measurements of this sort, carried out for different values of the barrier height at the surface, can be used to determine the energy distribution of the surface states. The interpretation of the data will be straightforward if only one discrete set of surface states dominates the absorption process. In this case two absorption bands with well-defined absorption edges may appear, one associated with electron transitions from occupied states into various levels in the conduction band, the other with transitions from the valence band into unoccupied states. The predominance of one band over the other will depend on the state of occupancy of the surface levels and can be controlled by varying the barrier height. The situation is considerably more complicated in practice, where several sets of surface states are usually present. The interpretation of the data in this case becomes quite involved, requiring careful analysis of many absorption spectra obtained at closely spaced values of barrier height[73]. Despite these difficulties, however, such measurements are particularly valuable since, besides demonstrating unambiguously the presence of surface states, they provide direct information on their energy position.

Due to the special characteristics peculiar to total internal reflection, the multiple reflection technique also constitutes a very sensitive tool in studying the adsorbed layer *outside* the surface proper[71, 75]. As predicted by Maxwell's equations and amply confirmed by experiment, light penetration into the rarer medium outside takes place under conditions of total reflection. Near grazing incidence, the penetration depth is only about $\lambda/10$, but it increases rapidly as the critical angle is approached. Any absorption in the rarer medium will make the reflection less than total. This effect can thus be used for measuring the absorption spectra of various adsorbed media. The variation of the penetration depth with angle of incidence points to the possibility of physically locating the adsorbed species. The method shows great promise for identifying foreign matter on the surface, even when present with monolayer coverage, as well as for gaining information on the nature of the bonding at the surface [76]. Investigation of adsorption characteristics on clean surfaces should prove particularly instructive[77].

6. Magnetic measurements

6.1. GALVANOMAGNETIC EFFECTS

Measurements of Hall effect and magnetoresistance have been used to estimate the surface mobility associated with carriers constrained to move in a potential well at the surface (Ch. 8, § 5 below). In general, the results

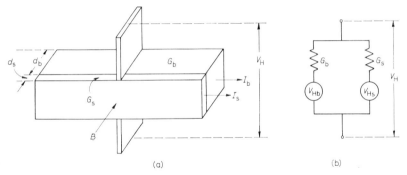

Fig. 7.9. A simplified representation of the bulk and surface contributions to the Hall effect.
(a) Hall configuration.
(b) Equivalent circuit used in calculating the resultant Hall coefficient.
(After Petritz, reference 78.)

reflect the contribution from both bulk and surface carriers. An oversimplified representation[78] of these contributions is shown in Fig. 7.9. The sample is considered as consisting of two parallel conductors, one corresponding to the space-charge layer (thickness d_s, conductance G_s), the other to the underlying bulk region (thickness d_b, conductance G_b). The two conductors will usually differ in both carrier density and mobility. The Hall voltage produced by a magnetic field applied normal to the surface is measured across the two side arms shown protruding from the sample.

It can be calculated from elementary circuit theory if one assumes that the two conductors are insulated from each other, except at the upper and lower ends where they are shortened by the Hall probes. The equivalent circuit applicable under these conditions is shown in Fig. 7.9b. The individual currents and Hall voltages are given by (see eq. (2.159))

$$I_b = \sigma_b d_b I/(\sigma_b d_b + \sigma_s d_s); \quad I_s = \sigma_s d_s I/(\sigma_b d_b + \sigma_s d_s);$$
$$V_{Hb} = R_{Hb} I_b B/d_b; \quad V_{Hs} = R_{Hs} I_s B/d_s; \quad (7.54)$$
$$I = I_b + I_s; \quad d = d_b + d_s.$$

The subscripts b and s refer to the bulk and the surface, respectively, σ_b and σ_s are the corresponding conductivities, R_{Hb} and R_{Hs} are the Hall coefficients, I_b and I_s are the currents, and B is the magnetic induction. An overall Hall coefficient R_H can be defined in terms of the measured open-circuit Hall voltage V_H:

$$R_H = V_H d/IB. \quad (7.55)$$

By reference to the equivalent circuit of Fig. 7.9b, it can readily be shown that R_H is given by

$$R_H = \frac{R_{Hb} \sigma_b^2 d_b + R_{Hs} \sigma_s^2 d_s}{\bar{\sigma}^2 d}, \quad (7.56)$$

where $\bar{\sigma} \equiv (\sigma_b d_b + \sigma_s d_s)/d$ is the average sample conductivity.

This approach is approximate in that it does not take into account the variation of carrier density with depth in the space-charge layer, nor does it include the possibility of circulating currents due to communication between the surface layer and the bulk. It does illustrate, however, how the surface contributes to the measured Hall effect. A more rigorous treatment of bulk and surface effects[78] will be discussed in detail in Ch. 8, § 4. At this point suffice it to say that such an analysis yields much the same expression for the Hall coefficient as (7.56).

The experimental procedure[79] consists in varying the surface barrier height and measuring the resulting change in sample conductance, Hall coefficient, and magnetoresistance. The last two are then plotted as a function of the fractional change in sample conductivity $\bar{\sigma}$, the minimum conductivity $\bar{\sigma}_{min}$ being used as a reference point. To maximize the surface contribution to the Hall coefficient (see eq. (7.56)) and to the magnetoresistance, high resistivity material is used (intrinsic or near intrinsic) and the sample thickness is made very small (hundredths to tenths of a millimetre).

A schematic diagram of the apparatus used by Zemel and Petritz[79] is shown in Fig. 7.10. The sample is enclosed in a chamber and placed between the pole pieces of an electromagnet. A bucking voltage is used to reduce the *IR* drop resulting from the misalignment of the voltage probes (one on either side of the sample). The misalignment permits the simultaneous measurement of conductance, Hall voltage, and magnetoresistance. The magnetic field is continually turned on, turned off, reversed, and turned off at intervals of 5 sec, while the gaseous ambient in the chamber is cycled.

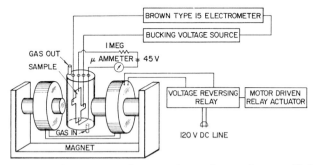

Fig. 7.10. Experimental arrangement for measuring surface conductance, Hall coefficient, and magnetoresistance as a function of barrier height. The barrier height is varied by a gaseous ambient cycle. (After Zemel and Petritz, reference 79.)

The voltage V_1 measured across the probes in the "off" intervals represents the change in surface conductance with ambient gas. The voltage V_2 measured for the magnetic field in a given direction is the sum of three components: the Hall voltage, the change in voltage drop due to the magnetoresistance, and V_1. In the opposite polarity of the field, the measured voltage V_3 is the same sum, but with the sign of the Hall voltage reversed. Thus the Hall voltage is given by $\frac{1}{2}(V_3 - V_2)$ and the magnetoresistance voltage by $\frac{1}{2}(V_2 + V_3) - V_1$.

Galvanomagnetic measurements of this sort can be advantageously applied to p–n–p (or n–p–n) surface channels[80]. Here, as in the channel conductance experiment (see Ch. 6, § 8), bulk effects are essentially absent. The Hall coefficient and the magnetoresistance thus yield directly the minority-carrier surface mobility in the inversion layer.

6.2. PARAMAGNETIC RESONANCE

Several workers[81–84] have applied electron or nuclear paramagnetic resonance techniques to the study of solid surfaces. In contrast to the case of the bulk, where such techniques have proved very valuable in characterizing various paramagnetic species[85], only a limited amount of information has been obtained so far on surface properties. The only paramagnetic centres that could be directly attributed to the surface were those introduced by severe mechanical damage (crushing, abrading, or sand blasting). Nevertheless, the potential interest for surface studies of several features of paramagnetic resonance warrants a brief discussion of the basic principles involved.

Paramagnetic resonance consists of the resonance absorption of electromagnetic radiation by paramagnetic species in a given material when subjected to a magnetic field. Such species are atoms, ions, nuclei, or any centres possessing a non-vanishing angular momentum and permanent magnetic moment. In the simplest case, the energy of a centre of angular momentum J is split by the field into equally spaced levels of energy $gM\mu_0 B$. Here g is the splitting factor, M the magnetic quantum number, and B the magnetic induction; μ_0 is the Bohr magneton for electron resonance and the nuclear magneton for nuclear resonance. In the former case, the magnetic moment associated with the orbital angular momentum is often "tied down" (quenched) by the lattice and is not free to orient itself in the magnetic field. Under these conditions only the spin can contribute significantly to the splitting of the levels, with J effectively equal to S, the spin angular momentum. In the presence of electromagnetic radiation of angular frequency ω, transitions may occur between neighbouring levels provided the resonance condition $\hbar\omega = g\mu_0 B$ is satisfied. For a magnetic field of several thousand oersted, ω is typically in the radio frequency range for nuclear resonance and in the microwave region for electron resonance (the Bohr magneton being about 2000 times larger than the nuclear magneton).

The experimental technique for measuring nuclear resonance is as follows: the sample is mounted within a coil whose axis is perpendicular to the direction of the applied magnetic field. The coil is part of a tuned circuit which is excited by a constant-current rf source of fixed frequency. At resonance, the absorption of the sample lowers the Q of the tuned circuit and thus the voltage across the coil. The measuring procedure consists of varying the magnetic field and noting the values at which power absorption takes place. For electron resonance, the rf coil is replaced by a resonant microwave cavity. Power absorption at resonance is detected either by transmission or by reflection techniques. Bridge methods are employed to improve sensitivity.

From the position and magnitude of the resonance absorption one may identify the species involved (through their g values) and determine their number. Two other quantities of interest are the spin–lattice and spin–spin relaxation times, T_1 and T_2. The former characterizes the return of the spin system to thermal equilibrium following the cessation of the external excitation. The latter is associated with the mutual interaction between various spins and is usually inversely proportional to the resonance width.

Measurements of surface effects are generally performed on crushed samples so as to increase the sensitivity. Walters[83] estimates that for electron resonance, as few as 10^{12} paramagnetic centres can be detected, while for nuclear resonance 10^{19} centres are required.

7. Noise

Electrical noise in an electronic conductor consists of random fluctuations about average in the current flowing through (or the voltage developed across) the conductor. It arises from the fact that the current is carried by discrete particles — electrons — whose number and velocity are apt to fluctuate due to the statistical nature of an ensemble of this sort. Thermal noise is a well-known example of fluctuations in carrier *velocity*, while shot noise in thermionic emission serves to illustrate fluctuations in the *number* of carriers. In semiconductors, noise may arise from minority-carrier injection or extraction at the contacts. Apart from this contribution, which can usually be reduced to negligible proportions by a proper choice of the experimental conditions, noise in semiconductors is associated mainly with fluctuations in the number of volume carriers[86-88].

One source of density fluctuations can be directly correlated with generation–recombination (G–R) processes in the bulk and at the surface. Even though under equilibrium conditions the *average* densities of free and trapped carriers are each maintained constant, statistical fluctuations between the two categories are bound to occur. For simple G–R processes characterized by a single time constant τ, the mean square noise current $\langle i^2 \rangle$ in a frequency interval Δf is given by

$$\langle i^2 \rangle = AI_{dc}^2 \frac{\tau}{1+\omega^2 \tau^2} \Delta f. \qquad (7.57)$$

Here I_{dc} is the dc current flowing through the sample and $\omega \equiv 2\pi f$ is the angular frequency at which the noise level is evaluated. The coefficient A depends on the specific G–R process considered — whether it involves transitions with recombination centres or traps, or band-to-band transitions. (The latter process is negligible in most semiconductors — see Ch. 2, § 7.2.) In the absence of significant trapping effects, τ usually represents the effective sample lifetime τ_{eff} as determined by bulk and surface recombination. As is seen from (7.57), the G–R noise is largest at $\omega \ll 1/\tau$ and falls off at higher frequencies. This feature can be used to separate the G–R component from the total noise.

The other important contribution to the overall noise is the $1/f$ noise. Its frequency dependence is given by

$$\langle i^2 \rangle \propto I_{dc}^n (1/f^m) \Delta f, \tag{7.58}$$

where m lies somewhere between 0.7 and 1.5 and n is around 2. This is usually the dominant noise current at low frequencies but its origin is not well understood. Accumulated evidence on germanium and silicon (see Ch. 9, § 1) shows that $1/f$ noise arises predominantly, if not entirely, from surface effects, mostly those associated with slow surface states.

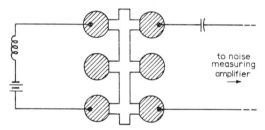

Fig. 7.11. Sample shape used in surface noise measurements. The large-area contacts on the side arms serve to minimize minority-carrier injection. (After Montgomery, reference 89.)

A typical experimental arrangement[89] for studying noise in semiconductors is shown in Fig. 7.11. Current is fed into the sample via the two outer electrodes on the left, the opposite electrodes being used to measure the noise signal. The middle electrodes (shown disconnected) can be employed to measure Hall effect. The Hall voltage, being inversely proportional to the carrier density, is another means of observing the noise. All six electrodes are applied on large sand-blasted side arms, which are an integral part of the sample. The low field and high recombination rates at the sand-blasted arms minimize the effect of minority-carrier injection (if present). This contribution is further suppressed by the large inductance in the bias circuit. The central part of the sample, where the noise level is to be investigated, may be etched, sand blasted, or otherwise treated. Provision can also be made to mount field-effect electrodes near the surface.

The noise spectrum is measured by means of a tuned amplifier whose resonance frequency can be varied over the range of interest. Since the noise signal is of the order of microvolts, rather stringent requirements must be satisfied by the amplifier, especially as regards its noise level. In order to determine the mean square noise current (or voltage), the amplified signal is fed into a long-time-constant detector. The probable relative error in the measured noise level is then given by $(2\Delta f \tau)^{-\frac{1}{2}}$, where Δf is the bandwidth of the tuned amplifier and τ the time constant of the detector. To show this, we recall that the output from the tuned amplifier corresponds to a wave of frequency f, amplitude-modulated at an average frequency of Δf. In other words, the average rate at which the output amplitude varies is $2\Delta f$. On the other hand, since the measuring time involved by the use of the detector is effectively τ, the total number N of independent observations embodied in the reading is $2\Delta f \tau$. The result quoted above now follows immediately from the fact that the noise obeys Poisson's distribution law, so that the relative probable error in the average value obtained by a series of N measurements is equal to $N^{-\frac{1}{2}}$.

It is evident that for adequate accuracy, both the detector time constant and the amplifier bandwidth should be made large. Good resolution of the noise spectrum, however, necessitates a small bandwidth, and a compromise between these two opposing requirements must be struck. There are no limitations of this sort on the detector time constant, and it is taken as long as feasible. The detector may be calibrated by a standard signal source connected in parallel to the sample. Either a sinusoidal signal generator or a known noise source (such as the thermal noise across a hot wire) can be used for this purpose.

References

[1] W. Shockley, *Electrons and Holes in Semiconductors* (Van Nostrand, New York, 1950).
[2] E. S. Rittner in *Photoconductivity Conference* (Edited by R. G. Breckenridge, B. R. Russell, and E. E. Hahn), (John Wiley, New York, 1956), p. 215.
[3] W. van Roosbroeck, Phys. Rev. **101** (1956) 1713.
[4] See, for example, R. Bray and A. Many in *Methods of Experimental Physics* (Edited by L. Marton), vol. 6B, *Solid State Physics* (Edited by K. Lark-Horovitz and V. A. Johnson), (Academic Press, New York, 1959), p. 78. (Review)
[5] H. Y. Fan, Theory of Photoconductivity. Unpublished report, June, 1952.
[6] A. Many, Proc. phys. Soc. London **B67** (1954) 9.
[7] J. R. Haynes and J. A. Hornbeck, Phys. Rev. **100** (1955) 606; J. A. Hornbeck and J. R. Haynes, Phys. Rev. **97** (1955) 311.

[8] R. L. Watters and G. W. Ludwig, J. appl. Phys. **27** (1956) 489.
[9] J. B. Cladis, C. S. Jones, and K. A. Wickersheim, Rev. sci. Instr. **27** (1956) 83.
[10] G. K. Wertheim and W. M. Augustyniak, Rev. sci. Instr. **27** (1956) 1062.
[11] D. Navon, R. Bray, and H. Y. Fan, Proc. Inst. Radio Engrs. **40** (1952) 1342.
[12] J. P. McKelvey and R. L. Longiny, J. appl. Phys. **25** (1954) 634.
[13] H. Suhl and W. Shockley, Phys. Rev. **76** (1949) 180.
[14] W. Shockley, G. L. Pearson, and J. R. Haynes, Bell System tech. J. **28** (1949) 344.
[15] H. Christensen, Proc. Inst. Radio Engrs. **42** (1954) 1371.
[16] A. R. Moore and W. M. Webster, Proc. Inst. Radio Engrs. **43** (1955) 427; Physica **20** (1954) 1046.
[17] J. R. Haynes and W. Shockley, Phys. Rev. **81** (1951) 835.
[18] F. S. Goucher, Phys. Rev. **81** (1951) 475.
[19] D. G. Avery and J. B. Gunn, Proc. phys. Soc. London **B68** (1955) 918.
[20] G. Adams, Physica **20** (1954) 1037.
[21] L. B. Valdes, Proc. Inst. Radio Engrs. **40** (1952) 1420.
[22] W. van Roosbroeck, J. appl. Phys. **26** (1955) 380.
[23] E. M. Pell, Phys. Rev. **90** (1953) 278.
[24] N. J. Harrick, Phys. Rev. **101** (1956) 491; **103** (1956) 1173; J. appl. Phys. **27** (1956) 1439.
[25] E. O. Johnson, J. appl. Phys. **28** (1957) 1349.
[26] I. F. Patai and M. A. Pomerantz, J. Franklin Inst. **252** (1951) 239.
[27] Lord Kelvin, Phil. Mag. [5] **46** (1898) 82.
[28] W. H. Brattain and C. G. B. Garrett in *Methods of Experimental Physics* (Edited by L. Marton), vol. 6B, *Solid State Physics* (Edited by K. Lark-Horovitz and V. A. Johnson), (Academic Press, New York, 1959), p. 136.
[29] W. H. Brattain and J. Bardeen, Bell System tech. J. **32** (1953) 1.
[30] F. G. Allen and G. W. Gobeli, Phys. Rev. **127** (1962) 150.
[31] W. H. Brattain and C. G. B. Garrett, Bell System tech. J. **35** (1956) 1019.
[32] E. O. Johnson, Phys. Rev. **111** (1958) 153.
[33] H. Dember, Z. Physik **32** (1931) 554, 856; **33** (1932) 207.
[34] F. Berz and D. C. Emmony, Phys. Letters **2** (1962) 197.
[35] C. G. B. Garrett and W. H. Brattain, Phys. Rev. **99** (1955) 376.
[36] G. W. Gobeli and F. G. Allen, Phys. Rev. **127** (1962) 141.
[37] F. G. Allen and G. W. Gobeli, *Proc. Int. Conf. Semiconductor Physics, Exeter*, 1962 (The Institute of Physics and the Physical Society, London, 1962), p. 818.
[38] J. van Laar and J. J. Scheer, Philips Res. Rep. **17** (1962) 101.
[39] J. van Laar and J. J. Scheer, *Proc. Int. Conf. Semiconductor Physics, Exeter*, 1962 (The Institute of Physics and the Physical Society, London 1962), p. 827.
[40] L. Apker and E. Taft, J. opt. Soc. Am. **43** (1953) 78.
[41] A. H. Sommer, Optica Acta **7** (1960) 121.
[42] E. O. Kane, Phys. Rev. **127** (1962) 131.
[43] W. E. Spicer, J. appl. Phys. **31** (1960) 2077.
[44] J. Eisinger, J. chem. Phys. **29** (1958) 1154.
[45] A. H. Sommer and W. E. Spicer in *Methods of Experimental Physics* (Edited by L. Marton), vol. 6B, *Solid State Physics* (Edited by K. Lark-Horovitz and V. A. Johnson), (Academic Press, New York, 1959), p. 376.
[46] J. S. Preston and G. W. Gordon-Smith, Brit. J. appl. Phys. **6** (1955) 329.
[47] K. J. Schmidt-Tiedemann in *Festkörperprobleme*, vol. I (Edited by F. Sauter), (Friedr. Vieweg, Braunschweig, 1962), p. 122. (Review)
[48] See, for example, J. Jamoshita in *Progress in Semiconductors*, vol. 4 (Edited by A. F. Gibson, F. A. Kröger, and R. E. Burgess), (Heywood, London, 1960), p. 63. (Review)

[49] See, for example, R. Stratton in *Progress in Dielectrics*, vol. 3 (Edited by J. B. Birks), (Heywood, London, 1961), p. 234. (Review)
[50] See, for example, R. H. Good and E. W. Müller in *Handbuch der Physik*, vol. 21 (Edited by S. Flügge), (Springer-Verlag, Berlin, 1956), p. 176. (Review)
[51] See, for example, A. G. Chynoweth in *Progress in Semiconductors*, vol. 4 (Edited by A. F. Gibson, F. A. Kröger, and R. E. Burgess), (Heywood, London, 1960), p. 95. (Review)
[52] N. Morgulis, Zhur. tekh. Fiz. **17** (1947) 983.
[53] R. Stratton, Proc. phys. Soc. London **B68** (1955) 746; Phys. Rev. **125** (1962) 67.
[54] R. H. Fowler and L. Nordheim, Proc. roy. Soc. London **A119** (1928) 173.
[55] L. Nordheim, Proc. roy. Soc. London **A121** (1928) 626.
[56] W. W. Dolan and W. P. Dyke, Phys. Rev. **95** (1954) 327.
[57] L. A. D'Asaro, J. appl. Phys. **29** (1958) 33.
[58] C. Zener, Proc. roy. Soc. London **A145** (1934) 523.
[59] W. Franz, Ergeb. exakt. Naturwiss. **27** (1953) 1; *Handbuch der Physik*, vol. 17, *Dielectric Breakdown* (Edited by S. Flügge), (Springer-Verlag, Berlin, 1956).
[60] W. Franz, Ann. Phys. **11** (1952) 17.
[61] A. G. Chynoweth, W. L. Feldmann, and R. A. Logan, Phys. Rev. **121** (1961) 684.
[62] A. von Hippel, E. P. Gross, J. G. Jelatis, and M. Geller, Phys. Rev. **91** (1953) 568.
[63] L. Esaki, Phys. Rev. **109** (1958) 603.
[64] R. A. Logan and A. G. Chynoweth, Phys. Rev. **131** (1963) 89.
[65] E. Harnik, Y. Goldstein, N. B. Grover, and A. Many, J. Phys. Chem. Solids **14** (1960) 193.
[66] A. Many and Y. Goldstein, *Proc. Int. Conf. Physics and Chemistry of Solid Surfaces*, Providence, 1964 (Surface Science **2** (1964) 114).
[67] R. Williams, Phys. Rev. **125** (1962) 850.
[68] A. G. Chynoweth and K. G. McKay, Phys. Rev. **106** (1957) 418.
[69] A. Many, J. Phys. Chem. Solids. To be published.
[70] H. J. G. Meyer, Phys. Rev. **112** (1958) 298.
[71] N. J. Harrick, Ann. N. Y. Acad. Sci. **101** (1963) 928. (Review)
[72] N. J. Harrick, J. Phys. Chem. Solids **14** (1960) 60.
[73] N. J. Harrick, Phys. Rev. **125** (1962) 1165.
[74] N. J. Harrick, J. Phys. Chem. Solids **8** (1959) 106.
[75] N. J. Harrick, J. phys. Chem. **64** (1960) 1110; Anal. Chem. **36** (1964) 188.
[76] L. H. Sharp, Proc. chem. Soc. London 461 (Dec. 1961).
[77] G. E. Becker and G. W. Gobeli, J. chem. Phys. **38** (1963) 2942.
[78] R. L. Petritz, Phys. Rev. **110** (1958) 1254.
[79] J. N. Zemel and R. L. Petritz, Phys. Rev. **110** (1958) 1263.
[80] N. St. J. Murphy and T. B. Watkins, *Proc. Int. Conf. Semiconductor Physics, Prague*, 1960 (Czechoslovak Academy of Sciences, Prague, 1961), p. 552.
[81] R. C. Fletcher, W. A. Yager, G. L. Pearson, A. N. Holden, W. T. Read, and F. R. Merritt, Phys. Rev. **94** (1954) 1392.
[82] G. Feher, Phys. Rev. **114** (1959) 1219.
[83] G. K. Walters, J. Phys. Chem. Solids **14** (1960) 43.
[84] G. K. Walters and T. L. Estle, J. appl. Phys. **32** (1961) 1854.
[85] See, for example, G. E. Pake, *Paramagnetic Resonance* (Benjamin, New York, 1962).
[86] A. van der Ziel, *Noise* (Prentice Hall, New York, 1954); *Fluctuation Phenomena in Semiconductors* (Academic Press, New York, 1959).
[87] K. M. van Vliet, Proc. Inst. Radio Engrs. **46** (1958) 1004. (Review)
[88] D. Sautter in *Progress in Semiconductors*, vol. 4 (Edited by A. F. Gibson, F. A. Kröger, and R. E. Burgess), (Heywood, London, 1960), p. 127.
[89] H. C. Montgomery, Bell System tech. J. **31** (1952) 950.

CHAPTER 8

TRANSPORT PROCESSES

Free carriers drifting along a filament of finite cross section are subject to scattering by the boundary surfaces in addition to the normal bulk scattering. This additional scattering will involve predominantly the carriers moving close to the surface and will generally reduce their mobility below that in the bulk. A study of the transport properties of such carriers can yield valuable information on the electronic structure of the surface. At the same time it is of considerable practical importance because of the vital role surface conductance plays in many of the electrical measurements on semiconductor surfaces. As we have seen in chapter 6, the surface conductance can provide one of the most accurate and convenient means of determining the barrier height, but a knowledge or at least a good estimate of the surface mobilities is first necessary. For these reasons the problem of surface transport has received a fair amount of attention in recent years, and in the present chapter we review the theoretical and experimental work on the subject.

The first calculations of surface mobility in semiconductors were carried out by Schrieffer [1], who extended the treatments of Fuchs [2] and Sondheimer [3] on metal films. The main additional feature introduced into the analysis was the incorporation of effects due to the surface potential barrier. (Such effects are insignificant in metals, where the thickness of the space-charge layer is of the order of an interatomic spacing.) Schrieffer's theory is based on rather special assumptions concerning the form of the potential barrier. As a result its range of validity is restricted to strong accumulation and inversion layers. Several workers [4] have since improved on the theory, which has also been extended to include galvanomagnetic effects. This work will be discussed in the first four sections of the chapter. In § 1 we define the pertinent surface parameters in terms of experimentally measurable quantities. Next we apply some elementary considerations in estimating the surface mobility. Sections 3 and 4 present a more rigorous analysis of surface transport processes, together with numerical solutions for the surface mobility as a function of barrier height. Most of the calculations reported so far consider a non-degenerate semiconductor characterized by spherical energy surfaces and a constant relaxation time. These conditions will be assumed to hold throughout our discussion. Attempts were also made [5,6] to treat the more general case of non-spherical energy surfaces, but the analysis becomes too involved and will not be considered here. In §5 we summarize the experimental data on surface mobility in germanium and silicon and compare the results with the theoretical calculations of the previous sections. The chapter concludes with a brief discussion of the present status of our knowledge of surface transport phenomena.

1. Diffuse and specular scattering

The two extreme ways in which the surface can behave towards free carriers striking it are as a completely diffuse (random) scatterer and as a specular (perfect) reflector. The former implies that the carriers emerge from the surface with a Maxwell–Boltzmann distribution, having lost all memory of their velocities prior to the collision. This type of scattering clearly leads to a reduction in mobility for the carriers drifting within a mean free path from the surface. Specular reflection, on the other hand, requires that only the momentum component normal to the surface change, the parallel components and the energy remaining constant. In the case of spherical constant-energy surfaces, the electrons (and holes) can be treated as free particles with a scalar effective mass. Reflection from a specular surface then results only in a reversal of the sign of the velocity component normal to the surface, the velocities parallel to the surface remaining unchanged. There can obviously not be any mobility reduction under these conditions. Once we depart from the assumption of spherical energy surfaces, however, the effective mass becomes a tensor quantity and the velocity and momentum are in general no longer parallel (see Ch. 2, §3). It has been estimated [5,6] that in this case there may be a mobility reduction even for specular scattering, such reduction being determined by the crystallographic orientation.

Specular reflection is the type of scattering one expects from an ideal surface. The disorder present on a non-ideal surface, on the other hand, will result in a measure of diffusivity, the exact extent of which should depend on the density and scattering cross sections of the defects that go to make up the disorder.

Any surface scattering that is not completely specular will result in a decrease in the conductance of a given sample. This reduction is appreciable only when the thickness of the sample is not too large compared to the mean free path of the carriers, thereby affording access to the surface to a significant fraction of the carriers. In thin samples it will be found useful to consider the average effect of surface scattering on the contribution of *all* the carriers to the current parallel to the surface. For this purpose, we define average electron and hole mobilities $\bar{\mu}_n$ and $\bar{\mu}_p$ as

$$\bar{\mu}_n = \frac{I_{nx}}{q\bar{n}\mathscr{E}_x 2d}; \quad \bar{\mu}_p = \frac{I_{px}}{q\bar{p}\mathscr{E}_x 2d}, \qquad (8.1)$$

where I_{nx} and I_{px} are the electron and hole currents (per unit width of a rectangular filament) set up by an electric field $\mathscr{E}_x$, $2d$ is the thickness, and $\bar{n}$ and $\bar{p}$ are the average electron and hole densities (given by the total number of carriers in the sample, divided by the volume).

While the average mobility is a useful concept for describing transport phenomena in thin films, it becomes inadequate in the case of thick samples. As the thickness increases, less and less of the bulk carriers are able to reach the surface before suffering a collision in the bulk. Thus surface scattering becomes increasingly less significant in the expression for the total sample conductance, and $\bar{\mu} \to \mu_b$. Here the effect of surface scattering is best seen by considering the contribution to the conductance of only those carriers that are in the space-charge region adjacent to the surface. This leads to a definition of the electron and hole surface mobilities μ_{ns} and μ_{ps} in terms of $\Delta\sigma$, the change in conductance (per square area) caused by the excess surface-carrier densities ΔN and ΔP (compare eq. (6.2), which is applicable in the case of no surface scattering):

$$\Delta\sigma = q(\mu_{ns}\Delta N + \mu_{ps}\Delta P). \qquad (8.2)$$

The surface mobilities so defined can also be expressed as

$$\mu_{ns} = \frac{\Delta I_{nx}}{q\Delta N \mathscr{E}_x}; \quad \mu_{ps} = \frac{\Delta I_{px}}{q\Delta P \mathscr{E}_x}, \qquad (8.3)$$

where ΔI_{nx} and ΔI_{px} are the increments (positive or negative) in the electron and hole currents for each surface (per unit of its width) with respect to their values at flat bands. The definition of the surface mobilities in (8.2) or (8.3) is not the only one possible [7,8]. It has the advantage, however, of being directly related to $\Delta\sigma$, which is a measurable quantity.

Since the contributions of the two carrier types to the total current are always additive, it will be sufficient to consider only one of them. Accordingly, all the expressions appearing in this chapter for the current and mobility will involve only one of the carrier types. Wherever there is no danger of ambiguity, we shall use the symbols I_x, ΔI_x, $\bar{\mu}$, and μ_s (with the subscripts n and p omitted) for either electrons or holes, as the case may be.

2. Simple considerations

In order to gain some insight into the main physical processes controlling the surface mobility, we present in this section a simplified treatment of carrier transport. A one-carrier system, corresponding to an extrinsic n-type sample, will be assumed throughout. We first estimate the average mobility in thin slabs and then extend the treatment to the calculation of surface mobility in accumulation and depletion layers of thick samples. Both completely diffuse and partially specular scattering are considered. Comparison of the surface mobility values derived by these simple considerations with those obtained numerically from the more rigorous analysis of § 3 below shows the agreement to be remarkably good. The former have the particular advantage that they are given in closed form and are thus very convenient to use in practice.

2.1. THIN SLABS

We consider a semiconductor sample bounded by the surfaces $z = 0$, $z = 2d$ and calculate the current flowing parallel to the surfaces as a result of an applied external field. The band edges are assumed to continue flat up to the surfaces ($V_s = 0$), the electron density n being uniform throughout the sample and equal to n_b. The effect of surface scattering will be introduced in the form of some average collision time τ_s, just as bulk scattering is characterized by the relaxation time τ_b (see Ch. 2, § 6.1). It is further assumed that these two processes, involving the bulk and the surface, act in parallel and independently in determining the average electron mobility, so that

$$\frac{1}{\bar{\tau}} = \frac{1}{\tau_s} + \frac{1}{\tau_b}, \qquad (8.4)$$

where $\bar{\tau}$ is the average collision time of the electrons in the sample. Another way of expressing (8.4) is to say that the probability per unit time that an electron be scattered and thus lose all memory of the previous action of the field is given by the sum of the scattering probabilities for the bulk and the surface taken separately.

For diffuse surface scattering, τ_s represents the average time an electron requires to collide with the surface towards which it is moving and thus lose all memory of its energy. As an estimate of τ_s we take the mean distance d of a carrier from the surface divided by the *unilateral* mean velocity $\bar{c}_z$,

$$\tau_s \approx d/\bar{c}_z = (d/\lambda)\tau_b. \tag{8.5}$$

The unilateral mean velocity is defined, analogously to the root mean square velocity (eq. (2.124)), as the average over the positive (or negative) velocity components c_z of all electrons; it is easily seen to be given by

$$\bar{c}_z = \sqrt{kT/2\pi m^*}, \tag{8.6}$$

where m^* is the scalar effective mass. The unilateral mean free path λ is related to $\bar{c}_z$ by means of the expression

$$\lambda = \tau_b \bar{c}_z = \tau_b \sqrt{kT/2\pi m^*} = \mu_b \sqrt{m^* kT/2\pi q^2}. \tag{8.7}$$

The average electron mobility $\bar{\mu}$ is taken as $\bar{\mu} = q\bar{\tau}/m^*$, in analogy with the corresponding expression for the bulk mobility $\mu_b = q\tau_b/m^*$ (eq. (2.114)). By using (8.4) and (8.5) we obtain

$$\frac{\bar{\mu}}{\mu_b} = \frac{1}{1+\lambda/d}. \tag{8.8}$$

It is seen that the average mobility decreases with decreasing sample thickness, as expected, while for $d \gg \lambda$ it approaches its value in the bulk. In the latter case, (8.8) can be written in the form $\bar{\mu}/\mu_b \approx (d-\lambda)/d$. In other words, instead of considering all the carriers as moving with an average mobility $\bar{\mu}$, one can look upon the sample as though the carriers within a distance λ of each of the surfaces have zero mobility while the carriers in the remaining part of the sample (of thickness $2d-2\lambda$) move with their bulk mobility μ_b.

For very thin slabs such that d is small compared to the effective Debye length L (see Ch. 4, § 2.2), the potential in the sample will be essentially constant regardless of whether $V_s = 0$ (as assumed above) or not. In both cases the electron density will be uniform, although for $V_s \neq 0$ its value will be different from n_b. This does not affect at all the derivation of (8.8), which is thus always valid for sufficiently thin slabs. For thick slabs, on the other hand, (8.8) is just a special case (flat bands), and a more general expression will be derived below.

2.2. THICK SAMPLES

The potential barriers at the two surfaces will be assumed symmetrical about the plane $z = d$. Consider first an accumulation layer. The electrons at the surface can be divided into two categories, *bounded* and *unbounded*. The former are constrained to move inside the potential well while the latter have energies above the well. The current increment ΔI_x in (8.3) consists not only of the contribution of the bounded electrons but also of that arising from the change in scattering conditions for the unbounded electrons close enough to the surface to be affected by it. The latter contribution is clearly negligible for strong accumulation layers or when the mean free path λ is small compared to the width of the well. Under these conditions the surface mobility μ_s associated with the excess surface-carrier density ΔN is given to a good approximation by the average mobility $\bar{\mu}$ defined in (8.1), but now corresponding to a thin slab whose thickness is no longer that of the sample but rather of the order of the width of the space-charge region. In order to estimate μ_s, we approximate the potential in the accumulation layer by a square-well potential and take for its width the effective charge distance L_c (Ch. 4, § 2.2). The carriers in the accumulation layer can then be looked upon as moving in a thin slab with one surface (the actual surface, $z = 0$) a diffuse scatterer and the other ($z = L_c$) a specular reflector. The mean distance of the carriers from the scattering surface is now L_c (and not half the sample thickness, as in the case of a slab in which *both* surfaces are diffuse scatterers). The discussion in the preceding subsection must be modified in another, much more important respect. The electrons in the accumulation layer are now accelerated towards the surface and their kinetic energy is increased due to the potential V through which they drop. As a result, the unilateral mean velocity $\bar{c}_z$ will be given by

$$\bar{c}_z = (\pi m^*)^{-\frac{1}{2}}\sqrt{\tfrac{1}{2}kT + q\bar{V}}, \tag{8.9}$$

where $\bar{V}$ is some average value of the potential barrier V. As an estimate of $\bar{V}$ we take $\tfrac{1}{2}V_s$, half the value of the barrier height. Substitution from (8.7) and (8.9) then yields the following expression for τ_s:

$$\tau_s \approx \frac{L_c \tau_b}{\lambda\sqrt{1+v_s}}. \tag{8.10}$$

The average collision time $\bar{\tau}$ of the bound electrons is again given by (8.4), but now the value of τ_s is as in (8.10) rather than in (8.5). The surface mobility is thus given by

$$\frac{\mu_s}{\mu_b} = \frac{1}{1+(\lambda/L_c)\sqrt{1+v_s}}. \tag{8.11}$$

Substituting Lv_s/F_s for L_c from (4.22), we have

$$\frac{\mu_s}{\mu_b} = \frac{1}{1+(rF_s/v_s)\sqrt{1+v_s}}, \qquad (8.12)$$

where

$$r \equiv \frac{\lambda}{L} = \mu_b \sqrt{\frac{m^*}{2\pi\kappa\varepsilon_0}} \sqrt{n_b+p_b}. \qquad (8.13)$$

A glance at Fig. 4.7 shows that F_s increases rapidly with v_s, so that μ_s is a decreasing function of barrier height. This is to be expected, since the carriers are constrained to move in ever narrower and deeper wells as the accumulation layer becomes stronger. Note that $\mu_s/\mu_b \to (1+r)^{-1}$ as $v_s \to 0$, implying that even for very slight curvature of the bands the surface mobility can differ from its bulk value.

In the derivation of (8.12) we have neglected the change in scattering conditions for the unbounded electrons. The circumstances under which this procedure is valid can be obtained from the following considerations. As was pointed out in the preceding subsection, under flat-band conditions the effect of surface scattering can be expressed by looking upon the carriers situated within a distance λ of each of the surfaces as though moving with zero mobility. In the presence of the potential barrier, the distance λ should be replaced by $\lambda(1+v_s)^{\frac{1}{2}}$ to allow for the increased unilateral mean velocity $\bar{c}_z$ (see eq. (8.9)). (We implicitly require that $\lambda < L$.) Our assumption amounts to neglecting $\mu_b n_b \lambda [(1+v_s)^{\frac{1}{2}} - 1]$ with respect to $\mu_s \Delta N$. Recalling that $\Delta N = n_b L G^+$ (see eq. (4.40)), we see that this condition is equivalent to

$$\frac{r}{G^+}(\sqrt{1+v_s}-1) \ll \frac{\mu_s}{\mu_b}. \qquad (8.14)$$

For strong accumulation layers, (8.14) is always satisfied, since $1/G^+ \ll 1$ and is much smaller than μ_s/μ_b, as can be seen by comparing (8.12) with the plot of G^+ given in Fig. 4.11. Such is no longer the case for weak accumulation layers, however, and only for r sufficiently small will the inequality (8.14) be maintained.

We shall now derive the surface mobility for depletion layers. Here only those electrons having sufficient energy to surmount the potential barrier are able to reach the surface. Such carriers, which are present with a density of approximately $n_b \exp v_s$, will be assumed to move with an average mobility $\bar{\mu}$ given by (8.8). All other carriers are specularly reflected at the potential barrier and are not expected to be affected by surface scattering. Under these assumptions we obtain for the current

$$I_x = 2[n_b d(1-e^{v_s})+\Delta N]q\mathscr{E}_x\mu_b + 2n_b d e^{v_s} q\mathscr{E}_x \bar{\mu}. \quad (8.15)$$

The square brackets represent the electrons moving with their bulk mobility. (Note that both ΔN and v_s are negative in depletion layers.) By subtracting from I_x its value at flat bands (obtained by equating v_s to zero in the above equation) and substituting for $\bar{\mu}$ from (8.8), we have (for each surface)

$$\Delta I_x = [n_b d(1-e^{v_s})+\Delta N]q\mathscr{E}_x\mu_b - n_b d(1-e^{v_s})q\mathscr{E}_x\mu_b(1+\lambda/d)^{-1}$$

$$= \Delta N q \mathscr{E}_x \mu_b \left[1 + \frac{\lambda n_b}{\Delta N(1+\lambda/d)}(1-e^{v_s})\right]. \quad (8.16)$$

For thick samples ($\lambda/d \ll 1$), the electron surface mobility can be expressed, with the help of (4.43), (8.3), and (8.13), in the form

$$\frac{\mu_s}{\mu_b} = 1 - \frac{r}{|G^-|}(1-e^{v_s}). \quad (8.17)$$

Thus μ_s is seen to approach μ_b for large (negative) values of v_s, as expected. At flat bands ($v_s \to 0$), $\mu_s/\mu_b \to 1-r$, as can easily be verified from the definition of G^- (eq. (4.45)). For small values of r, this result is the same as that obtained when flat-band conditions are approached from accumulation layers (eq. (8.12)).

In order to obtain the hole surface mobility in accumulation and depletion layers, it is necessary only to change the sign of v_s in (8.12) and (8.17).

2.3. PARTIALLY SPECULAR SCATTERING

The simple arguments presented above can readily be extended to the case where the scattering surface is partially diffuse, partially specular. We define ω as the probability that an electron reaching the surface be specularly reflected. Thus $\omega = 0$ corresponds to completely diffuse scattering, $\omega = 1$ to totally specular reflection.

The reciprocal of τ_s (see eq. (8.4)) expresses the probability that an electron be scattered by the surface in unit time. Thus, for diffuse scattering it also represents the probability that an electron lose all memory of its energy. For partially specular reflection, however, this probability must be modified to include the fact that of all the reflections taking place during unit time, only the fraction $(1-\omega)$ will lead to a loss of memory, so that

$1/\tau_s$ must be replaced by $(1-\omega)/\tau_s$. This is equivalent to replacing d in (8.5) by $d/(1-\omega)$. The average mobility for thin slabs (eq. (8.8)) then becomes

$$\frac{\bar{\mu}}{\mu_b} = \frac{1}{1+(1-\omega)\lambda/d}. \tag{8.18}$$

Identical reasoning for the case of thick samples leads to the replacing of L_c in (8.10) by $L_c/(1-\omega)$, so that the surface mobility for majority carriers in accumulation layers (eq. 8.12)) becomes

$$\frac{\mu_s}{\mu_b} = \frac{1}{1+(1-\omega)(rF_s/v_s)\sqrt{1+v_s}}. \tag{8.19}$$

A similar result is obtained for depletion layers (eq. (8.17)):

$$\frac{\mu_s}{\mu_b} = 1 - \frac{(1-\omega)r}{|G^-|}(1-e^{v_s}). \tag{8.20}$$

3. Calculations of average and surface mobilities

In this section we develop rigorous expressions for both the average and the surface mobilities for the case of diffuse (random) surface scattering, and present the numerical solutions in graphical form. Partially specular reflection is discussed briefly in the last subsection.

3.1. THE AVERAGE MOBILITY

We again consider a semiconductor slab bounded by the surfaces $z = 0$, $z = 2d$, and calculate the current set up by a small electric field $\mathscr{E}_x$ parallel to the surface. As in Ch. 2, § 6.2, we write down the steady-state Boltzmann transport equation

$$\boldsymbol{a} \cdot \operatorname{grad}_c f + \boldsymbol{c} \cdot \operatorname{grad}_r f = -(f-f_0)/\tau_b, \tag{8.21}$$

where f is the distribution function in the presence of the electric field, $\boldsymbol{a}$ is the acceleration of an electron due to the field, $\boldsymbol{c}$ is its velocity, and f_0 is the Maxwell–Boltzmann distribution function at thermal equilibrium. Obviously f will now be determined by both bulk and surface scattering. The former is represented by the (constant) bulk relaxation time τ_b in the right-hand side of (8.21), while the latter is introduced by appropriate boundary conditions imposed on f at the scattering surfaces. Equation (8.21) is solved by substituting $f = f_0 + \delta f$, where $\delta f = \delta f(\boldsymbol{c}, z)$ is a perturbing function (assumed small) which, in addition to the dependence on c (or k) as in the case of bulk scattering, is a function of z as well.

Consider first the case of flat bands ($v_s = 0$). Here $\mathbf{a} = (a_x, 0, 0)$, the acceleration component normal to the surface being zero, while $f_0 = n_b \frac{1}{2}(2\pi m^* kT/h^2)^{-\frac{3}{2}} \exp(-m^* c^2/2kT)$ and is independent of z (see eqs. (2.54), (2.66), (2.85), (2.89)). The assumption of diffuse scattering requires that the carriers leaving the surface be *randomly* distributed. Thus the boundary conditions are

$$\begin{aligned} \delta f(\mathbf{c}, 0) &= 0 \quad \text{for} \quad c_z \geq 0; \\ \delta f(\mathbf{c}, 2d) &= 0 \quad \text{for} \quad c_z \leq 0. \end{aligned} \tag{8.22}$$

It can be verified by direct substitution that by neglecting $a_x \partial(\delta f)/\partial c_x$ with respect to $c_z \partial(\delta f)/\partial z$ (this is justified, since $a_x = -(q/m^*)\mathscr{E}_x$, and $\mathscr{E}_x$ is taken sufficiently small), an approximate solution to the Boltzmann equation with these boundary conditions is obtained[7,9] in the form

$$\begin{aligned} \delta f &= \delta f_b [1 - e^{-z/c_z \tau_b}] \quad \text{for} \quad c_z \geq 0; \\ \delta f &= \delta f_b [1 - e^{-(z-2d)/c_z \tau_b}] \quad \text{for} \quad c_z \leq 0, \end{aligned} \tag{8.23}$$

where $\delta f_b = -q\tau_b \mathscr{E}_x c_x f_0/kT$ is the deviation from the thermal equilibrium distribution in the bulk (see eq. (2.137)). As we are dealing with the case of flat bands (c_z independent of z), the value of z/c_z or of $(z-2d)/c_z$ is just the time elapsed since the electron left the relevant surface. Thus the factors in square brackets in (8.23) are the probability that an electron having been scattered by the surface $z = 0$ (or $z = 2d$) suffer a bulk collision within the time z/c_z (or $(z-2d)/c_z$).

The electron current (per unit width) parallel to the surface is

$$I_x = -q \int_0^{2d} \int_{-\infty}^{\infty} \int_{-\infty}^{\infty} \int_{-\infty}^{\infty} c_x N(\mathbf{c}) \delta f \, dc_x \, dc_y \, dc_z \, dz, \tag{8.24}$$

where $N(\mathbf{c})$ is the density of states in velocity space and is given by $2(m^*/h)^3$ (see Ch. 2, § 3.2). Substituting for δf, we can integrate directly over all the variables but c_z. The introduction of the dimensionless energy parameter $\varepsilon \equiv m^* c_z^2/2kT$ then yields

$$I_x = 2n_b dq\mathscr{E}_x \mu_b [1 - (\lambda/d) + (\lambda/d)\Gamma_1(\lambda/d)], \tag{8.25}$$

where

$$\Gamma_1(\lambda/d) \equiv \int_0^\infty \exp\left[-\varepsilon - (d/\lambda)(\pi\varepsilon)^{-\frac{1}{2}}\right] d\varepsilon \tag{8.26}$$

and λ is the unilateral mean free path defined by (8.7). Using (8.1) we obtain for the average mobility $\bar{\mu}$

$$\frac{\bar{\mu}}{\mu_b} = 1 - \frac{\lambda}{d} + \frac{\lambda}{d}\Gamma_1\left(\frac{\lambda}{d}\right). \tag{8.27}$$

This function has been evaluated numerically[7,8] and is plotted in Fig. 8.1

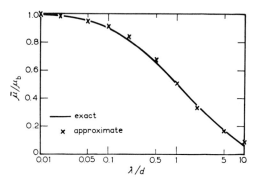

Fig. 8.1. Calculated average mobility for the case of flat bands (thin slabs) as a function of λ/d. (After Flietner, reference 8.) The crosses represent approximate values obtained from (8.8).

against λ/d. The crosses represent values calculated on the basis of the simplified analysis of the previous section (eq. (8.8)). They are seen to agree quite well with the results of the more rigorous treatment.

3.2. THE SURFACE MOBILITY

In order to calculate the surface mobility for the case of thick samples, it is necessary to extend the above treatment to include the influence of the potential barrier v. Two of the terms appearing in the Boltzmann transport equation (eq. (8.21)) are modified. First, the electron acceleration now has an additional component a_z due to the electric field $\mathscr{E}_z = -(kT/q)(dv/dz)$ associated with the potential barrier. Secondly, the expression for the equilibrium distribution function f_0 in the case of flat bands must be multiplied by the factor $\exp v$.

Consider first accumulation layers in n-type samples. The boundary condition for all electrons *leaving* the surface $z = 0$ is the same as before, namely

$$\delta f(c, 0) = 0 \quad \text{for} \quad c_z \geq 0. \tag{8.28}$$

It can be shown [9] that this condition determines δf uniquely at any point in the space-charge region for all *bounded* electrons ($m^*c_z^2/2kT \leq v$), the potential barrier acting as a specular reflector towards such electrons. For the unbounded electrons ($m^*c_z^2/2kT \geq v$) approaching the surface ($c_z \leq 0$), however, an additional condition is required. It is justified to assume that far removed from the surface ($z \to d$) these electrons have their bulk distribution — that is, $\delta f = \delta f_b$ for $z \to d$. It is again easily verified by direct substitution that by neglecting nonlinear terms, the solution to the transport equation subject to these boundary conditions is [9]

$$\delta f = \delta f_b \left[1 - \exp\left(-\frac{1}{\tau_b} \int_0^z \frac{dz}{c_z} \right) \right] \tag{8.29}$$

for the bounded electrons and

$$\delta f = \delta f_b, \quad c_z \leq 0;$$

$$\delta f = \delta f_b \left[1 - \exp\left(-\frac{1}{\tau_b} \int_0^z \frac{dz}{c_z} \right) \right], \quad c_z \geq 0 \tag{8.30}$$

for the unbounded electrons. The above integrals are to be evaluated with ε constant, where ε is a dimensionless energy parameter defined by $\varepsilon \equiv (m^*c_z^2/2kT) - v$. The solutions (8.29), (8.30) are similar to those for flat bands (eq. (8.23)). But as c_z is no longer independent of position, the time elapsed since the electron left the relevant surface must now be expressed in the form of an integral over z.

The electron current (per unit width) I_x is again given by (8.24). After integrating over c_y and c_x, we obtain

$$I_x = 2n_b q \mathscr{E}_x \mu_b (m^*/2\pi kT)^{\frac{1}{2}} \int_0^d \int_{-\infty}^{\infty} \left[1 - \exp \gamma \left(-\frac{1}{\tau_b} \int_0^z \frac{dz}{c_z} \right) \right] e^{-\varepsilon} dc_z \, dz, \tag{8.31}$$

where $\gamma = 0$ for $\varepsilon > 0$ and $c_z < 0$, and $\gamma = 1$ otherwise. The first term can be integrated once directly while the second term can be integrated after transforming the integration variables from (z, c_z) to (ε, c_z). We then obtain (using eq. (4.35))

$$I_x = 2n_b(d-\lambda)q\mathscr{E}_x \mu_b + 2\Delta Nq\mathscr{E}_x \mu_b [1 - (\lambda n_b/\Delta N)\Gamma_2(u_b, v_s, r)], \tag{8.32}$$

where

$$\Gamma_2(u_b, v_s, r) \equiv \int_{-v_s}^{0} e^{-\varepsilon} \left[1 - \exp\left(-\frac{1}{r\sqrt{\pi}} \int_{-\varepsilon}^{v_s} \frac{dv}{\sqrt{v + \varepsilon} \, F(u_b, v)} \right) \right] d\varepsilon \tag{8.33}$$

and F and r are given by (4.16) and (8.13), respectively. For the case of thick samples being considered, the first term in (8.32) corresponds to the current at flat bands (see eq. (8.25), where $\Gamma_1(\lambda/d) \to 0$ for $d \gg \lambda$). Thus the second term of this equation actually represents $2\Delta I_x$ (the current increment for both surfaces), and the electron surface mobility defined in (8.3) is given by

$$\frac{\mu_{ns}}{\mu_{nb}} = 1 - \frac{\lambda n_b}{\Delta N} \Gamma_2(u_b, v_s, r). \tag{8.34}$$

The analogous expression for the hole surface mobility in accumulation layers of a p-type sample follows from symmetry considerations and is obtained from (8.34) by replacing n_b, ΔN, u_b, and v_s by p_b, ΔP, $-u_b$, and $-v_s$, respectively. Thus

$$\frac{\mu_{ps}}{\mu_{pb}} = 1 - \frac{\lambda p_b}{\Delta P} \Gamma_2(-u_b, -v_s, r). \tag{8.35}$$

By substituting for ΔN or ΔP from (4.40) or (4.41), we can write for majority carriers (holes or electrons) in accumulation layers

$$\frac{\mu_s}{\mu_b} = 1 - \frac{r}{G^+} \Gamma_2(|u_b|, |v_s|, r). \tag{8.36}$$

(The function $\Gamma_2(|u_b|, |v_s|, r)$ is obtained from (8.33) by replacing the variables with their absolute values.) For extrinsic semiconductors ($|u_b| \geq 2$), the functions $F(|u_b|, v)$ and G^+ can be approximated by $F^+(v)$ and $F^+(v_s)$, respectively, and the mobility is no longer dependent on the bulk potential u_b:

$$\frac{\mu_s}{\mu_b} \approx 1 - \frac{r}{F^+(v_s)} \Gamma_2^+(v_s, r), \tag{8.37}$$

where

$$\Gamma_2^+(v_s, r) \equiv \int_{-|v_s|}^{0} e^{-\varepsilon} \left[1 - \exp\left(-\frac{1}{r\sqrt{\pi}} \int_{-\varepsilon}^{|v_s|} \frac{dv}{\sqrt{v+\varepsilon\, F^+(v)}}\right)\right] d\varepsilon. \tag{8.38}$$

The function μ_s/μ_b, as obtained[10] by numerical integration of (8.37), is plotted in Fig. 8.2 against $|v_s|$ for various values of the parameter r. Results for an intrinsic semiconductor ($u_b = 0$) are also included (dashed curves), and are seen not to differ much from those for extrinsic samples. The curves for $|u_b| = 0.4$ (not shown) lie approximately midway between the corresponding sets of curves.

When r is close to unity, we see that for small $|v_s|$ the function μ_s/μ_b decreases with decreasing $|v_s|$, which is somewhat surprising at first glance.

Such a behaviour is associated with the contribution of the unbounded electrons which, under these conditions (see eq. (8.14)), is not negligible compared to the contribution of the bounded electrons. For the same reason, μ_s/μ_b also exhibits an inverted cusp at $v_s = 0$ (not shown in the figure) as one goes from accumulation to depletion layers. The magnitude of the discontinuity in the derivative $d(\mu_s/\mu_b)/dv_s$ at $v_s = 0$ is given[11] by $\frac{1}{2}\pi r^3$. That

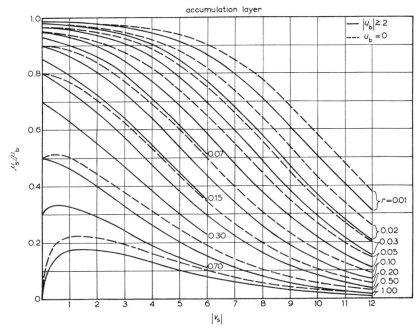

Fig. 8.2. Calculated surface mobility for majority carriers in accumulation layers as a function of $|v_s|$ for various values of the parameter r. The solid curves correspond to extrinsic semiconductors ($|u_b| \gtrsim 2$), the dashed curves to intrinsic semiconductors ($u_b = 0$). (After Goldstein et al., reference 10.)

the overall effect of these characteristics on the shape of the ΔI_x versus ΔN (or $\Delta\sigma$ versus v_s) curve is small is illustrated in Fig. 8.3 (calculated for an n-type sample with $r = 1$). The increase of μ_s/μ_b with v_s manifests itself in the almost indiscernible inflection point at $v_s \approx 1$ and zero slope at $v_s = 0$. It is obvious that there is little hope of detecting this behaviour experimentally.

As has been discussed in the preceding section, the relative contribution of the unbounded carriers is negligible near flat bands for $r \ll 1$ and in strong accumulation layers for r not necessarily small. Under these conditions the simple calculations of the surface mobilities (eq. (8.12)) should constitute a good approximation, and indeed the values so obtained are found to agree well (to within a few hundredths) with those calculated numerically (Fig. 8.2).

Equation (8.34) has been derived for accumulation layers on an n-type sample. As long as the excess electrons are constrained to move in a potential well, however, it is immaterial whether they are majority or minority carriers.

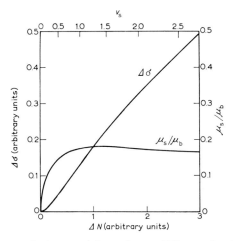

Fig. 8.3. The surface conductance and the surface mobility as a function of ΔN (or v_s) in weak accumulation layers of an n-type extrinsic semiconductor for the case $r = 1$.

Consequently, exactly the same expression (eq. (8.34)) applies to the case of electrons in an inversion or depletion layer of a p-type sample. Similarly, (8.35) is equally applicable to holes in both accumulation layers (p-type sample) and inversion or depletion layers (n-type sample). Thus for minority carriers in inversion or depletion layers we can write

$$\frac{\mu_s}{\mu_b} = 1 - \frac{r}{g^+ e^{2|u_b|}} \Gamma_2(-|u_b|, |v_s|, r), \qquad (8.39)$$

where the excess surface-carrier density has been expressed in terms of g^+, using (4.49) and (4.50).

The values of μ_s/μ_b for minority carriers in inversion layers, as obtained by numerical integration[12] of (8.39), are plotted in Figs. 8.4–8.6 as functions of

$|u_b|$ for various values of the parameter r. Each of the figures corresponds to two values of the parameter $|v_s|-2|u_b|$. This parameter, rather than $|v_s|$ itself, is the relevant quantity for minority carriers in inversion layers (see Ch. 4, § 3.2). The six values of $|v_s|-2|u_b|$ cover the range of practical interest, which extends from depletion layers where minority-carrier densities are

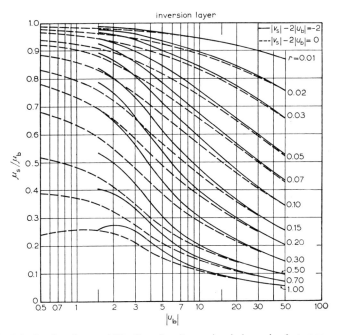

Fig. 8.4. Calculated surface mobility for minority carriers in inversion layers as a function of $|u_b|$ for various values of r. The solid curves refer to $|v_s|-2|u_b| = -2$ and the dashed curves to $|v_s|-2|u_b| = 0$. (In the former case, $|u_b| = 1$ corresponds to flat bands, where $\mu_s/\mu_b = 1-r$.) (After Grover et al., reference 12.)

as yet negligible and up to the strongest inversion layers normally attainable. The weak dependence of μ_s/μ_b on $|u_b|$ (approximately logarithmic) is obvious from the semilogarithmic plot used. Physically, this follows from the fact that the shape of inversion layers (particularly of strong ones — Fig. 8.6) is insensitive to sample resistivity. As in the case of majority carriers in accumulation layers, here too, for large values of r, the mobility decreases as flat-band conditions are approached (see Fig. 8.4).

Next we calculate the surface mobility for the carriers that are *repelled* from the surface. The treatment will be given in terms of electrons in depletion layers of an n-type sample. The expression for μ_s obtained in this manner will then be extended to cover the general case. The Boltzmann transport equation (eq. (8.21)) is again used as the starting point. The boundary conditions are exactly as for unbounded electrons in accumulation

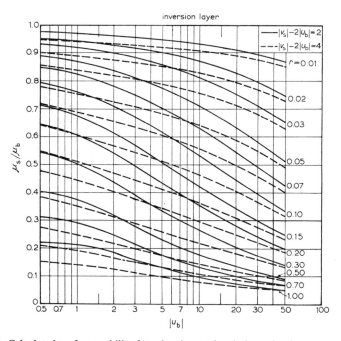

Fig. 8.5. Calculated surface mobility for minority carriers in inversion layers as a function of $|u_b|$ for various values of r. The two families of curves (solid and dashed) correspond to $|v_s|-2|u_b| = 2$ and $|v_s|-2|u_b| = 4$. (After Grover et al., reference 12.)

layers (eq. (8.28)). The solution for those electrons not possessing sufficient energy to reach the surface $(mc_z^2/2kT \leq |v_s-v|)$ is given by

$$\delta f = \delta f_b. \tag{8.40}$$

Physically, this solution expresses the fact that such electrons are in effect separated from the surface by the repulsive potential barrier. The solution for the carriers that do have sufficient energy to reach the surface $(mc_z^2/2kT \geq |v_s-v|)$, is

$$\delta f = \delta f_b \quad \text{for} \quad c_z \leq 0;$$

$$\delta f = \delta f_b \left[1 - \exp\left(-\frac{1}{\tau_b}\int_0^z \frac{dz}{c_z}\right)\right] \quad \text{for} \quad c_z \geq 0. \tag{8.41}$$

In order to calculate I_x, we substitute the expressions for δf and carry out the integrations. Due to the fact that now there are no bounded electrons, the expression for ΔI_x is considerably simplified, and one obtains for the

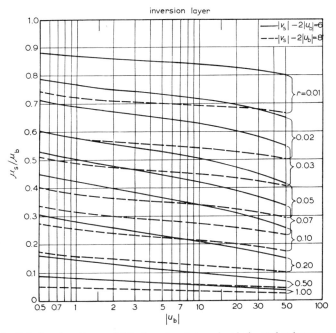

Fig. 8.6. Calculated surface mobility for minority carriers in inversion layers as a function of $|u_b|$ for various values of r. The two families of curves (solid and dashed) correspond to $|v_s| - 2|u_b| = 6$ and $|v_s| - 2|u_b| = 8$. (After Grover et al., reference 12.)

electron surface mobility in depletion layers of an n-type sample [9]

$$\frac{\mu_{ns}}{\mu_{nb}} = 1 - \frac{\lambda n_b}{\Delta N}(e^{v_s} - 1), \tag{8.42}$$

where both ΔN and v_s are negative. This equation is equally applicable to the case of electrons in an accumulation layer of a p-type semiconductor. The appropriate expression for holes is obtained as before by replacing n_b, ΔN, and v_s by p_b, ΔP, and $-v_s$, respectively. By substituting ΔN or ΔP from

(4.43) and (4.44), we obtain for majority carriers in depletion (or inversion) layers

$$\frac{\mu_s}{\mu_b} = 1 - \frac{r}{|G^-|}(1-e^{-|v_s|}). \qquad (8.43)$$

For minority carriers in accumulation layers, the use of (4.46) and (4.47) yields

$$\frac{\mu_s}{\mu_b} = 1 - \frac{r}{|g^-|e^{2|u_b|}}(1-e^{-|v_s|}). \qquad (8.44)$$

These expressions are essentially the same as (8.17), which was obtained so much more simply. We again observe that as $|G^-|$ and $|g^-|$ increase (with increasing $|v_s|$), μ_s approaches μ_b. On the other hand, as flat-band conditions are approached ($v_s \to 0$), in both cases

$$\mu_s/\mu_b = 1-r \quad \text{for} \quad v_s \to 0. \qquad (8.45)$$

An identical value is reached when flat-band conditions are approached from the opposite direction, as can be seen by taking the limit of (8.36) and (8.39) for $v_s \to 0$.

3.3. PARTIALLY DIFFUSE SCATTERING

Consider now the more general problem of partially diffuse scattering. To a linear approximation, the previous boundary conditions (eqs. (8.22), (8.28)) are then replaceable by[2, 3, 9]

$$\begin{aligned}\delta f(c_x, c_y, c_z, 0) &= \omega \delta f(c_x, c_y, -c_z, 0); \\ \delta f(c_x, c_y, c_z, 2d) &= \omega \delta f(c_x, c_y, -c_z, 2d),\end{aligned} \qquad (8.46)$$

where ω is the probability of specular reflection (see § 2.3). Solution of the Boltzmann transport equation subject to such boundary conditions yields[9] for the majority-carrier surface mobility in depletion layers

$$\frac{\mu_s}{\mu_b} = 1 - \frac{(1-\omega)r}{|G^-|}(1-e^{-|v_s|}). \qquad (8.47)$$

This expression is identical with (8.20), obtained on the basis of the simple considerations of § 2. For minority carriers in accumulation layers, μ_s is similarly given by

$$\frac{\mu_s}{\mu_b} = 1 - \frac{(1-\omega)r}{|g^-|e^{2|u_b|}}(1-e^{-|v_s|}). \qquad (8.48)$$

Just as in the case of completely diffuse scattering, the expression for μ_s for the bounded carriers[9] cannot be written in closed form. In the present instance, however, there are no numerical solutions available and we suggest the use of (8.19) as a substitute. That this approximation is adequate is supported by the good agreement obtained for the special case of flat bands from a comparison between several corresponding values of μ_s derived from (8.18) and from numerical calculations employing the rigorous expressions.

4. Galvanomagnetic surface effects

The contribution of surface scattering to the Hall effect and magnetoresistance can be derived by a solution of the Boltzmann equation as before, but now in the presence of a magnetic field. Such calculations have been carried out by Zemel[13] and by Amith[7] for semiconductors, along similar lines to those used by Sondheimer[3] for metals, and will be presented in § 4.1. An alternative approach has been developed by Petritz[14], who solved the problem by means of an effective mobility formalism. As will be seen in § 4.2, such a presentation makes it possible to express the galvanomagnetic quantities in an approximate way by means of the surface mobility μ_s associated with the conductivity processes discussed in the preceding sections.

4.1. SOLUTION OF TRANSPORT EQUATION

We consider the situation where in addition to the drift field there is also a magnetic field normal to the scattering surface. The Boltzmann transport equation (eq. (8.21)) is solved for an extrinsic semiconductor under conditions of diffuse surface scattering and constant τ_b. The treatment will be confined to the case of flat bands which, as discussed previously (§ 2.1), is an adequate representation of the situation in thin slabs ($d \ll L$). Here the boundary conditions are again given by (8.22), and the thermal equilibrium distribution function is the same as in § 3.1. The acceleration $a = (a_x, a_y, 0)$, however, is determined by the magnetic induction B_z and the Hall field $\mathscr{E}_y$ in addition to the electric field $\mathscr{E}_x$. As can be seen by substitution, the solution (for electrons) is given to a first approximation by[7,13]

$$\delta f = \delta f_b^H \{1 - e^{-z/\tau_b c_z}[\cos(\omega_0 z/c_z) + (Y/X)\sin(\omega_0 z/c_z)]\} \quad \text{for} \quad c_z \geqq 0;$$

(8.49)

$$\delta f = \delta f_b^H \{1 - e^{-(z-2d)/\tau_b c_z}[\cos\{\omega_0(z-2d)/c_z\} + (Y/X)\sin\{\omega_0(z-2d)/c_z\}]\}$$
$$\text{for} \quad c_z \leqq 0,$$

where

$$\omega_0 = qB_z/m^* = \mu_b B_z/\tau_b \tag{8.50}$$

is the cyclotron frequency, and

$$\begin{aligned} X &\equiv (\mathscr{E}_x - \omega_0\tau_b\mathscr{E}_y)c_x + (\mathscr{E}_y + \omega_0\tau_b\mathscr{E}_x)c_y; \\ Y &\equiv (\mathscr{E}_x - \omega_0\tau_b\mathscr{E}_y)c_y - (\mathscr{E}_y + \omega_0\tau_b\mathscr{E}_x)c_x. \end{aligned} \tag{8.51}$$

Here $\delta f_b^H = q\tau_b f_0 X/(1+\omega_0^2\tau_b^2)kT$ and is the bulk distribution function in the presence of the magnetic field. (Similar expressions have been derived by the simpler analysis of Ch. 2, §6.3.) The x component I_x of the current is obtained by evaluating the integral in (8.24), with δf taken from (8.49). The y component I_y is obtained similarly, but now replacing c_x by c_y. From the condition that $I_y = 0$ (corresponding to open-circuit Hall-voltage measurements — see Ch. 7, §6.1), one derives a relationship connecting $\mathscr{E}_x$ and $\mathscr{E}_y$. In the limit of small magnetic fields ($\omega_0^2\tau_b^2 \ll 1$), the various quantities associated with the Hall effect (defined analogously to those in Ch. 2, §6.3), are given by

$$R_H \equiv \frac{\mathscr{E}_y}{(I_x/2d)B_z} = -\eta(\lambda/d)\frac{1}{qn_b}; \tag{8.52}$$

$$\tan\theta \equiv \mathscr{E}_y/\mathscr{E}_x = -\eta(\lambda/d)\bar{\mu}B_z; \tag{8.53}$$

$$R_H\bar{\sigma} = -\eta(\lambda/d)\bar{\mu}. \tag{8.54}$$

Here R_H and θ are the Hall coefficient and Hall angle, $\bar{\sigma}$ is the average conductivity in the presence of surface scattering ($\bar{\sigma} = qn_b\bar{\mu}$), and

$$\eta(\lambda/d) \equiv \frac{1 - 2(\lambda/d) + 2(\lambda/d)\Gamma_1(\lambda/d) + \Gamma_3(\lambda/d)}{[1 - (\lambda/d) + (\lambda/d)\Gamma_1(\lambda/d)]^2}. \tag{8.55}$$

The function $\Gamma_1(\lambda/d)$ is given by (8.26), while

$$\Gamma_3(\lambda/d) \equiv \int_0^\infty (\varepsilon\pi)^{-\frac{1}{2}} \exp[-\varepsilon - (d/\lambda)(\varepsilon\pi)^{-\frac{1}{2}}]d\varepsilon. \tag{8.56}$$

It is seen that apart from the correction factor $\eta(\lambda/d)$, (8.52)–(8.54) are completely analogous to the corresponding expressions in the absence of surface scattering (eqs. (2.158)–(2.160)), provided that μ_b is replaced by $\bar{\mu}$ throughout.

The transverse magnetoresistance is expressed (as in the bulk — see eq. (2.161)) by the ratio $\Delta\rho/\rho B_z^2$, where $\Delta\rho$ is the change in the (average) resistivity ρ due to the presence of the magnetic field. This quantity is derived from the value of I_x and the relation between $\mathscr{E}_x$ and $\mathscr{E}_y$, and is given by

$$\Delta\rho/\rho B_z^2 = \mu_b^2 - \eta^2(\lambda/d)\bar{\mu}^2. \tag{8.57}$$

In the derivation of the first-order Hall effect parameters (eqs. (8.52)–(8.54)), one can suppose that the assumption of constant τ_b is justified. The difference between averaging τ^2 and averaging τ can be included by introducing into the expressions an appropriate coefficient which will probably not differ substantially from the corresponding bulk value (in the vicinity of unity — see Ch. 2, § 6.3). For the magnetoresistance, however, the assumption becomes questionable, since in the bulk a constant τ_b results in the vanishing of such higher-order effects, whereas in the presence of surface scattering, the solution of the Boltzmann transport equation based on this assumption (which is the only one available to date) yields a finite magnetoresistance.

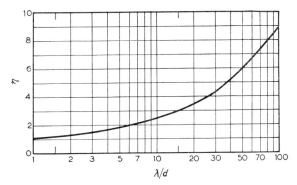

Fig. 8.7. The correction factor η for weak magnetic fields plotted as a function of λ/d. (After Amith, reference 7.)

The correction factor η appearing in (8.52)–(8.57) has been calculated by numerical integration[7] and is plotted in Fig. 8.7 as a function of λ/d. As expected, η approaches unity for thick samples ($\lambda/d \to 0$). (In this range the magnetoresistance given by (8.57) vanishes, as of course it should because of the constant τ_b.)

The above treatment has been extended to include two-carrier systems and the presence of potential barriers at the surface[7]. The mathematics becomes more complicated, however, and since there have been no exact numerical calculations of the integrals involved, we shall not give it here.

4.2. EFFECTIVE MOBILITY FORMALISM

In this subsection we treat the case of a thick sample and consider the modification in the galvanomagnetic quantities brought about by the presence of the space-charge region. A one-carrier system (electrons) will again be assumed. For weak magnetic fields, the expressions for the electron current density in the x direction (drift current) and in the y direction (Hall-field direction) can be written[14] in a form analogous to that in the bulk (see Ch. 2, § 6.3):

$$J_x(z) = qn\langle\mu(z)\rangle\mathscr{E}_x - qn\langle\mu^2(z)\rangle B_z\mathscr{E}_y - qn\langle\mu^3(z)\rangle B_z^2\mathscr{E}_x;$$
$$J_y(z) = qn\langle\mu(z)\rangle\mathscr{E}_y + qn\langle\mu^2(z)\rangle B_z\mathscr{E}_x - qn\langle\mu^3(z)\rangle B_z^2\mathscr{E}_y, \quad (8.58)$$

where $\langle\mu^n(z)\rangle$ is the average over *velocity* space of the n-th power of the mobility at a distance z from the scattering surface. (The reader should note that $\langle\mu^n(z)\rangle$ is a function of z and *not* the average over z that such a notation usually implies.) Within the bulk, where the effect of the surface can be neglected, these averages are given by (see also eqs. (2.156))

$$\langle\mu_b^n\rangle = \frac{q}{m^*}\frac{\langle E\tau^n\rangle}{\langle E\rangle}, \quad (8.59)$$

where E is the electron energy and $\tau(E)$ is the relaxation time associated with the bulk scattering process (considered in this formalism to be energy dependent). In the region near the surface, $\langle\mu^n(z)\rangle$ is less than $\langle\mu_b^n\rangle$ because of the effect of surface scattering.

The current components I_x and I_y (per unit width of the filament) are obtained by integrating (8.58) over z. It is then readily seen that the average conductivity $\bar{\sigma}$ and Hall coefficient R_H are given (up to order B_z) by

$$\bar{\sigma} \equiv \frac{I_x}{2d\mathscr{E}_x} = \frac{q}{2d}\int_0^{2d} n\langle\mu(z)\rangle\,\mathrm{d}z \quad (8.60)$$

and

$$R_\text{H} \equiv \frac{\mathscr{E}_y}{(I_x/2d)B_z} = -\frac{\dfrac{2d}{q}\displaystyle\int_0^{2d} n\langle\mu^2(z)\rangle\,\mathrm{d}z}{\left[\displaystyle\int_0^{2d} n\langle\mu(z)\rangle\,\mathrm{d}z\right]^2}. \quad (8.61)$$

Similarly, the magnetoresistance (to order B_z^2) is given by

$$\frac{\Delta\rho}{\rho B_z^2} = \frac{\int_0^{2d} n\langle\mu^3(z)\rangle dz}{\int_0^{2d} n\langle\mu(z)\rangle dz} - \frac{\left[\int_0^{2d} n\langle\mu^2(z)\rangle dz\right]^2}{\left[\int_0^{2d} n\langle\mu(z)\rangle dz\right]^2}. \qquad (8.62)$$

By means of the notation

$$\langle\mu^n(z)\rangle = \langle\mu_b^n\rangle + \Delta\langle\mu^n(z)\rangle, \qquad (8.63)$$

we can write for $\bar{\sigma}$, R_H, $\Delta\rho/(\rho B_z^2)$ the equations

$$\bar{\sigma} d/q = n_b\langle\mu_b\rangle d_b + (\Delta N + n_b d_s)\langle\mu_s\rangle + \Delta N\langle\mu_c\rangle; \qquad (8.64)$$

$$-R_H \bar{\sigma}^2 d/q = n_b\langle\mu_b^2\rangle d_b + (\Delta N + n_b d_s)\langle\mu_s^2\rangle + \Delta N\langle\mu_c^2\rangle; \qquad (8.65)$$

$$\frac{\bar{\sigma} d}{q}\left[\frac{\Delta\rho}{\rho B_z^2} + (R_H \bar{\sigma})^2\right] = n_b\langle\mu_b^3\rangle d_b + (\Delta N + n_b d_s)\langle\mu_s^3\rangle + \Delta N\langle\mu_c^3\rangle, \qquad (8.66)$$

where d_s is the effective width of the space-charge region. This width is not well defined; it is usually taken to be the effective Debye length L (see Ch. 4, § 2.2). The effective half width of the bulk d_b is defined as $d-d_s$, d being half the width of the sample. (We are considering the symmetrical case where the same potential barrier exists at both scattering surfaces.) In this treatment the n-th powers of the "surface mobility" $\langle\mu_s^n\rangle$ and the "correlation mobility" $\langle\mu_c^n\rangle$ are defined as

$$\langle\mu_s^n\rangle \equiv \langle\mu_b^n\rangle + (1/d_s)\int_0^d \Delta\langle\mu^n(z)\rangle dz; \qquad (8.67)$$

$$\Delta N\langle\mu_c^n\rangle \equiv \int_0^d [(n-n_b) - \Delta N/d_s]\Delta\langle\mu^n(z)\rangle dz. \qquad (8.68)$$

By subtracting the bulk contribution, we obtain from (8.64)–(8.66) that

$$(\bar{\sigma} - \sigma_b)(d/q) = \Delta N(\langle\mu_s\rangle + \langle\mu_c\rangle) - n_b d_s(\langle\mu_b\rangle - \langle\mu_s\rangle); \qquad (8.69)$$

$$-(R_H \bar{\sigma}^2 - R_{Hb}\sigma_b^2)(d/q) = \Delta N(\langle\mu_s^2\rangle + \langle\mu_c^2\rangle) - n_b d_s(\langle\mu_b^2\rangle - \langle\mu_s^2\rangle); \qquad (8.70)$$

$$\left\{\bar{\sigma}\left[\frac{\Delta\rho}{\rho B_z^2} + (R_H \bar{\sigma})^2\right] - \sigma_b\left[\left(\frac{\Delta\rho}{\rho B_z^2}\right)_b + (R_{Hb}\sigma_b)^2\right]\right\}(d/q)$$
$$= \Delta N(\langle\mu_s^3\rangle + \langle\mu_c^3\rangle) - n_b d_s(\langle\mu_b^3\rangle - \langle\mu_s^3\rangle), \qquad (8.71)$$

where

$$\sigma_b = qn_b\langle\mu_b\rangle; \tag{8.72}$$

$$-R_{Hb}\sigma_b^2 = qn_b\langle\mu_b^2\rangle; \tag{8.73}$$

$$\sigma_b\left[\left(\frac{\Delta\rho}{\rho B_z^2}\right) + (R_{Hb}\sigma_b)^2\right] = qn_b\langle\mu_b^3\rangle. \tag{8.74}$$

The expressions for systems of two or more carriers (including light holes) have also been developed along these lines[14]. Albers[15] has extended the formalism to the case of ellipsoidal energy surfaces.

The mobilities defined above have not been computed numerically, but by introducing a number of approximations, (8.69)–(8.71) can be expressed by means of the surface mobility μ_s calculated in § 3.2. One should first note that the quantity $n_b d_s$ is the density (per unit area) of carriers in the space-charge region for the case of flat bands. For accumulation and inversion layers, this density is usually small compared to the excess surface-carrier density ΔN and can therefore be neglected. As for the correlation mobility $\langle\mu_c^n\rangle$, we see from (8.68) that it vanishes for the case of a square potential well at the surface (since then $(n-n_b)d_s \equiv \Delta N$). In view of the fact that such a model yields a valid expression for the surface mobility (see § 2.2), setting $\langle\mu_c^n\rangle$ equal to zero in the general case would seem to be a reasonable approximation. Once the correlation mobility is neglected, a considerable simplification is obtained in (8.69)–(8.71). From (8.69) it follows that $\langle\mu_s\rangle$ is very nearly equal to the surface mobility μ_s as defined in (8.3). Moreover, one can reasonably assume that the averaging of the various powers $\langle\mu_s^n\rangle$ will result in correction terms not very different from those in the bulk — that is,

$$\frac{\langle\mu_s^n\rangle}{\mu_s^n} \approx \frac{\langle\mu_b^n\rangle}{\mu_b^n}. \tag{8.75}$$

Under these conditions (8.70) and (8.71) reduce to

$$-(R_H\bar{\sigma}^2 - R_{Hb}\sigma_b^2)(d/q) \approx \Delta N(\mu_s/\mu_b)^2\langle\mu_b^2\rangle; \tag{8.76}$$

$$\left\{\bar{\sigma}\left[\frac{\Delta\rho}{\rho B_z^2} + (R_H\bar{\sigma})^2\right] - \sigma_b\left[\left(\frac{\Delta\rho}{\rho B_z^2}\right)_b + (R_{Hb}\sigma_b)^2\right]\right\}(d/q) \approx \Delta N(\mu_s/\mu_b)^3\langle\mu_b^3\rangle. \tag{8.77}$$

(Equation (8.76) is essentially the same as (7.56) derived from the simple considerations of Ch. 7, § 6.1.)

It should be noted that even though the above assumptions appear sound, it has yet to be proved that (8.76) and (8.77) constitute a good approximation. The convenient form in which these expressions are presented, however, makes them very amenable to comparison with experiment. And indeed, almost all the galvanomagnetic data reported are interpreted on the basis of (8.76) and (8.77).

5. Experimental results

In this section we review the various experimental results reported to date on surface transport, and consider in detail their comparison with the theoretical calculations of the preceding sections. The key quantity in all the measurements is the surface conductance $\Delta\sigma$. The experimental conditions can usually be chosen so that the contribution from one carrier type is negligible. The measurement of $\Delta\sigma$ then yields the product of the surface mobility and the excess density of the dominant carrier at the surface. Additional, independent information is required, however, to separate the two unknowns. This has been derived from either field-effect or galvanomagnetic measurements, and the discussions that follow are divided accordingly.

5.1. CONDUCTIVITY MOBILITY

As has been pointed out in Ch. 6, § 6.3, the presence of fast surface states usually prevents an accurate determination of the surface mobility from field-effect data. A good estimate can only be obtained if the density of the states is sufficiently low or if their effect can be eliminated. In this case the excess surface density of the dominant carrier at the surface (ΔN or ΔP) is given by the known magnitude of the applied field, and the change in surface conductance is a direct measure of the surface mobility. To approach such ideal conditions one might attempt to reduce the density of fast states by means of suitable chemical treatments. Another, more effective method is provided by the pulsed field-effect measurements described in Ch. 6, § 6.3, where one tries to eliminate the effect of surface states rather than to reduce their density.

The earliest indication that a reduction in mobility takes place for carriers constrained to move in deep potential wells comes from dc field-effect measurements[16]. Subsequent measurements of this type, however, suggested that the mobility reduction is less than what was to be expected from completely diffuse surface scattering[17,18]. Figure 8.8 shows results obtained[18] on inversion layers of a thermally oxidized p-type germanium sample. The points represent the change in surface conductance (with respect to its minimum value) as a function of the total charge induced by the field. The solid and dashed curves correspond to the change in conductance calculated on the assumption that all the induced charge is mobile, the first on the basis of completely diffuse and the second on the basis of completely specular scattering. (Computed values of the surface potential u_s are marked off along the two curves.) Clearly, the horizontal distance between each of the experimental points and either of the calculated curves (depending on the type

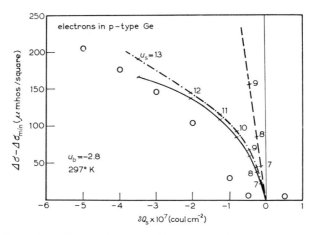

Fig. 8.8. Change in surface conductance with respect to the conductance minimum as a function of induced charge for inversion layers on a p-type germanium sample. The points refer to experimental data, the curves have been calculated on the basis of diffuse (solid line) and specular (dashed line) scattering models. The corresponding values of surface potential are marked off along the curves. (After Frankl and Feuersanger, reference 18.) The dash-dot curve is a recalculation of the solid curve using the value of 0.12 m for the electron conductivity effective mass.

of scattering assumed) corresponds to the fraction of the total charge that is immobile in the surface states. It is seen that the assumption of diffuse scattering (solid curve) gives an apparent trapped charge density that is a non-monotonic function of the surface potential, which is absurd, while the specular scattering model (dashed curve) leads to rather large surface state densities ($> 10^{12} \text{cm}^{-2}$), and is probably an exaggeration in the other direction. The surface studied is thus neither a completely diffuse nor a completely specular scatterer. It should be pointed out that the diffuse-scattering curve presented here has been calculated using a value of 0.25 m

for the electron conductivity effective mass. A better value would appear to be 0.12 m (Table 2.1, p. 41, and eq. (2.149)), and the solid curve has been replotted accordingly (dash-dot curve). The trapped charge is no longer a non-monotonic function of u_s but gives a more or less constant value over a large range of surface potential. A diffuse-scattering model would now imply that surface states are essentially absent near the conduction-band edge, contrary to what is usually found on germanium surfaces (see Ch. 9, § 2). The conclusion that a certain amount of specularity is necessary to interpret the results is

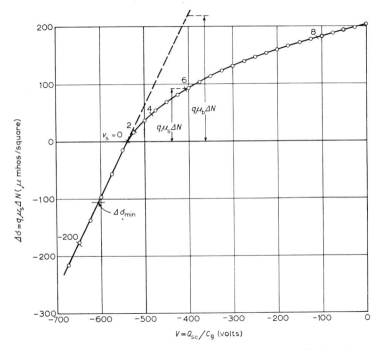

Fig. 8.9. Surface conductance as a function of the (negative) amplitude of the voltage pulse for an extrinsic n-type germanium sample at 80°K. The values of v_s derived from the induced charge are marked off along the curve. (After Many et al., reference 19.)

therefore probably still correct. The exact degree of specularity, however, cannot be determined from such measurements; for this the surface-state distribution would be required. The results thus yield only a lower bound for the surface mobility. The smaller the density of surface states on a given surface the closer does this lower bound come to representing the true situation.

We now turn to surface mobility measurements based on the pulsed field-effect technique (Ch. 6, §6.3). The experimental procedure is illustrated in Fig. 8.9, which corresponds to an n-type germanium sample at 80°K. The point at the extreme right represents the quiescent value of the surface conductance. Starting from that point (strong accumulation layer), negative voltage pulses of increasing amplitude are applied at the field plate. This amplitude serves as abscissa; as ordinate one plots the resulting change in surface conductance at the onset of the pulse, before any appreciable redistribution of the induced charge can take place between the majority-carrier band and the surface states. The minimum in conductance (obtained by ordinary dc field-effect measurements) is located on the curve and used to determine the point corresponding to flat bands ($v_s = 0$). The potential barrier at any other point along the curve is derived on the assumption that the entire induced charge appears in the conduction band and is given by $q\Delta N$ (see below). It is seen that for negative values of the potential barrier, the experimental points are colinear. The slope of the straight line is in remarkable agreement with that computed from the electron bulk mobility. Besides the obvious conclusion that $\mu_s = \mu_b$ for large negative v_s, these results prove that in this region there are no surface states characterized by release times smaller than the experimental resolution time. As the barrier moves towards accumulation-layer conditions, the surface mobility begins to decrease. For any particular v_s, the ordinate of the experimental curve is $\Delta\sigma = q\mu_s \Delta N$. The corresponding value of $q\mu_b \Delta N$ is the ordinate of the point having the same abscissa but lying on the extension of the straight line (dashed) representing $\Delta\sigma$ for $v_s < 0$. This value is just what $\Delta\sigma$ would be had the electrons continued to move with their bulk mobility. Thus μ_s/μ_b for any v_s is the ratio of the two ordinates at that point.

Measurements of this type have been carried out[19-22] on a large number of germanium samples of different impurity content. In each case the effect of both surface treatment and temperature was studied. Typical results of μ_s/μ_b are presented in Figs. 8.10–8.12. The data are shown only for $v_s \gtrsim 3$ (fairly strong accumulation layers), since in this range the steep dependence of ΔN on v_s permits an accurate determination of the barrier height. Moreover, the deduced values of μ_s are insensitive here to the exact location of the flat-band position ($v_s = 0$) in Fig. 8.9, such location being somewhat uncertain due to the slow variation of $\Delta\sigma$ with v_s in depletion layers (see, for example, Fig. 6.2). The results in Fig. 8.10 were obtained on a p-type[20] and an n-type[21] sample at 185°K, following two different etching procedures.

The effect of the etchants on the measured surface mobility is seen to be in opposite directions for the p and n samples. The theoretical curves were taken from Fig. 8.2 with r obtained from (8.13). The case of the hole surface mobility is somewhat more complicated due to the presence of light and heavy holes (see Ch. 2, §3.3). The surface conductances of both types of hole were calculated separately from the theoretical curves and were added after suitable weighting. For lack of better information, the density ratio of light to heavy holes at the surface was taken to be the same as that in the bulk (≈ 0.04).

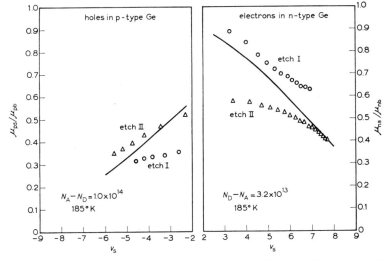

Fig. 8.10. Experimental results for the majority-carrier surface mobility in accumulation layers as a function of barrier height (together with the corresponding theoretical curves) in n- and p-type germanium samples following two different surface treatments. Etch I: CP-4A; Etch II: 1 part by volume of HF, 6 of HNO_3, 12 of CH_3COOH plus 1 mg KI for every 4ml HF.

The influence of doping is shown in Fig. 8.11 [19,20]. It will be noted that in both cases the effect of impurity concentration is in the direction predicted by the theory ($r \propto |N_D - N_A|^{\frac{1}{3}}$ for extrinsic samples), only somewhat smaller. Figure 8.12 illustrates the effect of temperature on the surface mobility [19,20]. The experimental points in the n-type sample are lower, the lower the temperature, as is to be expected from the increasing mean free path of the electrons as the temperature is decreased ($r \propto \mu_b$). A different behaviour, however, is indicated in the p-type sample. Here the results at the two

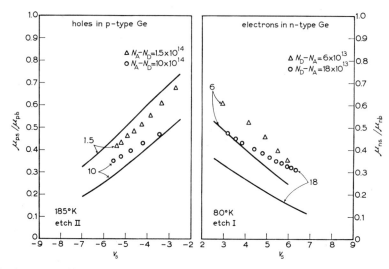

Fig. 8.11. Experimental results for the majority-carrier surface mobility in accumulation layers as a function of barrier height (together with the corresponding theoretical curves) in n- and p-type germanium samples of different impurity content.

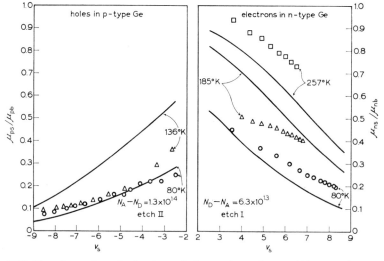

Fig. 8.12. Experimental results for the majority-carrier surface mobility as a function of barrier height (together with the corresponding theoretical curves) in n- and p-type germanium samples at different temperatures.

temperatures almost coincide, whereas the theoretical curves are of course different.

The data presented in Figs. 8.10–8.12 have several features in common. In many cases the experimental points lie above (or well above) the corresponding theoretical curve for completely diffuse scattering throughout the entire range of v_s. In the other cases, the deeper the accumulation layer the less is the separation between the points and the curve, until either the curve passes under the points or gives every indication that it is about to do so at values of v_s slightly beyond the range of measurement.

In the processing of the data, we have assumed that the experimental procedure employed eliminated effects arising from all surface states. Actually, the occurrence of any states with release times shorter than the experimental resolution time (for example, states close to the majority-carrier band edge) would result in the mobile charge being less than the total charge induced by the pulse. The effect of such states, if present, would then be to raise the actual surface mobility above its measured values and, on the other hand, to lower the values of $|v_s|$ below those calculated on the basis of no surface states. As v_s for deep accumulation layers depends only logarithmically on the induced mobile charge $q\Delta N$ (see Ch. 4, § 3.2), however, the deduced values of v_s would not be appreciably affected in this range[19]. The experimental curves thus come very near to representing the lowest possible limits to the true surface mobilities.

One can therefore conclude that some degree of specularity is necessary to account for the data obtained. The less-than-predicted sensitivity to such parameters as impurity concentration and temperature as well as the marked effect of etchants may very well be the results of the overriding influence of surface states and their variation with surface treatment.

A rough estimate of the surface mobility of *minority* carriers in inversion layers has been obtained[22] by the use of similar field-effect techniques, as explained in Ch. 6, § 6.4. Two quantities are measured at each barrier height: the steady-state change $\Delta\sigma - \Delta\sigma_{min}$ in surface conductance (with respect to its minimum), and the saturation value $(\delta\sigma)_{sat} = q(\mu_n + \mu_p)\Delta P$ of the transient change in sample conductance following the onset of a large positive pulse (an n-type sample is being considered). These two quantities are plotted in Fig. 8.13 for an n-type germanium sample at room temperature. In order to obtain the surface mobility, we should in principle take the ratio of the ordinate to the abscissa. Since the magnitude of the minimum conductance may be in large error, however, a correction is introduced (dashed curve) by extrapolating from the region of strong inversion layers. Such a correction does not appreciably affect the deduced values of surface mobility in that region, just as in the majority-carrier case the surface mobility in strong accumulation layers is insensitive to the exact location of the flat-band position. The values of v_s are derived from ΔP, as discussed in Ch. 6, § 6.4. The results of μ_{ps}/μ_p obtained in this manner are plotted in Fig. 8.14, together with the appropriate theoretical curve calculated from Figs. 8.4–8.6. Here again the experimental points represent the lowest possible limit to

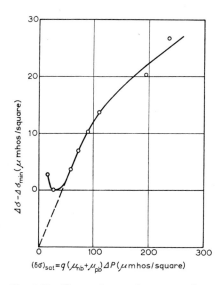

Fig. 8.13. Change in steady-state surface conductance with respect to the conductance minimum as a function of the transient conductivity change $(\delta\sigma)_{sat}$ for an n-type germanium sample at room temperature. The dashed curve corresponds to corrected values near the conductance minimum (as explained in text).

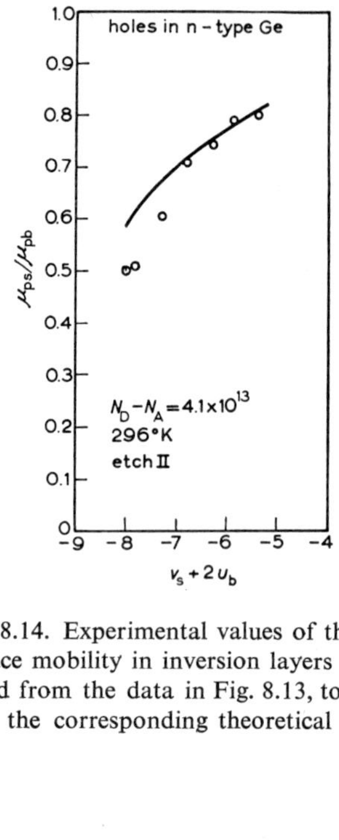

Fig. 8.14. Experimental values of the hole surface mobility in inversion layers as deduced from the data in Fig. 8.13, together with the corresponding theoretical curve.

the actual surface mobility. This limit is cruder than in the case of majority carriers, however, since in addition to possible contribution to ΔP from surface-state emission, a large error may be involved in the measurement of $\Delta\sigma - \Delta\sigma_{min}$ due to insufficient communication between the inversion layer and the underlying bulk (see Ch. 6, § 2).

5.2. HALL EFFECT AND MAGNETORESISTANCE

Galvanomagnetic data yield additional relationships between μ_s and ΔN (or ΔP), without having to resort to assumptions concerning surface states. As we have seen in the previous section, however, the theoretical expressions with which the experimental results are to be compared have so far been obtained with the aid of various approximations, some of which are hard to justify except as mathematical expediencies. Moreover, the galvanomagnetic quantities are not very sensitive to the exact degree of diffusivity of the scattering surface.

Most of the Hall effect and magnetoresistance measurements published so far have been performed on real surfaces (as in the case of conductivity mobility). Missman and Handler[23] are the only ones to report results on clean surfaces, and we begin by discussing these. The authors used a 20 ohm-cm n-type germanium sample cleaned by argon bombardment and annealed

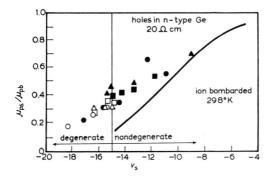

Fig. 8.15. Surface mobility as a function of barrier height for strong inversion layers on a clean germanium surface. The points refer to experimental data. (After Missman and Handler, reference 23.) The solid curve has been calculated on the basis of the diffuse scattering model.

(see Ch. 3, § 1.5). Simultaneous measurements of the conductance and Hall voltage were made as oxygen was slowly admitted into the system. It was found that the clean surface is associated initially with a degenerate inversion layer. As the oxygen coverage increases, the surface at first becomes more degenerate and then the direction of change in barrier height v_s is reversed and the surface conductance decreases. The lowest value reached by the surface conductance is assumed to correspond to flat bands. Both the sample conductance G and the Hall coefficient R_H at this point are taken to be equal to their respective bulk values (G_b and R_{Hb}); these are good approximations in the case of strong inversion layers. Experimental results for several different runs are shown in Fig. 8.15. The surface mobility is derived from the conductance and Hall coefficient by the use of (8.76) with $\langle \mu_b^2 \rangle$ set equal to μ_b^2. The barrier height v_s is obtained from the measured surface mobility and surface conductance. The solid curve is calculated using eq. (8.39) and Fig. 8.6. It lies slightly above the one computed by Missman and Handler using Schrieffer's theory[1]. The results are quite similar to the accumulation layer data presented in the previous subsection and again indicate less than completely diffuse surface scattering, especially for large values of $|v_s|$.

Aside from some magnetosurface studies on gold-doped germanium at

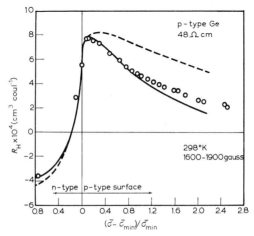

Fig. 8.16. Hall coefficient R_H as a function of the reduced conductivity change $(\bar{\sigma} - \bar{\sigma}_{min})/\bar{\sigma}_{min}$ for a near-intrinsic p-type germanium sample. The points refer to experimental data, the curves have been calculated on the basis of diffuse (solid line) and specular (dashed line) scattering models. (After Zemel and Petritz, reference 26.)

liquid air temperature showing greatly reduced hole mobilities[24], all measurements reported have been carried out in the vicinity of room temperature (in the case of germanium) or slightly above (silicon). Zemel and Petritz[25, 26] measured simultaneously the conductance, Hall coefficient, and magnetoresistance of near-intrinsic p-type germanium samples, using the techniques described in Ch. 7, §6.1. Figure 8.16 shows a plot of the Hall coefficient R_H as a function of the reduced conductivity change $(\bar{\sigma} - \bar{\sigma}_{min})/\bar{\sigma}_{min}$ obtained directly from the experimental data. A similar plot of the transverse magnetoresistance $\Delta \rho / \rho H^2$ (for low magnetic field) is shown in Fig. 8.17. In both figures the appropriate theoretical curves for completely diffuse scattering (solid lines) were obtained from (8.76) and (8.77), suitably modified to include holes[25]. The dashed curves, corresponding to total specular reflection, were obtained from the same equations with μ_s set equal to μ_b. The results indicate that diffuse rather than specular scattering is the dominant process at the surfaces studied.

Similar conclusions have been drawn by Albers[15] working on near-intrinsic, n-type germanium samples. Here, too, the Hall voltage, magneto-resistance, and conductivity were recorded simultaneously while ambients were employed to vary the surface potential. The bulk values of these quantities were determined from measurements on thick samples (to minimize the effect of the surfaces) neighbouring the actual slabs studied.

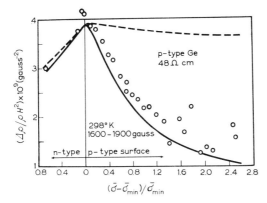

Fig. 8.17. Transverse magnetoresistance as a function of the reduced conductivity change $(\bar{\sigma}-\bar{\sigma}_{\min})/\bar{\sigma}_{\min}$ for a near-intrinsic p-type germanium sample. The points refer to experimental data, the curves have been calculated on the basis of diffuse (solid line) and specular (dashed line) scattering models. (After Zemel and Petritz, reference 26.)

Figure 8.18 shows a plot of the normalized Hall coefficient as a function of $(\bar{\sigma}-\bar{\sigma}_{\min})/\bar{\sigma}_{\min}$. The swing in conductivity corresponds to a variation in v_s of approximately -3 to $+6$. The appropriate theoretical curves for completely diffuse scattering (solid lines) were obtained from (8.65), suitably modified to include holes. The correlation mobility was neglected, and the approximation of (8.75) was used with μ_s taken from Schrieffer's calculations[1]. The two curves were chosen so that the roughly-known impurity concentration of the sample lies somewhere between them. That the theory is insensitive to the precise surface scattering mechanism assumed is seen by the proximity of the specular reflection curve (dashed line) to the other two.

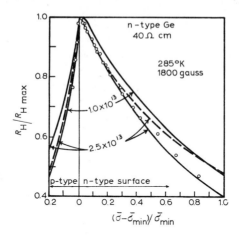

Fig. 8.18. Normalized Hall coefficient $R_H/R_{H\,\text{max}}$ as a function of the reduced conductivity change $(\bar{\sigma}-\bar{\sigma}_{\text{min}})/\bar{\sigma}_{\text{min}}$ for a near-intrinsic n-type germanium sample. The two solid curves represent theoretical values based on the diffuse scattering model and correspond to different estimates of the impurity concentration in the sample. The dashed curve is calculated from the specular scattering model. (After Albers, reference 15.)

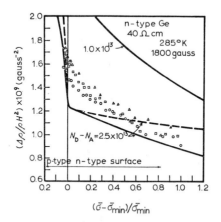

Fig. 8.19. Transverse magnetoresistance as a function of the reduced conductivity change $(\bar{\sigma}-\bar{\sigma}_{\text{min}})/\bar{\sigma}_{\text{min}}$ for a number of near-intrinsic n-type germanium samples. The two solid curves represent theoretical values based on the diffuse scattering model and correspond to different estimates of the impurity concentration in the samples. The dashed curve is calculated from the specular scattering model. (After Albers, reference 15.)

Figure 8.19 shows plots[15] of the transverse magnetoresistance $\Delta\rho/\rho H^2$ for three different samples. In this case the effect of the anisotropy of the conduction band has been included in the theory. Because of the uncertainty in the impurity concentration and its large effect on the theoretical curves, the interpretation of the results is not completely unambiguous. On the whole, however, a predominantly diffuse model is again suggested. This conclusion is supported by additional data obtained by Albers from longitudinal magnetoresistance measurements[15]. Because of the fact that the valence band is to a good approximation spherical (see Ch. 2, § 3.3), the hole contribution to the longitudinal magnetoresistance vanishes. This results in effect in a single-carrier system (electrons). Such measurements are therefore

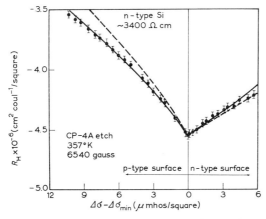

Fig. 8.20. Hall coefficient as a function of the change in surface conductance with respect to the conductance minimum for an n-type silicon sample. The points refer to experimental data, the curves have been calculated on the basis of diffuse (solid line) and specular (dashed line) scattering models. (After Coovert, reference 27.)

potentially very useful in yielding information concerning electron surface scattering, but at present the associated theory still awaits the refinements necessary to permit quantitative comparisons.

So much for germanium. The Hall coefficient and magnetoresistance of high-resistivity silicon samples (at temperatures slightly above room temperature) have been measured by Coovert[27]. The barrier was varied by means of an external dc field, the samples having been thermally oxidized to reduce shielding by slow surface states (see Ch. 9, § 2). Typical results for an n-type sample are shown in Figs. 8.20, 8.21, as a function of the surface conductance $\Delta\sigma$ measured with respect to its minimum value. The theoretical curves corresponding to diffuse surface scattering (solid lines) and to specular reflection (dashed lines) have been calculated from (8.76) and (8.77). The solid curve is seen to provide a somewhat better fit to the Hall data, while in the magnetoresistance case the results are inconclusive.

Murphy and Watkins[28] reported measurements on surface channels on p-n-p structures (Ch. 6, §8). Preliminary results indicate that the surface scattering of holes is partially specular, partially diffuse.

All the experiments described in the present section employed electric and/or magnetic measurements in order to elicit information concerning surface transport properties. Following a suggestion by Geballe[29] that the incorporation of thermal gradients might yield interesting information,

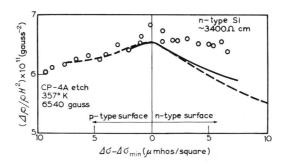

Fig. 8.21. Transverse magnetoresistance as a function of the change in surface conductance with respect to the conductance minimum for an n-type silicon sample. The points refer to experimental data, the curves have been calculated on the basis of diffuse (solid line) and specular (dashed line) scattering models. (After Coovert, reference 27.)

Zemel and Greene[30] measured the thermoelectric power of intrinsic germanium at room temperature. Their results agree remarkably well with the theoretical curve, but as the latter is totally insensitive to the type of surface scattering, no conclusions can be drawn regarding the applicability of either the diffuse or the specular model.

6. Concluding remarks

In this chapter we have reviewed the existing theoretical and experimental work on surface transport. We saw how it was possible to solve the Boltzmann transport equation both without and with a magnetic field under the assumption of constant bulk relaxation time. The calculations are considerably simplified if, in addition, the energy surfaces are taken to be spherical.

In practice these assumptions are usually adequate despite the fact that they do not represent the actual situation in semiconductors. The reason for this is that in all surface transport phenomena one is essentially dealing with changes corresponding to different conditions at the surface of the *same* sample, so that bulk effects are relegated to second order. Once the Boltzmann transport equation has been solved, the current can be calculated (usually by numerical integration) and in this way expressions obtained for the average and surface mobilities $\bar{\mu}$ and μ_s as a function of the scattering conditions at the surface. It should be remarked in this connection that the application to the scattering problem of very simple considerations has been successful in yielding results in good agreement with those obtained from the more rigorous treatments.

As for magnetosurface effects, here too expressions have been obtained connecting the galvanomagnetic parameters with the transport properties of the surface. These expressions, however, are considerably more complicated and so far no exact calculations have been carried out for the cases of practical interest. Approximations based on various assumptions concerning the shape of the potential barrier are available, but their accuracy is difficult to evaluate. Side by side with this approach, whose point of departure is a solution of the Boltzmann equation in the presence of a magnetic field, an effective mobility formalism has been developed with which it is possible to express approximately the galvanomagnetic quantities in terms of the conductivity mobility (for which numerical solutions are already available). In this treatment, as well, it is difficult to estimate the extent of the effect of the approximations involved.

The important advantage in the conductivity mobility results lies in their lending themselves to direct comparison with accurate theoretical calculations. One should recall, however, that these results yield only a lower, though certain bound to the conductivity mobility. How close this bound represents the actual situation depends on the distribution of the surface states, which are not taken into account in the evaluation of the data. Quite the contrary can be said concerning the galvanomagnetic parameters. Here the analysis of the measured quantities does not depend on assumptions concerning the surface states, but any accurate interpretation of the results obtained must await a more detailed theoretical treatment.

The general picture emerging from the information available at present indicates good agreement between the experimental data and the theoretical models. In most cases the scattering is definitely not completely diffuse, especially when the carriers are confined to deep wells, and some degree of specularity must be involved. Such information, however, is far from being capable of determining the dominant mechanism of surface scattering. Let us consider briefly the possible physical causes for the existence of non-specular reflection at the surface. It is reasonable to assume that an ideal surface having translational symmetry parallel to the surface will be a specular reflector at a temperature of absolute zero, just as a perfect crystal does not obstruct the motion of electrons moving through it. Every deviation from ideal conditions will of course give rise to a certain amount of diffusivity. At finite temperatures, phonons associated with surface waves (Rayleigh phonons) will appear in addition to the usual bulk phonons. Furthermore, structural defects arising from atomic disorder and impurities distributed randomly over the surface are able to serve as scattering centres that detract from the specular nature of the reflection. One should also take into consideration the electronic structure of such defects and impurities, since these are expected to scatter differently according as they are charged or neutral. Their charge condition will in turn depend on the position of the Fermi level at the surface (the surface potential).

In order to shed more light on the dominant mechanism in surface scattering, considerable work is still required. On the one hand, both the experimental and the theoretical tools available for surface transport studies have to be improved; on the other hand, the data must be intimately correlated with the different conditions prevailing at the surface. A relation has to be found between the scattering processes as manifested by the degree of diffusivity and the electronic and atomic structure of the surface. Studies of clean and nearly clean surfaces should prove very instructive in this respect.

References

[1] J. R. Schrieffer, Phys. Rev. **97** (1955) 641; *Semiconductor Surface Physics* (Edited by R. H. Kingston), (University of Pennsylvania Press, Philadelphia, 1957), p. 55.
[2] K. Fuchs, Proc. Cambridge phil. Soc. **34** (1938) 100.
[3] E. H. Sondheimer, Phys. Rev. **80** (1950) 401; *Advances in Physics* (Edited by N. F. Mott), (Taylor and Francis, London, 1952) vol. I, p. 1.
[4] See, for example, R. F. Greene, J. Phys. Chem. Solids **14** (1960) 291. (Review)
[5] F. S. Ham and D. C. Mattis, IBM J. Research Develop. **4** (1960) 143 ; see also *Electrical Properties of Thin Film Semiconductors*, Univ. Illinois Tech. Rep. No. 4 (1955), unpublished. (U.S. Army OOR, Contract DA-11-022-ORD-1001).
[6] P. J. Price, IBM J. Research Develop. **4** (1960) 152.
[7] A. Amith, J. Phys. Chem. Solids **14** (1960) 271.

[8] H. Flietner, Physica Status solidi **1** (1961) 483.
[9] R. F. Greene, D. R. Frankl, and J. N. Zemel, Phys. Rev. **118** (1960) 967.
[10] Y. Goldstein, N. B. Grover, A. Many, and R. F. Greene, J. appl. Phys. **32** (1961) 2540.
[11] R. F. Greene, Phys. Rev. **131** (1963) 592.
[12] N. B. Grover, Y. Goldstein, and A. Many, J. appl. Phys. **32** (1961) 2538.
[13] J. N. Zemel, Phys. Rev. **112** (1958) 762.
[14] R. L. Petritz, Phys. Rev. **110** (1958) 1254.
[15] W. A. Albers, Jr., J. Phys. Chem. Solids **23** (1962) 1249.
[16] J. Bardeen, R. E. Coovert, S. R. Morrison, J. R. Schrieffer, and R. Sun, Phys. Rev. **104** (1956) 47.
[17] M. F. Millea and T. C. Hall, Phys. Rev. Letters **1** (1958) 276.
[18] D. R. Frankl and A. Feuersanger, J. Phys. Chem. Solids **14** (1960), p. BD14 (Discussion following p. 225).
[19] A. Many, N. B. Grover, Y. Goldstein, and E. Harnik, J. Phys. Chem. Solids **14** (1960) 186.
[20] N. B. Grover and R. Oren, J. Phys. Chem. Solids **24** (1963) 693.
[21] N. B. Grover, Thesis, The Hebrew University of Jerusalem (1962).
[22] Y. Goldstein, Unpublished data (1962).
[23] R. Missman and P. Handler, J. Phys. Chem. Solids **8** (1959) 109.
[24] R. Cunningham and R. Bray, Bull. Am. phys. Soc. Ser. II **2** (1957) 348; R. Bray and R. Cunningham, J. Phys. Chem. Solids **8** (1959) 99.
[25] J. N. Zemel and R. L. Petritz, Phys. Rev. **110** (1958) 1263.
[26] J. N. Zemel and R. L. Petritz, J. Phys. Chem. Solids **8** (1959) 102.
[27] R. E. Coovert, J. Phys. Chem. Solids **21** (1961) 87.
[28] N. St. J. Murphy and T. B. Watkins, *Proc. Int. Conf. Semiconductor Physics, Prague*, 1960 (Czechoslovak Academy of Sciences, Prague, 1961), p. 552.
[29] T. H. Geballe, J. Phys. Chem. Solids **14** (1960) 72.
[30] J. N. Zemel and R. F. Greene, *Proc. Int. Conf. Semiconductor Physics, Prague*, 1960 (Czechoslovak Academy of Sciences, Prague, 1961), p. 549.

CHAPTER 9

THE ELECTRONIC STRUCTURE OF THE SURFACE

A great wealth of information on the electrical properties of semiconductor surfaces has been amassed during the last decade. Surface research has been the subject of five meetings [1-5] devoted entirely to its various aspects, as well as of special sessions in most of the recent international conferences on semiconductor physics [6-8]. Until several years ago, it was still possible to give adequate coverage to our knowledge of surface phenomena by means of general review articles [9-16]. The continuous accumulation of data, however, has made such comprehensive coverage increasingly more difficult, and review articles dealing with specific topics [17-22] have become more feasible. In the present chapter we shall attempt to strike a compromise between the scope of material covered and the degree of detail in its presentation. As regards the former, we shall limit the discussion mainly to single-crystal germanium and silicon surfaces. Although a large amount of work has been reported on surfaces of other semiconductors, most of it is still at the qualitative stage. The numerous studies of surface effects carried out primarily from the point of view of device performance will be omitted on similar grounds. In the presentation of the data, emphasis will be placed on those measurements that have either contributed directly to a better understanding of fundamental processes, or have lead to quantitative information on the surface structure. To do justice to all the work that falls into these two categories is still a formidable task, and no doubt much important material will receive less attention than it merits. In order to enable the reader to delve deeper into any topic that may be of specific interest to him, however, the discussion is accompanied by an extensive list of references.

The present chapter is divided into four sections. The first reviews the early work from which the basic model of a semiconductor surface evolved, as well as some of the later developments that have served to substantiate the various features of this model. Next we consider the detailed characteristics of the surface states on real surfaces, and their dependence on surface treatment. In § 3 we discuss the electronic structure of clean surfaces. The last section summarizes the present status of surface research and points out some of the important conclusions that may be drawn from the available data.

1. The surface of a semiconductor

The basic model of a semiconductor surface has emerged from studies made for the most part on germanium and silicon surfaces during the early 'fifties. This model applied fundamental theoretical concepts to describe and correlate the available experimental data, and was very successful in pointing to new relevant experiments. It is well established by now, and in fact was used extensively in many of our discussions in the preceding chapters. A simplified energy-level diagram of the surface, which embodies the main features of the model, is shown in Fig. 9.1. From the point of view of

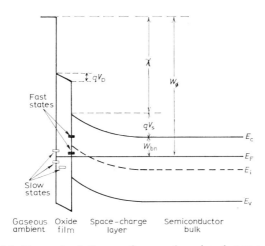

Fig. 9.1. Energy level diagram for a real semiconductor surface.

its electronic behaviour, the surface may be roughly divided into three separate regions: the space-charge layer, associated with surplus or deficit of free carriers, of thickness 10^{-5} cm or greater; the surface proper, consisting of the first few atomic planes of the semiconductor; and, in the case of a real surface, an adsorbed layer of foreign material. In a semiconductor such as germanium or silicon, this last consists mainly of an oxide film, usually 10–30 Å thick [23], together with various species adsorbed from the etchant and the surrounding ambient (see Ch. 3, § 4.1). The oxide film with its conduction and valence bands is shown next to the semiconductor, the midgaps of the two materials having been arbitrarily assumed to lie at the same energy level.

The fundamental variable characterizing the space-charge layer is the barrier height V_s (or surface potential ϕ_s) which, for given bulk conditions, determines uniquely the shape of the potential barrier and the carrier distribution in this region (see chapter 4). The surface proper and the adsorbed layer are usually the sites of localized electronic states — the surface states. These are known to be sensitive to mechanical and chemical treatment and to the gaseous or liquid environment to which the surface is exposed. As discussed in chapters 5–7, some sort of statistical equilibrium exists between the surface states and the semiconductor interior, and in the absence of an external field the overall charge in surface states is equal and opposite to that in the space-charge region. In germanium and silicon, the surface states can be divided into two distinct categories, according to whether the transition times between the states and the underlying bulk are very short (microseconds or less) or very long (seconds or more). The former, the *fast states*, are in intimate contact with the semiconductor bulk, and are probably located at, or very close to, the semiconductor/oxide interface. The latter, the *slow states*, are associated with the oxide and adsorbed species and are known to be distributed within the oxide film and/or on its outer surface. Another fundamental entity characterizing the surface is the work function W_ϕ. This is the sum of the energy parameter $W_{bn} \equiv E_c - E_F$, the barrier height $-qV_s$, and the electron affinity χ (see Ch. 4, § 1 and eq. (7.45)). For the sake of convenience, we have included in the affinity the potential drop $\pm qV_D$ across the oxide film, arising from a possible distribution of charge in the slow states. Under ordinary conditions, the thickness of the oxide film is quite small and this drop is negligible. A much more important contribution to the affinity may result from the alignment of adsorbed molecules having an electric dipole moment (see § 1.5 below).

In this section, the various features of the surface model are examined in the light of experiment. Here we shall be interested mainly in the fundamental electronic processes taking place at the surface. First we review the experimental data that serve to substantiate the properties of the space-charge layer as derived from a solution of Poisson's equation (see chapter 4). The general characteristics of the fast and slow states, and the interaction processes between the states and the underlying space-charge region, are considered in § 1.2. Two important processes of this sort, surface recombination and noise,

1.1. THE SPACE-CHARGE LAYER

The existence of a space-charge barrier layer at the free surface of a semiconductor was originally proposed by Bardeen [24]. Evidence for such a barrier first came from measurements of contact potential and rectification characteristics, as discussed in Ch. 1, § 1. Much more direct evidence was obtained by Brattain and Bardeen [25] from measurements of contact potential and surface photovoltage (Ch. 7, § 2) on etched germanium surfaces. An essential part of this work was the discovery that the barrier height can be varied in a reproducible way by exposing the surface cyclically to different gaseous ambients (the Brattain–Bardeen cycle — see Ch. 6, § 1). The observed range of variation in contact potential was about 0.5 V, one extreme obtained in ozone and corresponding to the highest germanium work function, the other in wet oxygen and corresponding to the lowest work function. The results were ascribed to the addition or removal by the cycling process of gas ions at the outer surface of the oxide film, and were interpreted in terms of the energy-level diagram in Fig. 9.1. The change in work function accompanying a change in surface ion charge falls partly across the space-charge region and partly across the oxide layer. The original estimate of Brattain and Bardeen [25] was that the change in qV_s amounted to only about 20 % of the observed change in W_ϕ. Later, however, Morrison [26, 27] measured changes in surface conductance in the same ambient cycle and concluded that a considerably larger fraction of the change in W_ϕ occurs across the barrier layer. This behaviour added more significance to contact-potential measurements as a tool for studying the characteristics of the space-charge region.

Brattain and Bardeen [25] found a direct correlation between the value of the work function and the surface photovoltage, in a manner consistent with the proposed surface model. The detailed theory of the surface photovoltage was worked out by Garrett and Brattain [28] and by Johnson [29, 30] (see Ch. 7, § 2). This theory was very successful in accounting for the experimental data and contributed greatly to a more quantitative understanding of the properties of the space-charge layer. Surface photovoltage measurements on germanium under small- and large-signal conditions are illustrated in Figs. 9.2 and 9.3, respectively. In both cases, the chopped light technique (see Ch. 7, § 2) has been used in order to eliminate the effect of slow states. In Fig. 9.2, taken from Brattain and Garrett [31,32], the ratio $\delta v_s/(\delta p_b/n_i)$ is plotted against surface potential u_s for an n-type ($u_b = 1.1$) and a p-type ($u_b = -2.9$) sample. The (dimensionless) surface photovoltage δv_s was obtained from δV_{cp} after subtracting the Dember potential (see (7.47)), while the excess carrier density δp_b was derived from changes in sample conductance. The surface potential was varied by the Brattain–Bardeen ambient cycle, and determined by surface conductance measurements. The right-hand branches represent negative values and the left-hand

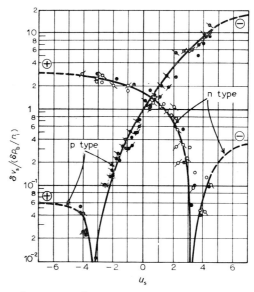

Fig. 9.2. Surface photovoltage δv_s (in units of fractional excess carrier density $\delta p_b/n_i$) *versus* surface potential u_s for an n- and a p-type germanium sample. Data from different runs are distinguished by modifications in symbols used for experimental points. Left- and right-hand branches represent positive and negative values, respectively. Theoretically calculated values for strong inversion and accumulation layers are indicated by dashed asymptotes. (After Brattain and Garrett, reference 31.)

branches, positive values. The surface photovoltage does not vanish at $u_s = u_b$ (where $v_s = 0$) because of the presence of (fast) surface states. For strong accumulation and inversion layers, however, it is independent of surface states, the asymptotic values being in good agreement with those predicted theoretically. In the former case (u_s large and positive in the n-type sample, large and negative in the p-type sample), we obtain from (4.66) that $\delta|v_s|/(\delta p_b/n_i) \approx -\exp(-|u_b|)$. A similar calculation for strong inversion layers yields the value $-\exp|u_b|$, provided δp_b is small not only with respect to $n_b + p_b$ (as required in the accumulation-layer case) but also with respect to the minority-carrier density. The theoretical values for the four branches are indicated by the dashed asymptotes. Johnson [30] measured the surface photovoltage δv_s as a function of light intensity. In Fig. 9.3, δv_s is plotted against the fractional change $\delta n_b/n_b$ in minority-carrier density (electrons) for a p-type sample. The circles were obtained when the surface was exposed to room air, the crosses and squares — under different

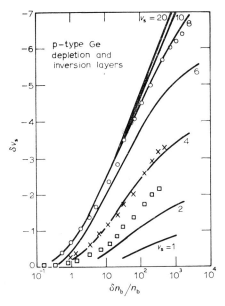

Fig. 9.3. Surface photovoltage δv_s as function of fractional change $\delta n_b/n_b$ in minority-carrier density for p-type germanium sample. The different points correspond to different flow rates of dry oxygen. The solid curves represent theoretical calculations for different values of the quiescent barrier height in the absence of surface states. (After Johnson, reference 30.)

flow rates of dry oxygen. The solid curves were calculated theoretically for different values of the quiescent barrier height on the assumption that surface states are absent (see Ch. 4, § 4). A good fit is obtained by assigning to the three sets of points the values 8, 4, and 3 for v_s. Particularly noteworthy is the confirmation of the theoretical expectation (see Ch. 7, § 2) that δv_s is insensitive to the presence of (fast) surface states, at least for sufficiently strong light.

The surface conductance $\Delta\sigma$ (Ch. 6, § 2) is another quantity that is intimately associated with the space-charge layer. The first reported measurements of this parameter were carried out by Clarke [33], who studied the variation in conductance of an etched germanium sample before and after exposure to oxygen. The direction of change in $\Delta\sigma$ indicated that oxygen adsorption introduced acceptor-like slow states which, as discussed in Ch. 4, § 1.3, act to bend the bands upward (that is, to decrease V_s). Mechanical damage to the surface produced by grinding, polishing, or sandblasting was also found to introduce a large density of acceptor-like states on germanium [34]: the sample conductance was observed to increase considerably upon sandblasting, while thermoelectric-power measurements with a hot point probe [35, 36] indicated the presence of a p-type layer at the damaged surface. This is consistent with Brattain and Bardeen's findings [25] that sandblasting increases the work function (V_s becomes large and negative). The thickness of the disturbed layer, as estimated from conductance [34] and surface recombination [37] measurements before and after controlled etching, ranges from 1 μ or less for fine polishes to 35 μ for heavy sandblasting. Damaged layers produced by ion bombardment exhibit a similar p-type character [38].

Considerably more quantitative information about the space-charge layer became available when surface-conductance measurements could be interpreted [27, 39-41] in terms of the conductance minimum $\Delta\sigma_{min}$. In order to obtain the large swing in barrier height usually necessary to include this point, use was made of an ambient gas cycle, strong electric fields normal to the surface, or some combination of the two. Typical results of low-frequency field-effect patterns (see Ch. 6, § 5) obtained on a p-type germanium sample by Montgomery and Brown [40] are shown in Fig. 9.4. Each pattern corresponds to different ambient conditions, as indicated. The horizontal axis is proportional to the charge induced at the semiconductor surface by the field, while the vertical axis represents the resulting change in sample conductance. A positive slope indicates hole conduction (accumulation layer) and a negative slope, electron conduction (inversion layer). The former corresponds to ozone and the latter to wet air, in agreement with the contact-potential measurements [25]. The minimum in conductance is clearly seen in the other four patterns, which correspond to intermediate values of V_s. Measurements of this sort established surface conductance as one of the most powerful tools for studying the space-charge layer. In fact, in nearly all subsequent surface work, the barrier height is derived from surface conductance.

The surface conductance of channels on p–n–p or n–p–n junction structures (see Ch. 6, § 8) has been studied extensively by Brown[42], by Kingston[43,44], and by Statz and co-workers[45,46]. The channel inversion layer can be produced by exposure to moist ambients or to dry ammonia in the case of a p-type base, and by oxidation in a mixture of wet oxygen and ozone

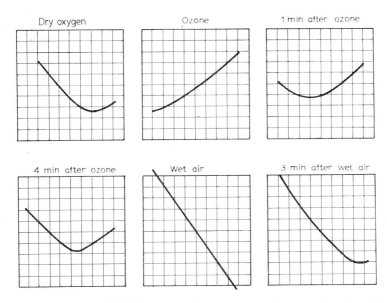

Fig. 9.4. Low-frequency field-effect patterns for p-type germanium sample under different gaseous ambients. Horizontal axis is proportional to charge induced at surface and vertical axis, to change in surface conductance. (After Montgomery and Brown, reference 40.)

in the case of an n-type base. Experimentally, the surface conductance is always observed to decrease with increasing reverse bias, in accordance with the theoretical predictions (Ch. 6, § 8). The change in conductance $\Delta\sigma$ upon application of the bias voltage V_a exhibits a transient behaviour, but for the moment we shall consider only the final, steady-state value, attained after the slow states have reached equilibrium with the interior (see § 1.2). Under these conditions Kingston[44] finds that for large applied biases, $\Delta\sigma$ varies very nearly as $|V_a|^{-1}$. Such behaviour can be understood if the slow states are assumed to be so dense as to anchor the Fermi level at the surface against any significant variation by the applied bias. Taking u_s to be constant and assuming $n_b(q|V_a|/kT)$ to be large compared to $n_i\exp|u_s|$, we can expand (6.69) and obtain for strong inversion layers on an n-type base

$$\Delta P \propto e^{|u_s|}[n_b(q|V_a|/kT+|u_b|-1)]^{-\frac{1}{2}} \approx e^{|u_s|}n_b^{-\frac{1}{2}}(q|V_a|/kT)^{-\frac{1}{2}}. \tag{9.1}$$

The observed $|V_a|^{-1}$ (rather than $|V_a|^{-\frac{1}{2}}$) dependence of $\Delta\sigma(\approx q\mu_{ps}\Delta P)$ on bias voltage could be quantitatively accounted for [44] by taking into consideration the decrease in surface mobility μ_{ps} with decreasing width of the inversion layer, according to Schrieffer's calculations [47] for diffuse surface scattering (see Ch. 8, § 3.2). Typical results of steady-state conductance *versus* applied bias obtained by Statz and co-workers [48] for a p-type layer on an n-type germanium base are shown in Fig. 9.5a. The four curves represent different oxidation pretreatments of the surface, the upper one corresponding to the longest exposure to wet oxygen and ozone. The surface potential ϕ_s *versus* V_a, as derived from the data of Fig. 9.5a with the use of (6.69) and Schrieffer's mobility calculations, is shown in Fig. 9.5b. The lack of depend-

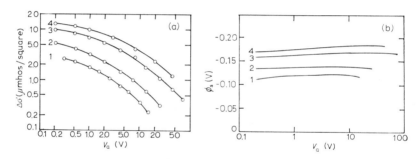

Fig. 9.5. Steady-state channel-conductance measurements on germanium p–n–p structure. The four curves in each figure represent different oxidation pretreatments.
(a) Surface conductance $\Delta\sigma$ *versus* applied bias V_a.
(b) Surface potential ϕ_s *versus* applied bias V_a, as derived from (a).
(After Statz et al., reference 48.)

ence of ϕ_s on V_a for a given condition of the surface confirms the presence of a large density of slow states and, at the same time, substantiates the basic assumptions underlying eq. (6.69). In other cases, however, a small decrease in $|\phi_s|$ was observed [48,49] with increasing $|V_a|$. This can be explained by the change in potential drop V_D across the oxide film (see Fig. 9.1) resulting from the increasing charge in the slow states. An oxide thickness of about 10 Å is required to account for the data.

The excess reverse current originating from conducting channels on the surface of a p–n junction has also received considerable attention [50–56]. Here again the results are, on the whole, consistent with theory. Anomalous effects associated with the presence of water and other condensable vapours will be considered in the following subsection.

An additional check on the theory was provided by measurements of the differential surface capacitance as a function of barrier height [57] (see Ch. 6, § 3). The results agreed well with theoretical calculations based on the values of barrier height derived from changes in surface conductance. More detailed data on surface capacitance were later obtained on semiconductor–electrolyte systems, as will be discussed in § 2.4 below.

The presence of a surplus or a deficit of carriers in the surface region was clearly demonstrated by Harrick [58-60], using absorption measurements on an infrared beam multiply reflected at the surface (see Ch. 7, § 5). This work will be discussed in §§ 1.2, 2.1. Similar measurements, bearing more directly on the space-charge layer, have been carried out by Seraphin and Orton [61]. These workers monitored the modulation in the reflected-beam intensity induced by a transverse ac field, as the surface was changed from p to n type by an ambient cycle or by a dc bias at the field plate. As expected, the change in absorption was found to follow both in magnitude and in sign the variation in field-effect mobility (eq. (6.50)): the modulation signal was in phase with the applied voltage for strong n-type layers, 180° out of phase for strong p-type layers. The signal decreased in depletion layers and its frequency doubled. The second harmonic appears because the field-effect mobility near the minimum changes sign within each cycle of the field.

A few special effects associated with the space-charge region may be mentioned. Longitudinal junctions in the space-charge layer can be produced by a field plate that covers only part of the sample [62]. Such surface junctions exhibit rectification and injection phenomena which are analogous to those occurring in bulk junctions. It has also been shown that a space-charge layer can be induced at the surface in a Hall-effect experiment [63, 64]. Here the Hall voltage alters the barrier height at the two surfaces parallel to the magnetic field.

We have discussed above only some of the many measurements that can serve to confirm the theoretically derived characteristics of the space-charge region. The self-consistency apparent in most of the quantitative studies of surface phenomena, in which these characteristics are used implicitly, has firmly established the basic assumptions underlying the theory.

1.2. FAST AND SLOW SURFACE STATES

Shockley and Pearson's field-effect measurements [65] (see Ch. 1, § 1) constituted the first direct evidence for the presence of surface states on a free germanium surface. That such states fall into two distinct categories, slow and fast, became apparent from the early work of Brattain and Bardeen [25]. In addition to the ambient-induced states, which must be of sufficient density to control the barrier height, it was necessary to assume a second type of states, ambient insensitive and much less dense, to account for the surface recombination data (see § 1.3 below). The large difference in capture times of the fast and slow states was demonstrated by Garrett and Brattain [28], who observed that the surface photovoltage disappeared after illumination for a period of seconds or minutes. This is the time required for the slow states to charge up and thus, because of their large density, to clamp the barrier height against any significant change. Slow drifts in contact potential [66] and in rectification characteristics [67] were also attributed to slow states. The most direct measurement of capture times, however, has been obtained from field-effect [39-41, 68-71] and channel conductance [44-46, 48, 72-74] experiments. The change in surface conductance $\Delta\sigma$ following a sudden application of the field is generally found to consist of two easily separable transients: a very fast relaxation (microseconds or less) from the initial instantaneous value to a new quasi-stationary value, followed by a much slower decay (seconds or longer) which brings $\Delta\sigma$ back almost to its original, field-free value. A similar decay (with the polarity of the change in $\Delta\sigma$ reversed) is observed when the field is switched off.

The fast relaxation has been studied by both the pulsed [68, 75] and the high-frequency [76-79] field-effect techniques (see Ch. 6, §§ 6, 7). The experimental behaviour can be well accounted for in terms of the interaction kinetics between the fast surface states and the underlying semiconductor bulk. In fact, the observed relaxation characteristics follow closely those predicted on the basis of the analyses of Ch. 6, §§ 6, 7 and need little further discussion. This is illustrated by Fig. 9.6, where the measured [76] frequency dependence of the in-phase field-effect mobility is shown for a p-type germanium sample under different ambient conditions. The similarity of the curves to those in Fig. 6.16 is apparent when one bears in mind that for a p-type sample, wet air gives rise to an inversion layer, ozone and dry oxygen to an accumulation layer. Measurements of this sort have also been extended to reduced temperatures [80-82]. Studies of the temperature dependence of the relaxation time have proved particularly useful in determining the energy and capture cross sections of the fast states, and will be discussed in detail in § 2.

An estimate of the density of the fast states was obtained by Brown [39] and Montgomery [40, 57] from low-frequency field-effect patterns of the type shown in Fig. 9.4. Bardeen and co-workers [41] studied the small-signal ac field effect in an ambient cycle, and expressed their results in terms of the

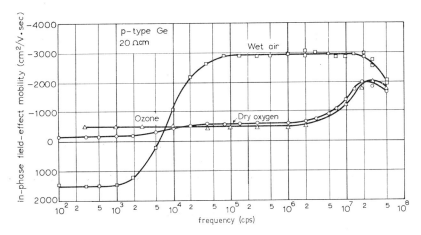

Fig. 9.6. In-phase field-effect mobility *versus* frequency for p-type germanium sample under different ambient conditions. (After Montgomery, reference 76.)

field-effect mobility. In both cases the period of the ac field was chosen long compared to the fast relaxation time but short compared to the slow relaxation time, as explained in Ch. 6, § 5. Similar measurements under quasi-stationary condition, but using channels, were performed by Statz and co-workers [45, 46, 48]. The density of fast states derived from such measurements is of the order of 10^{11} cm^{-2} for germanium and 10^{12} cm^{-2} for silicon. Their capture cross section for holes and electrons, as estimated from any of the theoretical expressions for the surface-state time constant or for the surface recombination velocity (Ch. 5, §§ 6, 7 and Ch. 6, § 7), is of the order of 10^{-15} cm^2. This value, which is comparable to that often quoted for bulk states, is strong evidence that the fast states are physically very close to the semiconductor, probably at its boundary with the oxide film. Such a location explains also the insensitivity of the fast states to ambients, the oxide film acting as a shield against external influences.

An altogether different character is exhibited by the slow states. The fact that the surface conductance eventually attains very nearly its value prior to the application of the field shows conclusively that the slow states are so dense as to shield the space-charge region almost completely from the external field. This makes an accurate determination of the slow-state density very difficult and so far only a lower limit to its value is available. Such a limit may be obtained if one assumes that the final, steady-state change in $\Delta\sigma$ is not zero but equal to the experimental accuracy, say 0.1 μmho. Reference to Fig. 6.1 and use of (5.48) then places the lower limit at about 10^{13} cm^{-2}. An estimate of the capture cross sections of the states based on this density and a time constant of seconds yields a value many orders of magnitude smaller than that for the fast states. This, combined with the observation that the slow relaxation is extremely sensitive to surface treatment and ambient, strongly suggests that the slow states are located in the oxide film and/or at its outer surface.

The process of charge transfer between the slow states and the semiconductor interior is far less understood than the parallel process involving fast states. In most cases the slow field-effect decay is non-exponential, the apparent time constant increasing with time. This behaviour is observed even when the applied fields are so low ($< 10^4$ V/cm) that V_s changes by less than 1 kT/q. In addition, the change in $\Delta\sigma$ at any instant of time is usually proportional to the applied field [70,71] and is symmetrical with respect to the field polarity. Such linearity is most strikingly demonstrated by the low-frequency response of the field effect: a sinusoidally-varying field gives rise to a sinusoidal variation in $\Delta\sigma$, of the same frequency and with no harmonic distortion [70]. At higher fields, departures from linearity are observed [71]. Two models have been proposed in an attempt to account for these observations. Morrison [71] assumes that the electron transfer to or from the slow states takes place by thermionic emission over the potential barrier presented by the forbidden gap of the oxide (see Fig. 9.1). The non-exponential decay is then attributed to a variation in surface potential with the changing occupation of the slow states. This model is similar to the one used in analysing the interaction process between the fast surface states and the space-charge region (see Ch. 5, § 6), except that here the electrons must surmount the additional potential barrier associated with the oxide film. Equation (5.57) applies equally well to the present case, with the capture probability K_n now corresponding to electron flow over the oxide barrier. Such a model accounts for the non-exponential decay at *high fields*. It does not, however, explain the low-field behaviour, since in this range (5.57) reduces to a linear differential equation (eq. (5.62)) and the small-signal treatment leading to a simple exponential decay applies.

The non-exponential behaviour at low fields has led Kingston and McWhorter [70] to propose that the wide range of time constants required to account for the decay curve arises not from a nonlinear process but from surface heterogeneity. The conductance of each element of area is assumed to decay exponentially, but the time constant fluctuates from one element to another over the surface. If the various decay characteristics are uncorrelated, their superposition would result in an overall decay that is not exponential, no matter how low the field is. Furthermore, this situation would give rise to the observed linearity of $\Delta\sigma$ with field. Kingston and McWhorter [70] measured the frequency response of the ac field effect in germanium over a wide range (10^{-2}–10^3 cps). The curves obtained differ for different treatments and ambients, but in all cases the response falls off gradually as the frequency is decreased over many decades, still continuing at 10^{-2} cps.

The distribution of time constants is derived from the data as follows. Consider first an element of area characterized by a simple exponential decay of time constant τ. The frequency response under these conditions has already been calculated for the analogous case involving fast states (Ch. 6, § 7). Since the density of states N_t is now very large, τ_{ss} and τ_1 in (6.53) and (6.54) are equal, to a good approximation, and (6.56) reduces to

$$\delta(\Delta N)_0 = \frac{i\omega\tau}{1+i\omega\tau}\frac{Q_{s0}}{q}, \tag{9.2}$$

where $\tau \equiv \tau_{ss} = \tau_1$. Averaging over all individual area elements of the surface, we obtain for the overall frequency response

$$\overline{\delta(\Delta N)_0} = \frac{\int_{\tau_1}^{\tau_2} g(\tau)i\omega\tau\,d\tau/(1+i\omega\tau)}{\int_{\tau_1}^{\tau_2} g(\tau)d\tau}\frac{Q_{s0}}{q}, \tag{9.3}$$

where $g(\tau)d\tau$ is the surface area associated with a time constant τ, and τ_1 and τ_2 are the limits of the distribution. A fit of (9.3) to the experimental data shows that the distribution $g(\tau)$ of time constants over the surface varies very nearly as $1/\tau$. The magnitudes of τ_1 and τ_2 depend on surface conditions, typical values being 10^{-2} and 10^2 sec, respectively. A similar distribution of time constants was inferred by Abkevich [83] and Kosman [84] from measurements of the slow relaxation in contact potential following the cessation of illumination. For freshly etched surfaces of both germanium and silicon [83], the distribution function has the form τ^{-m}, where m lies between 1 and 2.

After prolonged exposure to air, heating in vacuum, or electron bombardment, m becomes approximately equal to unity.

It is apparent that neither of the models considered above can account by itself for the observed relaxation characteristics, and possibly a combination of the two might be necessary. Morrison's barrier model may apply for high fields, while at low fields surface heterogeneity appears to play the dominant role. The mechanism of charge transfer between the slow states and the interior is likewise uncertain. The observed increase of the relaxation time with decreasing temperature [46, 71] suggests charge transfer *over* a potential barrier. The activation energy of 0.3–0.5 eV obtained for this process, however, is far too small to correspond to the oxide barrier; one would expect the barrier to be more like a few electron volts, in view of the value of 5.5 eV for the forbidden gap of germanium dioxide estimated by Papazian [85]. Phonon-assisted tunnelling or T–F emission (see Ch. 7, § 4) has been suggested by Statz et al.[46, 86] as a possible mechanism. A tunnelling process is also indicated by the measurements of Lasser and co-workers [87], who found that the relaxation time increases very rapidly with the thickness of the oxide layer. Oxide films 0.3 μ thick, for example, give immeasurably long decay times, and the surface potential becomes insensitive even to gaseous ambients as active as iodine vapour. These results indicate also that the slow states are located predominantly at the outer surface of the oxide.

Kingston and McWhorter [70] have shown that the observed $1/\tau$ distribution can be accounted for by a uniform distribution of barrier heights over the surface if electron transfer takes place by thermionic emission, and by a uniform distribution in the oxide film thickness if tunnelling through the barrier predominates. Other models of charge transfer, and more recent experimental data bearing on this question, will be discussed in § 2.3 below.

As pointed out above, the slow relaxation has been found to depend strongly on the surrounding ambient gas. Particularly pronounced is the effect of water vapour, especially in the case of germanium. Adsorption of water is accompanied by a drastic reduction in time constant, while desorption results in a large increase in time constant, the process being generally quite reversible [87]. Water vapour was found also to give rise to anomalous effects at the surfaces of p–n–p and p–n junction structures. For relative humidities greater than 50 % and at reverse bias voltages exceeding about 10 V, an additional surface conductance, well above the normal channel conductance, is usually observed [48, 50, 51, 55, 86, 88]. Similar effects were encountered in the presence of vapours of certain organic liquids [89–91].

Law and Meigs [51] attribute the leakage current to ionic conduction, while Statz and co-workers [48, 89] ascribe it to hole conduction in the adsorbed liquid film.

Another surface anomaly is the occurrence of surface breakdown in reverse-biased junctions [91, 92]. The breakdown field is a sensitive function of surface conditions and is usually lower than the threshold for bulk breakdown. Using a one-dimensional model that takes into account the charge and field distribution at the surface, Garrett and Brattain [92] were able to account for the main features of the breakdown phenomena.

There is a close similarity between the slow states observed on the free surface of germanium and silicon and those present on the surface when in contact with a metal electrode. Just as large a density of states is required here in order to explain the lack of dependence of the rectification characteristics on the work function of the metal. A lower limit for this density can readily be obtained [24] if one assumes that neither the semiconductor nor the metal surface changes structure upon contact. Referring to Fig. 4.4, we require that the barrier height V_s remain constant, say to within $1\,kT/q$, while the metal–semiconductor separation a is decreased to atomic distances, say 4 Å. Under these conditions, the field within the separation is very nearly equal to $(W_\phi - W'_\phi)/qa$. From Gauss's law we then obtain that the surface charge density on the metal or semiconductor is given by $|Q_s| = \varepsilon_0|W_\phi - W'_\phi|/qa$. On the semiconductor side, Q_s is composed of a change in surface-state charge density δQ_{ss} and a change in space-charge density δQ_{sc}. For the case of interest $|\delta Q_{sc}| \ll |\delta Q_{ss}|$, and if we consider a single discrete set of surface states at the Fermi level with density N_t, we may express Q_s in the form (see eqs. (5.44), (5.45))

$$|Q_s| \approx |\delta Q_{ss}| = \tfrac{1}{2} q N_t \frac{1 - \exp(-|\Delta v_s|)}{1 + \exp(-|\Delta v_s|)}, \qquad (9.4)$$

where $|\Delta v_s|$ is the change in the (dimensionless) barrier height brought about by the proximity of the metal. Thus

$$N_t = \frac{2\varepsilon_0 |W_\phi - W'_\phi|}{q^2 a} \frac{1 + \exp(-|\Delta v_s|)}{1 - \exp(-|\Delta v_s|)}. \qquad (9.5)$$

The contact potential $|W_\phi - W'_\phi|/q$ is of the order of 1V. Substituting 4 Å for a and unity for $|\Delta v_s|$, we obtain for N_t a value of $6 \times 10^{13}\,\mathrm{cm}^{-2}$. This value is much the same as that estimated above for the lower limit of the slow-state density.

The early measurements of rectification properties in germanium and silicon were performed mostly on point contacts. These are not very satisfactory because of complications due to forming and possible damage of the surface by the metal point. More reproducible results have been obtained for plated and evaporated contacts [93, 94]. The current–voltage characteristics of such contacts are much closer to those of an ideal rectifier (see Ch. 2, § 7.4), and enable reliable determination of the barrier height. Bocciarelli [95] and Hartig and Noyce [96] investigated the effect of different

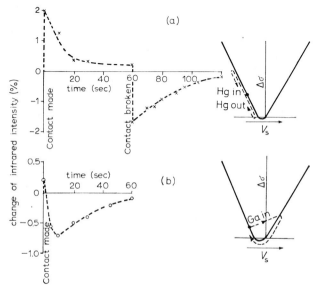

Fig. 9.7. Changes in intensity of infrared beam reflected from etched germanium sample, following contact with metal electrode. The accompanying changes in surface conductance are illustrated schematically in the diagrams at right of figure.
(a) Mercury electrode.
(b) Gallium electrode.
(After Harrick, reference 58.)

surface treatments and gaseous ambients on plated contacts. These studies supported the earlier conclusion that the rectification properties of contacts on germanium and silicon are chiefly a function of the semiconductor surface, and are relatively independent of the metal used.

Perhaps the most direct evidence for identifying the surface states involved in the metal–semiconductor contact with the slow states of the free surface has been provided by infrared absorption measurements (see Ch. 7, § 5).

Harrick[58,59] monitored the changes in intensity of an infrared beam reflected from an etched germanium surface, following contact with a mercury or gallium electrode. His results are shown in Fig. 9.7. The relaxation curves are seen to be very similar to those of the field effect, and can be accounted for by the same process of charge transfer between the slow states and the interior. Upon application of the metal electrode, most of the contact potential between the metal and the semiconductor falls across the space-charge region, resulting in a large change in V_s and hence in free-carrier absorption. The subsequent decay of the signal represents the return of the space-charge region to its original condition because of screening by slow states. A similar sequence of events occurs when the contact is broken. The sign of the change in absorption is in each case consistent with the polarity of the contact potential measured independently between the germanium sample and the metal electrode. The accompanying changes in surface conductance are illustrated schematically by the diagrams at the right-hand side of the figure.

1.3. SURFACE RECOMBINATION

The surface recombination velocity s is an extremely sensitive function of surface treatment. Values of s ranging from a hundred thousand centimetres per second on sandblasted surfaces to hundreds and even tens of centimetres per second on etched or otherwise treated surfaces have been reported for germanium [97-103] and silicon [104-106]. Systematic measurements of the effect of gaseous ambients on surface recombination on germanium, with the view of gaining some quantitative information on the surface states involved, were first carried out by Brattain and Bardeen[25]. It is from these studies that the basic model of surface recombination via fast surface states has evolved (see Ch. 5, § 7). The recombination centres were assumed to remain essentially unaltered during the ambient cycle, the recombination velocity being entirely controlled by the surface potential ϕ_s. In their original experiment, Brattain and Bardeen found little or no variation in s with ambient. This indicated that the recombination levels lie very close to the band edges. More careful measurements by Keyes and Maple[107] and by Stevenson and Keyes[108], using the photoconductance decay technique (Ch. 7, § 1.2), revealed that s may change by as much as a factor of two or three in the Brattain–Bardeen cycle. Larger changes in s have been reported subsequently[109,110]. Similar effects of gaseous ambients on surface recombination were observed for cadmium sulphide [111-113]. Here, however, changes in surface-state density may also play an important role.

Experiments in which surface recombination and surface conductance were measured simultaneously, so as to obtain a relation between s and ϕ_s, have been reported by Noyce [114] and by Stevenson [115] for germanium. The results were not in exact agreement with those expected from a simple recombination model (Ch. 5, § 7), but they could be explained qualitatively by a reasonable distribution in energy of active recombination centres. A qualitative agreement between experiment and theory was also obtained by Rzhanov and co-workers [116]. Attempts [97,108,117] to determine the energy of the recombination centres by measuring the temperature dependence of s have not been very successful, because possible changes in ϕ_s with temperature were not taken into account. The variation of s with bulk resistivity was studied by Schultz [118] and in more detail by Strikha [119]. Their results were in reasonably good agreement with the theoretically expected variation arising from the $(n_b + p_b)$ term in (5.83).

A more reliable check on the theory of surface recombination is provided by experiments in which the surface potential is varied by transverse fields, since any changes in the recombination centres due to ambients are avoided. Such measurements on germanium were first reported by Henisch et al.[120]. The surface recombination velocity was determined by the pulse-injection technique (see Ch. 7, § 1.2), and found to change by a factor as large as ten in the range of dc fields used. Subsequent measurements [121-125] revealed that following the application of the field, a slow relaxation in s usually takes place, similar to that observed with surface conductance. Both processes are of course associated with the relaxation of surface potential arising from the charging of slow states, and indeed good correlation was obtained [121,125] between the decay characteristics in the two cases.

Thomas and Rediker [126] combined an ambient cycle with a large transverse ac field, and obtained oscillograms of s *versus* induced charge that are analogous to the field-effect patterns shown in Fig. 9.4. Changes in s were derived from the variation in reverse current in a junction diode alloyed near the surface under study (see Ch. 7, § 1.2). The oscillograms for different ambients could be fitted together to form a smooth s curve, exhibiting a maximum and falling off for strong n- and p-type surfaces. Since the values of ϕ_s were not determined, however, the results could only serve as a qualitative confirmation of (5.83).

By the use of very high transverse fields (dc and ac), Many and co-workers [121,125] were able to swing the potential barrier over the entire range of interest, without recourse to an ambient gas cycle. The surface recombination velocity was determined by the pulse-injection technique, while the surface potential was derived from surface conductance measurements carried out simultaneously. Typical results [127] of s/s_{max} versus $u_s(\equiv q\phi_s/kT)$ obtained on an n-type germanium sample are shown in Fig. 9.8. The circles and dots

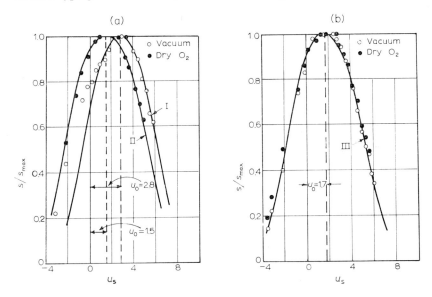

Fig. 9.8. Surface recombination velocity s (relative to its maximum value s_{max}) as function of surface potential u_s for an n-type germanium sample. The circles and dots represent measured values under vacuum and in dry oxygen atmosphere, respectively. The solid curves were calculated from (5.85). [I:$(E_t^f - E_i)/kT = 6.1$; II: $(E_t^f - E_i)/kT = 5.3$; III: $(E_t^f - E_i)/kT = 5.2$.]
(a) Shortly after etching.
(b) Two weeks later.
(After Many and Gerlich, reference 127.)

represent measured values in vacuum and in a dry oxygen atmosphere, respectively. The data in Fig. 9.8a were taken within 24 hours following etching, while those in Fig. 9.8b — two weeks later, when the surface had become stabilized. The solid curves in both figures were calculated from (5.85), with the parameters of the recombination centres having the values indicated. The remarkable agreement between experiment and theory provides very convincing evidence in favour of the Shockley-Read model in general (see Ch. 2,

§ 7.2), and of the basic assumptions underlying its extension to surface recombination in particular. Such excellent agreement is a consequence of the rather fortunate situation in germanium where, quite often, only one discrete set of fast states dominates the recombination process (see § 2 below). This lends confidence to the application of the theory, in its more general form, to determine the parameters of the recombination centres even when their distribution in energy is more complex.

The results discussed above correspond to the simplest case, in which the excess carrier density is small compared to the equilibrium *minority*-carrier density. At higher excitation levels, non-linear effects have been observed on both germanium [128, 129] and silicon [110, 130, 131]. In general s was found to increase with increasing excess-carrier density δp_b, but careful studies on silicon [131] revealed changes in either direction, depending on the surrounding ambient. Such effects can be satisfactorily accounted for by an appropriate modification [128, 129, 132] of the simple recombination model. The case of intermediate excitation levels (δp_b large compared to the minority-carrier density, but small compared to $n_b + p_b$) has been considered in detail in Ch. 5, § 7. It was shown there that the width of the s curve decreases with increasing $\delta p_b / n_i$, the main change occurring in the range of ϕ_s corresponding to depletion and inversion layers (see Fig. 5.7). In this range, the presence of a relatively small excess-carrier density is sufficient to give rise to a significant lowering of the barrier height, and thus to a corresponding change in s. Whether s will increase or decrease with $\delta p_b / n_i$ depends both upon the shape of the small-signal s curve and upon the quiescent value of the surface potential.

1.4. NOISE

Whereas surface recombination involves fast states only, noise is affected by both fast and slow states, depending on the frequency range considered. The fast states manifest themselves primarily in G–R noise (see Ch. 7, § 7), the dominant process at high frequencies. This type of noise has been studied in germanium and also in compound semiconductors such as cadmium and lead sulphide [133]. The experimental results can generally be well accounted for by the theoretical expressions (7.57), as regards the dependence of the noise power on frequency, temperature, and dc current. In germanium, surface treatment and ambient gas were shown to have considerable effect on the G–R noise spectrum [134–136], and the time constant τ derived on the basis of (7.57) could be reasonably identified with the sample effective lifetime τ_{eff} (see Ch. 7, § 1.1). Apart from this rather qualitative correlation between G–R noise and surface recombination, little more can be learned from the available data on the interaction process between the fast states and the bulk.

Considerably more information about germanium surfaces has been obtained from measurements of $1/f$ noise [133, 137]. The approximate $1/f$ law (see eq. (7.58)) has been found to hold over a remarkably wide frequency range [138–142]. It was observed [139] in germanium at frequencies as low as 10^{-4} cps. The upper limit of the spectrum usually extends to 10^2–10^4 cps, where it is masked by G–R noise [142–144]. A typical noise spectrum obtained by Mac Rae and Levinstein [144] on a gold-doped germanium filament at 90°K is shown in Fig. 9.9. The $1/f$ component dominates the spectrum at low

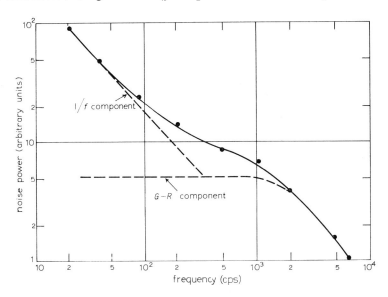

Fig. 9.9. Noise spectrum of gold-doped n-type germanium sample at 90°K. (After Mac Rae and Levinstein, reference 144.)

frequencies, while in the vicinity of 1 kc an appreciable contribution from G–R noise of the form $\tau/(1+\omega^2\tau^2)$ becomes apparent.

The $1/f$ noise power is usually found to be proportional to the square of the dc current I_{dc}, as in the case of G–R noise. This behaviour indicates that conductivity modulation arising from fluctuations in carrier density is chiefly responsible for the noise. Density fluctuations have been directly observed by Hall-effect measurements [145-147]. It should be pointed out, however, that deviations from the I_{dc}^2 law are sometimes found to occur, and noise power proportional to I_{dc}^3 and I_{dc}^4 have also been reported [141,148]. Brophy [148] has shown that such a higher-power dependence is often found in plastically deformed crystals.

There is extensive experimental evidence to show that $1/f$ noise, at least in germanium, is predominantly a surface phenomenon. Montgomery [138] reported that by etching a sandblasted germanium filament, the noise current was enhanced by a factor of ten. A similar enhancement was obtained by increasing the surface-to-volume ratio [149,150]. Changing the ambient gas has also been observed to have a drastic effect on the $1/f$ component [143,151]. Using combined measurements of noise and field effect, Mac Rae and Levinstein [144] found that the $1/f$ component is independent of barrier height when an accumulation layer is present at the surface, but increases rapidly as inversion-layer conditions are approached. Furthermore, the noise power associated with an accumulation layer was found to be temperature insensitive, while that associated with an inversion layer decreased rapidly as the temperature was reduced from 300°K to 200°K and then remained constant down to 90°K. Similar results were reported by Lauritzen and Gibbons [152], who employed heavy surface doping to control the barrier height. They observed an enhancement of a hundredfold in $1/f$ noise in inversion layers over that in accumulation layers. In contrast to these results, Noble and Thomas [153], using an ambient cycle to vary the barrier height, found that the noise attains a definite minimum when the surface conductance is a minimum, and increases as the surface is shifted towards either accumulation- or inversion-layer conditions — more so, however, in the latter case. The high noise power associated with inversion layers was also demonstrated by Komatsubara et al.[154], who studied the effect of transverse fields on the reverse current in germanium p–n junctions. Watkins [155] observed that avalanche breakdown in such junctions at high reverse bias is accompanied by a large enhancement in $1/f$ noise, part of which is probably due to surface breakdown at the junction region. An increase in $1/f$ noise with reverse bias was found by Karpov [156] even before breakdown, especially at reduced temperatures ($\approx -60°C$).

It should be noted that not all $1/f$ noise originates in the surface. Montgomery [138] observed an increase of noise in germanium with increasing bulk resistivity. Quenching from elevated temperatures and plastic deformation were also found to have a marked effect [142,150].

Several theories have been advanced to explain $1/f$ noise. In principle, a $1/f$ spectrum can be derived from a superposition of G–R noise terms of the form (7.57), with an appropriate distribution $g(\tau)$ of time constants over the sample [157]. By analogy with (9.3), we now write for the noise power

$$\langle i^2 \rangle \propto \Delta f \int_{\tau_1}^{\tau_2} g(\tau) \frac{\tau}{1+\omega^2 \tau^2} \, d\tau, \qquad (9.6)$$

where τ_1 and τ_2 are the limits of the distribution. It can readily be shown that for $g(\tau) \propto 1/\tau$, (9.6) results in a $1/f$ spectrum in the frequency range $1/\tau_2 \ll f \ll 1/\tau_1$. Thus the problem is reduced to finding a physical process that gives rise to a distribution function proportional to $1/\tau$ over an extremely wide range of time constants. It should be stressed that such a distribution cannot arise from transitions between the conduction (or valence) band and various traps in the *same* region — under these conditions the shortest time constant would dominate the noise spectrum, because it is the fastest process that determines the average lifetime of a free carrier.

As noted by Kingston and McWhorter [70] and by Morrison [71], the wide range of time constants characterizing $1/f$ noise in germanium is also apparent in the field-effect relaxation associated with the slow states. In fact, the $1/\tau$ distribution required to account for the $1/f$ component is just that derived [70] from the frequency response of the ac field effect. On this basis, McWhorter [158] has proposed that the same model of surface heterogeneity (see § 1.2 above) apply to both phenomena. The $1/f$ noise is thus attributed to changes in barrier height arising from fluctuations in the occupancy of slow states characterized by a $1/\tau$ distribution of time constants over the semiconductor surface. In the presence of an electric field along the filament, one has to consider as well conductivity modulation arising from injection and extraction of minority carriers between adjacent regions. Such effects were shown by McWhorter [158] to give a $1/f$ noise comparable in magnitude to that arising directly from barrier-height fluctuations. Depending on the surface conditions, one or the other process could easily dominate. In either case the calculated noise amplitude is of the same order of magnitude as that observed experimentally. Such agreement between theory and experiment is also apparent in the results of Noble and Thomas [153] mentioned earlier, although for inversion layers the measured amplitude is about ten times larger than predicted. Of particular interest is their observation that the noise attains a minimum at the conductance minimum. This indicates that $1/f$ noise is indeed caused by fluctuations in barrier height, since small changes in barrier height produce little or no effect on $\Delta\sigma$ at the conductance minimum. Montgomery[159], on the other hand, suggests that the increase in noise in accumulation layers may be due to a decrease in surface recombination velocity.

A $1/f$ spectrum may also be derived on the basis of Morrison's model [71] considered above (§ 1.2), without the necessity of assuming a distribution of time constants over the surface [160]. The difficulty [158] with this model, however, is that it requires fluctuations in barrier height much greater than 1 kT/q, the value to be expected on thermodynamic grounds. Furthermore, such fluctuations would produce a noise amplitude far too large to agree with experiment.

Although the process of charge-transfer between the slow states and the bulk has most of the qualitative characteristics necessary to account for the $1/f$ component in germanium, the role of the slow states in determining the noise has not been firmly established. It appears that not in all cases can the distribution of τ's deduced from the field effect account for the complete $1/f$ spectrum. Thus, Sochava and Mirlin [161] found little correlation between the frequency response of the field effect and the $1/f$ noise spectrum. Moreover, Mac Rae and Levinstein [144] have observed that the decay characteristics of the field effect are independent of those of $1/f$ noise measured simultaneously; the former are strongly temperature dependent while the latter are not. The insensitivity of $1/f$ noise to temperature has been observed by many workers and constitutes a serious criticism of any theory linking noise with the slow states. Measurements on clean silicon surfaces carried out by Mac Rae [162] indicate strongly that, at least on silicon, slow states are not the prime source of $1/f$ noise: removal of adsorbed gases by the cleaning process, which is known to eliminate slow states (see § 3 below), did not appreciably affect either the magnitude or the frequency dependence of the $1/f$ component. The presence or absence of an inversion layer at the surface, on the other hand, was found to have a strong effect on the noise, as in the case of germanium.

The large noise power observed under inversion layer conditions appears to be due to some process occurring in the space-charge layer. Avalanche breakdown in regions of microscopic irregularity, where the field is higher than average, has been proposed as a possible mechanism [152, 155, 156, 163]. Such breakdown would be more pronounced in inversion layers because of the high field in the inversion region. One cannot rule out the possibility, however, that the large noise under these conditions arises from the injection–extraction effects considered by McWhorter [158] and mentioned above.

An altogether different mechanism for $1/f$ noise, one involving ionic diffusion, has been proposed by a number of workers [164-166]. Bess [142] assumed that the basic noise event is the diffusion of an impurity atom along a dislocation line to or from the surface. In an n-type semiconductor, for example, a donor atom that migrates to the surface removes one mobile electron from the bulk, giving rise to conductance modulation. Bess showed that such a process could lead, under certain conditions, to a $1/f$ spectrum. It is difficult to see, however, how this model would account for the observed dependence of noise power on barrier height.

Summarizing, one can say that although it has been definitely established that $1/f$ noise in germanium is predominantly a surface effect, the controlling mechanism is still obscure. The various models outlined above seem to give the correct order of magnitude and frequency dependence of the noise, but none of them can account for the complex, and sometimes contradictory, experimental observations. Quite likely, a number of different mechanisms act simultaneously, one or the other predominating, depending on the prevailing surface conditions. Further experimental and theoretical work, on germanium as well as on other semiconductors, is required before a satisfactory physical understanding of $1/f$ noise and its relation to surface phenomena can be achieved.

1.5. ADSORPTION AND ELECTRICAL SURFACE PROPERTIES

The intimate relation between the ambient gas and the type of slow states induced makes it natural to identify the latter (or at least a portion of them) with atoms or molecules adsorbed at the surface from the gas phase. Only chemisorption (see chapter 3) is of interest in this context, since physical adsorption is unlikely to affect the electrical properties [167]. Field-effect and other measurements considered above provide ample evidence that charge transfer can take place between the slow states and the semiconductor interior. Such a charge-transfer process forms the basis of the chemisorption theories advanced by Aigrain and Dugas [168], by Hauffe and Engell [169], and by Weisz [170]. Consider a single electronegative molecule such as oxygen approaching the surface. If its electron affinity A is larger than the semiconductor work function W_ϕ, the molecule will tend to pick up an electron from the semiconductor and thereby become adsorbed at the surface. The difference $A - W_\phi$ represents the energy of adsorption and determines the position of the acceptor level introduced in this manner. As we have seen above, the presence of acceptor levels acts to bend the bands upwards. Thus, with increasing adsorption the work function increases until, at equilibrium, the Fermi levels of the semiconductor and the adsorbate are aligned. Beyond this point, charge transfer no longer leads to a decrease in the free energy of the system, and chemisorption stops. The formation of a barrier at the semiconductor surface affects both the equilibrium amount adsorbed and the rate of adsorption. Similar considerations apply in the case of donor states introduced by electropositive gases such as water vapour. Here an electron is transferred from the approaching molecule to the semiconductor. The energy of chemisorption is $W_\phi - I$, where I is the ionization energy of the adsorbate. Such adsorption leads to a lowering of the energy bands and a reduction in the work function. It should be noted that A and I are not be to taken literally as the affinity and ionization energy of the molecule in its free state — both quantities are expected to be greatly affected by the proximity of the substrate surface.

Electron transitions between acceptor or donor species and the semiconductor may occur by thermionic emission over the oxide barrier, by tunnelling through it, or by some combination of the two processes, as discussed in connection with the slow field-effect relaxation (§ 1.2). Here, as well, such processes would explain the slow change in barrier height following exposure of the surface to a given ambient.

Although the model just presented is no doubt an oversimplification of the actual situation, it illustrates clearly the fundamental processes involved in chemisorption. A number of workers [171-179] have developed these ideas and applied them to adsorption and catalysis in more complex systems, but it is beyond the scope of this book to consider such work in detail. Here we shall only touch on the important question concerning the relationship between the density of slow states and the quantity of adsorbed gas. The early theories [168-170] assumed that all of the adsorbed molecules are present as ions, so that neutralization of the molecule gives rise to desorption. Since the space-charge layer is hardly ever found to be degenerate when the surface is exposed to an ambient gas, it follows that the density of electronic charges is nearly always less than 10^{12} cm^{-2} (see Ch. 4, § 2). Thus, if there were only one type of adsorbed ions (positive *or* negative), the equilibrium coverage for chemisorption would be less than 0.1 %, the number of adsorption sites being of the order of 10^{15} cm^{-2}. Such low coverage has been questioned by Vol'kenshtein [179], who maintains that in reality many more molecules, both neutral and ionized, exist on the surface. On this premise, Vol'kenshtein distinguishes between two forms of chemisorption, "weak" and "strong", according to whether the adsorbed species is neutral or ionized, respectively. The actual chemical behaviour is not clear, but it is likely that, irrespective of whether neutral molecules are or are not adsorbed, very large densities (at least 10^{13}, probably 10^{15} cm^{-2}) of *both* positively charged donor and negatively charged acceptor ions are simultaneously present on the surface. Such a combination is consistent with the large overall *ionic* coverages to be expected for chemisorption and, at the same time, is able to account for the observations of both low *total* surface charge density and very strong anchoring of the barrier height. If, for instance, the Fermi level at the surface is controlled by a large density of donor levels in its vicinity, then a large density of charged acceptor levels must exist below the Fermi level so as to neutralize most of the positive charge of the ionized donors.

It is evident from the foregoing considerations that the variation in barrier height resulting, for example, from a change in ambient from ozone to wet nitrogen, should not be conceived of as a transition from purely acceptor-ion to purely donor-ion adsorption. Rather, changing the ambient shifts the energy of the adsorbed species and determines the relative predominance of the donor and acceptor states. In fact, one should consider not only states arising from the adsorbed molecules, but also those intrinsic to the surface and its oxide.

The experimental evidence for the validity of the charge-transfer model is extensive, but rather qualitative in nature. It consists essentially of measurements of changes in surface conductance and/or contact potential under various gaseous ambients, from which the sign of the surface charge and, in some cases, the rate of its build-up are determined. Most of these studies have been performed on compound semiconductors such as zinc and copper oxide [12, 179, 180]. The change in adsorption characteristics with illumination observed in a number of cases [179-182], offers another confirmation of the charge-transfer model. Hole–electron pairs, produced in the semiconductor by the light, alter the surface potential and surface carrier densities, so that a change in both the amount and the rate of adsorption is only to be expected. The theoretical aspects of the photoadsorptive effect have been considered in [183]. Similar considerations lead one to expect that electric fields applied normal to the surface would also affect its adsorption capacity [184], but so far this has not been demonstrated experimentally.

The adsorption kinetics in germanium under different conditions of ambient pressure and temperature has been studied by Lyashenko and Litovchenko[185]. The gases used were ethyl alcohol, acetone, benzene, carbon monoxide (donors), and oxygen (acceptor). In all cases the steady-state changes in contact potential and surface conductance were observed to increase monotonically with increasing pressure and (except for oxygen) with decreasing temperature. This can be accounted for by the enhanced rate of impingement as the pressure is raised, and by the reduced rate of desorption as the temperature is lowered. The time constant τ characterizing the attainment of steady-state conditions following gas admission was found to be inversely proportional to the square root of the pressure and to increase exponentially with decreasing temperature. The magnitudes of τ (minutes) and its activation energy (0.1–0.3 eV) are comparable to those observed in the slow field-effect relaxation (see § 1.2 above). Moreover, here as well, τ increases very rapidly with oxide-film thickness. Such behaviour constitutes strong evidence that charge transfer between the adsorbed molecules and the semiconductor interior does indeed play a dominant role in chemisorption.

The effect of water vapour on the surface properties of germanium has received much attention. While it is generally agreed upon that water forms donor-type slow states, the details of the process are not well understood. Water vapour would be expected to dissolve GeO_2 and also to promote the oxidation of insoluble GeO to soluble GeO_2. There appears to be little doubt that water penetrates the oxide layer [87,186], so that its effect on the electrical behaviour should depend on both the amount adsorbed and the initial structure of the oxide. From measurements of surface conductance as a function of relative humidity, Kingston [9,44] and Dorda [187] concluded that the principal increase in surface potential occurs during adsorption of the first and second monolayers. Their interpretation was based on Law's data [188] relating the number of adsorbed layers to the partial pressure of water vapour. To account for the relatively small increase in surface charge associated with the observed change in surface potential (from $-3 kT/q$ to $+5 kT/q$), Dorda [187] assumed that the donor levels lie well above the Fermi level and that it is mainly their density that increases with coverage. An energy distribution of this sort, however, cannot explain the strong anchoring of the surface potential observed in the field-effect experiment. It is more likely that at all times following water adsorption the Fermi level is clamped close to the energy of the donor states. (The large positive charge in these states is neutralized by occupied acceptor states, as explained above.) On this premise, the shift in surface potential results mainly from an increase in *energy* of the donors, although an increase in their density might also take place. A mechanism by which such a change in energy may come about has been proposed by Hutson [54]. Donor-type species are assumed to be present at the surface independently of the adsorbed water. Their energy, however, is strongly affected by the high dielectric constant of the water film, such that with increasing film thickness the donor levels continuously move up towards the conduction-band edge. Calculations based on this model were found to be in very good agreement with Kingston's data. In particular, the main change in energy was predicted to occur during adsorption of the first few monolayers of water, as observed. An altogether different behaviour, however, was reported by Kawasaki *et al.*[189], who observed saturation of the surface conductance after a coverage of only about one hundredth of a monolayer. The origin of the discrepancy between these results and those of Kingston and Dorda is not clear. The complexity of the effects produced by water adsorption is also apparent in the results reported by Fedorovich [190]. On prolonged heating of germanium at 100°C in a humid ambient, the water-induced slow states showed an acceptor-like character (*negative* charge added at the oxide layer); after cooling to room temperature, the normal donor behaviour was restored. Furthermore, it appears that under certain conditions water may introduce acceptor levels even at room temperature [191].

So far we have considered the effect of adsorption on the barrier height. This is reflected as a change in both the surface conductance and the work function. In the latter case, however, one has to consider also possible changes in the electron affinity χ (see Fig. 9.1). These may become quite large [192] if the adsorbed molecules have dipole moments aligned perpendicularly to the surface. To the approximation that such a dipole array can be represented as a perfect double layer, there will be no change in the electric field outside the layer, so that the barrier height will remain unaltered. A sharp jump in potential will occur, however, across the double layer, adding to or detracting from the micropotential associated with χ (depending on the polarity of the dipoles). The magnitude of this discontinuity in potential is given by $\sigma M/\varepsilon_0$, where σ is the number of aligned dipoles per unit area and M the dipole moment of each molecule. The latter will, in general, depend on the nature of both the adsorbant and the absorbate.

The above considerations are illustrated in Fig. 9.10, where conditions are shown before and after gas adsorption. The change in work function δW_ϕ is made up of a change in barrier height $q\delta V_s$ arising from positive adsorbed ions and a change in affinity $\delta\chi$ due to a layer of adsorbed dipoles. (The oxide film and the potential drop V_D associated with it have been omitted from the diagram.)

One way of distinguishing between the two contributions $q\delta V_s$ and $\delta\chi$, is to carry out simultaneous measurements of contact potential and surface conductance. Experiments of this sort discussed in § 1.1 show that none of the molecules adsorbed on etched germanium surfaces during the Brattain–Bardeen cycle are associated with a significant electric dipole. (A different

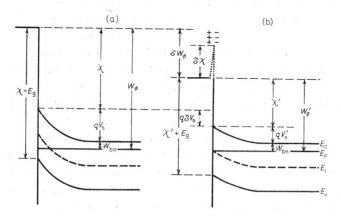

Fig. 9.10. Energy-level diagrams illustrating the effect on work function of the adsorption of positive ions and neutral polar molecules.
(a) Before adsorption.
(b) After adsorption.

behaviour is observed on clean surfaces — see § 3.6 below.) The same conclusion has been reached by Litovchenko and co-workers [185, 193] from measurements of the slow relaxation characteristics of contact potential and surface conductance, and by Sochanski [194] from studies of inversion channels in p–n junctions. The former measurements indicate that even polar molecules such as ethyl alcohol do not change χ. In another experiment, on the other hand, Sochanski [195] finds that long-term drifts in contact potential induced by corona discharge in air cannot be entirely accounted for by changes in V_s, suggesting that adsorption in this case involves neutral polar molecules as well. The possibility cannot be ruled out, however, that it is the affinity of the reference electrode rather than of the semiconductor that is affected by adsorption [193]. The use of a glass-coated reference electrode [185] appears to minimize such spurious effects.

Perhaps the most direct observation of changes in electron affinity is provided by studies of photoemission (see Ch. 7, § 3). For high-resistivity samples and non-degenerate surface conditions, the photoelectric threshold is insensitive to the barrier height, being simply equal to $\chi + E_g$. Extensive measurements of the effect of barium oxide, a polar molecule, on threshold in a variety of semiconductors have been carried out by Borzyak and co-workers [196]. Less than a monolayer coverage is usually sufficient to cause a drastic reduction in χ. In germanium, for example, one observes a decrease in affinity from about 4.2 eV for the untreated surface to 1.5 eV for a BaO-covered surface. A similar reduction is effected by cesium, as will be discussed in more detail in connection with clean surfaces (§ 3.6). Cesium adsorption on silicon [197-199] and on germanium [200] has been employed to obtain emission of hot electrons produced in the semiconductor by high fields.

2. Real germanium and silicon surfaces

As discussed in Ch. 3, § 4, a real surface is prepared by mechanical polishing and chemical etching, followed by exposure to room air or some other environment. In addition to an oxide film of variable thickness, structure, and water content, unknown traces of foreign impurities adsorbed from the etchant and the surrounding ambient are certain to be present at the surface. A real surface thus constitutes a rather complex system as regards both physical structure and chemical composition. Such complexity, however, is not reflected in its *electronic* structure. While the characteristics of the surface states depend on the particular treatment used, contamination and other uncontrollable factors inherent in the mode of preparation of the surface do not appear to play a dominant role. Good reproducibility is the rule rather than the exception, and results obtained in different laboratories under similar (but obviously far from identical) conditions are, on the whole, in satisfactory agreement.

The available experimental data on the density, energy distribution, and capture cross sections of the fast states, and the influence of surface treatment on these parameters, are reviewed in the first two subsections. Various effects related to the slow states are discussed in § 2.3. All three subsections pertain to real surfaces in contact with an ambient gas (or vacuum). In § 2.4 we consider surface states at the semiconductor–electrolyte interface.

2.1. CHARACTERISTICS OF FAST STATES ON ETCHED SURFACES

The structure of the fast states has no doubt received far more attention than any other aspect of the real surface. Rather than attempt a comprehensive discussion of the vast amount of data accumulated on the subject, we have chosen to select several typical examples to illustrate the methods used and the results obtained. The bulk of the data is then summarized in tabular form.

Field-effect measurements (under conditions in which the influence of slow states is eliminated) provide the most direct means of estimating the fast-state

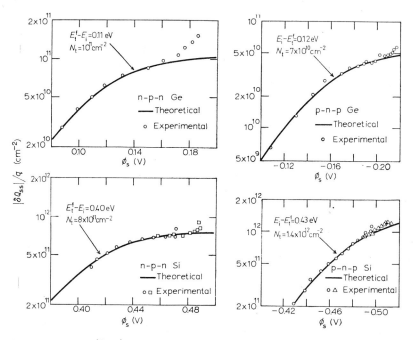

Fig. 9.11. Change $|\delta Q_{ss}|/q$ of trapped-carrier density in fast states on germanium and silicon surfaces *versus* surface potential ϕ_s, as derived from channel-conductance measurements on n–p–n and p–n–p structures. The solid curve in each case represents the occupation function for a single discrete set of states. (After Statz et al., references 46, 48, 86.)

density. They cannot, however, determine uniquely the energy distribution of the states, unless the distribution function is particularly simple or the swing in barrier height is sufficiently large. Brown[39,40], who initiated these measurements (see, for example, Fig. 9.4), found in fact that different distribution functions, ranging from a discrete scheme to a continuous one, could account equally well for the experimental data. While this behaviour is typical to most field-effect studies [41, 201-207], Statz and co-workers [46, 48, 86], using p–n–p and n–p–n channels, were able to obtain a fairly clear-cut characterization of the fast states near the band edges in both germanium and silicon. Their results are shown in Fig. 9.11, where the change $|\delta Q_{ss}|/q$ in carrier density trapped in fast states is plotted against the surface potential ϕ_s. (See Ch. 6, § 8 for a description of the experimental procedure employed.) In each case the experimental points can be fitted reasonably well to a theoretical curve (solid) representing the occupation function of a single discrete set of states (see eqs. (5.44), (5.45)). The values of the effective energy E_t^f and density N_t used in the construction of the curves are marked on the figure. It is seen that in both germanium and silicon a pair of surface states dominates the regions of the gap explored. The rise in $|\delta Q_{ss}|/q$ above the saturation values for large $|\phi_s|$, however, indicates the presence of additional states closer to the band edges. Since channel-conductance measurements are inherently limited to inversion layers, they are not suitable for exploring the central region of the gap. This region is best studied by the ordinary field-effect technique. The two methods thus complement one another in mapping out the surface-state distribution over the energy gap.

Considerably more information on the fast states is provided by combined measurements of field effect and surface recombination velocity s. This is particularly the case when only one set of surface states dominates the recombination process (see § 1.3 above). The analysis of the experimental data under these conditions is illustrated in Fig. 9.12. Here measured values [127] of s/s_{max} and $\delta Q_{ss}/q$, obtained on an etched germanium surface at two temperatures, are plotted against the (dimensionless) surface potential u_s. Also shown (Fig. 9.12b) is the observed temperature dependence of s_{max}. The solid curves in Fig. 9.12a represent the theoretical expression (5.85) for each temperature, the parameters of the recombination centres having been chosen so as to give the best fit to the data. Measurements at different temperatures are necessary in order to remove the ambiguity in the energy position of the recombination centres, as explained in Ch. 5, § 7. At the same time, they enable the determination of both the actual and the effective energy (E_t and E_t^f) of the recombination centres, and hence of the statistical weight factor g_1/g_0 (see eq. (2.74)). These objectives can be achieved in full only if the parameters of the recombination centres are temperature independent. That such is actually the case for the capture cross-section ratio A_p/A_n is immediately apparent from the constancy of u_0 in Fig. 9.12a (see eq. (5.82)). On this basis it is reasonable to suppose that A_p and A_n are each temperature independent. Combining the data of Figs. 9.12a and 9.12b, we then con-

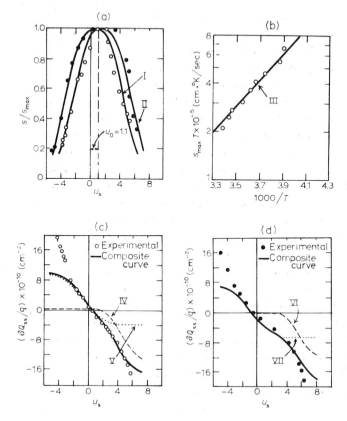

Fig. 9.12. Results obtained from combined measurements of surface recombination and field effect on n-type germanium sample.

(a) Surface recombination velocity s (relative to its maximum value s_{max}) as function of surface potential u_s at two temperatures $T_1 = 289°\text{K}$(o) and $T_2 = 264°\text{K}$(•). The solid curves were calculated from (5.85). [I: $(E_t^f - E_1)/kT_1 = 4.8$; II: $(E_t^f - E_1)/kT_2 = 5.8$.]

(b) Semilog plot of $s_{max}T$ versus $1000/T$. The slope of the straight line (III) yields $E_c - E_t = 0.16$ eV.

(c) Change $\delta Q_{ss}/q$ in trapped-carrier density in fast states versus u_s for $T = T_1$. Solid curve represents superposition of occupation functions (dashed and dotted curves) for two discrete sets of fast states:

$$\text{IV} \begin{cases} (E_t^f - E_1)/kT_1 = 4.8 \\ N_t = 1.2 \times 10^{11} \text{ cm}^{-2} \end{cases} \quad \text{V} \begin{cases} (E_t^f - E_1)/kT_1 = -1 \\ N_t = 1.4 \times 10^{11} \text{ cm}^{-2}. \end{cases}$$

(d) Similar to (c), but for $T = T_2$:

$$\text{VI} \begin{cases} (E_t^f - E_1)/kT_2 = 5.8 \\ N_t = 1.2 \times 10^{11} \text{ cm}^{-2} \end{cases} \quad \text{VII} \begin{cases} (E_t^f - E_1)/kT_2 = -1 \\ N_t = 1.4 \times 10^{11} \text{ cm}^{-2}. \end{cases}$$

(After Many and Gerlich, reference 127.)

clude that of the two possible values that can be assigned to E_t^f, the one above E_i (that is, closer to E_c) represents the true energy of the recombination centres. The linear relation between $\ln(s_{max}T)$ and $1/T$ suggests that E_t is essentially temperature independent. Comparison of the value for the activation energy $E_c - E_t$ derived from this relation, with that of $E_t^f - E_i$ deduced from the room-temperature s/s_{max} curve, indicates that g_1/g_0 is of the order of 10. A much larger value, however, is required to account for the observed width of the s/s_{max} curve at the lower temperature. This discrepancy may be due to the lower accuracy in the data for that temperature, as evidenced by the greater scatter of the experimental points. On the other hand, a slight increase in $E_t - E_i$ with decreasing temperature cannot be ruled out (see also § 2.2).

With the energy of one set of fast states accurately known, the analysis of the field-effect data in Figs. 9.12c and 9.12d becomes much more meaningful. First, the occupation function for the recombination centres is plotted (dashed curves). This obviously cannot account for the observed results, indicating that other sets of states, having a negligible effect on recombination, are also present at the surface. The effective energy of one of these sets (at about $1\,kT$ below E_i) can be determined quite unambiguously; its occupation function is shown by the dotted curve. The composite solid curve represents the combined effect of the two sets and accounts for the experimental data at both temperatures over a substantial range. The departures at extreme values of $|u_s|$ indicate the presence of additional states near the band edges.

Finally, the density of the recombination centres derived from the field-effect data can be used, in conjunction with the previously measured values of s_{max} and u_0, to determine the hole and electron capture cross sections. The values so obtained are $A_p \approx 10^{-14}$ and $A_n \approx 10^{-15}$ cm². The acceptor-like character of these states $(A_p > A_n)$ is quite general in germanium (see Table 9.1 below) and may arise, for example, from centres that are neutral when empty and negatively charged when occupied: due to Coulomb interaction, a negatively charged centre will have a larger cross section for holes than a neutral centre for electrons.

Similar studies have been carried out by Wang and Wallis[208], using an ambient cycle to vary the surface potential. Here the photoconductance response and the small-signal ac field effect in the dark and under illumination were measured as a function of u_s. The first measurement yields s (see eqs. (7.9), (7.25), (7.26)), the second — δQ_{ss}, and the third — ds/du_s. This last is of particular interest, since it provides a sensitive means of determining

A_p/A_n: for $s = s_{max}$, the derivative ds/du_s goes through zero, and this can be detected with high precision. Here, as well, the recombination data can be accounted for by a simple s/s_{max} curve (eq. (5.85)) while the field-effect results reveal two sets of surface states, one above and the other below E_i. Contrary to the conclusion reached above that only one of these sets contributes to recombination, Wang and Wallis maintain that *both* sets correspond to recombination centres having the same ratio A_p/A_n and density N_t. In other words, the two possible assignments for the energy of the recombination centres, considered above as *alternative*, are assumed here to be realized *simultaneously*. The latter situation, while possible in principle, is most unlikely, since it is improbable that two different sets would have identical cross sections and densities. Moreover, the measurements having been carried out at a single temperature, there is not sufficient basis for a decision one way or the other.

Zhuze et al.[209] have used direct display techniques to determine the variation of s and $\delta Q_{ss}/q$ with u_s. The former was derived from changes in sample resistance in a magnetic field [210] as u_s was varied by an ac field, the latter from field-effect patterns of the type shown in Fig. 9.4. The oscillograms representing s and $\Delta\sigma$ *versus* the charge Q_s induced at the surface are illustrated in Fig. 9.13 for a p-type germanium sample. The s and $\delta Q_{ss}/q$ curves obtained from these data are shown on the right-hand side of the figure, together with the appropriate theoretical curves. The surface structure is seen to be very similar to that observed on n-type samples (compare Fig. 9.12): the central region of the gap is dominated by two sets of states, only one of which contributes to recombination. A third set of levels (lying $6\ kT$ above E_i) is invoked to account for the field-effect results at large values of u_s.

The simple character of the recombination process apparent in the data considered above is revealed over and over again by many of the measurements done on germanium [211-216]. In quite a few instances, however, complex s curves have been reported, requiring the assumption of contributions from several sets of recombination centres, usually closely spaced in energy [217-222] or even continuously distributed [31, 32, 223, 224]. On most surfaces for which this behaviour is found, the density of states is relatively low ($1-3 \times 10^{10}$ cm^{-2}) compared to that observed on surfaces exhibiting a simple recombination character (10^{11} cm^{-2} or higher). This suggests the presence on all surfaces of a small ($\approx 10^{10}$ cm^{-2}) background distribution of recombination centres. Such a background would mask the predominance of a single set of centres in the former case, but would be hardly noticeable in the latter case. This is well demonstrated by Balk's measurements [224] on germanium samples etched in a weakly oxidizing mixture [102] of HF and H_2O_2. The surfaces so obtained have s values as small as 10–15 cm/sec, with correspondingly low densities of states. Essentially simple s/s_{max} curves emerge after subtraction from the measured results of a constant contri-

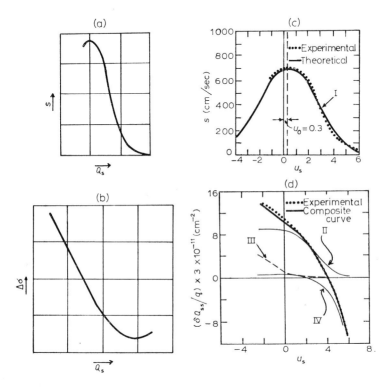

Fig. 9.13. Results obtained from combined measurements of surface recombination and field effect on p-type germanium sample.
(a) and (b) Oscillogram patterns representing surface recombination velocity s and surface conductance $\Delta\sigma$ as functions of charge Q_s induced at the surface.
(c) Surface recombination velocity s versus surface potential u_s, as derived from (a) and (b). Heavy curve was calculated from (5.83). [I: $(E_t^f - E_1)/kT = 3$.]
(d) Change $\delta Q_{ss}/q$ of trapped-carrier density in fast states versus surface potential u_s. Heavy curve represents superposition of occupation functions for three discrete sets of fast states:

II $\begin{cases} (E_t^f - E_1)/kT = 3 \\ N_t = 3 \times 10^{11}\ cm^{-2} \end{cases}$ III $\begin{cases} (E_t^f - E_1)/kT = -1 \\ N_t = 2 \times 10^{11}\ cm^{-2} \end{cases}$ IV $\begin{cases} (E_t^f - E_1)/kT = 6 \\ N_t = 8 \times 10^{11}\ cm^{-2} \end{cases}$

(After Zhuze et al., reference 209.)

bution, very likely associated with a small continuous distribution of recombination centres. One should not rule out the possibility, however, that some of the complex s curves reported originate from inhomogeneity in the surface structure ("patchiness") or in the applied transverse field. Either of these will result in an *apparent* spread of the surface states over a range of energies. The commonly-used techniques of effective-lifetime measurement, based as they are on the photoconductance response under *uniform* illumination, are unable to detect local variations in s. On the other hand, simple recombination characteristics have been obtained mostly by the pulse injection method, which is sensitive to inhomogeneity of either type. Any sample exhibiting surface patchiness or improper mounting can readily be rejected in this manner.

A *well-defined* contribution to recombination from sets of states other than the one dominating the central region of the gap has been observed in only a few cases [224, 225]. Here the s curve has the usual bell shape for moderate values of u_s, but for large positive [224] or negative [225] values, a second peak becomes apparent. The peak to the right ($u_s > 0$) originates from *acceptor*-like states lying close to E_c, the one to the left — from *donor*-like states close to E_v.

By comparing results obtained on many germanium samples, and under different surface treatments, Rzhanov [226] concludes that, quite generally, recombination levels in the upper region of the energy gap ($E_t^f - E_i > 0$) are acceptor-like, while those in the lower region are donor-like. Plotted on a semilog scale, A_p/A_n is found to increase roughly linearly from 10^{-4} to 10^{+6} as $(E_t^f - E_i)/kT$ increases from -8 to $+10$. Rzhanov suggests that this behaviour may arise either from differences in the charge condition of the recombination centres (see Ch. 5, § 8.1) or from the fact that all observed s curves are in effect unresolved composite curves resulting from two different sets of centres.

A number of interesting features are revealed by recombination studies at reduced temperatures. The position of s_{max} ($u_s = u_0$) has been observed to shift towards *higher* values of u_s for an n-type germanium sample [225] and towards *lower* values for a p-type sample [227] as the temperature is decreased from about 260°K. At first glance such behaviour would indicate that the cross-section ratio is temperature dependent, and indeed the measurements were initially [225] so interpreted. Rzhanov [228] attributed the change in A_p/A_n to the participation of excited states in the recombination process, in accordance with his theory discussed in detail in Ch. 5, § 8.2. Re-examination [229] of the data, however, shows that at reduced temperatures the excess-carrier density obtaining in the measurements ($\delta p_b/(n_b+p_b) \approx 5 \times 10^{-3}$) is no longer small compared to the minority-carrier density. These conditions (which are unavoidable if sufficient sensitivity is to be attained) give rise to a narrowing of the s curves and hence to an *apparent* shift in u_0 (see Fig. 5.7). The observed directions of change in u_0 for the n-type and the p-type samples are just those to be expected on this basis. It is thus obvious that at low temperatures u_0 cannot be used as a reliable measure of A_p/A_n. Quite likely, the temperature dependence of the cross sections reported by Litovchenko and Lyashenko [221] can be explained in a similar manner. Measurements at low temperatures may be further complicated by trapping effects (see Ch. 2, § 7.1) either in the bulk or at the surface [230, 231]. In silicon, surface trapping was observed even at room temperature [232-234], and has been very thoroughly investigated by Primachenko et al. [233].

Recombination studies on silicon are hampered by the limited range of u_s normally attainable (due to the high density of fast states), and by the lack of good communication between the surface and the bulk under inversion-layer conditions (see Ch. 6, § 2). Buck and McKim [235] and Harten [236] employed a combination of gaseous ambients and chemical pretreatments to effect the necessary swing in u_s. A rough estimate of the parameters of the recombination centres was obtained by piecing together the results for different pretreatments. Using transverse ac fields *in vacuo*, on the other hand, Snitko [237] was able to achieve sufficient swing in u_s to observe both s_{max} and $\Delta\sigma_{min}$. His results on n-type samples show clearly that the recombination centres involved are donor-like ($A_p < A_n$). In a more detailed investigation, Litovchenko and co-workers [238, 239] interpreted the complex s curves they observed in terms of three sets of recombination centres. The assumption of the existence of several other sets was found necessary to account for the field-effect data. It is rather questionable, however, whether under these conditions an analysis based on a discrete scheme can have much meaning.

Measurements of the temperature dependence of the field-effect relaxation constitute another powerful tool for studying the fast surface states. The large-signal pulsed field-effect technique, under conditions in which minority-carrier effects are absent, is particularly useful in this respect (see Ch. 6, § 6.2). Here the time constant τ_1 characterizing the carrier emission rate from the states into the majority-carrier band is measured as a function of temperature. Under the proper experimental conditions, τ_1 is given by $(\langle c \rangle A_n n_1)^{-1}$ for an n-type sample (see eq. (5.68)) and by $(\langle c \rangle A_p p_1)^{-1}$ for p-type, where n_1 or p_1 is a direct measure of the energy position of the states involved (eqs. (2.172), (2.173)). Typical results obtained on n- and p-type germanium samples are shown in Fig. 9.14. The dots cor-

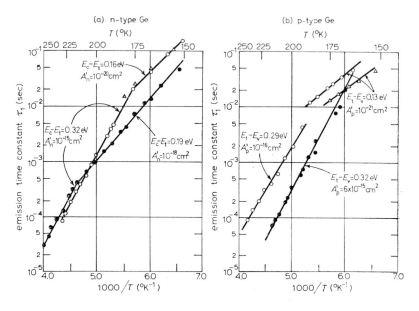

Fig. 9.14. Emission time constant τ_1 *versus* $1000/T$ for n- and p-type germanium samples. The dots correspond to data taken by Rupprecht (reference 240). The circles and triangles were obtained on different samples by Many et al. (reference 229).

respond to data taken by Rupprecht [240]; the circles and triangles were obtained by Many and co-workers [229]. It is seen that the results of the various measurements are on the whole in good agreement. At low temperatures shallow surface states with low majority-carrier capture cross sections dominate the thermal emission process, while at higher temperatures deep states with large cross sections take over. The energy levels of the various sets of states marked on the figure are derived from the slopes of the straight lines on the assumption that the cross sections are temperature independent. The T^2 dependence of the $\langle c \rangle N_c$ and $\langle c \rangle N_v$ terms (see eqs. (2.85), (2.86), (2.124)), however, has been taken into account. The effective cross sections $A'_n = (g_0/g_1)A_n$, $A'_p = (g_1/g_0)A_p$ are calculated from the room-temperature data, with N_c (or N_v) and $\langle c \rangle$ taken to be 10^{19} cm^{-3}

and 10^7 cm/sec, respectively. It will be recognized that the energy of the upper level (0.16–0.19 eV below E_c) matches that of the recombination centres derived from the data of Fig. 9.12, but the electron cross section is much smaller (10^{-20}–10^{-18} cm^2). The difference in cross sections is not surprising in view of the extreme sensitivity of this parameter to surface treatment (see Table 9.1, p. 390).

Pulsed field-effect measurements are of special interest in the case of silicon where, as has been pointed out above, field-effect and recombination studies are beset by many difficulties. Results obtained by Rupprecht[240,241] on a number of n- and p-type silicon samples are shown in Fig. 9.15. Here $(A_n/N_c)n_1$ and $(A_p/N_v)p_1$, as calculated from the measured emission time constants, are plotted against $1000/T$. The points correspond to surfaces

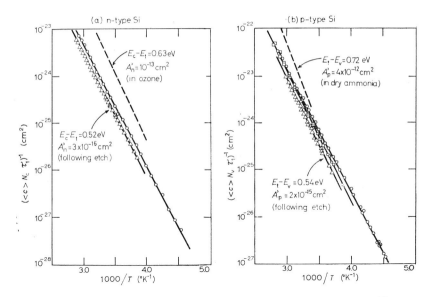

Fig. 9.15. Plots of $(\langle c \rangle N_c \tau_1)^{-1}$ and $(\langle c \rangle N_v \tau_1)^{-1}$ versus $1000/T$ for n- and p-type silicon samples, respectively. The dashed curves were obtained after exposure of the n-type samples to ozone and the p-type samples to dry ammonia. (After Rupprecht, references 240 and 241.)

etched in CP-4 or in the white etchant (see Ch. 3, § 4.2) and are associated with two discrete levels close to the midgap, $E_c - E_t = 0.52$ eV and $E_t - E_v = 0.54$ eV. In order to determine whether the two levels are present simultaneously, an n-type sample was exposed to ozone and a p-type sample to dry ammonia. The data obtained under these conditions are represented by the dashed lines, the activation energies now being 0.63 and 0.72 eV, respectively. Ozone exposure raises the bands and is therefore expected to empty the upper level. If the lower level remains occupied, electron emission from this level will control the relaxation process in the n-type sample. Similar conditions of occupancy may be expected when the p-type sample is exposed to ammonia. The controlling process here would then be hole emission from the upper level into the valence band. And, indeed, each corresponding pair of values (0.54 and 0.63 eV, 0.52 and 0.72 eV) adds up to approximately the band gap of silicon (at $T = 0°K$), indicating that the same two levels are present in both the n- and the p-type samples. This procedure also enables the determination of both the hole and the electron capture cross sections of each set of states. The values so obtained show the upper level to be acceptor-like, the lower to be donor-like.

The *small-signal* field-effect relaxation has been studied by Lindley and Banbury [78,82] and by Litovchenko and Lyashenko [221,238,239,242] using pulsed and ac techniques (see Ch. 6, §§ 6.4, 7). The surface-state time constant and its variation with quiescent barrier height and temperature were determined for single- as well as two-carrier systems. The latter was obtained either under inversion-layer conditions or in the presence of excess minority carriers produced by light. The relaxation process in this case involves the interaction of the surface states with both bands, and its analysis can lead, in principle, to an evaluation of the capture cross sections for both holes and electrons [82,243]. In practice, however, such an analysis is quite involved, and the results obtained are not as clear-cut as those derived from the simpler, large-signal relaxation process discussed previously.

Several workers have been able to detect optical transitions between surface states and the valence or conductance band (see Ch. 7, § 5). Spear [244] observed a photoconductive shoulder in germanium (at 80°K) below the fundamental absorption edge, in the photon-energy range between 0.55 and 0.7 eV. A study of the effect of sample thickness, chemical treatment, and low-energy electron bombardment indicated that this shoulder arises from transitions of electrons from the valence band into surface states lying near the conduction-band edge. Similar measurements have been carried out by Rzhanov and Plotnikov [228,245] following heating of a germanium sample in vacuum at various temperatures. The heat treatment used increased the density of surface states (see § 2.2 below) without affecting the bulk. A corresponding increase in photoconductivity (at 100°K) was observed, and after heating at 500°K a definite peak at 0.4 eV became apparent. The data were interpreted in terms of electron transitions from two discrete levels near the midgap into the conduction band.

Optical transitions involving surface states on silicon have been studied by Harrick [60] using absorption measurements of a multiply-reflected infrared beam under conditions of ac field modulation of the surface potential (see Ch. 7, § 5). For low fields, the only surface states that can contribute to the observed absorption modulation are those lying near the Fermi level, the occupation of all other states being practically unaffected by the field. The spectral dependence of the change in absorption due to such states (obtained after subtraction of the free-carrier absorption) is shown in Fig. 9.16. The

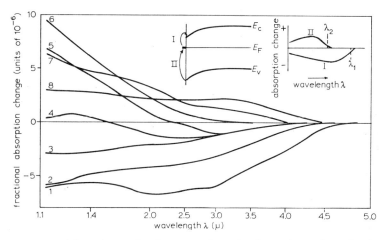

Fig. 9.16. Spectral dependence of surface-state absorption for various values of quiescent barrier height at silicon surface. Diagrams at top of figure illustrate optical transitions involved. (After Harrick, reference 60.)

various curves correspond to different values of the quiescent barrier height, starting from accumulation-layer (curve 1) and on to inversion-layer conditions (curve 8). Positive and negative values on the ordinate correspond, respectively, to enhancement and reduction in absorption during the *negative* cycles of the ac field (that is, when the electron occupation of the surface states is decreased). The shape of the absorption curves can be understood by reference to the schematic diagram at the top of the figure. For accumulator layers, transitions from surface states into the conduction band should predominate at long wavelengths, and a negative change in absorption is to be expected. For inversion layers, on the other hand, electron

TABLE 9.1

Summary of results obtained by different workers on the energy levels, densities, and capture cross sections of the fast states on real germanium surfaces

No.	Cryst. plane	Method*	Source	(1) $(E_t-E_i)/kT_0$ $[A_p(cm^2)]$	(1) $N_t(cm^{-2})$ $A_n(cm^2)]$	(2) $(E_t-E_i)/kT_0$ $[A_p(cm^2)]$	(2) $N_t(cm^{-2})$ $A_n(cm^2)$	(3) $(E_t-E_i)/kT_0$ $[A_p(cm^2)]$	(3) $N_t(cm^{-2})$ $A_n(cm^2)$	(4) $(E_t-E_i)/kT_0$ $[A_p(cm^2)]$	(4) $N_t(cm^{-2})$ $A_n(cm^2)]$
1		a (ambients)	40	>8	$\approx 10^{12}$	3	10^{11}	−1	10^{11}	<−6	$\approx 10^{12}$
2		a (ambients)	41	>5		......small continuous distribution......				−5.5	2×10^{11}
3		b (fields)	46, 48, 86			4	7×10^{10}			−5	5–10×10^{10}
4		a (ambients)	201	5.5	2×10^{11}	0.5	2×10^{10}	−1.5	3×10^{10}	−6	2×10^{11}
5		a (ambients)	202	5	10^{11}	......continuous distribution ($\approx 10^{11}$)......				−3.5	10^{11}
6	(110)	a (ambients)	203	7	10^{11}	5	3×10^{10}	−3	4×10^{10}	−7	10^{11}
7		a (ambients)	204	7.5	10^{11}	4.5	2×10^{10}	2, −2	10^{10}	−8	2×10^{11}
8		a+c+d (ambients)	31, 32, 57	>4	1–10×10^{11}	......continuous distribution ($\approx 10^{11}$)...... $[A_p \approx 6\times 10^{-15}, A_n \approx 4\times 10^{-17}]$		−1	1–10×10^{10}	<−4	1–10×10^{11}
9	(100)	a+d (fields)	127	>7		3–6(R)** $[4$–15×10^{-15} 1–$6\times 10^{-15}]$	1–4×10^{11}			<−6	
10	(111)	a+d (fields)	225	11(R) $[10^{-12}$	1×10^{11} $5\times 10^{-17}]$			0	10^{11}	−8(R) $[A_p/A_n \approx 10^{-7}$–$10^{-6}]$	>10^{11}
11	(110)	a+d (ambients)	208	>6	$\approx 10^{11}$	4(R) $[2\times 10^{-15}]$	4×10^{10} 2×10^{-16}	−1.5(R)	4×10^{10}	<−5	10^{11}

Ch. 9, § 2.1   REAL SURFACES   391

#	Orient.	Method	Ref.					
14		a+d (fields)	209			3–4.5(R) [2–200×10^{-15}]	6–30×10^{10} 1–30×10^{-15}	−1	10^{11}	
15	(110) (111)	a+d (fields)	214, 215		≈10^{12}	4.5–6(R) [2–10×10^{-14}]	2–6×10^{11} 5–30×10^{-16}	−2	1–3×10^{11}	
16		a+d (ambients)	78			4.7(R) [2–7×10^{-14}]	7×10^{10} 1–3×10^{-15}			
17	(110) (111)	a+d (fields)	249		8	4(R) [10^{-14}]	2×10^{10} 10^{-15}]	−1(R)	2×10^{10}	−5	10^{11}
18		a+d (ambients)	222		10^{11}	2.5(R) [2×10^{-16}]	3×10^{10} 10^{-17}]	−1.5(R) [3×10^{-17}]	2×10^{10} 2×10^{-16}]	−7.5	10^{11}
19	(100) (111)	a+d (fields)	216		7.5(R) [2×10^{-15}]	4.5–4.8(R) [1–2×10^{-15}]	4×10^{11} 1–2×10^{-17}]	−1	3×10^{11}	<−6	>4×10^{11}
20	(111)	a+d (fields)	224		>7 3×10^{11} 5×10^{-20}]	3–4(R) [1–5×10^{-15}]	2–4×10^{10} 1–8×10^{-16}]	−2	2–5×10^{10}	<−8(R) [$A_p/A_n \ll 1$]	
				>4×10^{11}							
				>9(R) [$A_p/A_n \gg 1$]							
21	(100)	a+d (fields) / e	229			6(R) [6×10^{-15}] 6 [ ]	5×10^{11} 2×10^{-16}] >10^{11} 10^{-20}]	0 0 [10^{-16}]	10^{11} >10^{11} 10^{-15}]	<−6 −7.5 [10^{-21}]	≈5×10^{11} >10^{11}
22		e	80			4 [ ]	≈10^{11} 4×10^{-15}]			−6 [2×10^{-13}]	≈10^{11}
23		e	240			5.5 [ ]	10^{-18}]				
24	(111)	e	229					0 [6×10^{-15}] 0 [1.5×10^{-14}]	10^{-15}]	−8 [10^{-21}]	3×10^{-13}]
25	(110)	a+d+f (fields)	220, 221		6(R) [2×10^{-14}]	3(R) [3×10^{-15}]	3×10^{10} 3×10^{-17}]	3×10^{-17}] −1.5(R) [10^{-17}]	3×10^{-17}] 2×10^{10} 10^{-16}]	−5(R) [3×10^{-19}]	2×10^{11} 3×10^{-15}]
				2×10^{11} 10^{-16}]							

* Method: a — field effect; b — channel field effect; c — surface photovoltage; d — surface recombination velocity; e — pulsed field effect; f — small-signal pulsed or ac field effect.
** The symbol (R) designates recombination centres.

TABLE 9.2

Summary of results obtained by different workers on the energy levels, densities, and capture cross sections of the fast states on real silicon surfaces

No.	Cryst. plane	Method*	Source	(1) $[E_t-E_1]/kT_0$ $[A_p(cm^2)]$	(1) $N_t(cm^{-2})$ $A_n(cm^2)]$	(2) $(E_t-E_1)/kT_0$ $[A_p(cm^2)]$	(2) $N_t(cm^{-2})$ $A_n(cm^2)]$	(3) $(E_t-E_1)/kT_0$ $[A_p(cm^2)]$	(3) $N_t(cm^{-2})$ $A_n(cm^2)]$	(4) $(E_t-E_1)/kT_0$ $[A_p(cm^2)]$	(4) $N_t(cm^{-2})$ $A_n(cm^2)]$
1		b (fields)	46, 86	**15**	8×10^{11}					-17	$\mathbf{1.5\times10^{12}}$
2		a (fields)	205			≈ 3	$\approx 10^{12}$				
3		a (fields)	206	≈ 11	$\approx 10^{12}$	≈ 0	$\approx 5\times10^{12}$			≈ -12	$\approx 3\times10^{12}$
4		a (fields)	207			≈ 5	$\approx 10^{12}$			<-10	$\approx 10^{12}$
5		d (fields)	248, 250	$\approx \pm 16$(R)**				small continuous distribution			
6		d (ambients)	235	≈ 10(R) $[A_p/A_n \approx 10^8]$							
7		e	240, 241			**4** $[4\times 10^{-12}$	$\approx 10^{12}-10^{13}$ $3\times 10^{-16}]$	-2 $[2\times 10^{-15}$	$\approx 10^{12}-10^{13}$ $10^{-13}]$		
8		e	251							-17 [	$10^{-25}]$
9		a+e (fields)	252	≈ 15	$\approx 10^{12}$	5.5; 0.5	2×10^{11}	**0** $[10^{-14}$	$10^{-12}]$		
10	(111)	a+d+f (fields)	220, 238 239	11(R) $[2\times 10^{-15}$	10^{12} $10^{-18}]$			-3(R) $[10^{-21}$	3×10^{11} $10^{-16}]$	-7(R) $[10^{-22}$	3×10^{12} $10^{-15}]$

* Method: a — field effect; b — channel field effect; d — surface recombination velocity; e — pulsed field effect; f — small-signal pulsed or ac field effect.
** The symbol (R) designates recombination centres.

transitions from the valence band into the surface states should be the controlling process (positive change in absorption). This behaviour is indeed exhibited by the data (curves 1 and 8). For intermediate values of barrier height, the two transition rates become comparable, and the absorption signal changes sign in different regions of the spectrum. Analysis of data on both n- and p-type samples indicates a large surface-state density ($\approx 10^{12}$ cm^{-2}) near the midgap and a decrease in density as the valence-band edge is approached. Similar studies, but not as detailed, have been carried out on germanium by Chiarotti and co-workers [246, 247]. On the assumption that the absorption observed is due to transitions from surface states into the conduction band, these authors conclude that the states involved lie 3, 5, and 6 kT below the midgap.

Tables 9.1 and 9.2 summarize the characteristics of the fast states on etched germanium and silicon surfaces as determined by different workers. The results refer to surfaces prepared by standard etching procedures, followed by "aging" in an ambient gas or in vacuum for at least several hours to attain stabilization. The most commonly used etchant was CP-4 or CP-4A, but other mixtures based on HNO_3, HF, and H_2O_2 were also employed. The first three columns of each table list the crystallographic plane of the surface (where known), the experimental method used, and the reference number(s) to the original article(s). The second column also indicates whether the main swing in barrier height was effected by ambients or by transverse fields. The sets of surface states found are grouped in four columns in order of decreasing energy in the gap. Each column lists the energy position $(E_t - E_i)/kT_0$ (where $T_0 = 300°K$) and the density N_t of the relevant set of states. Where measured, the hole and electron capture cross sections A_p and A_n are included in parentheses below the values for the energy and the density, respectively. Sets corresponding to recombination centres are designated by the symbol (R). In most cases, the energy values quoted represent the effective energies $E_t^f (\equiv E_t + kT \ln(g_0/g_1))$. The actual values E_t are available only for those sets whose energy is derived either from data that include the temperature dependence of the recombination characteristics (the recombination centres in experiments 9, 10, 15, 21 of Table 9.1) or from data based on the pulsed field effect (experiments 21–24 of Table 9.1 and 7–9 of Table 9.2). In the former instance, it is possible to determine both E_t and E_t^f and thus to estimate the statistical weight factor g_0/g_1 (see above). In the latter case the data yield $E_c - E_t$ (or $E_t - E_v$), and the energy values quoted in the tables were calculated on the assumption that E_i coincides with the midgap. Only the effective cross sections $A'_p \equiv (g_1/g_0)A_p$ and $A'_n \equiv (g_0/g_1)A_n$ can be obtained from such measurements.

Values printed in boldfaced type correspond to data that can be analysed unambiguously in terms of a single dominant set of states over an extended range. Such is the case for parameters determined from simple s/s_{max} curves (see, for example, Figs. 9.12, 9.13) or from pulsed field-effect results in which the linear relation extends over an appreciable temperature range (Figs. 9.14, 9.15). The spread in the boldfaced values in each experiment represents the range of variation of these parameters obtained on different samples and/or for different etchants and surrounding gaseous ambients. Results based on field-effect data (by themselves) or on complex s curves, where a unique resolution of the overlapping effects of different sets is much more difficult to attain, are obviously less accurate and yield only order-of-magnitude values.

Inspection of Table 9.1 shows that the grouping of the fast states in germanium into four discrete sets, each present with a density of the order of 10^{11} cm^{-2}, is consistent with most observations. The parameters of the two inner sets (Columns (2) and (3)) are known better than those of the outer ones, the former sets being within the normal range of the swing in surface potential. Of these two, only the one above E_i (Column (2)) appears to contribute to the recombination process. A similar contribution from the outer sets becomes apparent at extreme values of the surface potential.

The various results reported on germanium are in good agreement in view of the fact that the fast states are affected both by the initial treatment given to the surface and by the subsequent history (see following subsection). Such differences as are apparent, are usually within the range of values obtained by the same group of workers under different surface conditions. Changes in the fast state characteristics on surfaces of different crystallographic orientation, if present, also fall within these limits. A careful comparison of the different faces under identical surface treatments should prove highly instructive. The only significant variation from one experiment to the other is exhibited by the capture cross sections.

As can be seen from Table 9.2, the data on silicon surfaces is much less extensive. The surface-state structure does not appear to be very different from that in germanium — except for the density of states, which is about one order of magnitude greater.

2.2. EFFECT OF SURFACE TREATMENT ON FAST STATES

An extensive effort has been devoted to the study of the electrical structure of germanium and silicon surfaces under different surface conditions. One of the chief aims of such studies is to establish a definite correlation between the observed characteristics of the fast states and the particular treatment to which the surface is subjected. In this manner one can hope to gain insight into the nature and physico-chemical origin of the surface states involved. Changes in the condition of the surface can be produced either during the initial preparation stage or by various treatments of the finished surface. The former approach employs different chemical etchants, including those deliberately doped with impurities; the latter includes exposure to various ambients, heat treatment, and oxidation. The data available on the effect of these treatments on the fast-state parameters will be reviewed in the following paragraphs.

Chemical etching. Surprisingly enough, the basic structure of the fast states on stabilized surfaces of germanium (and possibly silicon) is insensitive to the etchant used. Almost all the common etchants give rise to four dominant sets with not-too-different energies and densities. This is apparent from Table 9.1, where similar structures are observed by most workers even though, if not always the etchant composition, the purity and mode of preparation certainly differ from one experiment to the other. Marked changes do occur, however, in the transition stage from a mechanically polished surface to a well-etched one. Measurements carried out by Harnik and Margoninski[253] on partially-etched surfaces show that with increasing etching time (in CP-4A) the energy, the density, and the cross-section ratio (A_p/A_n) of the recombination centres all decrease monotonically towards their final values on well-etched surfaces. Particularly noteworthy is the drastic reduction in A_p/A_n, from above 2×10^4 to 10^2.

Konorov and Romanov[254] measured the surface potential ϕ_s, the surface photovoltage δV_s, and the recombination velocity s in germanium *during* the etching process (while the sample was immersed in an H_2O_2 etchant). Three stages are apparent in their data. Initially, ϕ_s changes rapidly while δV_s and s remain practically constant. This is attributed to the gradual dissolution of the oxide (present on the surface from previous exposure to air), a process that does not penetrate to the fast states and so leaves them unaffected. The second stage contains the principal variation in δV_s and in s, with ϕ_s now changing only weakly, and is probably associated with the dissolution of the disturbed interface region. The final stage corresponds to the establishment of equilibrium dissolution and stable surface-state characteristics. Similar measurements were carried out [255] on etched germanium samples immersed in a nitric acid solution of varying concentration (1 − 10N). Beyond 7N (where passivation of the surface occurs [256]), s was found to decrease monotonically with increasing concentration. This was attributed to the increasing thickness of the oxide film.

A systematic comparison of the parameters of the recombination centres on germanium surfaces prepared by several different etchants has been undertaken by Wallis and Wang [218]. No significant change was found for surfaces etched in CP-4, in peroxide etchant (HF + H_2O_2), or in silver etchant (HNO_3 + HF + $AgNO_3$). A relatively small increase in the hole capture cross section A_p was observed for electrolytic etching (in KOH) and for an iodine etchant (a variant of CP-4 with iodine added). Margoninski [214] has attempted to correlate the characteristics of the recombination centres with the state of the oxide at the surface. A CP-4A etched germanium sample was alternately treated with concentrated HNO_3, which is an oxidizing agent, and with HF, which acts to dissolve the oxide. A very good reproducibility was obtained for each treatment. Except for a threefold increase in A_p observed on the HF treated surface, the parameters of the recombination centres in the two cases were equal to within the limits of the experimental accuracy. A similar insensitivity of the density of states to a wide variety of etchants is indicated by the somewhat less quantitative but more extensive studies performed by Morrison [257].

Impurities. In analogy with the situation in the bulk, one might expect impurities to play an important role in the electrical behaviour of the surface. This has been qualitatively demonstrated by Fistul and Andrianov [258] and by Ioselevich and Fistul [259], who found that traces of metallic impurities in the etchant gave rise to substantial changes in the surface conductance of germanium and silicon. In a more systematic study, Morrison [257] showed that the higher the purity of the etchant (CP-4 or mixtures of HF/HNO_3), the more n type is the germanium surface; admixtures of small amounts of copper, antimony, or silver in the etchant or rinse water made the surface more p type.

A detailed investigation of the effect of copper impurity on the structure of the fast states on germanium has been carried out by Boddy and Brattain [260] and by Frankl [261, 262]. The former workers employed a semiconductor–electrolyte system, and their work will be discussed in § 2.4. The etched surfaces studied by Frankl were first treated by HF to remove the oxide film, and then washed in 1 % KOH at 90°C. The latter procedure was found [261] to remove the hydrophobicity of the surface and make it sensitive to copper contamination. Measurements of s and δQ_{ss} versus u_s were then carried out [262] (a) with no further treatment, (b) following dipping in dilute $Cu(NO_3)_2$ solution (1 ppm), and (c) following immersion in 3 % KCN at 50°C (to remove any residual copper contamination). The copper treatment resulted in a considerable increase in s_{max} and in the trapped-charge density, whereas the opposite effect was produced by the KCN treatment; the difference in s_{max} amounted to a factor of 50. In all cases the s curves were bell-shaped but complex in character, indicating the presence of a background distribution of recombination centres superimposed on one dominant set. Although the reproducibility of the data was not always good, the results clearly show that the fast states on copper-treated surfaces are associated mostly with copper contamination. Further support for this conclusion was furnished by radioactive tracer analysis combined with electron microscopy and diffraction studies. The former shows that about 10^{14} cm^{-2} copper atoms are deposited by the copper treatment, and that the KCN removes all detectable traces of the metal. The latter measurements indicate that the copper is not deposited uniformly, but rather nucleates in the form of globules about 100 Å in diameter and 10^{11}–10^{12} cm^{-2} in density. Presumably the nucleation takes place on some sort of select sites, since the density of microcrystals is relatively fixed. This density is of just the same order of magnitude as that of the fast states measured electrically.

The fast-state structure of the copper-treated surface is very similar to that found on ordinary surfaces (Table 9.1). Contrary to most workers, however, Frankl [262] identifies the energy level of the recombination centres with the *lower* of the two possible values that can be deduced from the s curve (see § 2.1 above). Such an assignment appears to be somewhat more consistent with the field-effect results. It is rather doubtful, however, whether the available data permits one to make a decision either way. For one thing, it is well known that field-effect curves by themselves cannot usually be trusted to yield unique values for the energy levels, certainly not within the narrow limits that would make one choice preferable to the other. Moreover, and more important still, the recombination data yield complex s curves and are not very reproducible. Both factors necessarily detract (in different ways) from the accuracy of the energy values derived for the recombination centres.

Gaseous ambients. It is generally observed that the effect of ambients on the structure of the fast states is most pronounced during the early stages after etching. The surface is then highly unstable, with fairly rapid and irreversible changes in its electrical characteristics taking place continuously. Measurements under such conditions are difficult to perform, and only a few studies along these lines have been reported. Wang and Wallis [217] found a very large increase in the density of the inner sets of states following brief (1 sec) exposure to ozone of a freshly etched germanium surface. The density gradually recovered in dry oxygen and increased again in wet nitrogen. Recombination and field-effect data obtained on successive Brattain–Bardeen ambient cycles were analysed on the assumption that only the density of the inner states vary, the energy levels and the cross-section ratio A_p/A_n of the recombination centres remaining fixed. With this assumption, the latter parameters were estimated to be much the same as those for stable surfaces. In an attempt to slow down the rate of change in structure, Harnik and Margoninski [253] quickly inserted a freshly etched sample into a vacuum chamber, where all measurements were then carried out. Here again the parameters of the inner sets of states were found to be similar to those for stable surfaces, except that higher values were obtained for A_p/A_n. Exposure of the freshly etched surface to ozone resulted in a considerable decrease in A_p/A_n (from 2000 to 270 in a typical case). It is interesting to note that prior treatment with HNO_3, also an oxidizing agent, gave rise to an *increase* in A_p/A_n (from 2400 to 9900).

As the surface is aged for a period ranging from hours to days, depending on the ambient gas, its structure becomes stabilized and less sensitive to variations in ambient. One might then visualize the fast states as being "shielded" from the surrounding gases by the oxide film. Even then, however, reproducible changes upon exposure to different ambients persist. These are relatively small when the ambient is cycled between vacuum (10^{-5} mm Hg), oxygen, and nitrogen [127], but may become quite appreciable upon exposure to more active gases such as ozone and water vapour. Both these agents have a direct influence on the structure of the oxide, ozone acting to build it up, water vapour (in the case of germanium) to dissolve it. It is generally agreed upon that the former increases the density of the fast states [211,212,217,263] while the latter decreases it [202,222,249,263,264]. Margoninski and Farnsworth [215] found an enormous decrease in the hole capture cross section upon exposure to water-saturated oxygen or nitrogen, and the density of states appeared to increase somewhat. For such high humidity, Dorda [264] too observed an increase in density, which he attributed to the thorough dissolution of the oxide layer and the attendant formation of interface defects. For less humid atmospheres, on the other hand, an order-of-magnitude *decrease* in density is obtained. The observed rate of change of the density with time following exposure to a wet ambient can be fitted to a diffusion equation, with a diffusion constant of about 10^{-14} cm²/sec. From this Dorda concludes that the annihilation of the fast states (as evidenced by the decrease in their density) is effected by water molecules diffusing through the oxide film to within an atomic distance of the centres, at which proximity they are somehow able to neutralize the fast states [249, 265] (see below).

The marked influence of ozone and water vapour on the fast-state structure in germanium even in the case of stable surfaces, casts some doubt on the accuracy of results obtained by the use of the Brattain–Bardeen cycle. Variation of the barrier height by transverse fields is undoubtedly preferable but, as will be discussed below, this method too has its limitations.

Heat treatment. Several workers [86, 212, 218, 257] have observed an increase in the density of fast states on germanium after heating at relatively low temperatures (100–200°C). This phenomenon has been very extensively studied by Rzhanov and his co-workers [227, 228, 249, 263, 265–267]. Their experimental method consisted of displaying on a CRO the large-signal ac field effect in the dark and under illumination following different heat treatments. Typical results [266] of δQ_{ss} and s versus v_s obtained from such data for a sample heated *in vacuo* at various temperatures are shown in Figs. 9.17a and 9.17b. (Note that the variable used here is the barrier height v_s and not the surface potential u_s.) An increase in both the bell-shaped s curve and the slope of the δQ_{ss} curve is apparent for heating temperatures up to 500°K, followed by a decrease at higher temperatures. The corresponding variation [249] in s_{max} is shown by the semilog plot of Fig. 9.17c. Both temperature ranges are associated with approximately the same activation energy (0.15–0.2 eV). The extent to which the recombination data could be approximated by the simple relation (5.85) was found to vary from sample to sample. For some samples the fit was perfect; for others the s curves exhibited a complex character, suggesting a contribution to recombination from both discrete and continuously distributed states. An intermediate case [249] is shown in Fig. 9.17d. Unless the s/s_{max} curves are particularly complex, their general shape for different heating temperatures is well preserved. This indicates that only the densities of the recombination centres vary, their energy and cross sections remaining more or less fixed. Essentially the same behaviour was found when the sample was heated in oxygen and carbon monoxide [266], or even in liquids such as benzene and ether [265, 268], except that in the case of ether there is also a change in A_p/A_n (from 20 to 6000). It is thus apparent that the increase in fast-state density is asso-

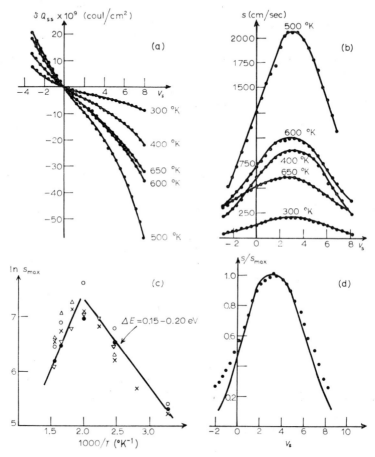

Fig. 9.17. Effect of heating *in vacuo* at various temperatures on the fast-state structure in germanium. (All data taken at room temperature.)
(a) Change δQ_{ss} in trapped-charge density in fast states *versus* barrier height v_s.
(b) Surface recombination velocity s *versus* barrier height v_s.
(c) Logarithm of s_{max} *versus* reciprocal of heating temperature.
(d) Typical results of s/s_{max} *versus* v_s for a heat-treated sample. Heavy curve was calculated from (5.85).

(After Rzhanov and co-workers, references 249 and 266.)

ciated mainly with the heating process itself and much less with the ambient in which the heating is performed. Results similar to those shown in Fig. 9.17 have also been obtained [269] on a freshly etched surface as the surrounding humidity content was progressively reduced, first by freezing out while the sample was kept in a nitrogen atmosphere, and then by pumping to successively lower pressures.

The experiments just discussed strongly indicate that the increase in the density of recombination centres is intimately connected with desorption of water from the surface. And indeed, exposure of the sample to a humid atmosphere destroys the previous effect of the heat treatment [228], the process being completely reversible. A similar reduction in density has been observed upon exposure of a heat-treated surface to vapours of other polar liquids, such as chlorobenzene and nitrobenzene [270]. The variation of the δQ_{ss} and s versus v_s curves with time of exposure indicates that in these cases the centres first lose their recombination properties (the cross sections decrease) and only after prolonged exposure does their density fall off as well.

The effect of ozone on the fast-state density has been found to depend on the condition of the surface. On a freshly etched surface, ozone gives rise to an increase in density [228, 267], just as heat treatment does below 500°K. In both cases, the removal of physically adsorbed water is probably the controlling mechanism. If, on the other hand, the etched sample has been first heat-treated at 500°C so as to become free of (physically) adsorbed water, ozone causes a *reduction* in surface-state density [267] similar to the reduction produced by heating at temperatures above 500°K (see Fig. 9.17). Rzhanov and co-workers [267] suggest that both these phenomena are associated with the removal from the surface of the more strongly bound (chemisorbed) water molecules. Their conclusion is supported by the observation that surfaces heat-treated at 750°K, and so presumably free of chemisorbed water [188], behave quite differently from those heated at lower temperatures: their fast-state density *increases* rather than decreases upon exposure to water vapour. It thus appears that physically and chemically adsorbed water molecules affect the density of surface states in opposite ways.

In order to account for the above observations, Rzhanov proposes that the germanium surface is inherently associated with some sort of defects (of density 10^{11}–10^{12} cm^{-2}) which are potential recombination (or trapping) centres. *Physically*-adsorbed water molecules, retained on the surface from the etching process or from other previous treatments, are assumed to neutralize most of these defects. Upon heating, water desorbs and the centres are activated to form fast states. The activation energy of 0.15–0.2 eV associated with this process (see Fig. 9.17c) corresponds to a heat of adsorption of about 4 kcal/mole, a reasonable value for physical adsorption and consistent with the reversibility of the process. The *reduction* in fast-state density following heat treatment at *above* 500°K is attributed to the removal of water molecules *chemisorbed at the surface defects;* such molecules are thus envisaged as constituting the activating species that give rise to the fast states. The difficulty with this explanation, however, is that the observed activation energy (≈ 0.2 eV) is too small to correspond to a chemisorption process.

The neutralizing action of water may arise [265] either from a saturation of the free bond of the centre by an H^+ or OH^- group, or from a drastic change in the parameters of the centre due to the high dipole field of the water molecule in its vicinity. The former mechanism is unlikely in view of the fact that chlorobenzene and nitrobenzene, which do not contain these groups, produce much the same effects as water. It thus appears that electrostatic interaction between physically adsorbed polar molecules and activated imperfection centres plays the dominant role in the neutralization process. Such interaction can decrease the capture cross sections of the recombination centres [271], as well as shift their energy to an ineffective position.

In silicon, heat treatment has been found [239] to *decrease* rather than increase the density of recombination centres. Furthermore, the change produced in this manner is irreversible — exposure to water vapour does not restore the original density prior to heating. This drastically different behaviour exhibited by silicon must somehow be connected with the insolubility of the silicon oxide.

Thermal oxidation. The remarkable stability of thermally-oxidized silicon surfaces and the obvious advantages in device performance that such stability offers, have prompted considerable activity in the study of the silicon/silicon oxide system. The first systematic investigation of this system was carried out by Atalla and co-workers [272]. Continuous and uniform oxide films, ranging in thickness from a few hundred angstroms to several microns, were obtained by heating etched silicon samples at a temperature of about 1000°C in an atmosphere of either oxygen or water vapour. If grown under carefully controlled conditions, the films are characterized by high resistivity ($\approx 10^{16}$ ohm · cm) and high dielectric strength ($\approx 5 \times 10^6$ V/cm). (They have a dielectric constant and refractive index [273] of 3.82 and 1.458, respectively.)

The oxide is usually amorphous, but prolonged annealing results in a definite crystalline structure [274]. In most cases the surface under the oxide is n type.

Atalla et al.[272] have found that thermal oxidation of the silicon surface completely eliminates slow states (see §2.3 below) and, at least in the case of zone-refined material, effects a marked reduction in the fast-state density. Field-effect measurements indicated the presence of two sets of fast states, at 0.3 eV above and 0.4 eV below the midgap, with densities of the order of 10^{10} and 10^{11} cm^{-2}, respectively. These are superimposed on a smaller, continuous distribution near the midgap. Champion [251], on the other hand, using pulsed field-effect techniques, observed a discrete set of states close to the midgap for an oxide 0.6 μ thick. Millea et al.[206] performed field-effect measurements at 100°C in order to eliminate the difficulties arising from poor communication between the bulk and the surface (see Ch. 6, § 2). They report some unusual effects for thermally oxidized surfaces that have not been observed by other workers.

The uniformity and high dielectric strength of the silicon oxide film makes feasible a new field-effect configuration, one in which the field electrode is evaporated directly on top of the oxide film. Such a configuration, referred to as the metal–oxide–semiconductor (MOS) structure, is obviously associated with a very large geometric capacitance C_g, quite often comparable to the surface capacitance C_s. It can therefore be used as a voltage-dependent capacitor element for parametric amplification [275, 276]. A much more important application, however, is the MOS field-effect transistor [277]. The relatively few studies reported so far of fast states on oxidized silicon surfaces employ mostly one or the other of these two structures.

The schematic representation of Fig. 6.5 (p. 221) can serve as well to illustrate the MOS capacitor element. One lead is connected to the evaporated field electrode (referred to as the gate), the other is ohmically bonded to the silicon sample. Measurements of differential capacitance *versus* applied dc bias have been carried out on such structures by a number of workers [273, 277, 278–281]. At low frequencies, the curves usually have an inverted bell shape, with the minimum somewhere in the depletion range (compare Fig. 6.6, p. 223). For large positive and negative bias voltages, corresponding to strong accumulation and inversion layers, the surface capacitance becomes large compared to the geometric capacitance and the curves level off at the latter value. At higher frequencies, the capacitance *versus* bias characteristics are generally altered considerably. Two entirely different mechanisms may then come into play. In the inversion-layer region, poor communication between surface and bulk can hamper the modulation of the surface charge by the ac signal, resulting in a reduction in the measured capacitance. The frequency response in this regime is controlled by minority-carrier recombination and generation processes (see Ch, 6, § 7). It was shown by Hofstein et al.[280] that extraneous effects originating in the region surrounding the gate may play an important role here. The other mechanism involves surface states. A reduction in capacitance, but now not necessarily confined to the inversion region, may take place with increasing frequency, as more and more surface states are no longer able to follow the applied signal. In principle the density, energy distribution, and time constants of the fast states may be derived from the bias and frequency dependence of the surface capacitance. Terman[278] and Lehovec et al.[279] conclude from such measurements that the fast-state distribution is quasi-continuous, with a density of the order of 10^{12} cm^{-2}. A similar density has been estimated by Gray[281] for silicon surfaces having very thin oxide films (40–60 Å). In this case, tunnelling between the surface states and the field electrode was clearly observed.

A theoretical analysis of the frequency response in terms of equivalent RC networks has been carried out by Terman[278] and in more detail by Lehovec and Slobodskoy[282]. Since too many unknowns are involved, however, interpretation of capacitance data by this or any other procedure cannot usually yield more than rough estimates for the surface-state parameters. Particular difficulties may be encountered in the inversion regime, where the not-too-well defined contribution from minority-carrier effects[280] (first mechanism discussed above) can introduce serious errors. Results quoted on the basis of capacitance data, in the absence of additional information derived from other types of measurements, should be considered with these limitations in mind.

A schematic diagram[277] of the MOS field-effect transistor is shown in Fig. 9.18a. The n-type inversion channel below the oxide layer is either formed during the oxidation process or else it can be induced by biasing the gate positively (enhancement mode). The diffused n+ regions on both sides of the gate ensure that the current flowing between the two ohmic contacts (source and drain) be confined to the inversion layer. (This is similar to the n–p–n channel discussed in Ch. 6, § 8.) In order to achieve high-frequency performance, the length of the channel is kept as short as possible (see Ch. 6, § 6.1). Current–voltage oscillograms taken by Hofstein and Heiman[283] on an n–p–n enhancement unit are shown in Fig. 9.18b. Here the current I_D through the channel is plotted against the source-to-drain voltage V_D for various values of the gate bias V_G. These characteristics are in very good agreement with those derived analytically by Shockley [284] for his unipolar field-effect transistor. The pentode-like shape of the curves can be understood from the following qualitative considerations. At low bias voltages, the n-type channel has not yet developed and the drain current is very small. At any given value of V_D, I_D increases monotonically with V_G as an increasingly greater density of electrons is induced in the channel. For $V_D \ll V_G$, the

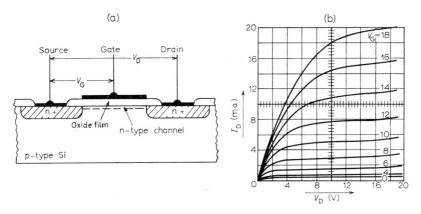

Fig. 9.18. MOS field-effect transistor.
(a) Schematic diagram of the device. (After Wallmark, reference 277.)
(b) Oscillograms of drain current I_D *versus* drain voltage V_D, with gate voltage V_G as parameter (enhancement unit). (After Hofstein and Heiman, reference 283.)

surface under the gate is essentially equipotential and the induced charge density between source and drain is uniform. As a result, I_D increases linearly with V_D. For larger values of V_D, the charge density in the channel is no longer uniform but decreases as the drain is approached, and the curves begin to bend. The current saturates when $V_D \gtrsim V_G$, since in this range the neighbourhood of the drain is completely depleted. The continuity of the current is then effected by electron injection into the depleted region, a space-charge limited process [285]. It can be shown [284] that in the absence of surface states the saturation value increases quadratically with bias voltage. In the linear regime, I_D is of course directly proportional to V_G. Both these relations are satisfied to a good approximation in the oscillograms of Fig. 9.18b.

Although the MOS transistor constitutes an ideal tool for studying the electronic structure of the surface, hardly any quantitative measurements of surface-state parameters have been reported so far. The linear dependence of the induced free-carrier density on gate voltage and the considerable swing attainable in barrier height suggest the absence of a large density of fast states on such structures. The density is probably of the order of 10^{11} cm^{-2}, the lowest value ever reported for silicon.

Heavily oxidized germanium surfaces have attracted much less attention, probably because the water solubility of the oxide precludes its use as a stabilizing agent. Measurements by Albers and Rickel [216] have shown that thermal oxidation drastically reduces the electron cross section of the recombination centres situated at 4 kT above E_i and, at the same time, introduces a new acceptor-like recombination level about 1 kT above E_i. The effect of the thickness and structure of the oxide film on surface recombination velocity in germanium has been studied by Berlaga et al.[286].

Coating. The thermally grown silicon oxide considered above is only one, even though undoubtedly the best, of many protective coatings devised for passivating the surface. To be effective, such coatings must be mechanically stable, uniform in structure (impermeable to gases), and free of mobile ionic impurities. More important still, they should not impair the electrical structure of the original surface. Most surface-passivation studies have been performed primarily from the standpoint of device performance. Quantitative measurements bearing directly on the characteristics of the fast states are available in only a few cases. Cullen, Amick, and Gerlich [287] have studied simultaneously the chemical and electrical properties of (111) germanium surfaces passivated by ethylation. Radioactive-tracer and mass-spectroscopy measurements indicated monolayer coverage, corresponding to one ethyl group ($-\mathrm{C_2H_5}$) bonded to each surface atom. The chemical stability of the ethylated surfaces on heating, exposure to air, or immersion in water and organic solvents suggests strong chemical bonding. A corresponding stability in the electrical structure is evidenced by the insensitivity of the barrier height and the fast-state characteristics to ambient conditions. The Fermi level at the surface is anchored at 4–5 kT below E_i, around which a fast-state density of the order of 5×10^{11} cm^{-2} is observed. This is somewhat greater than the densities found on ordinary surfaces. Similar stability has been achieved by the sulphurisation of germanium surfaces [288].

Bombardment. Studies of the effect of particle bombardment on the electronic structure of the surface can lead to a definite correlation between surface damage and surface states. So far, however, the available data is very limited. Gianola [289] has observed a hundredfold increase in the surface recombination velocity of silicon after bombardment with high-energy helium ions (3×10^4 eV). A similar increase was found by Dorda [290] following α-particle bombardment of germanium. From the temperature dependence of s, Dorda concluded that the recombination centres introduced in this manner lie close to one of the band edges. The effect of γ irradiation on surface recombination was studied by Komatsubara [291]. CRO patterns representing s as a function of induced charge were found to undergo marked changes by irradiation. Gamma and particle irradiation were also observed [292, 293] to enhance the leakage current in reverse-biased p–n junctions. This was shown [292] to arise from the accumulation on the surface of ions produced in the surrounding gas by the radiation.

In all these experiments, the depth of damage is considerable (0.6 μ for helium, 20 μ for α particles, and much larger for γ rays), greater in fact than the thickness of the space-charge layer. Accordingly, the localized levels that are produced cannot be considered as surface states in the usual sense. Low-energy ion bombardment, which affects only the first few atomic layers, should prove more illuminating.

Electric field and temperature. Several workers have reported changes in fast-state density produced by high transverse fields. Statz and co-workers [48], using channel conductance measurements on silicon, found a logarithmic increase in the density with time of application of a high reverse bias. This was attributed [86] to the drift of negative oxygen ions (or some other charged species) to or from the interface under the action of the high field ($\approx 10^6$ V/cm) across the oxide film. In the case of germanium, Rzhanov et al.[212] and Litovchenko and Lyashenko [203] have reported an increase in density at much lower fields (10^4–10^5 V/cm), which they explained on a similar basis. These effects are usually more pronounced in the presence of water vapour. Margoninski [294], on the other hand, finds no significant change in either the energy or the density of the fast states on germanium even at fields as high as 2×10^6 V/cm applied for 12 hours. Identical s/s_{max} curves were obtained before and after application of the field, and the δQ_{ss} versus u_s curve was only slightly shifted along the u_s axis. The discrepancy between these two results may be due to differences in surface treatment: whereas the surfaces that did exhibit changes were associated with a fast-state density of $2\text{--}5 \times 10^{10}$ cm^{-2}, those that did not contained about ten times as many states. Some support for this conjecture is provided by the measurements of Frankl[261], who found an increase in fast-state density with field for a KCN-treated surface (low density) but virtually no change for a copper-treated surface (high density). Balk and Peterson[224] observed a (reversible) increase of 20% in both s_{max} and the slope of the δQ_{ss} versus u_s curve due to a change in field from 2×10^4 to 8×10^4 V/cm. Here again, the surfaces studied were characterized by exceptionally low densities of fast states (see § 2.1 above).

There is some evidence that the energy and density of the fast states are temperature dependent[127, 212, 221, 224, 239, 295]. It has been suggested that variations in these parameters at different temperatures may be associated with changes in the amount of adsorbed water[227] or in the nature of the bond matching at the semiconductor/oxide interface[295]. It is rather difficult, however, to see how either of these mechanisms can account for the observed changes which, in some cases, occur over a temperature range of only a few tens of degrees.

2.3. SLOW STATES

The experimental data on the slow states accumulated since the early work reviewed in § 1.2 are far less extensive than for the case of fast states. The main effort in the later studies has been directed towards a better understanding of the charge-transfer mechanism between the slow states and the underlying semiconductor. To this end measurements were carried out under more carefully controlled conditions of the oxide film and the ambient gases. Although considerable progress has been achieved, some of the important questions in this area still remain open.

An indication as to how the energy bands of the oxide are aligned with respect to those of the semiconductor was obtained by studies of optical transitions between the slow states and the bulk. Bray and Cunningham[296] measured the photoconductance response of a high-resistivity gold-doped germanium sample at 80°K. Illumination at wavelengths shorter than the fundamental absorption threshold produced, in addition to the usual fast bulk photoconductivity, a very slowly decaying component. That such retentive conductance is a surface effect requiring the presence of an oxide film, is evidenced by its complete absence on surfaces freshly reduced in

HF. Field-effect measurements showed that the higher the retentive conductance, the more p type the surface became — from which it was concluded that the effect of illumination is to increase the electron density in the slow states. The magnitude of the retentive conductance was found to be a sensitive function of wavelength. The strongest response has a threshold energy of about 3 eV and a peak near 3.4 eV. A much weaker response occurs at longer wavelengths, peaking at approximately 2 eV and decreasing monotonically in the infrared. The strong response was attributed to the excitation of hole–electron pairs at the germanium surface with enough energy to enable the electrons to surmount the oxide barrier and fall into the slow states on the outer surface of the oxide film. This would require the conduction-band edge of the oxide to be about 3 eV above the valence-band edge of the germanium. The weaker response, which is characterized by a faster recovery time, was ascribed to transitions into internal oxide states close enough to the bulk germanium to communicate by tunnelling. The retentive conductance built up in either of these spectral ranges can be quenched by subsequent illumination in two other regions. One, corresponding to photon energies higher than about 4 eV, was attributed to excitation of electrons trapped in the slow states back to the germanium bulk. The other quench occurs at a narrow band centred about a photon energy of approximately 2.2 eV. This may be associated with the excitation of free holes at the germanium surface into the valence band of the oxide, where the holes recombine with electrons trapped in slow states. The complete spectral response then suggests a value of about 5.2 eV for the forbidden gap of the oxide, in good agreement with Papazian's estimate [85] of 5.5 eV (see § 1.2).

Abkevich [297] arrived at somewhat different conclusions from measurements of the retentive change in contact potential following short-wave illumination. In both germanium and silicon this change corresponds to an increase in work function, indicating that here as well illumination excites electrons from the semiconductor bulk into the slow states. The spectral response of the retentive component exhibits an approximately exponential rise as the photon energy is increased up to a value of 3.15–3.45 eV. (The spread represents different samples.) Thereafter, the response rises more slowly or even levels off completely. The energy at the inflexion point was interpreted as the height of the barrier between the bulk and the slow states, while the response below this point was attributed to the exponential increase in tunnelling probability *through* the barrier with increasing energy of the excited electrons. A fit of the data to a tunnelling formula of the type (7.53) yields for the width of the barrier values that range from 6 to 8 Å and are independent of whether the surface is freshly etched or aged. Since the thickness of the oxide film on an aged surface must exceed 10–30 Å, it was concluded that the slow states involved lie *inside* the oxide film rather than on its outer surface. The results of Abkevich are in agreement with those of Bray and Cunningham [296] as regards the position of the conduction-band edge of the oxide, but they are at variance as far as the details of the spectral response are concerned. Evidence for optical transitions from the semiconductor bulk into the slow states has also been obtained by Surduts [298] from measurements of the spectral response of the surface photovoltage δV_s. At long wavelengths, δV_s exhibits its normal behaviour (see, for example, Figs. 9.2, 9.3). Below about 4000 Å, however, a rapid change takes place in δV_s, indicating the formation of a p-type space-charge layer, presumably as a result of the increased electron occupation of the slow states.

It has been fairly well established that the slow relaxation in both germanium [299, 300] and silicon [301] practically disappears after evacuation to pressures of 10^{-6} mm Hg or less. This substantiates the previously reached conclusion that most of the slow states on ordinary surfaces consist of ions adsorbed from the surrounding ambient gas. Heating in vacuum [266] greatly enhances the relaxation time, while exposure to vapours of water or other polar liquids [270] effects a drastic reduction in time constant. As has been mentioned in § 1.2, the slow relaxation is essentially absent on heavily oxidized surfaces of germanium and, except for water vapour, the barrier height is insensitive to the gaseous ambient [87]. This is much more pronounced in the case of thermally grown oxides on silicon. Atalla [272] has shown that ac field-effect patterns in the presence of a superimposed dc bias remain unaltered even if the bias is maintained for months. Moreover, the surface is entirely unaffected by ambient gases, including even ozone and water vapour. It appears that the extremely low rate of charge transfer through the thick oxide film completely inhibits gas chemisorption and thus the formation of slow states on the outer oxide film (see § 1.5). The fact that the quiescent value of the surface potential depends on the growth conditions of the oxide, however, strongly indicates the presence of another type of slow states, one associated with impurities or imperfections within the oxide film [280]. The negligible influence of such states on the field effect must be attributed to their very poor communication with the silicon bulk.

The original observation [70] that the slow field-effect decay in germanium is non-exponential even under small-signal conditions (see § 1.2) has been substantiated by the very careful measurements carried out by Koc [302]. An empirical relation of the form

$$\delta\sigma = \delta\sigma_0 e^{-(t/\tau)^a} \tag{9.7}$$

was found to provide a good description of the surface-conductance relaxation process. The index a has an approximately constant value of 0.6, while the "time constant" τ decreases with increasing temperature as $\exp(\Delta E/kT)$. Relation (9.7) remains valid upon admixture of wet air, but τ drops sharply. With prolonged exposure to the wet air, τ recovers slowly — approximately logarithmically with time [303]. This last effect is ascribed to the hydration of the oxide proposed by Wallmark and Johnson [186]. The activation energy ΔE is likewise a function of humidity [304], ranging from 0.8 eV *in vacuo* to 0.2 eV in moist air at 200–250 mm Hg. These values agree fairly well with those obtained previously (§ 1.2). Koc [305] attempts to explain the empirical relation (9.7) in terms of electron diffusion through the oxide film. For small t, the exponential function in (9.7) can be approximated by $1-(t/\tau)^{\frac{1}{2}}$, which is just the form to be expected on the basis of such a model. It is difficult to see, however, how diffusion can be significant when the field in the oxide is 10^4 V/cm or higher.

Similar results have been reported by Pilkuhn [306], who was also able to fit his experimental data to a relation of the form (9.7). The index a, however, was found to depend on the water content of the oxide, its value varying between 0.3 and 1.0. The latter value, corresponding to a truly exponential decay, obtains for high humidity. The activation energy ΔE ranges between 0.25 and 0.8 eV, depending on the nature of the oxide layer: surfaces with thick oxides are associated with larger activation energies than freshly etched ones. Contrary to Koc's observations [302], ΔE was found to be independent of water content. The slow recovery of the time constant with prolonged exposure to water vapour reported by Koc [303] was observed here in the case of thick oxide films and found to be completely reversible. Hydrophobic surfaces [262] (freshly etched with an after-treatment of HF), on the other hand, showed very small changes in time constant upon exposure to water vapour, as expected. A more or less self-consistent model, which assumes the charge transfer between the slow states and the bulk to be of an ionic rather than electronic nature, has been advanced by Pilkuhn [306]. The slow decay in the charge initially induced in the space-charge region is attributed to the drift of ions in the oxide, a process that gradually polarizes out the external field. On this basis, the time constant τ should be inversely proportional to the mobility and to the density of the ions. The mobility is expected to be proportional to the number of vacancies in the oxide and to vary with temperature as $\exp(-\Delta E/kT)$, where ΔE is the activation energy for ion migration. And indeed, the observed activation energy of 0.25–0.8 eV is not unreasonable for such a process. The sharp drop in τ upon exposure to water vapour may be interpreted as due to an increase in the density of ions (possibly OH^-), its subsequent recovery as due to a decrease in the number of vacancies resulting from the hydration of the oxide. The proposed model thus accounts for the main features of the experimental data, and it is quite possible that ion migration may be one of the mechanisms of charge transfer. In fact, essentially the same mechanism has been invoked to explain changes in fast-state density under high fields (see § 2.2 above). The model, however, cannot explain the non-exponential character of the relaxation for very small signals, nor is it compatible with Koc's findings [304] that the activation energy varies with the water content of the oxide.

An altogether different mechanism of charge transfer, consisting of ion migration *outside* the semiconductor rather than through the oxide layer, has been proposed by Lyashenko and Chernaya [299]. Field emission of electrons from the metal or semiconductor surface was assumed to ionize the interjacent gas. The ions, in turn, accumulate on the semiconductor surface and gradually shield the interior from the external field. In support of this assumption, good correlation was observed between the decay characteristics of the measured ion current and those of the slow field effect. It is doubtful, however, whether such a mechanism is in general significant. Relaxation of the contact potential following illumination or of the field effect in p–n–p channels, for example, certainly cannot be explained on this basis. Rather, the results stress the necessity of avoiding sharp protrusions in the field-effect setup so as to eliminate any possible field emission.

It is well established by now that, at least under high fields, tunnelling through the barrier presented by the oxide is the controlling mechanism in the charge transfer process between the slow states and the bulk. Early indications for this conclusion came from two different phenomena observed on germanium surfaces. One, the so-called storage or accumulation effect [203, 212, 215, 307], consists of a shift in the quiescent surface potential upon the application of a large ac field. This shift, and the change in slow-state occupation that it implies, take place almost instantaneously — much faster than the normal relaxation process characterizing the charge transfer at low fields. The storage is more pronounced, and its onset occurs at lower fields, the higher the humidity of the ambient gas. The other phenomenon pointing to a tunnelling mechanism was observed [213] in pulsed field-effect measurements at reduced temperatures (see Ch. 6, § 6.2). For low pulse amplitudes the relaxation is controlled by thermal emission from fast states, as discussed in the previous subsections. At large amplitudes or for strong inversion layers, however, a much faster relaxation sets in which completely masks the thermal emission process. Such field-enhanced relaxation cannot be explained in terms of any of the known processes involving the fast surface states.

Dorda [308] has shown that the storage effect is a consequence of the unequal rates of tunnelling through the oxide film for the two polarities of the applied ac field. The field dependence of the charge accumulated in the slow states, was studied under ac [308], dc, and pulsed [309] fields. The results obtained could be well accounted for by a tunnelling expression of the type (7.53), appropriately modified [308] to take into account the fact that here tunnelling must proceed partly through the forbidden gap of the oxide and partly through the forbidden gap of the semiconductor. On this basis, Dorda concluded that the slow states involved lie 0.02 to 0.05 eV below the germanium conduction-band edge.

Measurements of the storage effect are rather indirect, and their analysis is quite involved. Pulsed field-effect studies of the type mentioned above, on the other hand, bring out the characteristics of the tunnelling process much more clearly. Results obtained in this manner [310] on n- and p-type germanium at several reduced temperatures are shown in Fig. 9.19. The ordinate represents the rate of increase in the excess surface density of majority carriers (ΔN in n type, ΔP in p type) just following the onset of the pulse, and is a direct measure of the initial rate of carrier emission from surface states into the majority-carrier band. The abscissa represents the reciprocal of the field $\mathscr{E}_s$ at the surface, again at the onset of the pulse. The strong field-dependence of the emission rate at high fields is apparent from the steepness of the curves in that range. A fit of the data to the tunnelling expression (7.53) shows that the surface states involved lie 0.03 eV below E_c for the n-type sample (electron tunnelling) and 0.023 eV above E_v for the p-type sample (hole tunnelling), in agreement with Dorda's results [308, 309]. The shallowness of these levels rules out the possibility that they correspond to fast states, since such states would release carriers by thermal emission much more readily than by tunnelling. Taking $E_c - E_t = 0.03$ eV, one calculates from (2.173) and (5.68) a value that is less than 10^{-10} sec for the thermal emission time at 200°K, whereas the observed time constant is many orders of magnitude larger and even at the highest applied field it is several microseconds. Thus the surface states involved must be physically separated from the semiconductor bulk, so that despite their shallowness in energy they cannot maintain under normal conditions (low fields) good electrical communication with the bulk. Only the slow states, situated as they are within the oxide layer or on its outer surface, are able to meet this requirement.

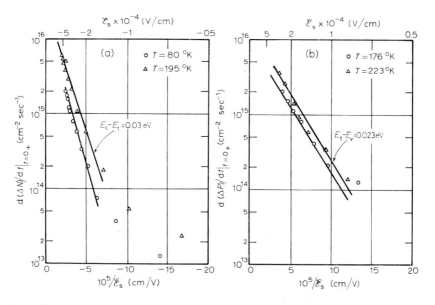

Fig. 9.19. Semilog plot of initial rate of increase in excess majority-carrier surface density as function of reciprocal electric field at surface, for an n- and a p-type germanium sample. (After Many and Goldstein, reference 310.)

Anomalous effects associated with the slow states or with some other charged species in the oxide film have been reported by several workers (see also § 1.2). In germanium, Lyashenko and Chernaya [311] observed very slow drifts in the field effect under high vacuum conditions. These could not be explained in terms of a simple charge-transfer process between the slow states and the bulk. Field desorption of ions from the surface was proposed as a possible mechanism. In silicon, high transverse fields ($\gtrsim 10^6$ V/cm) were found [301] to cause an upward displacement of the entire ac field-effect pattern, including the value of the conductance minimum $\Delta\sigma_{min}$. This indicates the presence of an additional conductance over and above that of the space-charge layer. After the removal of the field, it required several days for the original value of $\Delta\sigma_{min}$ to be restored. It was suggested that the additional conductance is due to mobile ions produced on the oxide film by the high field. Clear-cut evidence for the motion of charge on the oxide surface has been obtained by Shockley *et al.*[312] from contact-potential measurements on silicon p–n junctions coated with thermally-grown oxide films. The surface-charge distribution under different conditions of bias and ambient was scanned by moving the reference electrode (a gold probe) along the surface. A reverse bias was found to give rise to a gradual redistribution of the surface charge; wet air (100 % humidity) accelerated the process. The form of the redistribution is just that to be expected from drift of charged species on the oxide surface under the action of the field at the junction. The enhanced charge migration in the presence of water vapour is consistent with the common observation of leakage current under these conditions (see §1.2). In contrast, adsorption of certain amines was found [313] to inhibit the leakage current in germanium p–n junctions. This was attributed to the bonding of the amines with surface ions, thereby reducing their mobility.

2.4. SEMICONDUCTOR–ELECTROLYTE INTERFACE

The main difference between the "dry" surfaces considered so far and a surface in contact with an electrolyte lies in the latter being the boundary between *two conducting* phases. The semiconductor–electrolyte (S–E) system offers a number of advantages in the experimental study of the electronic structure of the surface. Under favourable conditions, changes in potential between the semiconductor and the electrolyte fall mostly or even entirely across the space-charge layer of the former, while the measured capacitance represents to a very good approximation the semiconductor surface capacitance (see Ch. 6, §§ 3, 6.1). Difficulties with this system arise from the incomplete blocking of the interface to current flow. The interfacial current hampers the electrical measurements and complicates the interpretation of the data. At the same time, it is usually accompanied by chemical reactions which, in turn, continuously alter conditions at the surface. Such reactions and the various electrochemical effects associated with them [314–317] will not be dealt with here. Rather, we shall be concerned mainly with the electronic processes at the semiconductor side of the system, those involving the surface states and the underlying space-charge region. In the study of these processes, attempts are usually made to minimize chemical changes by keeping the interfacial current low or, if unfeasible, by using high-speed measurements.

There are several points of similarity between the semiconductor–electrolyte and the semiconductor–metal systems (see Ch. 4, §§ 1.2, 1.3). Here, too, when the two phases are brought into contact, charge transfer takes place until, at equilibrium, a potential difference (the so-called Galvani potential) is established between the semiconductor bulk and the interior of the solution. The Galvani potential is distributed across three adjacent layers: the space-charge region of the semiconductor, the Helmholtz region encompassing the geometric interface, and the space-charge region of the electrolyte (the Gouy layer). The Helmholtz region is formed by adsorbed ions, and its thickness is comparable to the radius of the solvation shells surrounding the ions. This layer can accommodate very strong electrical fields, and the micropotential drop across it constitutes a significant portion of the Galvani potential. The Gouy layer is similar to the semiconductor space-charge layer, but because of the high ion density it is usually much thinner, becoming practically nonexistent at concentrations larger than about 0.1 N. In such cases all excess charge on the solution side can be considered to be in the compact Helmholtz region, and the Galvani potential is then just the sum of the micropotential drop V_H across the Helmholtz layer and the barrier height V_s at the semiconductor surface. Most of the electrical measurements are carried out under these conditions.

The rectification properties of the S–E system can be understood from the following considerations [318]. The electrons in the electrolyte are on negative ions, and their energy is usually located below the conduction- or even the valence-band edge of the semiconductor. Hence, when an n-type semiconductor is passing anodic current, the electrons attached to these ions cannot enter the conduction band of the semiconductor, and the interface is blocking with respect to electron flow. The current can then consist only of holes flowing from the valence band into the solution, conditions being analogous to those obtaining in a reverse-biased p–n junction (see Ch. 2, § 7.4). In the other polarity (the forward direction), electrons can easily flow from the conduction band into the solution, and a large current results. Actually, the situation is more complex since electrochemical reactions usually play a dominant role in such processes.

In a dry surface, the barrier height can be varied by the use of gaseous ambients and transverse fields. The analogous methods for a wet surface are variation of the composition of the electrolyte and application of bias voltages between the semiconductor and the electrolyte. These two procedures, as applied to an n-type silicon electrode, are illustrated in Fig. 9.20, taken from Harten [319,320]. In Fig. 9.20a, measured values of surface conductance $\Delta\sigma$ (points) are plotted against the electrode potential V_E (that is, against the voltage between a calomel reference electrode and the semiconductor bulk). Here V_E was varied by the addition of different amounts of cerium ammonium sulphate to the electrolyte (sulphuric acid). The solid curve represents the theoretically calculated relation between $\Delta\sigma$ and the surface potential ϕ_s (see Ch. 6, § 2). The V_E and ϕ_s scales (upper and lower abscissae) are the same except for a constant displacement of one with respect

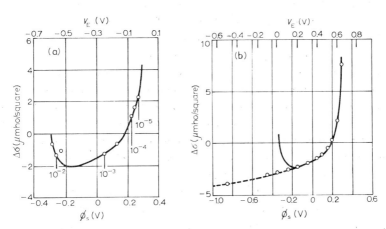

Fig. 9.20. Surface conductance $\Delta\sigma$ as function of electrode potential V_E (upper abscissa) and surface potential ϕ_s (lower abscissa) for n-type silicon electrode ($\rho = 250$ ohm · cm). Solid curves represent theoretical relation between $\Delta\sigma$ and ϕ_s.
(a) V_E varied by addition of different amounts of cerium ammonium sulphate to electrolyte (H_2SO_4). Numbers near points indicate molar concentrations of salt.
(b) V_E varied by external voltage.
(After Harten, references 319 and 320.)

to the other. The good correspondence between the points and the curve shows that changes in V_E introduced by different salt concentrations are precisely reflected as changes in ϕ_s (that is, $\delta V_E = \delta\phi_s$). A similar behaviour is apparent in the data of Fig. 20b, where an external bias was used to vary the electrode potential. Here, however, the measured values coincide with the theoretical curve only in the range of accumulation and depletion layers. At negative biases, holes can flow readily from the semiconductor into the electrolyte, so that an inversion layer cannot form. Instead, a Schottky barrier, in which both majority and minority carriers are depleted, is established, and is maintained up to very large reverse biases. In this range, $\Delta\sigma$ should decrease approximately parabolically with decreasing barrier height (see eqs. (4.26), (4.43), (4.57)), as is indeed observed. Such good agreement between experiment and theory in the Schottky-barrier regime is also revealed by surface capacitance measurements on electrodes of zinc oxide [321], cadmium sulphide [322], germanium, and silicon [323, 324]. In most of these cases the variation of measured capacitance with applied bias follows closely over an extended range of bias the theoretical variation of space-charge capacitance with barrier height (see eqs. (4.22), (4.23), (4.26)). This behaviour indicates the absence of active surface states in the regions of the gap explored (see below). If the semiconductor electrode is sufficiently thin, its carriers can be depleted completely by the application of high reverse biases [325].

The potential drop V_H across the Helmholtz layer is unmeasurable. Its variation δV_H under different surface conditions, however, can be determined by comparing the accompanying changes in electrode potential and surface potential. The measurements discussed above show that V_H is not appreciably affected by the salt concentration in the electrolyte or by the applied bias, the variation in V_E being completely taken up by the space-charge region. In germanium, on the other hand, changes in V_H are usually found to occur as either the pH of the electrolyte [326-328] or the applied bias [328-330] is varied. In the former case, V_H is displaced by about 60 mV per decade change in hydrogen-ion concentration. In the latter, δV_H is a function of the amount of charge that flowed through the system. High-speed measurements [331] were found to minimize the change in V_H. Brattain and Boddy [328-330] have suggested that the displacement in V_H might be due to changes in the surface dipole associated with adsorbed water molecules, envisaged as attached to the germanium atoms via hydroxyl groups. The structure of the adsorbed layer, and hence the degree of dipole orientation, would be expected to be strongly affected by the current flow. Changes in V_H observed for different crystallographic planes (under identical conditions) have been explained on a similar basis [330]: the surface dipole associated with the oriented water molecules varies from one plane to the other because of the different densities of adsorption sites in each case (see Ch. 3, § 3.2). Another contribution to the variation of V_H with applied bias would be expected [332] to arise from the charging of surface states (if present). Pleskov and Krotova [333] have explained their surface-conductance measurements on germanium on this basis. In Brattain and Boddy's experiments [329, 330], the only surface states that could affect V_H were slow states, since the surfaces studied were shown to be devoid of a significant density of fast states (see below). No such clear-cut distinction is apparent in the data of Pleskov and Krotova [333]. One should also keep in mind that a sizable fraction of the applied voltage can be accommodated in the oxide film if its thickness exceeds a few monolayers [320]. This effect would be very difficult to distinguish from that due to surface states.

Surface capacitance measurements have been very widely used in the study of the S–E system. They represent in effect the analogue of field-effect studies on dry surfaces. The inverted-bell shape of the differential capacitance *versus* bias curves, first reported by Bohnenkamp and Engel [334], has been observed repeatedly in germanium [260, 326, 328–331, 335, 336] and in silicon [335, 337]. In general, the measured capacitance is larger than the calculated space-charge capacitance, indicating a contribution from surface states. This is supported by the observed [334–338] decrease in capacitance with increasing frequency of the ac signal (or, in the case of pulse measurements, with decreasing duration of the pulse). In some cases [260, 328, 337, 339], the surface states manifest themselves as a well-defined peak in capacitance, the measured curves having the same general form as the theoretical curve illustrated in Fig. 6.6 (p. 223).

Surface recombination processes at the S–E interface have been found to be very similar to those on dry surfaces. Bell-shaped curves of s *versus* applied bias were first observed by Harten [340] on silicon electrodes. In a later paper [320], the surface conductance was measured as well in order to determine the surface potential ϕ_s. The plot of s as a function of ϕ_s for an n-type sample revealed the predominance of a single set of recombination centres, situated at about 0.1 eV above the midgap. The width of the s curve, however, was much smaller than that in the corresponding case of germanium. This has been attributed [21] to the fact that the excess-carrier density used in the measurement was necessarily larger than the equilibrium minority-carrier density since the latter has an extremely low value in silicon. As we have seen in §§ 1.3, 2.1, such conditions can give rise to a considerable narrowing of the (small-signal) s curve. Detailed measurements carried out by Memming [341] on silicon electrodes have reconfirmed this behaviour.

Considerable progress in our understanding of the electrical properties of the S–E interface has been achieved by Brattain and Boddy, who succeeded first in producing an essentially trap-free germanium surface [329, 342] and then in deliberately introducing surface states under controlled conditions [260, 343]. The trap-free surface was obtained in a carefully purified solution of 0.1M K_2SO_4 kept at pH 7.4 by a phosphate buffer; a purified helium or nitrogen atmosphere was maintained over the electrolyte. In order to eliminate highly adsorptive impurities, the solution was gettered by stirring with finely crushed germanium for several hours before use. Anodic etching *in situ* was employed to remove the first few (or more) atomic layers of the crystal.

The trap-free character of surfaces prepared in this manner was established from combined measurements of differential surface capacitance c_s, small-signal surface photovoltage δv_s, surface conductance $\Delta\sigma$, and surface recombination velocity s — all as a function of electrode potential V_E. With the assumption that $\delta V_E = \delta\phi_s$, the measured c_s *versus* V_E curve matches the theoretically calculated c_{sc} *versus* ϕ_s curve over an extended range of bias, apart from a vertical displacement of the two curves by a factor of 1.3. This last was interpreted as the "roughness factor" — that is, the amount by which the actual electrode surface is greater than the geometric surface — in agreement with Law's estimate [167] derived from gas-adsorption data. The surface-photovoltage curves are similar to those obtained for dry surfaces (Fig. 9.2), but now agreement with theory extends over almost the entire range of surface potential studied. (The measurements were taken under pulsed-light conditions in order to suppress slow changes in surface potential caused by light-sensitive interface reactions.) Less accord between theory and experiment is exhibited by the surface-conductance data. This has been attributed to parallel conduction through the electrolyte and to variation in potential along the electrode introduced by the measuring current [344]. The effective lifetime was found to be essentially constant over the accessible range of electrode potential and equal to the bulk lifetime, clearly indicating that surface recombination is negligible. From these data, Brattain and Boddy conclude that the density of surface states must be less than 3×10^9 cm^{-2} within about 8 kT of either side of the midgap. Similar conditions hold [328] over a range of pH values extending from 4.5 to 11.5. Outside this range, surface states of unidentified origin were detected, even when special precautions were taken in purifying the electrolyte. It was suggested that these states may be associated with some structural defects intrinsic to the surface. Trap-free conditions probably correspond to most of the free bonds of the surface atoms, including those associated with such defects, being saturated by hydroxyl groups [326].

The trap-free surface in contact with an electrolyte is no doubt the best configuration for checking the theoretically derived characteristics of the space-charge region. At the same time, it is particularly suitable for investigating the surface states introduced by various impurities. The effects of copper, silver, and gold ions on the electronic structure have received particular attention [260, 343]. In all three cases, minute traces ($\approx 10^{-7}$ mole/litre) of the contaminants are sufficient to give rise to surface states with densities of the order of 10^{11} cm^{-2}. Typical results of differential capacitance and surface recombination obtained by Boddy and Brattain [260] on a p-type germanium electrode before and after the addition of copper nitrate to the highly purified electrolyte are shown in Fig. 9.21. The data are plotted against the dimensionless surface potential u_s. (In the original paper, the barrier height V_s was used as abscissa.). The values of u_s (or V_s) were derived from surface-conductance measurements, the presence of surface states making capacitance and photovoltage data inadequate for this purpose. In order to circumvent the difficulties arising from the parallel electrolyte conduction mentioned above, use was made of an empirical relation between the *measured* conductance and the surface potential obtained for the uncontaminated (trap-free) surface.

The additional capacitance introduced by different concentrations of cupric ion can be obtained by subtracting the space-charge capacitance (lower curve in Fig. 9.21a) from the total measured capacitance (upper curves), and is plotted in Fig. 9.21b for one of the concentrations used. For values of u_s within several units on either side of zero, the results can be accounted for by a pair of discrete sets of surface states 2.2 kT below and 1.2 kT above E_i, as demonstrated by the good fit between the experimental points and the solid theoretical curve. The latter represents the combined effect of the two sets (dashed curves), as calculated from (6.28). An additional capacitance is also apparent at more extreme values of u_s, but its measurement is rendered difficult because of the relatively large space-charge capacitance in these regions. In Fig. 9.21c, the reciprocal of the sample effective lifetime is plotted against u_s for the same cupric ion concentrations. The corresponding plot of s/s_{max} *versus* u_s derived from these measurements is shown in Fig. 9.21d. The solid curve represents the theoretical expression (5.85), and corresponds to a set of recombination centres situated at either 4.8 kT or 1.2 kT above E_i. In the case of dry germanium surfaces, the experi-

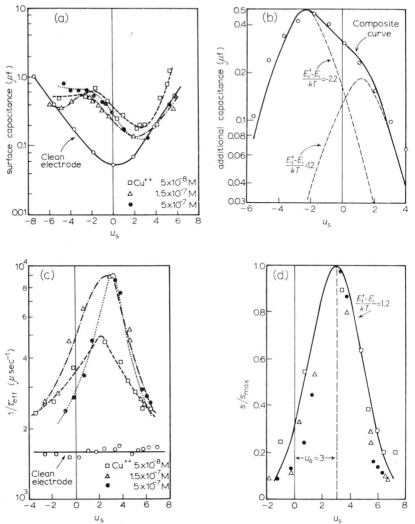

Fig. 9.21. Combined measurements of surface capacitance and surface recombination velocity as functions of surface potential u_s for p-type germanium electrode before and after addition of copper nitrate to electrolyte.

(a) Differential surface capacitance.
(b) Differential surface-state capacitance.
(c) Reciprocal of sample effective lifetime τ_{eff}.
(d) Surface recombination velocity s (relative to its maximum value $s_{\max}$). Solid curve represents theoretical expression (5.85).

(After Boddy and Brattain, reference 260.)

mental evidence pointed to the former value as the proper choice (see § 2.1 above). Boddy and Brattain, on the other hand, favour the alternative assignment, since the value 1.2 kT matches that of one of the energy levels deduced from the capacitance data. They point out, however, that the sensitivity of their capacitance measurements for large u_s (where c_{sc} is relatively high) is not sufficient to rule out the presence of surface states at 4.8 kT above E_i. In fact, the data in this range do exhibit a capacitance in excess of that of the space-charge layer. Thus, the question as to whether the set at 1.2 kT above E_i does or does not contribute to the recombination process cannot be answered unequivocally from the available data. All that can be definitely stated is that the set at 2.2 kT below E_i is ineffective in recombination.

Surface states with similar characteristics are formed when silver or gold ions are added to the electrolyte [343]. Here, as well, a pair of levels dominates the central region of the gap and an acceptor-like recombination level controls the recombination process. The latter was identified, as before, with the upper member of the pair of central levels. The various surface-state parameters obtained by averaging over several experiments on both n- and p-type electrodes, are shown in Table 9.3.

TABLE 9.3
Parameters of surface states formed on germanium electrodes
by copper, silver, and gold ions
(After Brattain and Boddy, reference 343.)

Ion	$\dfrac{(E_{t1}^f - E_i)}{kT}$	N_{t1} (cm^{-2})	$\dfrac{(E_{t2}^f - E_i)}{kT}$	N_{t2} (cm^{-2})	A_p (cm^2)	A_n (cm^2)
Cu^{++}	-1.9	1.0×10^{11}	1.2	1.9×10^{10}	6.0×10^{-14}	1.5×10^{-16}
Ag$^+$	-4.1	1.9×10^{11}	-0.55	1.0×10^{11}	2.0×10^{-14}	7.4×10^{-17}
Au^{+++}	-0.9	1.8×10^{11}	2.5	4.5×10^{10}	4.1×10^{-15}	1.4×10^{-16}

Brattain and Boddy's results are very similar to those obtained by Frankl [262] on copper-contaminated dry surfaces (see § 2.2 above). In the present case as well, the density of states is of the order of 10^{11} cm^{-2} even though there is evidence [345] of monolayer coverage (that is, 10^{15} atoms/cm^2) by the metal impurity at the concentrations used. In fact, little or no correlation exists between the state density and the concentration of metal atoms adsorbed at the surface, as has been directly demonstrated by Memming [346] using radioactive-tracer techniques. It appears then that only those atoms deposited (in the form of clusters) at select sites are effective. This suggests once again that the surface states are associated with interface defects activated by the metal ions. The fact that the parameters of the states formed by copper, silver, and gold are somewhat different (see Table 9.3) may be due to the different environment of the defect site in each case.

That the mechanism of defect activation is by no means a simple one is revealed by Memming's measurements on the effect of the electrolyte acidity[346] and of illumination [347] on the surface-state formation by metal ions. Raising the pH of the electrolyte from 5 to 9 was found to increase the surface-state density by about a hundredfold without affecting to any appreciable extent the amount of metal deposited at the germanium electrode. In the case of extrinsic p-type samples ($\approx$ 1 ohm · cm) and low copper-ion concentrations ($\lesssim 10^{-6}$ M), illumination was observed to enhance the rate of surface-state formation, again with no noticeable change in the amount of metal deposition. One must conclude from these measurements that the extent to which the defect centres are activated depends not so much on the mere presence of the metal species as on the *conditions* under which they are deposited at the surface. Why large pH values, and sometimes illumination as well, should promote the activation process is not clear.

Studies of the effects of other species on the surface-state structure of germanium electrodes are less quantitative in nature. Pleskov [348] observed an increase in surface recombination velocity on germanium following adsorption of electrolytically-discharged hydrogen. This was attributed to the penetration of hydrogen atoms into the germanium lattice so as to form recombination centres. Using a modified field-effect experiment, Green et al.[349] concluded that the surface states associated with adsorbed hydrogen are very dense ($\gtrsim 10^{14}$ cm^{-2}) and are situated 0.22 eV above the midgap. An entirely different interpretation of the role of hydrogen has been put forward by Harvey [350]. Harvey considers the increase in surface recombination velocity by the passage of cathodic current as being due, at least in part, to the removal of oxygen by electrolytic reduction. This conclusion is based on the observation [350,351] that dissolved oxygen gives rise to a decrease in s while its removal (by prolonged bubbling of nitrogen) results in an increase in s. Furthermore, no change in s was detected when an oxygen-free solution had been saturated with hydrogen gas. Harvey[350] suggests that, if anything, adsorption of atomic hydrogen gives rise to a *decrease* in the density of recombination centres. These results are very difficult to reconcile with the extremely large surface-state density observed by Green et al.[349] under what appear to be similar conditions.

Krotova and Pleskov [336] employed a KBr solution in methylformamide

Ch. 9, § 3   CLEAN SURFACES   425

as the electrolyte in contact with the germanium electrode. This has the advantage that electrochemical reactions are essentially absent. Capacitance and photovoltage measurements suggest the presence of surface states (density 10^{11}–10^{12} cm^{-2}) well removed from the midgap. An acceptor-like recombination level near the midgap is also indicated.

3. Clean surfaces

Of the various types of clean surfaces described in Ch. 3, §§ 1–3, those produced by argon bombardment and subsequent annealing and by cleavage in ultrahigh vacuum have proved to be the most suitable for electrical measurements. In both cases the surfaces obtained are characterized by an ordered lattice structure and are practically free of adsorbed foreign matter. Accordingly, one would expect the study of their electrical properties to provide direct and basic information on the electronic behaviour of the surface. So far, however, the data available on clean surfaces are not nearly as detailed as those on real surfaces. This is due in part to the many experimental difficulties encountered in producing and handling clean surfaces. With the continuous advance in high-vacuum technology, these difficulties are gradually being overcome, and more and more of the measuring procedures developed for real surfaces are becoming adaptable to clean surfaces. Progress is now impeded not so much by technique as by the inherent properties of the clean surface: in both germanium and silicon the surface states are so dense that they firmly anchor the Fermi level at the surface. As a consequence, the large swing in barrier height essential for exploring the surface-state distribution cannot be realized, and little is known about the characteristics of the states apart from rough estimates of their density and energy position. This is in marked contrast to the case of fast states on real surfaces, the situation being in fact rather analogous to that of the slow states.

The various electrical measurements used in the study of clean surfaces are outlined in § 3.1. The results obtained from such measurements on germanium and silicon surfaces produced by bombardment and annealing and by cleavage are reviewed in the next four subsections. In § 3.6 we consider the effect of gas adsorption on the electrical properties of the surface.

3.1. EXPERIMENTAL PROCEDURES

Electrical measurements on clean surfaces fall into two main categories. The first encompasses transport parameters such as surface conductance, field-effect mobility, and surface recombination velocity (see Ch. 6, §§ 2, 7 and Ch. 7, § 1). Since transverse fields are ineffective in varying the surface potential, recourse is usually had to ambient gases. One or more of the parameters listed above is first measured on the clean surface. Next, oxygen or some other gas is introduced into the high-vacuum system and the resulting changes are followed as a function of time. Such a procedure cannot yield any direct information on the states originally present on the clean surface since these are drastically altered by gas adsorption. The main purpose served by the ambient gas is to drive the surface to the point of minimum conductance and thus permit the absolute determination of the *initial* surface conductance $\Delta\sigma$ — the conductance associated with the *clean* surface (prior to the admission of the gas). In calculating $\Delta\sigma$ one should eliminate from the data any changes in conductance induced by the gas at the *uncleaned* (or uncleaved) faces of the sample. Such changes may be estimated from a control run before applying the cleaning procedure. Obviously, accurate results can only be obtained if the contribution of the unclean faces to the overall sample conductance is small compared to that of the clean surface under study. The use of ambient gases can be dispensed with in the case of clean surfaces on p–n–p or n–p–n junction structures, since there $\Delta\sigma$ is available without recourse to the conductance minimum (see Ch. 6, § 8). The measured value of $\Delta\sigma$ can be used to calculate the surface potential u_s of the clean surface, the surface mobilities being estimated theoretically (see Ch. 8, § 3.2) or else obtained directly from Hall-effect measurements (see Ch. 8, § 5.2). Using the theoretical calculations of Ch. 4, § 2, one can then derive the space-charge density $q(\Delta P - \Delta N)$ which, from the neutrality condition, must be equal (and opposite in sign) to the density of charge stored in the surface states.

All measurements of field-effect mobility reported so far have been carried out at sufficiently low frequencies for the surface states to be in complete equilibrium with both bands (see Ch. 6, §§ 5,7). Slow states have never been observed on clean surfaces, and equilibrium conditions usually obtain for any frequency from zero (dc) to several kilocycles. Throughout this range $\mu'_{fe} = \mu_{fe}$, and we can rewrite (6.50) in the form

$$\mu_{fe} = -\frac{d(\Delta\sigma)}{dQ_s} = -\frac{d(\Delta\sigma)/du_s}{(dQ_{sc}/du_s)+(dQ_{ss}/du_s)}. \tag{9.8}$$

(The minus sign enters because here Q_s denotes the charge density at the semiconductor surface.) The sign of the field-effect mobility indicates immediately whether the surface is n or p type, while its magnitude provides an estimate of the surface-state density near the Fermi level. The derivatives $d(\Delta\sigma)/du_s$ and dQ_{sc}/du_s for any given value of the surface potential u_s can readily be calculated from the equations and graphs presented in Ch. 4, §§ 2, 3 and Ch. 6, § 2. Having determined u_s and μ_{fe} experimentally, one can thus obtain dQ_{ss}/du_s, the rate of change with surface potential of the charge trapped in surface states. A few limiting cases, useful in the analysis of field-effect data on clean surfaces, will now be considered.

For $|\mu_{fe}|$ very small compared to the free-carrier surface mobilities, $|dQ_{sc}/du_s| \ll |dQ_{ss}/du_s|$, and unless $\Delta\sigma$ is too close to $\Delta\sigma_{min}$ we obtain from (9.8)

$$\frac{d(Q_{ss}/q)}{du_s} \approx -\frac{1}{q\mu_{fe}} \frac{d(\Delta\sigma)}{du_s}. \tag{9.9}$$

For strong p-type surfaces we have $|d(\Delta N)/du_s| \ll |d(\Delta P)/du_s|$ and $\Delta\sigma \approx Q_{sc}\mu_{ps} \approx q\Delta P\mu_{ps}$ (see Ch. 4, § 3 and eq. (8.2)), so that (9.8) reduces to

$$\frac{d(Q_{ss}/q)}{du_s} \approx -\left(\frac{\mu_{ps}}{\mu_{fe}} + 1\right)\frac{d(\Delta P)}{du_s}. \tag{9.10}$$

Equation (9.10) assumes a particularly simple form when the surface is not degenerate (but still strongly p type). In this range ΔP varies with u_s very nearly as $\exp(-\tfrac{1}{2}u_s)$ (see eqs. (4.56), (4.59)), so that $d(\Delta P)/du_s \approx -\tfrac{1}{2}\Delta P$. We then have

$$\frac{dQ_{ss}}{du_s} \approx \tfrac{1}{2}\Delta\sigma\left(\frac{1}{\mu_{fe}} + \frac{1}{\mu_{ps}}\right). \tag{9.11}$$

Since the derivative dQ_{ss}/du_s is usually available for just a single value of u_s (corresponding to the clean surface), an estimate of the surface-state density is possibly only if special assumptions are made regarding the energy distribution of the states. The simplest assumption is that of a continuous distribution, in which case

$$\frac{d(Q_{ss}/q)}{du_s} = -kT\overline{N}_t, \tag{9.12}$$

where $\overline{N}_t$ is the surface-state density (per unit energy) near the Fermi level. Both the magnitude and the sign of the total charge Q_{ss} stored in the surface states will depend on the position of the Fermi level at the surface with respect to the so-called neutral level E_n. The latter is defined [24] as the energy at which the Fermi level must intersect the surface in order for the overall charge in the surface states to be zero. For a *uniform* distribution ($\overline{N}_t$ = const) beginning several kT below E_F, we obtain (see similar calculation in Ch. 4, § 5.2)

$$Q_{ss}/q = -(\Delta P - \Delta N) = \overline{N}_t(E_n - E_F). \tag{9.13}$$

For the case in which the continuous distribution degenerates into a single discrete level coinciding with the Fermi level, we obtain from (5.44) and (5.45)

$$\frac{d(Q_{ss}/q)}{du_s} = -\tfrac{1}{4} N_t, \tag{9.14}$$

where N_t is the total density of surface states. Comparison of (9.12) and (9.14) shows that the continuous and discrete distributions lead to the *same* value of μ_{fe} provided that $N_t = 4kT\overline{N}_t$.

The second category of clean-surface studies consists of work-function and photoelectric-yield measurements (see Ch. 7, §§ 2, 3). Such experiments are usually carried out as a function of either gas adsorption or bulk resistivity. The former procedure is of interest as a means of investigating adsorption processes, mainly through their effect on the electron affinity. The latter procedure, which is really an extension of the one employed in the early studies of real surfaces (see Ch. 1, § 1), has proved particularly useful in estimating the surface-state distribution on clean surfaces. By using samples of different resistivities, from degenerate p type to degenerate n type, the Fermi level in the bulk can be moved from below E_v to above E_c. In the absence of surface states, the work function W_ϕ would change by the same amount. In the presence of surface states, on the other hand, W_ϕ will in general vary less rapidly, due to the tendency of the states to clamp the Fermi level, and a potential barrier will form at the surface.

The surface potential for each doping may be obtained by combining work function and photoelectric threshold data. On the assumption that photo-emission from the valence band rather than from surface states dominates the yield near threshold, the threshold energy will be equal to $\chi + E_g$ provided that two conditions are satisfied: E_F must lie well inside the band edges, and the band bending over the escape depth of the emitted electrons must be negligible (see Ch. 7, § 3). As we shall see in §§ 3.3, 3.5, this last condition is satisfied for high resistivity samples of both germanium and silicon.

Once measured, $\chi + E_g$ will remain constant for all other dopings, except for small variations at heavy dopings. The surface potential can then be derived from the relation (see, for example, Fig. 9.10, p. 376)

$$q\phi_s = \chi + E_g - W_\phi - (E_i - E_v). \tag{9.15}$$

For each pair of (ϕ_s, ϕ_b) values, one can calculate the space-charge density (see Ch. 4, § 2) and hence the charge density Q_{ss} in the surface states. In this manner one obtains a plot of Q_{ss} *versus* ϕ_s over the range of surface-potential values covered by the variation in bulk doping. It should be noted, however, that whereas *changes* in ϕ_s can usually be measured with an accuracy of several millivolts, the *absolute* value of ϕ_s may be in error by as much as one or two tenths of a volt. This error is incurred partly in the derivation of W_ϕ from the contact potential, which involves the uncertainty in the work function of the reference electrode (measured photoelectrically), and partly in the determination of the photoelectric threshold.

Analysis of the photoelectric-yield *versus* wavelength curves can provide information on the transition processes dominating the various spectral ranges explored: whether they involve purely bulk, barrier-assisted, or surface-state excitation (see Ch. 7, § 3). Such an analysis, however, relies heavily on the theoretical models assumed, and the interpretation of the data is not always unambiguous. It is particularly difficult to distinguish between valence-band emission and surface-state emission, and in fact different workers have given one or the other interpretation to essentially the same data (see § 3.5). As will be shown below, this uncertainty is reflected mainly in the value derived for the neutral level E_n, but it does not affect too much the general conclusions that can be drawn from the data concerning the surface-state distribution about E_n.

3.2. BOMBARDMENT–ANNEALED GERMANIUM

It is generally agreed upon that the bombardment–annealed germanium surface is strongly p type, irrespective of bulk resistivity. Surface-conductance measurements [352-355] yield a value of about 10^{12} cm^{-2} for the excess hole density ΔP in the space-charge layer, from which one can infer the presence of an (approximately) equal density of occupied surface states near the valence-band edge. Upon the introduction of oxygen, $\Delta\sigma$ shows a small increase followed by a decrease towards a value below that of the clean surface. These changes are mostly irreversible, as to be expected from a chemisorption process. The initial increase in $\Delta\sigma$ indicates that the surface first becomes more strongly p type, and only with prolonged exposure to oxygen does it gradually move towards intrinsic conditions (see § 3.6 below). These conclusions are supported by Hall effect data [356], as discussed in Ch. 8, § 5.2.

All recombination velocity measurements on the clean surface yield much the same results, the values quoted for s lying within fairly narrow limits (250–600 cm/sec). The effect of oxygen adsorption on s appears to depend on the annealing treatment given to the surface. Samples that have been annealed for short periods near 500°C exhibit a slow rise in s as the exposure time to oxygen is increased [352, 357, 358]. In the case of (100) surfaces annealed over prolonged periods (15 hours) at temperatures of 550°C or higher, little or no change is observed [355, 359] even at oxygen pressures as high as 5×10^{-5} mm Hg. For (111) surfaces treated similarly, on the other hand, s is found [355] first to decrease and then to increase upon further adsorption. The minimum in s occurs at about the same oxygen pressure at which $\Delta\sigma$ attains its maximum value (see Fig. 9.22 below).

There is far less accord among the various measurements of field-effect mobility, values of $|\mu_{fe}|$ reported by different workers [352-355, 358, 360] ranging from 20 to 500 cm^2/V·sec. The field-effect mobility is the most direct measure of the surface-state density, and the large spread in its values is a strong indication that the different surfaces investigated were far from identical. One is led to suspect that the degree of cleanliness (and possibly the extent of annealing) varied over fairly wide limits from one sample to the other. And indeed, more recent investigations [361] have shown that unless special precautions are taken, boron originating in the glass envelope may contaminate the surface by introducing a p-type film several microns deep (see § 3.4 below). This hitherto unsuspected effect may well have been present in some of the earlier experiments, particularly those in which the samples were initially outgassed at elevated temperatures [358, 360]. Ion bombardment at high argon pressures [358] may be another source of surface contamination. It appears [362] that under these conditions, some of the germanium atoms removed by the bombardment are scattered by the gas back onto the surface.

The different character of clean (111) and (100) surfaces was clearly demonstrated by Forman [354]. For (111) surfaces, the conductance minimum $\Delta\sigma_{min}$ could actually be reached (in wet nitrogen), as was evidenced also by the change in sign in field-effect mobility, leading to a value of about 130 μmho/square for the conductance of the clean surface. The barrier height estimated from $\Delta\sigma$ by the use of Schrieffer's mobility calculations was $-12\,kT/q$, corresponding to a p-type degenerate surface ($E_v - E_F \approx 1\,kT$). In the case of (100) surfaces, $\Delta\sigma_{min}$ could not be reached but $\Delta\sigma$ was shown to be at least 400 μmho/square. The (100) surface is thus more strongly degenerate than the (111) surface, indicating a larger density of occupied surface states in the first case. A possible explanation for this anisotropy [354] is that the (100) surface has a larger density of surface states than a (111) surface because of the greater number of dangling bonds (see Ch. 3, § 3.2).

Forman's results are substantiated by the more extensive measurements of Margoninski [355], who studied the variation in both $\Delta\sigma$ and s with oxygen adsorption. Margoninski, in contrast to Forman, did not use a dielectric spacer in the field-effect configuration, a precaution that eliminates any possible contamination of the clean surface by another solid in contact with it. Moreover, special care was taken to avoid boron contamination, a problem of which Forman was not aware. Using strong transverse fields after the samples had been exposed to room air, Margoninski was able to observe the conductance minimum for both (111) and (100) surfaces. Typical results obtained on (111) surfaces of an n-type sample are shown in Fig. 9.22.

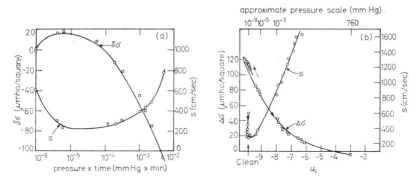

Fig. 9.22. Effect of oxygen adsorption on conductance $\Delta\sigma$ and on recombination velocity s of clean (111) germanium surface.
(a) s and change in $\Delta\sigma$ *versus* product of oxygen pressure and exposure time.
(b) s and $\Delta\sigma$ *versus* surface potential u_s.
(After Margoninski, reference 355.)

In Fig. 9.22a, s and $\Delta\sigma$ are plotted against the product of oxygen pressure and exposure time. (The changes in $\Delta\sigma$ are related to an arbitrary zero before the admission of oxygen.) The two quantities (measured on a different sample) are plotted in Fig. 9.22b as a function of surface potential u_s, the latter having been derived by reference to $\Delta\sigma_{min}$ and with the assumption of negligible surface scattering ($\mu_{ps} = \mu_p$). On this basis, the surface potential at the clean surface was estimated at $-10\ kT/q$ (p type but not degenerate). Following oxygen adsorption, u_s first decreases slightly and then increases steadily towards zero. A similar behaviour is exhibited by the (100) surface, but u_s at the clean surface is more negative ($-14\ kT/q$), corresponding to a stronger inversion layer (in this case, degenerate). Actually, however, the neglect of surface scattering in the derivation of u_s is not justified (see Ch. 8, § 5). The use of surface rather than bulk mobility will shift the calculated values of u_s to the left—the larger $|u_s|$, the greater the shift. The surface potentials for the clean (111) and (100) surfaces are thus more likely to be $-12\ kT/q$ (slightly degenerate) and $-17\ kT/q$ (highly degenerate), respectively.

The drop in the recombination velocity s on the (111) surface during the first stage of adsorption is probably associated with the initial increase in $|u_s|$, which limits still further the flow of electrons to the surface. No quantitative conclusions as to the characteristics of the recombination centres involved can be drawn from the data, however, since adsorption affects not only the surface potential but also, and mainly, the density of surface states (see § 3.6 below). A rough estimate [355] of the density of recombination centres may be obtained if one assumes the centres to be near the Fermi level and to have normal cross sections ($\approx 10^{-15}\ cm^2$). Taking $s \approx 500\ cm/sec$ and $u_s \approx -12$, one derives from (5.83) a density of the order of $10^{14}\ cm^{-2}$, which is comparable to the density of surface states indicated by the field-effect data (see below).

The values obtained by Margoninski [355] for ΔP and μ_{fe} were $5 \times 10^{11}\ cm^{-2}$ and $-80\ cm^2/V\cdot sec$ on (111) surfaces, and $2.4 \times 10^{12}\ cm^{-2}$ and $-165\ cm^2/V\cdot sec$ on (100) surfaces. On the assumption of a continuous distribution of surface states near E_F, the density $\bar{N}_t$ for the two orientations was estimated at 2.4×10^{14} and $5 \times 10^{14}\ cm^{-2}eV^{-1}$, respectively. These estimates are based on (9.10) and (9.12) but with the surface mobility μ_{ps} set equal to the bulk mobility μ_p. This does not make a great deal of difference because the effect of using μ_{ps} rather than μ_p in (9.10) would be partially compensated for by the larger value of ΔP to be derived on this basis from the conductance data. In fact, for a non-degenerate surface, $\bar{N}_t$ is practically independent of the value assumed for the surface mobility, as can be seen from (9.11). Since $|\mu_{fe}|$ is small compared to μ_{ps}, $\bar{N}_t$ depends essentially only on $\Delta\sigma$ and μ_{fe}, both of which are directly measurable quantities not involving μ_{ps}

explicitly. Referring to (9.13), we see that an increase in $|\phi_s|$ by about 0.01 V is sufficient to reduce to zero the charge stored in the surface states. This places the neutral level E_n at less than 1 kT below the Fermi level at the surface — that is, at approximately 1 kT above and 4 kT below E_v on the (111) and (100) surfaces, respectively.

Some support for the presence of a continuous distribution of surface states is provided by Handler and Portnoy's measurements [362] of the temperature dependence of $\Delta\sigma$ and μ_{fe} on clean (100) surfaces. Since the conductance minimum was not reached, only approximate values for $\Delta\sigma$ could be obtained: at each temperature, $\Delta\sigma$ was taken as the difference between the conductance of the clean surface and that following oxygen adsorption, the latter being assumed to lie close to $\Delta\sigma_{min}$. Both $\Delta\sigma$ and μ_{fe} were found to be almost independent of temperature, varying by a factor of two over the range 77–300°K. This is just what might be expected from a uniform distribution of surface states, as can be seen from (9.11)–(9.13). These results should be treated with some caution, however, since the $\Delta\sigma$ values estimated by Handler and Portnoy (150–200 μmho/square) are considerably smaller than those (600–700 μmho/square) obtained by the more quantitative measurements of Margoninski [355]. Whether this discrepancy is due to the procedure used here in estimating $\Delta\sigma$ or to boron contamination (the samples were outgassed at 500–600°C prior to bombardment) is hard to say.

A two-dimensional band model for the surface states on clean surfaces has been advanced by Handler [362, 363]. The unsaturated or dangling bonds of the surface atoms (see Ch. 5, § 2.2) are assumed to give rise to two adjacent surface bands, one lying below the valence-band edge, the other overlapping the energy gap. The lower band is filled at all times, and its occupation (by the unpaired electrons) does not lead to any surface charge. The upper band, on the other hand, is nearly empty, and the occupation of its acceptor-like states (by bulk electrons) is responsible for the observed p-type space-charge layer. The density of states $\overline{N}_t$ in such a two-dimensional band can be readily derived if one assumes circular energy surfaces. A calculation similar to that performed in the analogous case of a spherical three-dimensional band (see Ch. 2, § 3.2) then yields for $\overline{N}_t$ the constant value

$$\overline{N}_t = 4\pi m^*/h^2, \tag{9.16}$$

corresponding to a *uniform* distribution. Nothing is known about the effective mass m^* in the surface band. Taking the free-electron value, one obtains $\bar{N}_t \approx 4 \times 10^{14}$ cm^{-2}eV^{-1}, in good agreement with the experimental estimates quoted above. Unfortunately. the available data is too limited to enable a closer examination of the proposed model. In fact, one does not even know for certain whether the actual distribution is really continuous, let alone uniform. In the case of cleaved silicon surfaces, it appears rather that the Fermi level is anchored between two surface-state bands (or levels) several kT apart (see § 3.5 below). It should be noted that a spreading of the surface states into one or more continuous bands is to be expected on the basis of their large density alone, and can provide no indication as to their origin.

Conductance in such surface-state bands is possible in principle, but one would expect the mobility to be so much less than for the free carriers that the latter would predominate. And indeed, most surface conductance measurements can be fully accounted for in terms of free carrier transport in the space-charge region (see also § 3.5). Some anomalies in surface conductance and Hall effect observed by Kobayashi and co-workers [364] at very low temperatures (10–20°K) were at first attributed to conductance in surface states. However, the surfaces studied were cleaned by heating (at 800–850°C) and were probably contaminated by boron. Upon repetition of the measurements by the same group of workers [365] on *bombardment*-annealed surfaces, most of the anomalous effects disappeared, the results obtained being consistent with the accepted picture of the clean germanium surface.

As was pointed out in Ch. 4, § 2.2, the use of classical methods in calculating the shape of the potential barrier and the carrier distribution in the space-charge region may not be valid for deep potential wells such as those encountered on clean surfaces. The effective charge distance L_c in these layers is typically 100 Å or smaller (see Fig. 4.9), which means that an appreciable fraction of the excess carriers are constrained to move in a potential well that is less than 100 Å wide. When the width at an energy about 1 kT above the bottom of the well is equal to or less than the wavelength of the carriers, quantization effects become important and the carriers in this region can no longer be considered as quasi-free particles. The uncertainty principle prevents the carriers from approaching too close to the surface, so that their lowest permissible energy is somewhat removed from the band edge. More precisely, the carriers near the bottom of the well are forced into a number of discrete states, and only at higher energies will this discrete distribution merge with the normal quasi-continuum. As a result, one should expect a reduction in the number of available states near the surface, and consequently a wider potential barrier than that derived by classical theory. This requires a modification of the F and G functions (see Ch. 4, §§ 2, 3) for steep barrier wells. So far, however, no detailed calculations have been carried out along these lines, and in the processing of transport data on clean surfaces one is forced to resort to the classical expressions.

The broadening of the potential barrier will also reduce the effect of surface scattering and thus raise the surface mobility in deep wells above its calculated value (see Ch. 8, §§ 2, 3). A comparatively simple mathematical procedure for estimating the correction involved on the basis of the available classical calculations has been described by Greene [366]. It appears quite likely that the relatively high values of surface mobility observed in strong accumulation layers (see Ch. 8, § 5.1) and more recently [367] in strong inversion layers are a result of such quantization effects rather than of incomplete diffusivity of the surface. Evidence for quantization of the *light*-hole states in the space-charge layer of a clean (111) germanium surface has been obtained by Handler and Eisenhour [368] from measurements of Hall effect and magnetoresistance as functions of magnetic field intensity. The measurements were carried out first on the clean surface ($u_s \approx -11$) and then for successively larger coverages of adsorbed water vapour, which gradually reduced the band bending towards flat-band conditions. On the assumption that in the range of magnetic field investigated only the light-hole mobility is large enough to be dependent upon field intensity, Handler and Eisenhour were able to obtain the surface mobility of the light and heavy holes *separately*, both as functions of surface potential. The variation of the heavy-hole mobility was found to be consistent with the diffuse-scattering model. The light-hole mobility, however, appears to *increase* with increasing $|u_s|$, saturating at a value about four times that in the bulk. Analysis of the data shows also that as the well becomes deeper, the ratio of light- to heavy-hole density *decreases*, its value at the clean surface ($\approx 10^{-4}$) being less than a hundredth of that in the bulk. These results were attributed to quantization of the light-hole states in the space-charge region. Since the light holes have an effective mass about eight times smaller than the heavy holes, their behaviour should be affected more strongly. In comparison, one may neglect quantization effects as far as the heavy holes are concerned. To this approximation, the shape of the potential barrier is identical with that derived classically (see Fig. 4.6, p. 144), the heavy holes constituting the major source of charge in the space-charge region. Solution of a one-dimensional Schrödinger equation based on this potential then shows that the first allowed state for the light holes at the clean surface ($u_s \approx -11$) lies $4\,kT$ below the valence-band edge, the next state lies $1.3\,kT$ lower, and so on down into the continuum. Most of the light holes occupy the first level, so that the ratio of light- to heavy-hole density in the well should be roughly e^{-4} times that in the bulk. As the band bending decreases (the well becomes wider), the first allowed state approaches the valence-band edge and the density ratio increases. Both these predictions are borne out by the experimental data. The anomalously high surface mobility of the light holes, however, could not be explained as readily. It was suggested [368] that the observed change in mobility is a result of either a decrease in the effective mass due to the lifting of the valence-band degeneracy at $k = 0$ or an increase in the scattering time owing to the reduced number of final states.

Measurements of contact potential and photoelectric yield are in essential agreement with the transport data discussed above. The first attempt to determine the work function of clean germanium surfaces was made by Apker et al.[369] using mostly evaporated films. Their value of 4.78 eV for W_ϕ was later confirmed by Dillon and Farnsworth [370] for bombardment–annealed surfaces, and shown to be largely insensitive to bulk doping and surface orientation. Upon exposure to oxygen, W_ϕ first increased by about 0.2 eV and then decreased slightly with further adsorption. The *direction* of the change in W_ϕ with time is the same as that of $|\phi_s|$ derived from the surface-conductance data (see above). The *magnitude* of the change is quite different, however, indicating that the electron affinity varies as well during the adsorption process (see § 3.6 below). Dillon and Farnsworth [370] also measured the photoelectric threshold E_T as a function of oxygen adsorption. In the clean condition, E_T was found to be approximately equal to W_ϕ. On the assumption that $E_T = \chi + E_g$ (see § 3.1 above), one concludes that the clean surface is strongly p type, with the Fermi level lying close to the valence-band edge.

The insensitivity of the work function to crystal orientation and bulk doping was also observed by Allen and Fowler [371]. The contact potentials measured between clean (100), (110), and (111) faces of a germanium crystal and the *same* reference electrode did not differ by more than 0.06 eV. Variations in the bulk Fermi level by as much as 0.5 eV, achieved by varying either the doping concentration or the ambient temperature, did not produce changes in work function exceeding the experimental uncertainty of ± 0.05 eV. The strong anchoring of the Fermi level at the surface indicated by these results sets a lower limit of about 7×10^{13} cm^{-2}eV^{-1} for the density of surface states near the Fermi level. A similar density was inferred by Fowler [372] from measurements on a clean (110) surface of a p–n junction. The difference in contact potential between the heavily doped p and n sides amounted to only 0.015 eV, compared to a value of 0.34 eV that this difference would have assumed in the absence of surface states.

3.3. CLEAVED GERMANIUM

All measurements on cleaved germanium (and silicon) surfaces are of necessity restricted to the (111) faces, the cleavage plane of the crystal. Cleaved surfaces have the important advantage of being clean, at least initially, to a degree that cannot be surpassed by any other method of preparation. Their main disadvantage is that they do not represent the equilibrium configuration of the surface, which can be approached only after prolonged annealing at elevated temperatures (see Ch. 3, § 3.1). Moreover, the process of cleavage may introduce plastic deformations resulting in severe local damage near the surface [373].

Transport measurements carried out by Palmer et al.[300] indicate that the cleaved germanium surface is not appreciably different in structure from the bombardment–annealed surface. Here too the surface is found to be strongly p type ($\Delta P \approx 10^{12}$ cm^{-2}) and to have a field-effect mobility of 50–150 cm^2/V·sec. Also, on exposure to oxygen, the surface initially becomes slightly more p type and then changes continuously towards an n-type configuration. More quantitative results were obtained [374] by channel conductance measurements on cleaved p–n–p structures. As expected, cleavage produced on the n-type base the strong inversion layer characteristic of the clean surface. The density of surface states estimated from the bias-dependence of the channel conductance was somewhat lower than that observed on annealed surfaces.

An altogether different picture of the cleaved surface emerges from the measurements of Banbury and co-workers [375, 376]. Their surface-conductance data indicate that the surface is only slightly p type ($u_s \approx -2$), with a field-effect mobility of 200–300 cm^2/V·sec. On near-intrinsic n-type samples the conductance minimum is clearly seen in the ac field-effect pattern, being close to the field-free point. From the swing obtained in barrier height (up to 0.2 V), it was estimated that for a surface-state distribution concentrated near the centre of the gap, the density did not exceed 10^{12}cm^{-2}, which is comparable to the fast-state density on real surfaces. These rather surprising results however, should be considered, with some reservations. In the earlier experiments [375], the separation between the field-effect plate and the cleaved surface was 0.3 mm — comparable to the width and the thickness of the sample. As a result, the stray capacitance to the *uncleaved*, contaminated faces was not negligible in comparison with the capacitance to the cleaved surface, so that the field effect could have been strongly influenced by the contaminated faces. This deficiency was corrected in the later work of Banbury et al.[376] by the use of a thin mica spacer, but then the intimate contact of the dielectric with the cleaved surface might very well have been a source of contamination.

Both Palmer [300, 374] and Banbury [376] have noted that during the cleavage process gas is evolved from the moving parts, giving rise to momentary pressure bursts of up to 10^{-8}–10^{-7} mm Hg. Although this gas was shown [376] to be composed mainly of carbon monoxide, which is relatively non-reactive with germanium and silicon, possible contamination by some minor constituents cannot be completely ruled out.

Gobeli and Allen [377] have carried out a detailed study of photoelectric emission and work function over a wide range of bulk doping. Their results on the photoelectric yield Y *versus* photon energy $h\nu$ are shown in Fig. 9.23a.

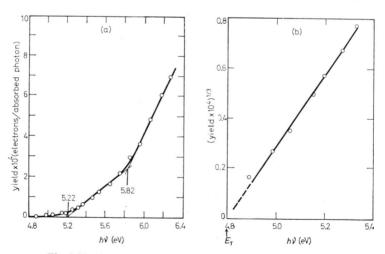

Fig. 9.23. Photoelectric yield from cleaved germanium surface.
    (a) Yield *versus* photon energy $h\nu$.
    (b) Cube root plot of low-energy tail.
(After Gobeli and Allen, reference 377.)

The yield curve was found to be almost independent of bulk doping, indicating that the effect of band bending on photoemission is negligible. And indeed, the width of the space-charge layer was shown to be large compared to the electron escape depth (20–30 Å) for all dopings used, so that the bands were practically flat over the depth below the surface from which the photoelectrons originated, irrespective of the position of the Fermi level at the surface. The yield in the low photon-energy tail region fits a cube power law as shown in Fig. 9.23b, and has an extrapolated intercept of $E_T = 4.80$ eV. The cube power law is in agreement, within experimental accuracy, with the $\frac{5}{2}$ law predicted for an indirect volume or surface-state excitation process (see Table 7.1, p. 282). Evidence has been obtained [378], however, that the distribution of the emitted electrons from both germanium and silicon depends upon the direction of polarization of the light, the effect extending well into the tail region. This indicates that the transverse momentum of the excited electron must be conserved as the electron crosses the surface barrier, which appears to rule out the indirect process since phonon scattering would be expected to randomize the distribution. Gobeli and Allen suggest that transitions in the low photon-energy range occur sufficiently close to the surface that essentially only *normal* momentum transfer (to the surface barrier) takes place during the act of absorption. The normal momentum is altered by an arbitrary amount, so that optical transitions from arbitrary valence-band states up to the top of the band at $k = 0$ are possible. On the assumption that surface-state emission is relatively unimportant, the extrapolated threshold $E_T = 4.80$ eV is then equal to $\chi + E_g$ (just as in the case of indirect excitation). The linear component at higher photon energies (with threshold at 5.2 eV) is attributed to direct optical transitions in the bulk, with *no* scattering of the emitted electrons either in the bulk or at the surface. This range and beyond it need not concern us since they provide little direct information on the surface proper.

The contact-potential data reveal no work-function changes greater than the experimental uncertainty of ± 0.04 eV, for a variation in bulk Fermi level from 0.15 eV below E_v (highly degenerate p-type samples) to 0.07 eV below E_c (strongly n type). This behaviour is similar to that exhibited by the annealed germanium surface (see § 3.2 above), and sets much the same lower limit to the surface-state density. The average value of W_ϕ (4.80 eV) is equal, within experimental uncertainty, to the threshold E_T, indicating that there is appreciable emission from filled levels immediately below the Fermi level at the surface. If surface-state emission does not play a significant role, as is deemed likely in view of the polarization evidence discussed above, this means that the Fermi level at the surface lies either close to or below the valence-band edge. How far below cannot be determined from these measurements, since valence-band states above E_F are empty and so do not contribute to photoemission. On the assumption that at the surface $E_F - E_v \approx \pm 1\,kT$ and that E_F is anchored between two groups of surface states several kT apart (see § 3.5), the total density of states was estimated to be at least 1.4×10^{13} cm^{-2}. If, on the other hand, $E_F - E_v \approx -5\,kT$, then a calculation based on a continuous distribution of surface states near E_F yields the value 4×10^{13} cm^{-2}eV^{-1} for the state density.

After the annealing of one of the p-type samples at 450°K, W_ϕ dropped from 4.79 to 4.61 eV, while E_T fell only from 4.81 to 4.74 eV. The change in E_T was tentatively attributed to a change in electron affinity resulting from the rearrangement of the surface atoms (see Ch. 3, § 3.1). In any case, it appears that annealing shifts the Fermi level to a position about 4 kT *above* E_v, making the surface considerably less p type, In view of the uncertainty involved in the *absolute* determination of W_ϕ (see § 3.1 above), this result should be considered to be in good agreement with those obtained by Forman [354] and by Margoninski [355] from the more direct measurement of surface conductance (see § 3.2 above). Much more significance, however, should be attached to the observed *change* in Fermi-level position and hence in surface-state structure between the cleaved and the annealed surface, such change being measurable with a high degree of accuracy.

3.4. BOMBARDMENT–ANNEALED SILICON

In the case of silicon, high-temperature heating (without ion bombardment) is sufficient to remove the oxide and other adsorbed foreign matter from the surface (see Ch. 3, § 1.4). Although surfaces prepared in this manner are atomically clean in a two-dimensional sense, they are usually characterized by highly conductive p-type films extending several microns into the bulk. The presence of such films was suspected by Allen [379] in his early contact-potential studies. Later investigations [361] established that these films are formed by boron contamination originating in the borosilicate glass (Pyrex) used in most vacuum systems. The boron is transferred onto the surface through volatilization of boron oxide by the water vapour present in the system during the mounting of the sample and the initial pump down. Upon heating (above 1000°C for silicon, 850°C for germanium), the deposited boron diffuses inwards to form the p-type film. The film thickness was estimated by measuring the sample conductance while successively larger amounts of material were removed from the surface by etching. Direct evidence for the presence of boron on heat-treated silicon surfaces has been obtained by radioactive-tracer techniques [380]. Dillon and Oman [381] have shown that the p-type film on both germanium and silicon surfaces can be avoided entirely if the sample is ion-bombarded *prior* to heating. Such bombardment removes the superficial boron layer before it is able to diffuse into the bulk. Care should be exercised to eliminate any traces of water vapour and to avoid bombardment of the glass walls, as both of these may lead to boron re-contamination. If the ion bombardment is carried out after the heating, then much heavier sputtering is of course necessary [361].

The adsorption characteristics of boron-free bombardment–annealed silicon surfaces have been studied in detail by Law [382-384] (see § 3.6 below). Both a uniform-resistivity sample (orientation unspecified) and a p–n junction [384] (orientation (110)) were employed. Surface-photovoltage measurements indicated that in all cases the clean surface was p type. The surface conductance was insensitive to oxygen adsorption, but decreased monotonically on exposure to *atomic* hydrogen, saturating at monolayer coverage. For the uniform-resistivity [383] sample (280 ohm·cm p type), the overall drop in conductance amounted to 30 μmho/square, from which it was inferred that $\Delta\sigma_{min}$ was at least that much below the clean-surface conductance. On this basis the clean surface is fairly strongly p type, with the Fermi level at the surface lying less than 0.12–0.14 eV above the valence-band edge. A similar conclusion was reached from the conductance data for the lower-resistivity p and n sides of the p–n junction [384]. The field-effect mobility was not measured and no estimates were given for the surface-state density.

Law's conclusion that the clean silicon surface is strongly p type is not confirmed by the more recent measurements of Heiland and Lamatsch [385]. These workers used thin cylindrical filaments (0.3 mm diameter) of 1000 ohm·cm p-type silicon. A large-diameter coaxial metal grid served both as the anode during argon-bombardment cleaning and as the field electrode in the field-effect experiment. Here, as well, all necessary precautions were taken to prevent boron contamination. Measurements of field-effect mobility and surface conductance were carried out at 89°C, the lowest temperature at which satisfactory performance was obtained from the molybdenum contacts employed. In order to achieve a swing in barrier height sufficient to include the point of minimum conductance, weak argon bombardment was used, a treatment known to bend the bands downwards [365,382]. It was established in this manner that the clean surface has a low p-type conductance of 1–2 μmho/square, corresponding to a Fermi-level position of 0.3 eV above the valence band edge (weak accumulation layer). This is to be contrasted with the value of 0.12–0.14 eV estimated by Law on the basis of the much larger surface conductance observed by him [383,384]. The field-effect mobility at the clean surface was determined as 0.1 cm^2/V · sec. Using (9.9), Heiland and Lamatsch derived from their data a value of 1.5×10^{15} cm^{-2}eV^{-1} for the surface-state density assuming the states to be continuously distributed, and a value of 10^{14} cm^{-2} assuming the states to have a discrete energy near the Fermi level. At first glance one might wonder how a field-effect mobility three orders of magnitude smaller than that observed on clean germanium surfaces gives rise to a surface-state density only ten times larger. The reason for this is that comparable changes in barrier height result in a much smaller change in $\Delta\sigma$ at the near-intrinsic silicon surface than at the degenerate germanium surface. Adsorption of oxygen was found to increase the field-effect mobility by a factor of 20 without affecting significantly the surface conductance. This means that the density of surface states was decreased by about the same factor.

The use of a cylindrical filament has the obvious disadvantage that it does not provide a well-defined crystallographic orientation for the surface. Its advantage lies in the fact that the *entire* filament surface is clean. Spurious effects that may arise from the uncleaned faces of the sample when using ordinary, planar geometry are thus completely eliminated. This feature is an invaluable asset in the case of silicon, where both the field-effect mobility and the surface conductance have extremely small values. In fact, it is rather doubtful whether such low values can be reliably measured by any other means. It appears likely therefore that the large surface conductance reported by Law [383,384] represents the contribution of the unclean rather than the clean surface.

Contact potential measurements carried out by Dillon and Farnsworth [386] show that in silicon, as opposed to germanium, the work function varies significantly with crystallographic orientation. Although at the time these workers were not aware of the danger of boron contamination, they ion-bombarded the samples *before* annealing, so that presumably their surfaces were clean. The values of W_ϕ reported for the (100), (110), and (111) faces were 4.82, 4.70, and 4.67 eV, respectively. This decreasing sequence in the magnitude of W_ϕ is in the same order as that of the corresponding surface densities of dangling bonds, and may suggest that the unfilled orbitals are partially responsible for the surface dipole [353]. In all three cases, the work function was found to be insensitive to the type of heat treatment given to the surface, even though different treatments resulted in different lattice structures [387].

Considerable effort has been devoted to the investigation of field emission from clean silicon tips [388, 389] (see Ch. 7, § 4). Most of these studies were concerned with the current–field characteristics at different temperatures and bulk dopings. Although attempts were made to obtain some information on the surface states, the results so far are inconclusive. Even the source of the emitted electrons and the role of the surface states in shielding the space-charge layer from the external field are still undetermined.

3.5. CLEAVED SILICON

Transport measurements performed by Palmer et al.[390] suggested a slightly n-type surface on both p- and n-type samples, with a field-effect mobility of about 50 cm^2/V·sec. Neither μ_{fe} nor $\Delta\sigma$ showed any detectable changes following exposure of the cleaved surface to oxygen even at atmospheric pressure. As a result, no information could be gained about the surface potential. In order to overcome this difficulty, use was made of cleaved p–n–p and n–p–n junction structures. In neither case was the channel conductance greater than 0.01 μmho/square, indicating that there was no inversion layer present at the base region. From this behaviour, Palmer et al. concluded that the Fermi level at the surface lies near the centre of the gap. As has been pointed out by Handler[391], however, a Fermi-level position anywhere within about ± 10 kT of the centre of the gap could equally well account for the absence of surface conductance. The resistivities of the n- and p-type base regions were 20 and 80 ohm·cm, corresponding to bulk potentials u_b of $+11$ and -11, respectively. Accordingly, the excess density of *minority* carriers is negligibly small for $v_s \gtrsim -20$ in the n-type base and for $v_s \lesssim 20$ in the p-type base. (See the $\Delta\sigma$ versus v_s plots in Fig. 6.2, p. 214, which are somewhat similar to those applicable in the present case.) Thus, no channel conductance would be detectable when the surface potential $u_s(\equiv v_s+u_b)$ lies in the range from -9 to $+9$. (The presence of diffuse scattering will extend these limits somewhat.) Throughout this range, the sign of the field-effect mobility should be controlled by the majority carrier — positive for n-type samples, negative for p-type. The positive sign observed by Palmer et al.[390] on uniform-conductivity samples of *both* resistivity types cannot therefore be explained. More surprising still is the large field-effect mobility (50 cm^2/V·sec), since it implies a surface-state density as low as 10^{12} cm^{-2} eV^{-1}. Even larger mobilities have been reported by Kawaji and Takishima [392] for surfaces cleaved from 9.5 ohm·cm silicon. It is very likely that in both experiments the field-effect data were distorted by capacitive coupling between the field plate and the *uncleaved* faces of the sample (see § 3.3 above). Such spurious effects would be far more difficult to avoid in silicon than in germanium because of the much smaller conductance of the silicon surface.

The results on cleaved silicon junction structures indicate that conductance in surface-state bands, if present, must be smaller than 0.01 μmho/square. On the assumption that the density of the states contributing to such conductance is comparable to the density of surface atoms, one concludes [391] that the average surface-state mobility does not exceed 10^{-4} cm^2/V·sec. The possibility does exist, however, that most of the occupied states are in filled bands. In this case only the relatively few electrons in the nearly empty upper band can contribute to the surface conductance, and a considerably higher upper limit for their mobility would have to be set. This mobility, rather than its average value, is of interest since it is mainly the occupation of the nearly empty band that would be expected to vary in the field-effect experiment.

An elegant technique for the absolute determination of the surface conductance, which does not necessitate the use of either gaseous ambients or p–n–p structures, has been described by Handler [393]. A semiconductor wafer provided with two evaporated contacts (front and back) is cleaved in such a manner that the cleavage plane intersects the contacts at right angles. Since the total contact area remains unchanged, the difference between the conductance of the severed parts and the sample conductance before cleavage should yield the surface conductance of the two cleaved faces. Handler [393] reports that experiments of this sort, performed on p-type silicon wafers of different resistivities, revealed the transition in $\Delta\sigma$ from positive to negative that is expected to occur as the Fermi level crosses the neutral point of the surface states. The zero of $\Delta\sigma$ (zero barrier height) was obtained for a bulk potential $q\phi_b$ of about -0.23 eV, which means that the neutral level lies below the midgap by the same amount.

The most extensive information to date on the electronic structure of the clean surface comes from the contact-potential and photoemission measurements of Allen and Gobeli on cleaved silicon [394–396]. In contrast to the case of cleaved germanium, appreciable shifts in the position of the Fermi level at the surface were obtained in the range of bulk dopings explored [395].

The observed variation of the work function W_ϕ with bulk potential ϕ_b is shown by the heavy curve in Fig. 9.24a. The full circles correspond to data taken directly across p–n junctions and, hence, were given the most weight in determining the relative variation in W_ϕ between the highly p- and the

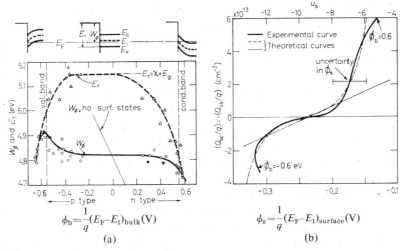

Fig. 9.24. Results obtained from combined measurements of contact potential and photo-electric emission on cleaved silicon surfaces.
(a) Work function W_ϕ and photoelectric threshold E_T *versus* bulk potential ϕ_b. Straight line represents predicted variation of W_ϕ in the absence of surface states. (Energy-level diagrams for degenerate p-type and n-type and for flat-band conditions are shown at the appropriate ϕ_b values.)
(b) Density of electrons Q_{ss}/q trapped in surface states *versus* surface potential ϕ_s. Heavy curve obtained by analysis of data in (a). Other two curves show fit to linear and to sinh relations (see text).
(After Allen and Gobeli, reference 395.)

highly n-type dopings. It is seen that while the Fermi level in the bulk moves 1.2 eV from degenerate n to degenerate p, W_ϕ changes at the extremes only from 4.76 to 4.92 eV. This once again demonstrates the strong clamping action of the surface states. (The straight line represents the path W_ϕ would have taken were there no surface states.) The sharp drop in W_ϕ at the p-type extreme was attributed to a decrease in the electron affinity χ, since no real distribution of surface states could cause ϕ_s to vary in the opposite direction from ϕ_b. The photoelectric threshold data in Fig. 9.24a (triangles) were obtained by extrapolating to zero the yield *versus hv* plots measured on the same samples. It is seen that the threshold E_T depends strongly on bulk doping but that throughout the high-resistivity range it is nearly constant and about 0.3 eV higher than the work function. The spectral-yield curve in this range is similar in shape to that observed in germanium (Fig. 9.23), being characterized by a cube (or $\frac{5}{2}$) power law at low photon energies and a sharp linear rise at higher energies. Here as well, the extrapolated threshold ($E_T = 5.15$ eV) was identified with $\chi + E_g$, so that the position of the Fermi level at the surface was fixed at 0.3 eV above the valence-band edge. As long as the bulk doping is not extreme the Fermi level is very close to E_n, the neutral level of the surface states. It coincides with E_n for $\phi_b \approx -0.22$ V ($\rho \approx 200$ ohm · cm, p type), when the charge in both the surface states and the space-charge region vanishes (flat-band conditions).

At large dopings E_T falls, and it approaches W_ϕ at both n and p extremes. This is as expected, since when the valence or conduction band crosses the Fermi level, electrons can be emitted from filled states right up to the Fermi level (see energy-level diagrams at top of Fig. 9.24a). The gradual decrease in E_T in going towards the p-type side can be quantitatively explained [396] by taking into account the effect of band bending (see Ch. 7, § 3), the fit between theory and experiment yielding a value of 20–30 Å for the electron escape depth. The slower decrease in E_T on the n side, however, begins long before emission should be possible from filled states in the conduction band. This effect was therefore attributed to emission from surface states that are progressively filled as the bulk doping becomes more strongly n type.

The next step in the analysis is to determine the surface potential ϕ_s at each bulk doping. The detailed shape of the E_T versus ϕ_b curve does not enter into the calculations; only the straight horizontal part of the curve, corresponding to the high-resistivity range, is required to determine $\chi + E_g$. On the assumption that χ and E_g remain constant for all other resistivities as well, ϕ_s can be derived from W_ϕ by the use of (9.15). Referring to Fig. 9.24a, we then see that ϕ_s varies from -0.30 to -0.14 V in going from extreme p- to extreme n-type doping. The corresponding variation in Q_{sc}/q or $-Q_{ss}/q$, obtained by the procedure described in § 3.1, is shown by the heavy curve in Fig. 9.24b. The most striking feature of the data is the very large range covered by Q_{ss}/q, from -6×10^{13} to about $+2 \times 10^{13}$ cm^{-2}. This change in Q_{ss}/q is more than two orders of magnitude greater than could possibly be induced by transverse fields. The inaccuracy in the ϕ_s position of the entire Q_{ss}/q versus ϕ_s curve is ± 0.1 V, incurred mostly in the absolute determination of W_ϕ. This inaccuracy is reflected in the value derived for E_n. It does not appreciably affect the *shape* of the curve, however, the estimated uncertainty in the relative value of ϕ_s being only ± 0.025 V. Thus, fairly accurate information can be obtained on the surface-state distribution around E_n, since the analysis of the data relies on the shape rather than on the absolute position of the curve.

The linear and dashed curves in Fig. 9.24b represent two assumed surface-state distributions. The straight line corresponds to a uniform distribution near the neutral level E_n, with $\overline{N}_t = 1.8 \times 10^{14}$ cm^{-2} eV^{-1} (see eq. (9.13)). It is seen that at best only a poor fit can be made to the central portion of the curve by such a distribution, the deviations at higher dopings being well outside the experimental uncertainty. A much better fit over the entire range of dopings is provided by the dashed curve, which corresponds to two surface-state bands on either side of E_n. The state densities $\overline{N}_{t1}$ and $\overline{N}_{t2}$ in the two bands are taken to be uniform, while the bottom E_{t1} of the upper band and the top E_{t2} of the lower band are assumed to be removed from the Fermi level by at least several kT at all bulk dopings. Under these conditions, one can readily show [395] that (see eqs. (5.44), (5.45))

$$Q_{ss}/q = -2N_0 \sinh [u_s + (E_i - E_n)/kT], \quad (9.17)$$

where

$$N_0 = kT\overline{N}_{t1} e^{-(E_{t1} - E_n)/kT} = kT\overline{N}_{t2} e^{-(E_n - E_{t2})/kT}. \quad (9.18)$$

The same expression (9.17) is obtained for the limiting case in which the bands degenerate into two discrete levels at energies E_{t1} and E_{t2} (densities N_{t1}, N_{t2}). In this case, however,

$$N_0 = N_{t1} e^{-(E_{t1} - E_n)/kT} = N_{t2} e^{-(E_n - E_{t2})/kT}. \quad (9.19)$$

The best fit for the data in Fig. 9.24b is obtained for $(E_i - E_n)/kT = 9.7$ and $N_0 \approx 1.2 \times 10^{12}$ cm^{-2}. Since the hyperbolic sine relation (9.17) will not result for this model unless E_{t1} and E_{t2} are at least 1–2 kT beyond the largest excursion of $q\phi_s$ (± 0.1 eV), the smallest value of $\overline{N}_{t1}$ and $\overline{N}_{t2}$ (or N_{t1} and N_{t2}) that will fit the data is given by 3×10^{15} cm^{-2} eV^{-1} (or 2×10^{14} cm^{-2}). On the assumption that the total density of states does not exceed the density of surface atoms (8×10^{14} cm^{-2} for the (111) face of silicon), one concludes that the width of each band must be less than about 0.16 eV. As the band edges E_{t1} and E_{t2} move beyond 0.1 eV from E_n, $\overline{N}_{t1}$ and $\overline{N}_{t2}$ will increase exponentially, and in order to maintain the constant density of 8×10^{14} cm^{-2}, the width of the bands must diminish. The largest possible separation between E_{t1} and E_{t2} obtains when the bands degenerate into two discrete levels, in which case $E_{t1} - E_n = E_n - E_{t2} \approx 0.15$ eV.

It is apparent that the exact form of the distribution function cannot be determined on the basis of the available data. All that can be stated definitely is that the surface-state density is relatively low near the neutral point and rises rapidly below and above this point. Further information may be gained by measuring the temperature dependence of the work function and through it of the neutral level E_n. Referring to (9.18) we see, for example, that E_n will depend on temperature unless $\overline{N}_{t1} = \overline{N}_{t2}$.

The photoemission properties of cleaved silicon have also been studied by van Laar and Scheer [397,398]. For high-resistivity p-type samples, the entire spectral yield could be represented by the relation

$$Y \propto (h\nu - 4.85)^{\frac{5}{2}} + 28.3(h\nu - 5.36)^{\frac{3}{2}}. \tag{9.20}$$

These results differ from those of Allen and Gobeli in two respects: the threshold of the low energy tail is 4.85 eV instead of 5.15 eV, and the steep rise at higher energies is characterized by a $\frac{3}{2}$ power law rather than a linear one. The two research teams differ also in their interpretation of the data. Van Laar and Scheer [398] ascribe the low-energy tail to emission from surface states, the threshold for this process (4.85 eV) being identified with the work function. This assignment is based essentially upon the fact that the observed $\frac{5}{2}$ power law is just what might be expected for emission from a surface-state band extending above and below the Fermi level at the surface (see Table 7.1, p. 282). In support of their conclusion, Scheer and van Laar [399] point out that the $\frac{5}{2}$ power law holds for quite a number of other semiconductors, independently of surface treatment. The second term in (9.20) was attributed [397,398] to emission from the valence band (presumably by indirect optical transitions), and the threshold of 5.36 eV was taken to represent $\chi + E_g$. A strong case against such an assignment is provided by the polarization effects observed by Gobeli et al.[378], which clearly prove the existence of unscattered electrons in photoemission (see § 3.3 above). The predominance of direct transitions at photon energies above 5.4 eV is also evidenced by the linear relation that characterizes all the yield curves of Allen and Gobeli [395,396] in this energy range. It has been suggested [395] that the $\frac{3}{2}$ power law observed by van Laar and Scheer is a result of imperfect cleavage, which might be expected to add curvature to their yield curves. And indeed, a fractured rather than cleaved surface was found to yield poor and non-reproducible results [400].

Van Laar and Scheer [397,398] did not carry out contact-potential measurements, so that no direct information on the surface-state distribution could be gained. An estimate of the position of the neutral level was obtained, however, by an analysis of the effect of band bending on photoemission for different p-type bulk dopings. It was concluded on this basis that the cleaved surface is near-intrinsic and that the electron escape depth is about 20 Å, in fairly good agreement with the results of Allen and Gobeli [395].

Allen and Gobeli [400] have studied the changes in work function, photoelectric yield, and emitted electron-energy distribution introduced by annealing of the cleaved silicon surface. The measurements were performed on 200 ohm·cm p-type samples, which should be close to the flat-band condition in the cleaved state (see above). After annealing, the work function dropped from 4.8 to 4.6 eV and the low-energy photoelectric threshold fell from 5.15 to about 4.6 eV. The results for the cleaved-and-annealed surface were found to be similar to those obtained on (111) bombardment–annealed surfaces, as expected in view of the identical lattice structures observed for the two types of surfaces (see Ch. 3, § 3.1). The important thing to note, however, is that annealing decreased $W_\phi - E_T$ from about 0.3 eV to near zero. One possible explanation is that the valence-band edge has been drawn up to the Fermi level at the surface (ϕ_s changed from -0.23 to -0.55 V). This implies a decrease in electron affinity of 0.55 eV, which is hard to explain by surface-atom rearrangement, considering that the greatest change in W_ϕ observed [386] between different crystal faces of clean silicon amounts to less than 0.2 eV. Moreover, degenerate surface conditions are definitely ruled out by Heiland's results [385] ($\phi_s \approx -0.25$ V), which were obtained by the more accurate method of surface conductance. It appears more likely, therefore, that no large change in ϕ_s has occurred upon annealing and that the photoemission near threshold ($E_T \approx 4.6$ eV) arises from a high-density band of surface states that has moved up to the Fermi level from below. On this basis, one should also associate the 5.15-eV threshold for the cleaved surface with emission from surface states (lying 0.1–0.2 eV above E_v) rather than with valence-band emission, in agreement with van Laar and Scheer's view [398]. Such a re-interpretation of the data would shift the position of the neutral level 0.1–0.2 eV upwards but, as was pointed out above, it would not appreciably alter the conclusions discussed there concerning the distribution of surface states about E_n.

3.6. EFFECT OF ADSORPTION ON ELECTRICAL PROPERTIES

The clean surface constitutes an ideal medium for studying the relation between the chemical and the electrical properties of the surface. Adsorption processes can be investigated over the entire range from zero to equilibrium coverage, with only one type of adsorbed species involved throughout. One is thus dealing here with a relatively simple, two-component system, so that observed changes in electrical properties can be directly correlated with the specific adsorbate used. Moreover, by measuring these changes as a function of surface coverage, one may gain some insight into aspects of the adsorption process not revealed by ordinary chemical measurements. The coverage at each point can be estimated from the product of gas pressure and exposure time if the sticking coefficient is known or determined directly by the flash filament technique (see Ch. 3, § 2.5).

As expected, no changes in the electrical characteristics of the clean surface were observed following exposure to inert gases such as nitrogen and hydrogen [382, 386, 401, 402]. Chemisorbed gases, on the other hand, usually alter both the surface-state structure and the electron affinity. In the study of these effects, oxygen has received particular attention. As discussed in Ch. 3, § 3.2, the reaction of oxygen with clean germanium and silicon surfaces proceeds in two well-defined stages: a rapid adsorption up to a coverage of about one monolayer, and a much slower adsorption leading to the formation of one or two additional monolayers. These two adsorption phases are clearly apparent in most surface-conductance and work-function data. In Fig. 9.22, for example, the initial decrease in the surface potential ϕ_s is associated with the first adsorption phase while the subsequent slow increase is associated with the second phase, the minimum of ϕ_s corresponding more or less to the completion of the first monolayer. This behaviour is quite generally observed in germanium [300, 354–356, 362], indicating that at low coverages oxygen enhances the overall acceptor character of the surface states while above a monolayer coverage it introduces donor-like states. Opinions vary, however, as to the *magnitude* of the changes in ϕ_s. Thus, estimates for the initial decrease in ϕ_s range from less than [355] 0.01 V to about [356] 0.1 V. Hardly any quantitative results have been reported for the variation of ϕ_s at larger coverages. On bombardment–annealed surfaces the point of minimum conductance was never reached, but on cleaved germanium surfaces [300] prolonged exposure to oxygen was observed to transform the surface from p type all the way to n type. In the case of silicon, oxygen adsorption was found to have little or no effect on the surface conductance [383–385, 390]. This is not too surprising, because ϕ_s must change appreciably before a change in conductance of the near-intrinsic silicon surface can be detected (see § 3.4 above).

Apart from the rather qualitative results mentioned above, very little information is available on the surface-state structure at different oxygen coverages. The adsorbate may very well change the density of the states, displace their energy position, or affect both simultaneously — one does not know at present. All that has been definitely established is that *prolonged* exposure to oxygen reduces the density of the states by one to two orders of magnitude [300, 374, 385, 403], the final density being characteristic of the fast states on real (etched) surfaces. The two types of oxidized surfaces appear to be very similar, except possibly in the structure of their oxides. It has been demonstrated [355] that under comparable conditions (vacuum of 2×10^{-5} mm Hg), slow field-effect relaxations were entirely absent on a clean-and-oxidized germanium surface whereas they were pronounced, as usual, on an etched surface. The absence of slow states in the former case was attributed to the fact that the oxide film had grown in the presence of carefully dried oxygen and so was free of occluded water molecules. Slow relaxation effects have been observed, however, on oxidized surfaces of cleaved germanium [300] and bombardment–annealed silicon [385], but in neither case was it indicated whether special precautions had been taken to eliminate traces of water vapour from the oxygen used.

The distinctly different nature of the adsorption processes below and above monolayer coverage is brought into sharp focus by contact-potential measurements. In germanium the work function was observed [370] to increase steadily from 4.78 to 4.96 eV during adsorption of the first oxygen monolayer, and to decrease only slightly with further adsorption, approaching a value of 4.90 eV at equilibrium coverage. That the second adsorbed layer is much less strongly bound to the surface than the first monolayer was demonstrated by the fact that on removal of the oxygen from the system, W_ϕ increased back to 4.96 eV and maintained this value even after prolonged pumping. The clean-surface value (4.78 eV) could only be restored by heating at 500°C in high vacuum, a procedure known to regenerate the clean surface by germanium oxide evaporation (see Ch. 3, § 1.4). It should be noted that the change in work function of about 0.2 eV induced in the initial adsorption phase is due mainly to an increase in electron affinity, the change in surface potential amounting to at most [356] 0.1 V and probably much less [355] (see, for example, Fig. 9.10).

The effect of oxygen on the work function of silicon is very similar [383, 386], except that beyond monolayer coverage adsorption appears [383] to increase W_ϕ slightly rather than decrease it. Adsorption tends to mask the differences in work function exhibited in the clean condition by different crystallographic faces [386] (see § 3.4 above). Starting from the clean surface, the change δW_ϕ was found [383] to increase *linearly* with coverage, reaching a value of about 0.35 eV at monolayer coverage. Here $|\delta\phi_s|$ appears to be even smaller than in the case of germanium, and to a good approximation $\delta W_\phi \approx \delta\chi$. This is further supported by the observation [395] that the work function and photoelectric threshold increased by approximately the same amount following oxygen adsorption. Using the relation $\delta\chi/q = \sigma M/\varepsilon_0$ (see § 1.5), one concludes that the average dipole moment M of an adsorbed oxygen atom is 0.12 debye units (1 debye unit = 10^{-18} esu) and is oriented with its negative end pointing outwards from the surface.

Sparnaay and co-workers [404] have shown that a number of other gases and vapours (such as CO_2, H_2O, NH_3, CH_3OH) affect the conductance of cleaned germanium samples in much the same way as oxygen does — increasing $\Delta\sigma$ at low pressures and decreasing it at high pressures. The behaviour of the various gases differed mainly in the pressure at which the maximum in $\Delta\sigma$ occurred and in the extent of irreversibility of the change in $\Delta\sigma$ on removal of the ambient gas. The clean surfaces studied were prepared in the following manner [405]. Thick cylindrical filaments were "burned off" in low-pressure oxygen atmosphere (at 700°C) until their diameter was reduced to 0.1 mm or less. The system was then evacuated to a pressure of 10^{-9} mm Hg and the filament cooled to room temperature. The only indication that the surfaces prepared in this manner were clean [406] was that application of the same treatment to germanium crystals crushed in air resulted in surfaces with adsorption properties similar to those reported for crystals crushed in high vacuum (see Ch. 3, §§ 1.2, 3.2). However, the fact that such surfaces were found to be near-intrinsic [406,407] rather than degenerate p type (as invariably observed on bombardment–annealed germanium) casts some doubt as to their cleanliness.

Several workers have studied the effect of *atomic* hydrogen on the surface conductance [382-384,402], the work function [382,383,386], and the photoelectric threshold [401] of clean surfaces. On both germanium and silicon, molecular hydrogen is adsorbed to a negligible extent, if at all. In the presence of a hot filament, on the other hand, atomic hydrogen is produced, and adsorption up to monolayer coverage can readily be obtained [408]. For bombardment–annealed germanium surfaces, adsorbed hydrogen acts as an acceptor in that it increases the p-type conductance of the surface [402]. The opposite effect has been observed on clean silicon surfaces [382-384], where hydrogen appears to introduce donor-like surface states. Here no change was detected in $\Delta\sigma$ up to a fractional coverage θ of 0.2 of a monolayer, after which $\Delta\sigma$ decreased steadily as θ approached unity (monolayer coverage). The overall increase in surface potential was estimated at 0.08 V. As pointed out in § 3.4, however, such conductance data on clean silicon surfaces are open to question because of the difficulty in eliminating the contribution of the uncleaned faces of the sample. Work-function measurements [383] are free of this limitation, and are therefore of more significance. In contrast to the case of oxygen adsorption, where the work function increases monotonically up to monolayer coverage ($\theta = 1$), atomic hydrogen *decreases* W_ϕ in the range $0 \leq \theta \lesssim 0.35$ and *increases* it in the range $0.35 \lesssim \theta \leq 1$. The changes in W_ϕ amount to -0.1 and $+0.2$ eV for $\theta = 0.35$ and $\theta = 1$, respectively. On the assumption that $\delta W_\phi \approx \delta\chi$ throughout, one concludes that the dipole associated with an adsorbed hydrogen atom is oriented with its positive end pointing *outwards* at low coverages ($\theta < 0.35$) and *inwards* at higher coverages, the average dipole moments being comparable in the two ranges ($M \approx 0.10$–0.15 debye units).

In order to explain this rather remarkable behaviour, Law [383] has postulated the existence of two possible sites for hydrogen adsorption, one (α) located slightly above and the other (β) slightly below the surface plane of the silicon atoms. The hydrogen atoms in both sites are assumed to be present as protons, having donated their electrons to surface states. In this manner, two arrays of oppositely oriented dipoles are formed, with the protons constituting the positive ends and the electrons trapped in surface states the negative ends. The initial adsorption stage ($\theta \lesssim 0.35$) is ascribed to the filling of the α sites, which would account for the initial decrease in electron affinity. At higher coverages, the β sites begin to be occupied and the affinity increases. This range is characterized by a much smaller sticking coefficient [408], as to be expected for adsorption in deep-lying sites: hydrogen atoms have to surmount a potential barrier before they are able to penetrate the lattice and occupy the β sites. It is by no means obvious, however, why the transition between the two types of adsorption should occur at $\theta \approx 0.35$.

The extremely small value observed for the dipole moment indicates that the situation is more complex than implied by the simple model discussed above. Even if the protons in the two adsorption sites are assumed to be as close to the surface states as 1 Å, the resulting dipole moment should be 5 debye units, nearly fifty times *larger* than the observed value. Since a negligible number of electrons are donated to the space-charge region (as evidenced by the small change in surface conductance), one is forced to conclude that most of the adsorbed atoms are in the neutral state. Such a situation is highly improbable in view of the fact that hydrogen is chemisorbed at the surface. A more likely possibility is that nearly all the hydrogen atoms do donate their electrons to surface states, but that their location and hence the orientation of the resulting dipoles are largely random. The observed dipole moment would then correspond to the relatively *small* average component of the dipoles in a direction normal to the surface. On this premise, adsorption proceeds *simultaneously* at both α and β sites (which are now "smeared" on both sides of the surface plane), with a slight preference for α-type sites at low coverages ($\theta < 0.35$) and for β-type sites at higher coverages.

By far the largest change in electron affinity of the clean surface has been observed for cesium adsorption. In silicon, both the work function [395] and the photoelectric threshold [397,409-411] are reduced to a value as low as ≈ 1.5 eV, most of the change occurring during the initial adsorption of less than a monolayer [395]. In this process the electron affinity drops considerably and, at the same time, the surface becomes n-type degenerate with its Fermi level lying at or above the conduction-band edge [409]. It appears then that cesium atoms donate electrons both to the surface states and to the space-charge region, the former process probably being responsible for the decrease in affinity and the latter giving rise to the strong bending of the bands downwards. Assuming that the conduction-band edge does not lie too far below the Fermi level at the surface, one concludes that the electron affinity of the cesium-coated surface is not appreciably larger than 1.5 eV, corresponding to a change $\delta\chi$ of about -2.5 eV with respect to the clean condition. On this basis, the average dipole moment associated with an adsorbed cesium atom is approximately 1 debye unit. This is nearly ten times greater than the average dipole moment of an adsorbed hydrogen atom and suggests a larger degree of dipole alignment normal to the surface.

4. Summary and conclusions

The experimental data reviewed in this chapter provide a fairly clear and self-consistent picture of the germanium and silicon surface. Most of the observed phenomena can be accounted for by means of straightforward phenomenological theories based on a relatively simple model, and comparison between theory and experiment yields detailed information on the various parameters characterizing the surface structure. Analysis of the data almost invariably relies on the properties of the space-charge region as derived from a solution of Poisson's equation. These properties have been confirmed independently by a variety of experiments, the most direct of which consist of measurements on trap-free germanium surfaces in contact with an electrolyte.

The major part of the data on the electronic behaviour of the surface pertains to the fast states on real surfaces. Surface recombination and pulsed field-effect studies have established that the fast-state distribution is essentially discrete. These studies, combined with low-frequency field-effect measurements, reveal the presence of four dominant sets of states: an acceptor-like set near the conduction-band edge, a donor-like set close to the valence-band edge, and two sets in the central region of the gap. Of the last two, the one just above the midgap consists of acceptor-like recombination centres and, at least in the case of germanium, it controls the recombination process over an extended range of surface potential. Considerable information is now available on the energies, densities, and capture cross sections of the various sets. Systematic measurements of these parameters under different surface conditions reveal that the real surface is characterized by a more or less fixed fast-state structure — a structure that can be modified but not drastically altered by different surface treatments. This is very strikingly demonstrated by the narrow limits between which the number of fast states varies: $\approx 10^{10}-10^{11}$ cm^{-2} for germanium, $\approx 10^{11}-10^{12}$ cm^{-2} for silicon. The occurrence of such low densities, corresponding to only a small fraction of the density of surface atoms ($\approx 10^{15}$ cm^{-2}), indicates a nearly perfect matching between the semiconductor and its oxide, with the fast states representing the relatively small number of interface sites where mismatch or incomplete bonding exists between the semiconductor and oxide lattices. Impurities and lack of stoichiometry in the first few atomic layers of the oxide may provide further sources of fast states.

The insensitivity of the fast-state structure to the condition of the oxide, as well as the results of heat treatment and metal contamination, suggests that the interface defects are characteristic of the semiconductor surface itself rather than of the oxide film. It appears that the fast-state density is increased or decreased by the activation or neutralization of *pre-existing* defect centres rather than by actual changes in the density of defect sites. In this model, a centre is activated if the energy of the fast state associated with it is brought inside the forbidden gap; it is neutralized if its energy level is shifted into an ineffective position. Quite likely, the same type of defect centre is involved in the heat-treatment and the metal-contamination experiments. In the former, excess oxygen or chemisorbed water molecules appears to be the activating agent, while physically adsorbed water acts as the neutralizing species. In the latter, metal microcrystals have definitely been identified as the activating agent. The trap-free surface in contact with an electrolyte may very well correspond to the limiting case in which neutralization of the defect centres (possibly by hydroxyl groups) is nearly perfect. In both treatments the density of fast states "saturates" at much the same maximum value (10^{11}–10^{12} cm^{-2}), which probably represents the number of (fixed) defect sites. The increase in the number of fast states under high applied fields, observed on surfaces of *low* rather than high density of states, can be accounted for by a similar process of saturation. A whole range of intermediate conditions can be envisaged in which only partial changes in the parameters of the fast states take place. This would explain, for example, the observation that certain surface treatments markedly alter the energy and cross sections of the states without significantly affecting their density.

The nature and origin of the surface defects giving rise to the fast states are not known. The persistently occurring density of 10^{11}–10^{12} cm^{-2} does not represent any of the known structural imperfections (such as dislocations) on germanium and silicon surfaces. It may be associated with point defects, with strains, or with monatomic steps—none of which are easily detectable. Steps of only a few atomic layers have been observed on cleaved surfaces (see Ch. 3, § 1.1), and are quite likely present on etched surfaces as well. The latter do not correspond exactly to a principal crystallographic plane, and one would expect steps to occur at the intersections of high-index planes.

As regards the slow states, it has been fairly well established that most of them originate in ions adsorbed from the ambient gases, but very little is known about their density and energy distribution. The strong clamping action of the slow states apparent in the dc field effect sets a lower limit of about 10^{13} cm^{-2} to their density, while their poor communication with the semiconductor interior indicates that they must be well removed from the semiconductor/oxide interface. The manner in which such communication is established is not properly understood. Several mechanisms of charge transfer have been proposed, of which phonon-aided field emission through the oxide film seems to be consistent with most of the observations. Direct evidence for the predominance of field emission under high transverse fields has been obtained in the case of germanium. The slow states involved in this process lie within a few hundredths of an electron volt of the band edges.

The structure of the slow states is intimately linked with chemical problems such as adsorption and catalysis. Although mostly of a qualitative nature, the experimental data bearing on this question leave little doubt that charge transfer through the oxide film is the controlling process in chemisorption. Thus, for thick oxide films and dry ambient gases, where the charge transfer is reduced to negligible proportions, hardly any adsorption appears to take place and no ambient-induced slow states are formed. Slow states arising from lattice imperfections and/or occluded impurities in the oxide film have been detected under these conditions. In the case of oxidized surfaces of clean germanium, the absence of water vapour prevents the formation of slow states even though the oxide film consists of only one or two monolayers. This observation, as well as many others, indicates strongly that water vapour plays the dominant if not the decisive role in the charge transfer process.

TABLE 9.4

Neutral level E_n and density $\overline{N}_t$ of surface states on clean surfaces

Type of clean surface	Cryst. plane	Method *	$E_n - E_v$ (eV)	$\overline{N}_t$** (cm^{-2}eV^{-1})
Bombardment–annealed Ge	(111)	a	+0.025	2.4×10^{14}
	(100)	a	−0.1	5×10^{14}
Cleaved Ge	(111)	b	< 0	$> 4 \times 10^{13}$
Bombardment–annealed Si	—	a	0.3	1.5×10^{15}
Cleaved Si	(111)	b	0.3	$> 3 \times 10^{15}$

* Method: a — surface conductance and field-effect mobility
            b — work function and photoelectric threshold
** Densities quoted correspond to assumption of continuous distribution near E_n, except in the case of cleaved silicon where they refer to two surface-state bands on either side of E_n.

Measurements of surface states on clean surfaces have so far been successful only in determining the neutral level of the surface states and the lower bound to their density. The most reliable information on these two quantities is listed in Table 9.4. For moderately doped samples, the neutral point E_n lies very close to the Fermi level, and surface-conductance measurements provide the most accurate means for its determination. With the exception of cleaved silicon surfaces, the available data provide no indication as to the form of the surface-state distribution function, and the densities quoted correspond to the simplest assumption — that of a continuous distribution near the neutral level. The lower bounds to the *total* density of surface states are obtained if all states are assumed to be located at the Fermi level. In the units used, the numerical values of these bounds are about ten times smaller than the corresponding values of $\overline{N}_t$ listed in the Table (see eqs. (9.12), (9.14)). Much more detailed results are available on cleaved silicon surfaces, where an appreciable swing in surface potential could be effected by varying the bulk doping. Measurements of work function and photoelectric threshold as functions of bulk potential have clearly indicated in this case the presence of two surface-state bands (or levels) on either side of E_n, with a minimum density of 3×10^{15} cm^{-2}eV^{-1} (or 2×10^{14} cm^{-2}). The exact position of E_n could not be determined, however, and it would be desirable to supplement such studies by surface-conductance measurements.

It is interesting to note that the clean germanium and silicon surfaces are markedly different in their surface-state structure. The former is p-type degenerate (the (100) face more so than the (111) face) with its neutral level lying close to the valence-band edge, whereas the latter is near-intrinsic with its neutral level well above E_v. Differences have been detected also between cleaved and annealed surfaces, as to be expected in view of their different lattice structures, but sufficient data are not available to permit a quantitative comparison of the two types of clean surfaces.

On both germanium and silicon, the surface-state densities are within an order of magnitude of the surface-atom densities. The mere presence of such a high density makes it unlikely that the surface states arise from structural imperfections such as monatomic steps. The occurrence of roughly one state per surface atom points rather to the possibility that they correspond to Tamm- or Shockley-like states associated with the termination of the periodic potential at the surface. Apart from these considerations, however, little can be said at present about the physical origin of the surface states. Neither the theoretical calculations nor the experimental data are sufficiently detailed to establish a definite correlation between the predicted and observed states. All surface-state calculations so far have been carried out on the basis of simplifying assumptions. In particular, they treat an ideal surface having a lattice structure identical to that of the parallel crystallographic planes in the bulk. The displacement of the surface atoms from their regular sites that is known to occur on semiconductor surfaces will no doubt alter significantly the results of the theoretical analyses. The pairing of surface atoms, for example, suggests that most if not all of the dangling bonds saturate each other. With present knowledge of atomic arrangements at the surface, one would hope that further theoretical work will provide more quantitative predictions of the surface-state structure. On the experimental side, it would be highly desirable to combine electrical measurements with structural studies so as to determine the relation between surface states and lattice structure. Unfortunately, the strong clamping action of the surface states makes a *complete* determination of their energy distribution by field-effect and contact-potential measurements rather hopeless. Other experiments, such as infrared absorption and paramagnetic resonance, may prove more successful in this respect.

The surface states are influenced by and play a dominant role in chemisorption. The drastic reduction in surface-state density accompanying the transition from a clean to a real surface shows that the conditions necessary for the formation of surface states are largely obviated by oxygen adsorption. Very likely, the adsorbate saturates most of the dangling bonds and, as a consequence, shifts the surface atoms (partially or completely) into their regular lattice sites. Both these interrelated processes are probably responsible for the altered surface-state structure. Chemisorption on the clean surface of either oxygen or hydrogen is not accompanied by a significant change in space charge, indicating that on the semiconductor side it is mainly the surface states that are involved in the charge transfer process between the adsorbate and the adsorbant. The formation by this process of surface dipoles between adsorbed ions and oppositely charged surface states is consistent with the observed changes in electron affinity. More extensive studies of this sort, in which electrical and chemical characteristics are systematically investigated on one and the same surface, would no doubt lead to a new level of understanding of both the clean and the real surface.

References

[1] R. H. Kingston (Editor), *Semiconductor Surface Physics* (University of Pennsylvania Press, Philadelphia, 1957).
[2] J. N. Zemel (Editor), *Semiconductor Surfaces* (Pergamon Press, Oxford, 1960); J. Phys. Chem. Solids **14** (1960).
[3] A. N. Frumkin (Editor), *Semiconductor Surface Physics* (U.S.S.R. Academy of Science Press, Moscow, 1962) [translation: Consultants Bureau, New York, 1964].
[4] F. R. Eirich and M. C. Johnstone (Editors), *Conference on Clean Surfaces*, Ann. N. Y. Acad. Sci. **101** (1963).
[5] H. C. Gatos (Editor), *Proc. Int. Conf. Physics and Chemistry of Solid Surfaces*, Providence, 1964; Surface Science **2** (1964).

[6] *Proc. Int. Conf. Semiconductor Physics, Rochester*, 1958 (Pergamon Press, Oxford, 1958); J. Phys. Chem. Solids **8** (1959).
[7] *Proc. Int. Conf. Semiconductor Physics, Prague*, 1960 (Czechoslovak Academy of Sciences, Prague, 1961).
[8] *Proc. Int. Conf. Semiconductor Physics, Exeter*, 1962 (Institute of Physics and Physical Society, London, 1962).
[9] R. H. Kingston, J. appl. Phys. **27** (1956) 101.
[10] J. Bardeen in *Semiconductors and Phosphors* (Edited by M. Schön and H. Welker), (Friedr. Vieweg, Braunschweig, 1958).
[11] A. Many, J. Phys. Chem. Solids **8** (1959) 87.
[12] J. T. Law in *Semiconductors* (Edited by N.B. Hannay), (Reinhold, New York, 1959), p. 676.
[13] M. Lax in reference 7, p. 484.
[14] P. Handler in *The Surface Chemistry of Metals and Semiconductors* (Edited by H. C. Gatos), (John Wiley, New York, 1960), p. 54.
[15] T. B. Watkins in *Progress in Semiconductors* (Edited by A. F. Gibson, F. A. Kröger, and R. E. Burgess), (Heywood, London), vol. V (1960), p. 1.
[16] A. Frova and A. Stella, Nuovo cimento Suppl. **22** (1961) 517.
[17] G. Heiland, Fortschr. Physik **9** (1961) 393.
[18] H. Flietner, Phys. Status solidi **2** (1961) 221.
[19] A. R. Plummer in *The Electrochemistry of Semiconductors* (Edited by P. J. Holmes), (Academic Press, New York, 1962), p. 61.
[20] G. Heiland in *Festkörperprobleme* (Edited by F. Sauter), (Friedr. Vieweg, Braunschweig), vol. III (1964), p. 125.
[21] H. U. Harten in *Festkörperprobleme* (Edited by F. Sauter), (Friedr. Vieweg, Braunschweig), vol. III (1964), p. 81.
[22] Review articles in reference 4, Suppl.
[23] R. J. Archer, J. electrochem. Soc. **104** (1957) 619.
[24] J. Bardeen, Phys. Rev. **71** (1947) 717.
[25] W. H. Brattain and J. Bardeen, Bell System tech. J. **32** (1953) 1.
[26] S. R. Morrison, J. phys. Chem. **57** (1953) 860.
[27] J. Bardeen and S. R. Morrison, Physica **20** (1954) 873.
[28] C. G. B. Garrett and W. H. Brattain, Phys. Rev. **99** (1955) 376.
[29] E. O. Johnson, RCA Rev. **18** (1957) 556.
[30] E. O. Johnson, Phys. Rev. **111** (1958) 153.
[31] W. H. Brattain and C. G. B. Garrett, Bell System tech. J. **35** (1956) 1019; C. G. B. Garrett and W. H. Brattain, *ibid.* **35** (1956) 1041.
[32] C. G. B. Garrett, W. H. Brattain, W. L. Brown, and H. C. Montgomery in reference 1, p. 126.
[33] E. N. Clarke, Phys. Rev. **91** (1953) 756; **95** (1954) 284; Ann. N. Y. Acad. Sci. **58** (1954) 937; Sylvania Technol. **7** (1954) 102.
[34] E. N. Clarke and R. L. Hopkins, Phys. Rev. **91** (1953) 1566.
[35] J. W. Granville and C. A. Hogarth, Proc. phys. Soc. (London) **B64** (1951) 488.
[36] S. G. Kalashnikov and Y. E. Pokrovskii, Zhur. tekh. Fiz. **22** (1952) 883.
[37] T. M. Buck and F. S. McKim, J. electrochem. Soc. **103** (1956) 593.
[38] W. D. Cussins, Proc. phys. Soc. (London) **B68** (1955) 213.
[39] W. L. Brown, Phys. Rev. **100** (1955) 590.
[40] H. C. Montgomery and W. L. Brown, Phys. Rev. **103** (1956) 865.
[41] J. Bardeen, R. E. Coovert, S. R. Morrison, J. R. Schrieffer, and R. Sun, Phys. Rev. **104** (1956) 47.
[42] W. L. Brown, Phys. Rev. **91** (1953) 518.

[43] R. H. Kingston, Phys. Rev. **93** (1954) 346; **94** (1954) 1416.
[44] R. H. Kingston, Phys. Rev. **98** (1955) 1766.
[45] G. A. deMars, H. Statz, and L. Davis, Jr., Phys. Rev. **98** (1955) 539; H. Statz, L. Davis, Jr., and G. A. deMars, *ibid.* **98** (1955) 540.
[46] H. Statz, G. A. deMars, L. Davis, Jr., and A. Adams, Jr., Phys. Rev. **101** (1956) 1272.
[47] J. R. Schrieffer, Phys. Rev. **94** (1954) 420.
[48] H. Statz, G. A. deMars, L. Davis, Jr., and A. Adams, Jr. in reference 1, p. 139.
[49] R. C. Sirrine, as quoted by J. R. Schrieffer in reference 1, p. 55.
[50] H. Christensen, Proc. Inst. Radio Engrs. **42** (1954) 1371; Phys. Rev. **96** (1954) 827.
[51] J. T. Law, Proc. Inst. Radio Engrs. **42** (1954) 1367; J. T. Law and P. S. Meigs, J. appl. Phys. **26** (1955) 1265.
[52] A. L. McWhorter and R. H. Kingston, Proc. Inst. Radio Engrs. **42** (1954) 1376.
[53] E. N. Clarke, Phys. Rev. **99** (1955) 1899.
[54] A. R. Hutson, Phys. Rev. **102** (1956) 381.
[55] A. R. F. Plummer, Proc. phys. Soc. (London) **B69** (1956) 539.
[56] J. I. Carasso and I. Stelzer, J. chem. Soc. (1956) 3726.
[57] W. L. Brown, W. H. Brattain, C. G. B. Garrett, and H. C. Montgomery in reference 1, p. 111.
[58] N. J. Harrick in reference 6, p. 106.
[59] N. J. Harrick in reference 2, p. 60.
[60] N. J. Harrick, Phys. Rev. **125** (1962) 1165
[61] B. O. Seraphin and D. A. Orton, J. appl. Phys. **34** (1963) 1743.
[62] J. D. Nixon and P. C. Banbury, Proc. phys. Soc. (London) **B70** (1957) 481.
[63] E. Aerts, S. Amelinckx, and J. Vennik, J. Electronics and Control **7** (1959) 497.
[64] N. G. Einspruch, J. Phys. Chem. Solids **23** (1962) 1743.
[65] W. Shockley and G. L. Pearson, Phys. Rev. **74** (1948) 232.
[66] G. W. Pratt, Jr. and H. H. Kolm in reference 1, p. 297.
[67] M. Kikuchi and T. Onishi, J. phys. Soc. Japan **9** (1954) 130; M. Kikuchi, *ibid.* **9** (1954) 665.
[68] G. G. E. Low, Proc. phys. Soc. (London) **B68** (1955) 10, 1154.
[69] V. I. Lyashenko and R. O. Litvinov, Ukrayin. fiz. Zhur. **1** (1956) 143.
[70] R. H. Kingston and A. L. McWhorter, Phys. Rev. **103** (1956) 534.
[71] S. R. Morrison, Phys. Rev. **102** (1956) 1297; in reference 1, p. 169.
[72] T. B. Watkins, Proc. phys. Soc. (London) **B69** (1956) 1353.
[73] R. E. Burgess, Brit. J. appl. Phys. **8** (1957) 62.
[74] S. Müller, Z. Naturforsch. **12a** (1957) 112.
[75] P. C. Banbury, G. G. E. Low, and J. D. Nixon in reference 1, p. 70.
[76] H. C. Montgomery, Phys. Rev. **106** (1957) 441.
[77] A. E. Yunovich, Zh. tekh. Fiz. **27** (1957) 1707 [translation: Soviet Phys. — Tech. Phys. **2** (1957) 1587]; Fiz. tver. Tela **1** (1959) 908 [translation: Soviet Phys. — Solid State **1** (1959) 829]; A. E. Yunovich in reference 3, p. 88; A. E. Yunovich and V. I. Tikhonov in reference 3, p. 97.
[78] D. H. Lindley and P. C. Banbury, Proc. phys. Soc. (London) **B74** (1959) 395.
[79] D. R. Frankl, J. electrochem. Soc. **109** (1962) 608.
[80] G. Rupprecht, Phys. Rev. **111** (1958) 75.
[81] S. R. Morrison, Phys. Rev. **114** (1959) 437.
[82] D. H. Lindley and P. C. Banbury in reference 2, p. 200.
[83] I. I. Abkevich, Dokl. Akad. Nauk **127** (1959) 1199 [translation: Soviet Phys. — Doklady **4** (1960) 899].
[84] M. S. Kosman and I. I. Abkevich, Fiz. tver. Tela **1** (1959) 378 [translation: Soviet Phys. — Solid State **1** (1959) 341].

[85] H. A. Papazian, J. appl Phys. 27 (1956) 1253.
[86] H. Statz, G. deMars, L. Davis, Jr., and A. Adams, Jr., Phys. Rev. 106 (1957) 455.
[87] M. Lasser, C. Wysocki, and B. Bernstein, Phys. Rev. 105 (1957) 491; in reference 1, p. 197.
[88] R. O. Litvinov and T.-L. Hsü in reference 3, p. 165; Radiotekh. i Elektron. 7 (1962) 1030 [translation: Radio Eng. electron. Phys. 7 (1962) 965].
[89] W. T. Eriksen, H. Statz, and G. A. deMars, J. appl. Phys. 28 (1957) 133.
[90] W. T. Eriksen, J. appl. Phys. 29 (1958) 730.
[91] H. Statz and G. A. deMars, Phys. Rev. 111 (1958) 169.
[92] C. G. B. Garrett and W. H. Brattain, J. appl. Phys. 27 (1956) 299.
[93] W. E. Bradley et al., Proc. Inst. Radio Engrs. 41 (1953) 1702.
[94] J. B. Gunn, Proc. phys. Soc. (London) B67 (1954) 409.
[95] C. V. Bocciarelli, Physica 20 (1954) 1020.
[96] P. A. Hartig and R. N. Noyce, J. appl. Phys. 27 (1956) 843.
[97] W. N. Reynolds, Proc. phys. Soc. (London) B66 (1953) 899.
[98] J. P. McKelvey and R. L. Longini, J. appl. Phys. 25 (1954) 634.
[99] A. Many, Proc. phys. Soc. London B67 (1954) 9.
[100] N. Holonyak, Jr. and H. Letaw, Jr., J. appl. Phys. 26 (1955) 355.
[101] T. M. Buck and W. H. Brattain, J. electrochem. Soc. 102 (1955) 636.
[102] S. G. Ellis, J. appl. Phys. 28 (1957) 1262.
[103] F. D. Rosi, RCA Rev. 19 (1958) 349.
[104] A. R. Moore and H. Nelson, RCA Rev. 17 (1956) 5.
[105] C. A. Hogarth, Proc. phys. Soc. (London) B69 (1956) 791.
[106] V. A. Petrusevich, Fiz. tver. Tela 1 (1959) 1695 [translation: Soviet Phys. — Solid State 1 (1960) 1549].
[107] R. J. Keyes and T. G. Maple, Phys. Rev. 94 (1954) 1416(A).
[108] D. T. Stevenson and R. J. Keyes, Physica 20 (1954) 1041.
[109] G. Adam, Z. Naturforsch. 12a (1957) 574.
[110] A. H. Benny and F. D. Morten, Proc. phys. Soc. (London) 72 (1958) 1007.
[111] R. H. Bube, J. chem. Phys. 21 (1953) 1409; Phys. Rev. 101 (1956) 1668.
[112] B. Seraphin, Ann. Physik Leipzig 13 (1953) 198.
[113] S. H. Liebson, J. electrochem. Soc. 101 (1954) 359; J. chem. Phys. 23 (1955)1732.
[114] R. N. Noyce, Meeting of the Electrochemical Society, Cincinnati, May 1955.
[115] D. T. Stevenson, Phys. Rev. 98 (1955) 1566 (A).
[116] A. V. Rzhanov, I. G. Neizvestnyi, and V. V. Roslyakov, Zhur. tekh. Fiz. 26 (1956) 2142 [translation: Soviet Phys. — Tech. Phys. 1 (1956) 2081].
[117] Y. Kanai, J. phys. Soc. Japan 9 (1954) 292.
[118] B. H. Schultz, Physica 20 (1954) 1031.
[119] V. I. Strikha in reference 3, p. 123.
[120] H. K. Henisch, W. N. Reynolds, and P. M. Tipple, Physica 20 (1954) 1033; Proc. phys. Soc. (London) B68 (1955) 353.
[121] A. Many, Y. Margoninski, E. Harnik, and E. Alexander, Phys. Rev. 101 (1956) 1433, 1434.
[122] J. D. Nixon and P. C. Banbury, Proc. phys. Soc. (London) B69 (1956) 487.
[123] M. Kikuchi, J. Phys. Soc. Japan 12 (1957) 756.
[124] V. I. Lyashenko, Zhur. tekh. Fiz. 27 (1957) 1613 [translation: Soviet Phys. — Tech. Phys. 2 (1957) 1496].
[125] A. Many, E. Harnik, and Y. Margoninski in reference 1, p. 85.
[126] J. E. Thomas, Jr. and R. H. Rediker, Phys. Rev. 101 (1956) 984.
[127] A. Many and D. Gerlich, Phys. Rev. 107 (1957) 404.
[128] A. Wolska in reference 7, p. 510; Acta. phys. Polon. 23 (1963) 103.
[129] A. V. Rzhanov and I. A. Arkhipova, Fiz. tver. Tela 3 (1961) 1954 [translation: Soviet

Phys. — Solid State **3** (1962) 1424].
[130] H. M. Bath and M. Cutler, J. Phys. Chem. Solids **5** (1958) 171.
[131] V. A. Petrusevich and T. N. Lobanova, Fiz. tver. Tela **3** (1961) 3546 [translation: Soviet Phys. — Solid State **3** (1962) 2575].
[132] G. C. Dousmanis, J. Appl. Phys. **30** (1959) 180.
[133] D. Sautter in *Progress in Semiconductors* (Edited by A. F. Gibson, F. A. Kröger, and R. E. Burgess), (Heywood, London), vol. IV (1960), p. 125. (Review).
[134] D. Sautter and K. Seiler, Z. Naturforsch. **12a** (1957) 489.
[135] J. E. Hill and K. M. van Vliet, J. appl. Phys. **29** (1958) 177.
[136] K. Komatsubara, J. phys. Soc. Japan. **13** (1958) 1409.
[137] K. M. van Vliet, Proc. Inst. Radio Engrs. **46** (1958) 1004. (Review).
[138] H. C. Montgomery, Bell System tech. J. **31** (1952) 950.
[139] B. V. Rollin and I. M. Templeton, Proc. phys. Soc. (London) **B66** (1953) 259; **B67** (1954) 271.
[140] D. Baker, J. appl. Phys. **25** (1954) 922.
[141] F. J. Hyde, Proc. phys. Soc. (London) **B66** (1956) 242.
[142] L. Bess, Phys. Rev. **103** (1956) 72.
[143] T. G. Maple, L. Bess, and H. E. Gebbie, J. appl. Phys. **26** (1955) 490.
[144] A. U. Mac Rae and H. Levinstein, Phys. Rev. **119** (1960) 62.
[145] J. J. Brophy and N. Rostoker, Phys. Rev. **100** (1955) 754.
[146] L. Bess, J. appl. Phys. **86** (1955) 1377.
[147] J. J. Brophy, Phys. Rev. **106** (1957) 675.
[148] J. J. Brophy, Phys. Rev. **115** (1959) 1122.
[149] J. J. Brophy, J. appl. Phys. **29** (1958) 1377.
[150] J. J. Brophy in *Solid State Physics in Electronics and Telecommunications* (Edited by M. Désirant and J. L. Michiels), (Academic Press, London), vol. I (1960), p. 548.
[151] G. L. Pearson, H. C. Montgomery, and W. L. Feldmann, J. appl. Phys. **27** (1956) 9.
[152] P. O. Lauritzen and J. F. Gibbons, J. appl. Phys. **33** (1962) 758.
[153] V. E. Noble and J. E. Thomas, Jr., J. appl. Phys. **32** (1961) 1709.
[154] K. Komatsubara, Y. Inuishi, H. Edagawa, and T. Shibaike, J. phys. Soc. Japan, **15** (1960) 1713.
[155] T. B. Watkins, Proc. phys. Soc. (London) **73** (1959) 59.
[156] Yu. S. Karpov, Fiz. tver. Tela **3** (1961) 1571 [translation: Soviet Phys. — Solid State **3** (1961) 1141].
[157] J. Bernamont, Proc. phys. Soc. (London) **49** (1937) 138.
[158] A. L. McWhorter in reference 1, p. 207.
[159] H. C. Montgomery, J. appl. Phys. **33** (1962) 2143.
[160] S. R. Morrison, Phys. Rev. **99** (1955) 1904.
[161] L. S. Sochava and D. N. Mirlin, Fiz. tver. Tela **2** (1960) 23 [translation: Soviet Phys. — Solid State **2** (1960) 18].
[162] A. U. Mac Rae, J. appl. Phys. **33** (1962) 2570.
[163] R. L. Petritz in reference 1, p. 226.
[164] J. M. Richardson, Bell System tech. J. **29** (1950) 117.
[165] R. L. Petritz, Phys. Rev. **87** (1952) 189(A).
[166] L. Bess, Phys. Rev. **91** (1953) 1569.
[167] J. T. Law, J. phys. Chem. **59** (1955) 543; J. T. Law and E. E. François, *ibid.* **60** (1956) 353.
[168] P. Aigrain and C. Dugas, Z. Elektrochem. **56** (1952) 363.
[169] K. Hauffe and H. J. Engell, Z. Elektrochem. **56** (1952) 366; **57** (1953) 762; H. J. Engell in *Halbleiterprobleme* (Edited by W. Schottky), (Fried. Vieweg, Braunschweig), vol. I (1954), p. 176; K. Hauffe, Angew. Chem. **67** (1955) 189.
[170] P. B. Weisz, J. chem. Phys. **20** (1952) 1483; **21** (1953) 1531.

[171] K. Hauffe in reference 1, p. 259.
[172] H. J. Krusemeyer and D. G. Thomas, J. Phys. Chem. Solids **4** (1958) 78.
[173] T. J. Gray and S. D. Savage, Discussions Faraday Soc. No. 28 (1959) 159.
[174] V. B. Sandomirskii, Izv. Akad. Nauk, Ser. Fiz. **21** (1957) 211; Sh. M. Kogan and V. B. Sandomirskii, Dokl. Akad. Nauk **127** (1959) 377; V. B. Sandomirskii and Sh. M. Kogan, Zhur. fiz. Khim. **33** (1959) 1709.
[175] I. Stelzer, J. Electronics and Control **8** (1960) 39.
[176] C. G. B. Garrett, J. chem. Phys. **33** (1960) 966.
[177] O. Jäntsch, J. Phys. Chem. Solids **21** (1961) 33.
[178] J. R. Macdonald and C. A. Barlow, Jr., J. chem. Phys. **39** (1963) 412.
[179] F. F. Vol'kenshtein, *The Electronic Theory of Catalysis on Semiconductors* (Pergamon Press, Oxford, 1963), and references cited therein.
[180] G. Heiland, Discussions Faraday Soc. No. 28 (1959) 168; G. Heiland, E. Mollwo, and F. Stöckmann in *Solid State Physics* (Edited by F. Seitz and D. Turnbull), (Academic Press, New York), vol. **8** (1959), p. 193. (Review).
[181] A. Kobayashi and S. Kawaji, J. phys. Soc. Japan **10** (1955) 270; J. chem. Phys. **24** (1956) 907.
[182] R. H. Bube, J. chem. Phys. **27** (1957) 496.
[183] F. F. Vol'kenshtein and I. V. Karpenko in reference 3, p. 79; J. appl Phys. **33** Suppl. No. 1 (1962) 460.
[184] F. F. Vol'kenshtein and V. B. Sandomirskii, Dokl. Akad. Nauk **118** (1958) 980.
[185] V. I. Lyashenko and V. G. Litovchenko, Zhur. tekh. Fiz. **28** (1958) 447, 454 [translation: Soviet Phys. — Tech. Phys. **3** (1958) 422, 429]; V. G. Litovchenko, V. I. Lyashenko, and O. S. Frolov in reference 3, p. 103.
[186] J. T. Wallmark and R. R. Johnson, RCA Rev. **18** (1957) 512.
[187] G. Dorda in reference 7, p. 506; Czech. J. Phys. **B10** (1960) 820.
[188] J. T. Law, J. phys. Chem. **59** (1955) 67.
[189] K. Kawasaki, K. Kanou, and Y. Sekita, J. phys. Soc. Japan **14** (1959) 233.
[190] Yu. V. Fedorovich, Fiz. tver. Tela **3** (1961) 2885 [translation: Soviet Phys. — Solid State **3** (1962) 2107].
[191] R. Kh. Burshtein, L. A. Larin, and G. F. Voronina, Dokl. Akad. Nauk **133** (1960) 148; R. Kh. Burshtein, L. A. Larin, and S. I. Sergeev in reference 3, p. 20.
[192] Sh. M. Kogan and V. B. Sandomirskii, Izv. Akad. Nauk, Otd. khim. Nauk No. 10 (1959) 1854.
[193] V. G. Litovchenko, O. S. Frolov, and S.-M. Bao, Fiz. tver. Tela **4** (1962) 833 [translation: Soviet Phys. — Solid State **4** (1962) 613].
[194] J. Sochanski, Phys. Status solidi **2** (1962) 1312, 1317.
[195] J. Sochanski, Phys. Status solidi **1** (1961) 317.
[196] P. G. Borzyak and O. G. Sarbei, Ukrayin. fiz. Zhur. **1** (1956) 395; P. G. Borzyak, L. S. Miroshnichenko, and O. G. Sarbei, Fiz. tver. Tela **2** (1960) 314 [translation: Soviet Phys. — Solid State **2** (1960) 284]; P. G. Borzyak in reference 3, p. 570.
[197] J. A. Burton, Phys. Rev. **108** (1957) 1342.
[198] R. J. Hodgkinson, Solid-State Electronics **5** (1962) 269.
[199] E. A. Davies, J. Phys. Chem. Solids **25** (1964) 201.
[200] Yu. K. Pozhela and V. I. Shilal'nikas, Fiz. tver. Tela **4** (1962) 1601; *ibid.* **5** (1963) 730 [translation: Soviet Phys. — Solid State **4** (1962) 1173; *ibid.* **5** (1963) 532].
[201] A. E. Yunovich and B. G. Anochin, Nauchn. Dokl. Vysshei Shkoly, Fiz. mat. Nauki, No. 5 (1958) 176.
[202] H. Flietner, Ann. Physik Leipzig **3** (1959) 414.
[203] V. G. Litovchenko and V. I. Lyashenko, Fiz. tver. Tela **1** (1959) 1609 [translation: Soviet Phys. — Solid State **1** (1960) 1470].

[204] A. Balzarotti, G. Chiarotti, and A. Frova, Nuovo cimento **26** (1962) 1205.
[205] N. G. Einspruch, J. electrochem. Soc. **108** (1961) 1164; Phys. Status solidi **2** (1962) 188.
[206] M. F. Millea, T. C. Hall, and J. O. Kopplin, J. Phys. Chem. Solids **23** (1962) 611.
[207] D. Gerlich, J. Phys. Chem. Solids **23** (1962) 837.
[208] S. Wang and G. Wallis, Phys. Rev. **105** (1957) 1459.
[209] V. P. Zhuze, G. E. Pikus, and O. V. Sorokin, Fiz. tver. Tela **1** (1959) 1420 [translation: Soviet Phys. — Solid State **1** (1959) 1302].
[210] V. P. Zhuze, G. E. Pikus, and O. V. Sorokin, Zhur. tekh. Fiz. **27** (1957) 1167 [translation: Soviet Phys. — Tech. Phys. **2** (1957) 1061].
[211] A. V. Rzhanov, Yu. F. Novototskii-Vlasov, and I. G. Neizvestnyi, Zhur. tekh. Fiz. **27** (1957) 2440 [translation: Soviet Phys. — tech. Phys. **2** (1957) 2274].
[212] A. V. Rzhanov, N. M. Pavlov, and M. A. Selezneva, Zhur. tekh. Fiz. **28** (1958) 2645 [translation: Soviet Phys. — Tech. Phys. **3** (1958) 2419].
[213] E. Harnik, Y. Goldstein, N. B. Grover, and A. Many in reference 2, p. 193.
[214] Y. Margoninski, J. chem. Phys. **32** (1960) 1791.
[215] Y. Margoninski and H. E. Farnsworth, Phys. Rev. **123** (1961) 135.
[216] W. A. Albers, Jr. and A. M. Rickel, J. electrochem. Soc. **109** (1962) 582.
[217] S. Wang and G. Wallis, Phys. Rev. **107** (1957) 947.
[218] G. Wallis and S. Wang, J. electrochem. Soc. **106** (1959) 231.
[219] V. E. Primachenko, V. G. Litovchenko, V. J. Lyashenko, and O. V. Snitko, Ukrayin. fiz. Zhur. **5** (1960) 345.
[220] V. I. Lyashenko, O. V. Snitko, and V. G. Litovchenko in reference 7, p. 515.
[221] V. G. Litovchenko and V. I. Lyashenko, Fiz. tver. Tela **3** (1961) 61, 73 [translation: Soviet Phys. — Solid State **3** (1961) 44, 53].
[222] G. Dorda, Czech. J. Phys. **B11** (1961) 406.
[223] A. V. Rzhanov, Fiz. tver. Tela **4** (1962) 1279 [translation: Soviet Phys. — Solid State **4** (1962) 937].
[224] P. Balk and E. L. Peterson, J. electrochem. Soc. **110** (1963) 1245.
[225] A. Many, Y. Goldstein, N. B. Grover, and E. Harnik in reference 7, p. 498.
[226] A. V. Rzhanov, Fiz. tver. Tela **1** (1959) 522 [translation: Soviet Phys. — Solid State **1** (1959) 469].
[227] A. V. Rzhanov, N. M. Pavlov, and M. A. Selezneva, Fiz. tver. Tela **3** (1961) 832 [translation: Soviet Phys. — Solid State **3** (1961) 607].
[228] A. V. Rzhanov in reference 3, p. 70; in reference 8, p. 673.
[229] A. Many, Y. Goldstein, and O. Brafman. Unpublished data.
[230] V. G. Litovchenko and V. I. Lyashenko, Fiz. tver. Tela **4** (1962) 1985 [translation: Soviet Phys. — Solid State **4** (1963) 1455].
[231] D. H. Lindley and P. C. Banbury. Proc. phys. Soc. (London) **82** (1963) 422.
[232] R. Solomon, J. appl. Phys. **31** (1960) 1791.
[233] V. E. Primachenko, V. G. Litovchenko, and V. I. Lyashenko, Fiz. tver. Tela **4** (1962) 2925 [translation: Soviet Phys. — Solid State **4** (1963) 2145].
[234] M. B. Hall, J. appl. Phys. **34** (1963) 1003.
[235] T. M. Buck and F. S. McKim, J. electrochem. Soc. **105** (1958) 709.
[236] H. U. Harten, Philips Res. Rep. **14** (1959) 207, 346; *Solid State Physics in Electronics and Telecommunications* (Edited by M. Désirant and J. L. Michiels), (Academic Press, London), vol. I (1960), p. 624.
[237] O. V. Snitko, Fiz. tver. Tela **1** (1959) 980 [translation: Soviet Phys. — Solid State **1** (1959) 898].
[238] V. G. Litovchenko and O. V. Snitko, Fiz. tver. Tela **2** (1960) 591 [translation: Soviet Phys. — Solid State **2** (1960) 554].
[239] V. G. Litovchenko and V. I. Lyashenko, Fiz. tver. Tela **5** (1963) 3207 [translation:

Soviet Phys. — Solid State **5** (1964) 2347].
[240] G. Rupprecht in reference 4, p. 960.
[241] G. Rupprecht in reference 2, p. 208.
[242] V. G. Litovchenko and V. I. Lyashenko, Fiz. tver. Tela **2** (1960) 1592 [translation: Soviet Phys. — Solid State **2** (1961) 1442].
[243] V. G. Litovchenko, Surface Science **1** (1964) 291.
[244] W. E. Spear, Phys. Rev. **112** (1958) 362.
[245] A. V. Rzhanov and A. F. Plotnikov, Fiz. tver. Tela **3** (1961) 1557 [translation: Soviet Phys. — Solid State **3** (1961) 1130].
[246] G. Chiarotti, G. del Signore, A. Frova, and G. Samoggia, Nuovo cimento **26** (1962) 403
[247] G. Samoggia, Nuovo cimento **29** (1963) 336.
[248] G. C. Dousmanis, Phys. Rev. **112** (1958) 369.
[249] A. V. Rzhanov, Yu. F. Novototskii-Vlasov, I. G. Neizvestnyi, S. V. Pokrovskaya, and T. I. Galkina, Fiz. tver. Tela **3** (1961) 822 [translation: Soviet Phys. — Solid State **3** (1961) 600].
[250] G. C. Alexandrakis and G. C. Dousmanis, J. Appl. Phys. **34** (1963) 3077.
[251] J. A. Champion, Proc. phys. Soc. (London) **79** (1962) 662.
[252] B. R. Marathe, Proc. phys. Soc. (London) **79** (1962) 503.
[253] E. Harnik and Y. Margoninski, J. Phys. Chem. Solids **8** (1959) 121.
[254] P. P. Konorov and O. V. Romanov, Fiz. tver. Tela **4** (1962) 1655 [translation: Soviet Phys. — Solid State **4** (1962) 1215].
[255] P. P. Konorov and O. V. Romanov, Fiz. tver. Tela **5** (1963) 3039 [translation: Soviet Phys. — Solid State **5** (1964) 2225].
[256] M. C. Cretella and H. J. Gatos, J. electrochem. Soc. **105** (1958) 487.
[257] S. R. Morrison in reference 2, p. 214.
[258] V. I. Fistul and D. G. Andrianov, Dokl. Akad. Nauk SSSR **130** (1960) 374.
[259] M. L. Ioselevich and V. I. Fistul, Fiz. tver. Tela **3** (1961) 1132 [translation: Soviet Phys. — Solid State **3** (1961) 822].
[260] P. J. Boddy and W. H. Brattain, J. electrochem. Soc. **109** (1962) 812.
[261] D. R. Frankl, J. electrochem. Soc. **109** (1962) 238.
[262] D. R. Frankl, Phys. Rev. **128** (1962) 2609.
[263] A. V. Rzhanov, Yu. F. Novototskii-Vlasov, and I. G. Neizvestnyi, Fiz. tver. Tela **1** (1959) 1471 [translation: Soviet Phys. — Solid State **1** (1960) 1349].
[264] G. Dorda, Czech. J. Phys. **B13** (1963) 272.
[265] A. V. Rzhanov and I. G. Neizvestnyi, Fiz. tver. Tela **3** (1961) 3317 [translation: Soviet Phys. — Solid State **3** (1962) 2408].
[266] A. V. Rzhanov, Yu. F. Novototskii-Vlasov, I. G. Neizvestnyi, T. I. Galkina, and S. V. Pokrovskaya in reference 7, p. 503.
[267] Yu. F. Novototskii-Vlasov and A. V. Rzhanov in reference 5, p. 93.
[268] I. G. Neizvestnyi in reference 3, p. 51.
[269] A. V. Rzhanov and I. A. Arkhipova, Fiz. tver. Tela **4** (1962) 1274 [translation: Soviet Phys. — Solid State **4** (1962) 934].
[270] Yu. F. Novototskii-Vlasov and M. P. Sinyukov in reference 3, p. 45.
[271] Yu. A. Kurskii, Fiz. tver. Tela **4** (1962) 2620 [translation: Soviet Phys. — Solid State **4** (1963) 1922].
[272] M. M. Atalla, E. Tannenbaum, and E. J. Scheibner, Bell System tech. J. **38** (1959) 749; M. M. Atalla in *Properties of Elemental and Compound Semiconductors* (Edited by H. C. Gatos), (Interscience, New York, 1960), p. 163.
[273] J. L. Sprague, J. A. Minhan, and O. J. Wied, J. electrochem. Soc. **109** (1962) 94.
[274] H. Edagawa, Y. Morita, S. Maekawa, and Y. Inuishi, Japan. J. appl. Phys. **2** (1963) 765.
[275] W. G. Pfann and C. G. B. Garrett, Proc. Inst. Radio Engrs. **47** (1959) 2011.

[276] J. L. Moll, I.R. E. Wescon Conf. Rec., Part 3, Electron. Dev. (1959) 32.
[277] J. T. Wallmark, RCA Rev. **24** (1963) 641. (Review).
[278] L. M. Terman, Solid-State Electron. **5** (1962) 285.
[279] K. Lehovec, Solid-State Electron. **6** (1963) 536; K. Lehovec, A. Slobodskoy, and J. L. Sprague, Phys. Status solidi **3** (1963) 447.
[280] S. R. Hofstein, K. H. Zaininger, and G. Warfield, Proc. Inst. Elect. Electron. Engrs. **52** (1964) 971.
[281] P. V. Gray, Phys. Rev. Letters **9** (1962) 302.
[282] K. Lehovec and A. Slobodskoy, Solid-State Electron. **7** (1964) 59.
[283] S. R. Hofstein and F. P. Heiman, Proc. Inst. Elect. Electron. Engrs. **51** (1963) 1190.
[284] W. Shockley, Proc. Inst. Radio Engrs. **40** (1952) 1363.
[285] A. Rose, *Concepts in Photoconductivity and Allied Problems* (Interscience, New York, 1963).
[286] R. Ya. Berlaga, L. P. Bol'shakov, P. P. Konorov, and M. I. Rudenok, Fiz. tver. Tela **5** (1963) 2991 [translation: Soviet Phys. — Solid State **5** (1964) 2189].
[287] G. W. Cullen, J. A. Amick, and D. Gerlich, J. electrochem. Soc. **109** (1962) 124; J. A. Amick, G. W. Cullen, and D. Gerlich, *ibid.* **109** (1962) 127; D. Gerlich, G. W. Cullen, and J. A. Amick, *ibid.* **109** (1962) 133.
[288] V. F. Synorov, V. V. D'yakov, and L. I. Bobrova in reference 3, p. 159.
[289] U. F. Gianola, J. appl. Phys. **28** (1957) 868.
[290] G. Dorda, Czech. J. Phys. **8** (1958) 181.
[291] K. Komatsubara, J. phys. Soc. Japan **14** (1959) 383; **17** (1962) 62.
[292] D. S. Peck, R. R. Blair, W. L. Brown, and F. M. Smits, Bell System tech. J. **42** (1963) 95.
[293] D. R. Kerr, Proc. Inst. Elect. Electron. Engrs. **51** (1963) 1142.
[294] Y. Margoninski, Phys. Rev. **121** (1961) 1282.
[295] H. Flietner and G. Hundt, Phys. Status solidi **3** (1963) 108.
[296] R. Bray and R. W. Cunningham in reference 6, p. 99.
[297] I. I. Abkevich, Fiz. tver. Tela **1** (1959) 1676 [translation: Soviet Phys. — Solid State **1** (1960) 1532].
[298] A. Surduts, Compt. rend. **251** (1960) 2665.
[299] V. I. Lyashenko and N. S. Chernaya, Fiz. tver. Tela **1** (1959) 1005 [translation: Soviet Phys. — Solid State **1** (1959) 921].
[300] D. R. Palmer, S. R. Morrison, and C. E. Dauenbaugh in reference 2, p. 27.
[301] V. G. Litovchenko and O. V. Snitko, Fiz. tver. Tela **2** (1960) 815 [translation: Soviet Phys. — Solid State **2** (1960) 748].
[302] S. Koc, Czech. J. Phys. **B11** (1961) 193.
[303] S. Koc, Czech. J. Phys. **B11** (1961) 287.
[304] S. Koc, Appl. Phys. Letters **4** (1964) 151.
[305] S. Koc, Phys. Status solidi **2** (1962) 1304.
[306] M. H. Pilkuhn, J. appl. Phys. **34** (1963) 3302.
[307] S. Koc, Czech. J. Phys. **B11** (1961) 289; **B13** (1963) 781.
[308] G. Dorda, Phys. Status solidi **3** (1963) 1318.
[309] G. Dorda, Phys. Status solidi **5** (1964) 107; G. Dorda and S. Koc in reference 5, p. 120.
[310] A. Many and Y. Goldstein in reference 5, p. 114.
[311] V. I. Lyashenko and N. S. Chernaya, Fiz. tver. Tela **1** (1959) 878 [translation: Soviet Phys. — Solid State **1** (1959) 799].
[312] W. Schockley, H. J. Queisser, and W. W. Hooper, Phys. Rev. Letters **11** (1963) 489; W. Schockley, W. W. Hooper, H. J. Queisser, and W. Schroen in reference 5, p. 277.
[313] G. A. Kataev, V. A. Presnov, E. N. Batueva, Yu. G. Kataev, and L. L. Lyuze in reference 3, p. 151.
[314] J. F. Dewald in *Semiconductors* (Edited by N. B. Hannay), (Reinhold, New York,

1959), p. 727.
[315] M. Green in *Modern Aspects of Electrochemistry* (Edited by J. O'M. Bockris), (Butterworth, London), No. 2 (1959), p. 343; *Solid State Physics in Electronics and Telecommunications* (Edited by M. Désirant and J. L. Michiels), (Academic Press, London), vol. I (1960), p. 619.
[316] H. Gerischer in *Advances in Electrochemistry and Electrical Engineering* (Edited by P. Delahay), (Interscience, New York, vol. **1** (1961), p. 139.
[317] D. R. Turner in *The Electrochemistry of Semiconductors* (Edited by P. J. Holmes), (Academic Press, London, 1962), p. 155.
[318] W. H. Brattain and C. G. B. Garrett, Physica **20** (1954) 885; Bell System tech. J. **34** (1955) 129.
[319] H. U. Harten, Z. Naturforsch. **16a** (1961) 1401.
[320] H. U. Harten, Z. Naturforsch. **16a** (1961) 459.
[321] J. F. Dewald, Bell System tech. J. **39** (1960) 615; in reference 2, p. 155.
[322] A. Many, J. Phys. Chem. Solids **26** (1965) 587.
[323] H. Gobrecht and O. Meinhardt, Ber. Bunsengesell. phys. Chem. **67** (1963) 142.
[324] R. Memming, Philips Res. Repts. **19** (1964) 323.
[325] R. Williams, Phys. Rev. **123** (1961) 1645.
[326] M. Hofmann-Perez and H. Gerischer, Z. Elektrochem. **65** (1961) 771.
[327] H. U. Harten and R. Memming, Phys. Letters **3** (1962) 95.
[328] P. J. Boddy and W. H. Brattain, J. electrochem. Soc. **110** (1963) 570.
[329] W. H. Brattain and P. J. Boddy, J. electrochem. Soc. **109** (1962) 574.
[330] W. H. Brattain and P. J. Boddy, Proc. Natl. Acad. Sci. U. S. **48** (1962) 2005.
[331] P. J. Boddy and W. J. Sundburg, J. electrochem. Soc. **110** (1963) 1170.
[332] M. Green, J. chem. Phys. **31** (1959) 200.
[333] Yu. V. Pleskov and M. D. Krotova in reference 8, p. 807; M. D. Krotova and Yu. V. Pleskov, Phys. Status solidi **2** (1962) 4411.
[334] K. Bohnenkamp and H. J. Engell, Z. Elektrochem. **61** (1957) 1184.
[335] P. T. Wrotenbery and A. W. Nolle, J. electrochem. Soc. **109** (1962) 534.
[336] M. D. Krotova and Yu. V. Pleskov, Phys. Status solidi **3** (1963) 2119.
[337] R. M. Hurd and P. T. Wrotenbery in reference 4, p. 876.
[338] H. Gobrecht and O. Meinhardt, Ber. Bunsengesell. phys. Chem. **67** (1963) 151.
[339] H. Gobrecht, O. Meinhardt, and B. Reinicke, Ber. Bunsengesell. phys. Chem. **67** (1963) 492.
[340] H. U. Harten, Proc. Inst. Elect. Engrs. **106B** (1959) 906; reference 2, p. 220.
[341] R. Memming, Surface Science **1** (1964) 88.
[342] P. G. Boddy and W. H. Brattain in reference 4, p. 683.
[343] W. H. Brattain and P. J. Boddy in reference 8, p. 797.
[344] W. W. Harvey, J. electrochem. Soc. **109** (1962) 638; reference 4, p. 904.
[345] V. S. Sotinkov and A. S. Belanovskii, Zhur. fiz. Khim. **35** (1961) 509 [translation: Russ. J. phys. Chem. **35** (1961) 249].
[346] R. Memming in reference 5, p. 436.
[347] R. Memming, Phys. Letters **7** (1963) 89.
[348] Yu. V. Pleskov, Doklady Akad. Nauk SSSR **126** (1959) 111 [translation: Proc. Acad. Sci. U.S.S.R., Phys. Chem. Sect. 1959, **126** (1961) 381]; in reference 7, p. 573.
[349] M. Green, V. Jendrasic, and J. McBreen, J. Phys. Chem. Solids **24** (1963) 701.
[350] W. W. Harvey in reference 2, p. 82.
[351] W. W. Harvey and H. C. Gatos, J. appl. Phys. **29** (1958) 1267.
[352] S. H. Autler, A. L. McWhorter, and H. A. Gebbie, Bull. Am. phys. Soc. **1** (1956) 145.
[353] P. Handler in reference 1, p. 23.
[354] R. Forman, Phys. Rev. **117** (1960) 698.

[355] Y. Margoninski, Appl. Phys. Letters **2** (1963) 143; Phys. Rev. **132** (1963) 1910.
[356] R. Missman and P. Handler in reference 6, p. 109.
[357] J. T. Law and C. G. B. Garrett, J. appl. Phys. **27** (1956) 656.
[358] S. Wang and G. Wallis, J. appl. Phys. **30** (1959) 285.
[359] H. H. Madden and H. E. Farnsworth, Phys. Rev. **112** (1958) 793.
[360] V. I. Lyashenko, N. S. Chernaya, and A. B. Gerssiniov. Fiz. tver. Tela **2** (1960) 2421 [translation: Soviet Phys. — Solid State **2** (1961) 2158].
[361] F. G. Allen, T. M. Buck, and J. T. Law, J. appl. Phys. **31** (1960) 979.
[362] P. Handler and W. M. Portnoy, Phys. Rev. **116** (1959) 516.
[363] P. Handler in reference 2, p. 1.
[364] A. Kobayashi, Z. Oda, S. Kawaji, H. Arata, and K. Sugiyama, J. Phys. Chem. Solids **14** (1960) 37; Z. Oda, K. Sugiyama, H. Arata, and A. Kobayashi, in reference 7, p. 544.
[365] A. Kobayashi, K. Sugiyama, H. Arata, and Z. Oda, J. phys. Soc. Japan **16** (1961) 2481.
[366] R. F. Greene in reference 5, p. 101.
[367] N. St. J. Murphy in reference 5, p. 86.
[368] P. Handler and S. Eisenhour in reference 5, p. 64.
[369] L. Apker, E. Taft, and J. Dickey, Phys. Rev. **74** (1948) 1462.
[370] J. A. Dillon, Jr. and H. E. Farnsworth, J. appl. Phys. **28** (1957) 174.
[371] F. G. Allen and A. B. Fowler, J. Phys. Chem. Solids **3** (1957) 107.
[372] A. B. Fowler, J. appl. Phys. **30** (1959) 556.
[373] D. R. Frankl, J. appl. Phys. **34** (1963) 3514.
[374] D. R. Palmer, S. R. Morrison, and C. E. Dauenbaugh, Phys. Rev. **129** (1963) 608.
[375] G. A. Barnes and P. C. Banbury, Proc. phys. Soc. (London) **71** (1958) 1020; in reference 6, p. 111.
[376] P. C. Banbury, E. A. Davies, and G. W. Green in reference 5, p. 813.
[377] G. W. Gobeli and F. G. Allen in reference 5, p. 402.
[378] G. W. Gobeli, F. G. Allen, and E. O. Kane, Phys. Rev. Letters **12** (1964) 94; E. O. Kane, *ibid.* **12** (1964) 97.
[379] F. G. Allen in reference 6, p. 119.
[380] G. Busch, H. Schade, A. Gobbi, and P. Marmier, J. Phys. Chem. Solids **23** (1962) 513.
[381] J. A. Dillon, Jr. and R. M. Oman in reference 7, p. 533.
[382] J. T. Law in reference 2, p. 9.
[383] J. T. Law, J. appl. Phys. **32** (1961) 600.
[384] J. T. Law, J. appl. Phys. **32** (1961) 848.
[385] G. Heiland and H. Lamatsch in reference 5, p. 18; H. Lamatch, Phys. Status solidi **9** (1965) 119.
[386] J. A. Dillon, Jr. and H. E. Farnsworth, J. appl. Phys. **29** (1958) 1195.
[387] H. E. Farnsworth, R. E. Schlier, and J. A. Dillon, Jr. in reference 6, p. 116.
[388] F. G. Allen in reference 4, p. 850. (References).
[389] G. Busch and T. Fischer, Phys. Kondens. Materie **1** (1963) 367. (References).
[390] D. R. Palmer, S. R. Morrison, and C. E. Dauenbaugh, Phys. Rev. Letters **6** (1961) 170.
[391] P. Handler, Phys. Rev. **126** (1962) 971.
[392] S. Kawaji and Y. Takishima, Surface Science **1** (1964) 119.
[393] P. Handler, Appl. Phys. Letters **3** (1963) 96.
[394] G. W. Gobeli and F. G. Allen in reference 2, p. 23.
[395] F. G. Allen and G. W. Gobeli, Phys. Rev. **127** (1962) 150; in reference 8, p. 818; G. W. Gobeli and F. G. Allen in reference 4, p. 647.
[396] G. W. Gobeli and F. G. Allen, Phys. Rev. **127** (1962) 141.
[397] J. van Laar and J. J. Scheer, Philips Res. Repts. **17** (1962) 101.

[398] J. van Laar and J. J. Scheer in reference 8, p. 827.
[399] J. J. Scheer and J. van Laar, Phys. Letters **3** (1963) 246.
[400] F. G. Allen and G. W. Gobeli, J. appl. Phys. **35** (1964) 597.
[401] J. Eisinger, J. chem. Phys. **30** (1959) 927.
[402] G. Heiland and P. Handler, J. appl. Phys. **30** (1959) 446.
[403] A. Kobayashi and S. Kawaji, J. phys. Soc. Japan **12** (1957) 1054; S. Kawaji, *ibid.* **15** (1960) 95.
[404] M. J. Sparnaay in reference 4, p. 973; A. H. Boonstra, J. van Ruler, and M. J. Sparnaay, Proc. K. Ned. Akad. Wetensch. **B66** (1963) 64.
[405] M. J. Sparnaay and J. van Ruler, Physica **27** (1961) 153.
[406] M. J. Sparnaay, A. H. Boonstra, and J. van Ruler in reference 5, p. 56.
[407] A. H. Boonstra, J. van Ruler, and M. J. Sparnaay, Proc. K. Ned. Akad. Wetensch. **B66** (1963) 70.
[408] J. T. Law, J. chem. Phys. **30** (1959) 1568.
[409] J. J. Scheer, Philips Res. Repts. **15** (1960) 584.
[410] R. E. Simon and W. E. Spicer, Phys. Rev. **119** (1960) 621.
[411] W. E. Spicer and R. E. Simon, Phys. Rev. Letters **9** (1962) 385; J. Phys. Chem. Solids **23** (1962) 1817.

AUTHOR INDEX

ABKEVICH, I. I., (83) 359; (84) 359; (297) 409
ADAM, G., (20) 272; (109) 363
ADAMS, A., Jr., (50) 250–254; (46) 353; (48) 354; (46) 356; (48) 356; (46) 357; (48) 357; (46) 360; (48) 360; (86) 360; (48) 361; (46) 378; (48) 378; (86) 378; (46) 390; (48) 390; (86) 390; (46) 392; (86) 392, 399; (48) 407; (86) 407
ADAMS, W. G., (3) 2
AERTS, E., (63) 355
AIGRAIN, P., (9) 15; (36) 244; (168) 371
ALBERS, W. A., Jr., (15) 328, 339–341; (216) 382, 391, 406
ALEXANDER, E., (23) 227; (121) 364, 365
ALEXANDRAKIS, G. C., (250) 392
ALLEN, F. G., (10) 92; (24) 94; 95, 100; (44) 100, 101; (24) 102; (30) 275, 279; (36) 279; (37) 279, 281; (36) 283, 284; (30) 285; (36) 285; (37) 285; (361) 430; (371) 436; (377) 438; (378) 439; (361) 440; (379) 440; (361) 441; (388) 443; (394) 444; (395) 444; (396) 444, 446; (395) 447, 448; (396) 448; (400) 448; (395) 449; (400) 449; (395) 452, 454
ALLEN, J. W., (96) 120
ALPERT, D., (1) 90, 104; (58) 104
AMELINCKX, S., (63) 355
AMICK, J. A., (287) 406
AMITH, A., (7) 306, 313, 314, 323, 325
ANDRIANOV, D. G., (258) 396
ANOCHIN, B. G., (201) 378, 390
ANTONČÍK, E., (10) 172, 173
APKER, L., (40) 280; (369) 436
ARATA, H., (364) 434; (365) 434, 442
ARCHER, R. J., (23) 347
ARKHIPOVA, I. A., (30) 196; (129) 366; (269) 401
ARTMAN, K., (12) 173, 174, 180

ATALLA, M. M., (87) 119; (272) 402, 403, 410
AUGUSTYNIAK, W. M., (10) 268
AUTLER, S. H., (352) 429, 430
AVERY, D. G., (19) 272

BAHADIN, K., (46) 101
BAKER, D., (140) 367
BALK, P., (224) 382–384, 391, 408
BALZAROTTI, A., (204) 378, 390
BANBURY, P. C., (7) 92; (9) 92; (28) 229; (62) 355; (75) 356; (78) 356; (82) 356; (122) 364; (231) 385; (78) 388; (82) 388; (78) 391; (375) 437; (376) 437; (375) 438; (376) 438
BAO, S.-M., (193) 376
BARDEEN, J., (13) 5; (16) 6, 7; (21) 7; (18) 132; (20) 134; (25) 156; (27) 194, 195; (1) 209; (25) 227; (29) 275; (16) 329; (10) 346; (24) 349; (25) 349; (27) 349; (25) 352; (27) 352; (41) 352; (25) 356; (41) 356, 357; (24) 361; (25) 363; (41) 378, 390; (24) 428
BARDSLEY, W., (116) 120, 123
BARLOW, C. A., JR., (178) 372
BARNES, G. A., (7) 92; (9) 92; (375) 437, 438
BATH, H. M., (11) 220; (49) 250; (51) 250; (130) 366
BATTERMAN, B. W., (99) 121
BATUEVA, E. N., (313) 415
BECKER, G. E., (77) 295
BECKER, J. A., (54) 103; (55) 103
BELANOVSKII, A. S., (345) 423
BELL, R. L., (116) 123
BENNY, A. H., (110) 363, 366
BENZER, S., (11) 5
BERLAGA, R. Ya., (286) 406
BERNAMONT, J., (157) 369
BERNARD, M., (34) 201

Numbers in parentheses are reference numbers. References are listed at the end of each chapter, on the following pages: 13–14, 89, 124–127, 163–164, 208, 254–256, 301–303, 344–345, 459–470.

AUTHOR INDEX

BERNSTEIN, B., (87) 360, 374, 410
BERNSTEIN, R. B., (85) 119
BERZ, F., (31) 201; (42) 244; (34) 277
BESS, L., (142) 367; (143) 367; (142) 368; (143) 368; (146) 368; (142) 371; (166) 371
BETHE, H., (4) 3; (10) 16
BIERIG, R., (60) 104
BILLIG, E., (103) 121
BLAIR, R. R., (292) 407
BLOCH, F., (5) 3; (12) 18, 19
BOBROVA, L. I., (288) 406
BOCCIARELLI, C. V., (95) 362
BODDY, P. J., (17) 224; (18) 225; (17) 233; (260) 369; (328) 418; (329) 418; (330) 418; (331) 418; (260) 419; (328) 419; (329) 419; (330) 419; (331) 419; (260) 420; (328) 420; (329) 420; (342) 420; (343) 420; (260) 421; (343) 421; (260) 422; (343) 423
BOHNENKAMP, K., (16) 224, 233; (334) 419
BOL'SHAKOV, L. P., (286) 406
BONCH-BRUEVICH, V. L., (41) 244
BOONSTRA, A. H., (404) 452; (406) 452; (407) 452
BORZYAK, P. G., (196) 377
BRADLEY, W. E., (93) 362
BRAFMAN, O., (229) 384, 386, 391
BRATTAIN, W. H., (16) 6; (19) 6; (20) 6; (16) 7; (21) 7; (107) 122; (9) 128, 141, 154; (25) 156; (27) 194, 195; (1) 209; (5) 210; (8) 212; (10) 220; (5) 223; (17) 224; (18) 225; (17) 233; (28) 274; (29) 275; (31) 275, 278; (35) 279; (25) 349; (28) 349; (31) 350; (32) 350; (25) 352; (57) 355; (25) 356; (28) 356; (57) 357; (92) 361; (25) 363; (101) 363; (31) 382; (32) 382; (31) 391; (32) 391; (57) 391; (260) 396; (318) 416; (328) 418; (330) 418; (260) 419; (328) 419; (329) 419; (330) 419; (260) 420; (328) 420; (329) 420; (342) 420; (343) 420; (260) 421; (343) 421, 423
BRAUN, F., (1) 2
BRAY, R., (22) 7; (4) 264; (11) 268; (24) 338; (296) 408, 410
BRENNAN, D., (63) 105, 114

BRODER, J. D., (12) 92, 93, 108
BROOKS, H., (15) 71
BROPHY, J. J., (145) 368; (147) 368; (148) 368; (149) 368; (150) 368
BROWN, W. L., (4) 210; (10) 220; (20) 225; (4) 227; (20) 227; (44) 250; (32) 350; (39) 352; (40) 352, 353; (42) 353; (57) 355; (39) 356; (40) 356; (39) 357; (40) 357; (57) 357; (39) 378; (40) 378; (32) 382, 390; (40) 390; (57) 390; (292) 407
BRUNAUER, S., (17) 93; (19) 93
BUBE, R. H., (111) 364; (182) 373
BUCK, T. M., (78) 117, 120; (37) 352; (101) 363; (235) 385, 392; (361) 430, 440, 441
BURGER, R. M., (29) 95; (37) 95; (29) 96; (37) 97
BURGESS, R. E., (9) 15; (73) 356
BURITZ, R. S., (58) 104
BURSHTEIN, R. Kh., (191) 375
BURTON, J. A., (197) 377
BUSCH, G., (380) 441; (389) 443

CALLEN, E., (19) 132
CAMP, P. R., (100) 121
CARASSO, J. I., (56) 354
CHAMPION, J. A., (251) 392, 403
CHERNAYA, N. S., (299) 410, 412; (311) 413; (360) 430
CHIAROTTI, G., (204) 378, 390; (246) 396
CHRISTENSEN, H., (47) 250; (15) 272; (50) 354, 360
CHURCHMAN, A. T., (102) 121
CHYNOWETH, A. G., (51) 286; (61) 290; (64) 291; (68) 292
CLADIS, J. B., (9) 268
CLARK, J. C., (120) 124
CLARKE, E. N., (33) 352; (34) 352; (53) 354
CONWELL, E. M., (7) 15
COOVERT, R. E., (16) 329; (27) 341, 342; (41) 352, 356, 357, 378, 390
COUTTS, M. D., (11) 92
CRETELLA, M. C., (91) 120; (256) 396
CROCKER, A. J., (115) 122
CUBICCIOTTI, D., (85) 119
CULLEN, G. W., (287) 406

Numbers in parentheses are reference numbers. References are listed at the end of each chapter, on the following pages: 13–14, 89, 124–127, 163–164, 208, 254–256, 301–303, 344–345, 459–470.

AUTHOR INDEX 473

CUNNINGHAM, R. W., (24) 338; (296) 408, 410
CUSSINS, W. D., (38) 352
CUTTLER, M., (11) 220; (49) 250; (51) 250; (130) 366

D'ASARO, L. A., (57) 289
DASH, W. C., (97) 121; (117) 124
DAUENBAUGH, C. E., (8) 92; (300) 410; (374) 437, 438; (390) 443, 450; (374) 451
DAVIDOV, B., (10) 4; (25) 7; (3) 128
DAVIES, E. A., (199) 377; (376) 437, 438
DAVIS, L., Jr., (46) 250; (50) 250-254; (45) 353; (46) 353; (48) 354; (45) 356; (46) 356; (48) 356; (45) 357; (46) 357; (48) 357; (46) 360; (48) 360; (86) 360; (48) 361; (46) 378; (48) 378; (86) 378; (46) 390; (48) 390; (86) 390; (46) 392; (86) 392, 399; (48) 407; (86) 407
DAVISON, C. J., (36) 96
DAY, R. E., (3) 2
DELL, R. M., (22) 93, 114
del SIGNORE, G., (246) 393
deMARS, G. A., (46) 250; (50) 250-254; (45) 353; (46) 353; (48) 354; (45) 356; (46) 356; (48) 356; (45) 357; (46) 357; (48) 357; (46) 360; (48) 360; (86) 360; (89) 360, 361; (91) 361; (46) 378; (48) 378; (86) 378; (46) 390; (48) 390; (86) 390; (46) 392; (86) 392, 399; (48) 407; (86) 407
DEMBER, H., (33) 276
DENNIS, L. M., (21) 93
DEWALD, J. F., (14) 224; (314) 415; (321) 418
DICKEY, J., (369) 436
DILLON, J. A., Jr., (42) 100; (370) 436; (381) 441; (386) 442; (387) 443; (386) 449, 450; (370) 451; (386) 451, 452
DOLAN, W. W., (56) 288
DORDA, G., (187) 374; (222) 382, 391, 398; (264) 398, 399; (290) 407; (308) 413; (309) 413
DOUSMANIS, G. C., (10) 128, 139; (132) 366; (248) 390, 392; (250) 392
DUGAS, C., (168) 371

DUNCAN, R. C., (10) 128, 139
DUNLAP, W. C. JR., (2) 15
DUSHMAN, S., (2) 90, 104
D'YAKOV, V. V., (288) 406
DYKE, W. P., (56) 288

EBHARDT, R., (12) 128; (52) 250
EDAGAWA, H., (154) 368; (274) 403
EDELMAN, S., (119) 124
EHRENBERG, W., (3) 15; (39) 97
EINSPRUCH, N. G., (64) 355; (205) 378, 392
EISENHOUR, S., (368) 435, 436
EISINGER, J., (24) 94, 95, 100, 102; (56) 103, 104; (69) 113; (44) 285; (401) 450, 452
ELLIS, R. C., (113) 122
ELLIS, S. G., (101) 121; (105) 121; (102) 363, 383
EMMETT, P. H., (17) 93; (19) 93; (20) 93
EMMONY, D. C., (34) 277
ENGELL, H. J., (16) 224, 233; (169) 371; (334) 419
ERIKSEN, W. T., (89) 360; (90) 360; (89) 361
ESAKI, L., (63) 291
ESTLE, T. L., (84) 297

FAN, H. Y., (5) 264; (11) 268
FARNSWORTH, H. E., (3) 90; (29) 95; (30) 95, 96; (37) 95; (29) 96; (35) 96; (30) 97; (37) 97; (3) 99; (41) 99; (42) 100; (35) 102; (30) 106; (41) 106; (30) 107, 109; (41) 109; (3) 110; (41) 112, 113; (30) 117; (215) 382, 391, 398, 412; (359) 430; (370) 436; (386) 442; (387) 443; (386) 449, 450; (370) 451; (386) 451, 452
FAUST, J. W., Jr., (93) 120; (114) 120; (93) 121; (114) 122, 123
FEDOROVICH, Yu. V., (190) 375
FEHER, G., (82) 297
FELDMANN, W. L., (61) 290; (151) 368
FEUERSANGER, A., (18) 329, 330
FINGERLAND, A., (4) 165
FISCHER, T., (389) 443
FISHER, R. B., (89) 119
FISTUL, V. I., (9) 215; (258) 396; (259) 396
FLETCHER, R. C., (81) 297
FLIETNER, H., (11) 128, 141; (8) 306, 314;

Numbers in parentheses are reference numbers. References are listed at the end of each chapter, on the following pages: 13-14, 89, 124-127, 163-164, 208, 254-256, 301-303, 344-345, 459-470.

AUTHOR INDEX

(18) 346; (202) 378, 390, 398; (295) 408
FORMAN, R., (354) 429–431, 440, 450
FOWLER, A. B., (371) 436; (372) 437
FOWLER, R. H., (54) 287
FRANÇOIS, E. E., (167) 371, 420
FRANKL, D. R., (13) 128; (23) 139; (13) 141;
(9) 313, 315, 321–323; (18) 329, 330; (79) 356; (261) 396; (262) 396; (261) 397; (262) 397; (261) 408; (262) 411, 423; (373) 437
FRANKS, J., (102) 121
FRANZ, W., (59) 289; (60) 289
FRENKEL, J., (7) 3
FROLOV, O. S., (185) 374, 376; (193) 376
FROVA, A., (16) 346; (204) 378, 390; (246) 393
FUCHS, E., (112) 122
FUCHS, K., (2) 304, 322

GALKINA, T. I., (249) 391, 398, 399; (266) 399; (249) 400; (266) 400; (266) 410
GARRETT, C. G. B., (107) 122; (9) 128, 141, 154; (5) 211; (8) 212; (10) 220; (5) 223; (39) 244, 247; (28) 274; (31) 275, 278; (35) 279; (28) 349; (31) 350; (32) 350; (57) 355; (28) 356; (57) 357; (92) 361; (176) 372; (31) 382; (32) 382; (31) 390; (32) 390; (57) 390; (275) 403; (318) 416; (357) 430
GATOS, H. C., (4) 92; (14) 93, 95; (79) 117; (86) 119; (91) 120; (256) 396; (351) 424
GEACH, G. A., (102) 121
GEBALLE, T. H., (29) 342
GEBBIE, H. E., (143) 367, 368; (352) 429, 430
GELLER, M., (62) 290
GEORGE, T. H., (29) 95; (37) 95; (29) 96; (37) 97
GERISCHER, H., (316) 415; (326) 418, 419, 421
GERLICH, D., (27) 228; (127) 365; (207) 378; (127) 379, 380, 390; (207) 392; (127) 398; (287) 406; (127) 408
GERMER, L. H., (36) 96; (38) 97
GERSSINIOV, A. B., (360) 430
GIANOLA, U. F., (289) 407
GIBBONS, J. F., (152) 368, 370
GIBSON, A. F., (9) 15

GOBBI, A., (380) 441
GOBELI, G. W., (10) 92, (67) 106, 109–113, 117; (30) 275, 279; (36) 279; (37) 279, 281; (36) 283, 284; (30) 285; (36) 285; (37) 285; (77) 295; (377) 438; (378) 439; (394) 444; (395) 444; (396) 444, 446; (395) 447, 448; (396) 448; (400) 448; (395) 449; (400) 449; (395) 452, 454
GOBRECHT, H., (323) 418; (338) 419; (339) 419
GOERING, H. L., (6) 92
GOLDSTEIN, Y., (14) 128; (15) 128; (22) 139; (22) 226; (30) 229; (31) 235; (30) 239; (32) 241; (65) 291; (66) 291; (10) 317; (12) 318–321; (19) 331, 332; (22) 332, 335; (213) 382; (225) 384; (229) 385, 386; (127) 390; (225) 390; (229) 391; (310) 413
GOOD, R. H., (50) 286
GOODWIN, E. T., (8) 172; (17) 174, 177
GORDON-SMITH, G. W., (46) 286
GOUCHER, F. S., (18) 272
GRANVILLE, J. W., (35) 352
GRAY, P. V., (281) 403, 404
GRAY, T. J., (173) 372
GREEN, G. W., (376) 437, 438
GREEN, M., (13) 92, 93; (15) 93; (23) 93; (25) 94; (13) 113; (15) 113; (13) 114; (15) 114; (70) 114; (75) 115; (15) 116; (75) 116; (76) 116, 117; (17) 128, 145; (315) 415; (332) 418; (349) 424
GREENE, R. F., (15) 128; (23) 139; (15) 141; (4) 304; (9) 313; (9) 315; (11) 317; (9) 321–323; (30) 342; (366) 435
GRIMLEY, T. B., (5) 165
GROSCHWITZ, E., (12) 128, (52) 250
GROSS, E. P., (62) 290
GROVER, N. B., (14) 128; (22) 139; (22) 226; (30) 229; (31) 235; (30) 239; (32) 241; (65) 291; (10) 317; (12) 318–321; (19) 331, 332; (20) 332; (21) 332–335; (213) 382; (225) 384, 390
GUNN, J. B., (19) 272; (94) 362

HAGSTRUM, H. D., (24) 94; (26) 94; (24) 95; (26) 95; (33) 95; (24) 100; (48) 101;

Numbers in parentheses are reference numbers. References are listed at the end of each chapter, on the following pages: 13–14, 89, 124–127, 163–164, 208, 254–256, 301–303, 344–345, 459–470.

AUTHOR INDEX

(49) 101; (50) 101; (24) 102; (50) 102; (51) 102; (52) 102; (53) 102; (26) 113
HALL, M. B., (234) 385
HALL, T. C., (17) 329; (206) 378, 392, 403
HAM, F. S., (5) 305
HANCE, F. E., (21) 93
HANCOCK, R. D., (119) 124
HANDLER, P., (77) 117; (23) 337, 338; (14) 346; (353) 429, 430; (356) 430; (362) 430, 433; (363) 433; (368) 435, 436; (353) 443; (391) 443, 444; (393) 444; (356) 450; (362) 450; (402) 450; (356) 451; (402) 452
HANEMAN, D., (9) 92; (27) 94; (32) 95; (27) 96, 106; (32) 106; (27) 108; (68) 108; (27) 109; (32) 109; (27) 110; (68) 110; (27) 111; (68) 111; (27) 113; (32) 113; (27) 117
HANNAY, N. B., (8) 15
HARNIK, E., (22) 139; (29) 194; (23) 227; (24) 227; (30) 229; (31) 235; (30) 239; (32) 241; (65) 291; (19) 331–333; (121) 364; (125) 365; (213) 382; (225) 384, 390; (253) 395, 398
HARRICK, N. J., (34) 243; (24) 273; (71) 292; (24) 293; (72) 293; (73) 293; (74) 294; (71) 294; (73) 294); (75) 294; (58) 355; (59) 355; (60) 355; (58) 362, 363; (59) 363; (60) 389
HARTEN, H. U., (15) 224; (21) 346; (236) 385; (319) 417; (320) 417; (327) 418; (21) 419; (320) 419; (340) 419
HARTIG, P. A., (96) 362
HARTMAN, C. D., (38) 97; (54) 103
HARVEY, W. W., (19) 225; (344) 420; (350) 424; (351) 424
HAUFFE, K., (169) 371; (171) 372
HAYNES, J. R., (35) 243; (7) 268; (14) 272; (17) 272
HAYWARD, D. V., (63) 105, 114
HEIDENREICH, R. D., (94) 120
HEIL, O., (18) 6
HEILAND, G., (77) 117; (17) 346; (20) 346; (180) 373; (385) 441, 449, 450; (402) 450; (385) 451; (402) 452
HEIMAN, F. P., (13) 223; (283) 404, 405

HEINE, V., (11) 172
HEITLER, W., (11) 18
HENISCH, H. K., (6) 128; (120) 364
HILL, J. E., (135) 366
HIPPLE, J. A., (57) 104
HOBSON, J. P., (2) 90, 104; (62) 105
HOBSTETTER, J. N., (17) 51
HODGKINSON, R. J., (198) 377
HOFMANN-PEREZ, M., (326) 418, 419, 421
HOFMEISTER, E., (12) 128
HOFSTEIN, S. R., (13) 223; (280) 403, 404; (283) 404, 405; (280) 410
HOGARTH C. A., (35) 352; (105) 363
HOLDEN, A. N., (81) 297
HOLMES, P. J., (5) 92; (80) 118; (81) 118; (80) 120; (81) 120; (95) 120; (80) 121; (81) 121; (118) 124
HOLONYAK, N., Jr., (100) 363
HOOPER, W. W., (312) 415
HOPKINS, R. L., (34) 352
HORNBECK, J. A., (7) 268
HROSTOWSKI, H. J., (22) 185
HSÜ, T.-L., (88) 360
HUNDT, G., (295) 408
HURD, R. M., (337) 419
HUTSON, A. R., (54) 354, 375
HYDE, F. J., (141) 367, 368

INUISHI, Y., (154) 368; (274) 403
IOFFE, A. F., (5) 15
IOSELEVICH, M. L., (259) 396

JAMOSHITA, J., (48) 286, 290, 291
JÄNTSCH, O., (177) 372
JELATIS, J. G., (62) 290
JENDRASIC, V., (349) 424
JOHNSON, E. O., (25) 273; (32) 275, 277, 279; (29) 349; (30) 349, 351
JOHNSON, R. R., (186) 374, 411
JONES, C. S., (9) 268

KAFALAS, A. J., (13) 92, 93, 113, 114
KALASHNIKOV, S. G., (21) 185; (36) 352
KANAI, Y., (117) 364
KANE, E. O., (42) 282, 283, 285; (378) 439
KANOU, K., (189) 375

Numbers in parentheses are reference numbers. References are listed at the end of each chapter, on the following pages: 13–14, 89, 124–127, 163–164, 208, 254–256, 301–303, 344–345, 459–470.

KARPENKO, I. V., (183) 373
KARPOV, Yu. S., (156) 368, 370
KATAEV, G. A., (313) 415
KAWAJI, S., (181) 373; (364) 434; (365) 442; (392) 444; (403) 451
KAWASAKI, K., (189) 375
KELVIN, Lord, (27) 274
KERR, D. R., (293) 407
KEYES, R. J., (28) 194; (107) 363; (108) 363, 364
KIKUCHI, M., (67) 356; (123) 364
KINGSTON, R. H., (8) 128, 154; (2) 210; (7) 212; (45) 250; (48) 250; (9) 346; (43) 353; (44) 353, 354; (52) 354; (44) 356; (70) 356, 358–360, 369; (9) 374; (44) 374; (70) 410
KLEIN, D. L., (106) 122
KOBAYASHI, A., (181) 373; (364) 434; (365) 434, 442; (403) 451
KOC, S., (302) 411; (303) 411; (304) 411; (305) 411; (304) 412; (307) 412; (309) 413
KOGAN, Sh. M., (174) 372; (192) 375
KOLB, G. A., (106) 122
KOLM, H. H., (66) 352
KOMATSUBARA, K., (136) 366; (154) 368; (291) 407
KONOROV, P. P., (254) 395; (255) 396; (286) 406
KOPPLIN, J. O., (206) 378, 392, 403
KORNELSEN, E. V., (2) 90, 104
KOSMAN, M. S., (84) 359
KOUTECKÝ, J., (2) 165; (4) 165; (2) 171; (2) 174; (13) 174; (14) 174; (15) 174; (16) 174; (13) 181; (14) 182; (18) 182
KOZLOVSKAYA, V. M., (83) 118
KRÖGER, F. A., (9) 15
KRONIG, R. de L., (14) 19
KROTOVA, M. D., (333) 419; (336) 419, 424
KRUSEMEYER, H. J., (172) 372
KURSKII, Yu. A., (271) 402

LAGRENAUDI, J., (36) 244
LAMATSCH, H., (385) 441, 449–451
LANDER, J. J., (40) 97, 98; (66) 106; (67) 106; (66) 108, 109; (67) 109, 110; (66) 111, 112; (67; 112; (66) 113; (67) 113; (66) 117; (67) 117
LANDSBERG, P. T., (72) 114; (32) 201
LARIN, L. A., (191) 375
LASSER, M., (87) 360, 374, 410
LAST, J. T., (23) 186
LAURITZEN, P. O., (152) 368, 370
LAVINE, M. C., (79) 117; (86) 119
LAW, J. T., (24) 94, 95, 100, 102, (56) 103; 104; (69) 113; (73) 114; (84) 119; (12) 346; (51) 354, 360; (167) 371; (12) 373; (188) 374, 401; (167) 420; (357) 430; (361) 430, 440, 441; (382) 441; (383) 441; (384) 441; (382) 442; (383) 442; (384) 442; (382) 450; (383) 450, 451; (382) 452; (383) 452; (384) 452; (408) 452; (383) 453; (408) 453
LAX, M., (35) 204; (13) 346
LEHOVEC, K., (279) 403, 404; (282) 404
LETAW, H., Jr., (100) 363
LEVINSTEIN, H., (144) 367, 368, 370
LIANDRAT, G., (36) 244
LIBERMAN, I. A., (25) 94; (75) 115, 116
LIEBSON, S. H., (113) 364
LIGENZA, J. R., (74) 115
LINDLEY, D. H., (78) 356; (82) 356; (231) 385; (78) 388; (82) 388; (78) 391
LIPMANN, A. B., (6) 170
LITOVCHENKO, V. G., (185) 374, 376; (193) 376; (203) 378; (219) 382; (220) 382; (221) 382, 385; (230) 385; (233) 385; (238) 385; (239) 385; (221) 388; (238) 388; (239) 388; (242) 388; (243) 388; (203) 390; (220) 391; (221) 391; (220) 392; (238) 392; (239) 392, 402; (203) 407; (221) 408; (239) 408; (301) 410; (203) 412; (301) 414
LITVINOV, R. O., (69) 356; (88) 360
LOBANOVA, T. N., (131) 366
LOGAN, R. A., (61) 290; (64) 291
LONDON, F., (11) 18
LONGINY, R. L., (12) 268; (98) 363
LOW, G. G. E., (24) 191; (26) 227–229; (28) 229; (68) 356; (75) 356
LUDWIG, G. W., (8) 268
LYASHENKO, V. I., (69) 356; (124) 364;

Numbers in parentheses are reference numbers. References are listed at the end of each chapter, on the following pages: 13–14, 89, 124–127, 163–164, 208, 254–256, 301–303, 344–345, 459–470.

(185) 374, 376; (203) 378; (219) 382; (220) 382; (221) 382, 385; (230) 385; (233) 385; (239) 385; (221) 388; (239) 388; (242) 388; (203) 390; (220) 391; (221) 391; (220) 392; (239) 392, 402; (203) 407; (221) 408; (239) 408; (299) 410, 412; (203) 412; (311) 414; (360) 430
LYUZE, L. L., (313) 415

MACDONALD, J. R., (26) 158; (178) 372
MAC RAE, A. U., (144) 367, 368, 370; (162) 370
MADDEN, H. H., (359) 430
MAEKAWA, S., (274) 403
MANY, A., (14) 128; (15) 128; (22) 139; (29) 194; (23) 227; (24) 227; (27) 228; (30) 229; (31) 235; (30) 239; (32) 241; (4) 264; (6) 267; (65) 291; (66) 291; (69) 292; (10) 317; (12) 318–321; (19) 331–333; (11) 346; (121) 364; (125) 365; (127) 365, 379; (213) 382; (225) 384; (229) 385, 386; (127) 390; (225) 390; (229) 391; (310) 413; (322) 418
MAPLE, T. G., (107) 363; (143) 367, 368
MARATHE, B. R., (252) 392
MARGONINSKI, Y., (29) 194; (23) 227; (24) 227; (121) 364; (125) 365; (214) 382; (215) 382; (214) 391; (215) 391; (253) 395; (214) 396; (215) 398; (253) 398; (294) 407; (215) 412; (355) 429–433, 450, 451
MARMIER, P., (380) 441
MARSH, J. B., (3) 90, 99, 110
MATTIS, D. C., (5) 305
MAUE, A. W., (7) 172
MAXWELL, K. H., (15) 93, 113, 114, 116; (76) 116, 117
MCBREEN, J., (349) 424
MCDOUGALL, J., (24) 145
MCKAY, K. G., (68) 292
MCKELVEY, J. P., (12) 268; (98) 363
MCKIM, F. S., (37) 352; (235) 385, 392
MCWHORTER, A. L., (48) 250; (52) 354; (70) 356, 358–360, 369; (158) 369–371; (70) 410
MEHL, W., (11) 92

MEIGS, P. S., (84) 119; (51) 354, 360
MEINHARDT, O., (323) 418; (338) 419; (339) 419
MEMMING, R., (324) 418; (327) 418; (341) 419; (346) 423, 424; (347) 424
MERRITT, F. R., (81) 297
MEYER, H. J. G., (70) 292
MEYERHOF, W. E., (12) 5
MICHEL, W., (110) 122
MILLEA, M. F., (17) 329; (206) 378, 392, 403
MINHAN, J. A., (273) 402, 403
MIRLIN, D. N., (161) 370
MIROSHNICHENKO, L. S., (196) 377
MISSMAN, R., (23) 337, 338; (356) 430, 450, 451
MITCHELL, E. W., J., (9) 92
MOLL, J. L., (276) 403
MOLLWO, E., (180) 373
MONTGOMERY, H. C., (4) 210; (10) 220; (4) 227; (37) 244, 248, 249; (89) 300; (32) 350; (40) 352, 353; (57) 355; (40) 356; (76) 356; (40) 357; (57) 357; (76) 357; (138) 367, 368; (151) 368; (159) 370; (40) 378; (32) 382, 390; (40) 390; (57) 390
MOORE, A. R., (16) 272; (104) 363
MORGULIS, N., (52) 287
MORITA, Y., (274) 403
MORRISON, J., (40) 97, 98; (66) 106; (67) 106; (66) 108, 109; (67) 109, 110; (66) 111, 112; (67) 112; (66) 113; (67) 113; (66) 117; (67) 117
MORRISON, S. R., (21) 226; (25) 227; (16) 329; (26) 349; (27) 349, 352; (41) 352, 356; (71) 356; (81) 356; (41) 357; (71) 358, 360, 369, 370; (160) 370; (41) 378, 390; (257) 396, 399; (374) 437, 438; (390) 444, 450; (374) 451
MORTEN, F. D., (110) 363, 366
MOTT, Sir N. F., (9) 4; (2) 128
MÜLLER, E. W., (43) 100; (45) 101; (46) 101; (47) 101; (50) 286
MÜLLER, S., (74) 356
MURPHY, N. St. J., (80) 297; (28) 342; (367) 435

Numbers in parentheses are reference numbers. References are listed at the end of each chapter, on the following pages: 13–14, 89, 124–127, 163–164, 208, 254–256, 301–303, 344–345, 459–470.

AUTHOR INDEX

NAVON, D., (11) 268
NEIZVESTNYI, I. G., (116) 364; (211) 382; (249) 391; (211) 398; (263) 398; (249) 399; (263) 399; (265) 399; (266) 399; (268) 399; (249) 400; (266) 400; (265) 402; (266) 410
NELSON, H., (104) 363
NEUSTADTER, S. F., (8) 128, 154; (7) 212
NEWMAN, R. C., (81) 118, 121
NIXON, J. D., (28) 229; (62) 355; (75) 356; (122) 364
NOBLE, V. E., (153) 368, 369
NOLLE, A. W., (335) 419
NORDHEIM, L., (54) 287; (55) 287, 288
NOVOTOTSKII-VLASOV, Yu, F., (211) 382; (249) 391; (211) 398; (263) 398; (249) 399; (263) 399; (266) 399; (267) 399; (249) 400; (266) 400; (267) 401; (270) 401; (266) 410; (270) 410
NOYCE, R. N., (96) 362; (114) 364

ODA, Z., (364) 434; (365) 434, 442
OHL, R. S., (34) 95
OMAN, R. M., (381) 441
ONISHI, T., (67) 356
OREN, R., (20) 332, 333
ORTON, D. A., (61) 355
OSTAPKOVICH, P. L., (122) 124

PAKE, G. E., (85) 298
PALMER, D. R., (8) 92; (300) 410, 437; (374) 437; (300) 438; (374) 438; (390) 443; (300) 450; (390) 450; (300) 451; (374) 451
PANKOVE, J. I., (108) 122
PAPAZIAN, H. A., (85) 360, 409
PATAI, I. F., (26) 274
PAVLOV, N. M., (212) 382; (227) 384; (212) 391, 398, 399; (227) 399; (212) 407, 408; (227) 408; (212) 412
PEARSON, G. L., (17) 6; (3) 210; (14) 272; (81) 297; (65) 355; (151) 368
PECK, D. S., (292) 407
PELL, E. M., (23) 273
PENNEY, W. G., (14) 19
PETERSON, E. L., (224) 382–384, 391, 408

PETRITZ, R. L., (78) 295, 296; (79) 296, 297; (14) 323, 326, 328; (16) 338; (25) 339; (26) 339; (163) 370; (165) 371
PETRUSEVICH, V. A., (106) 363; (131) 366
PFANN, W. G., (275) 403
PIKUS, G. E., (209) 382; (210) 382; (209) 383, 391
PILKUHN, M. H., (306) 411
PIRANI, M., (2) 90, 104
PLESKOV, Yu. V., (333) 419; (336) 419, 424; (348) 424
PLOTNIKOV, A. F., (245) 388
PLUMMER, A. R., (19) 346
PLUMMER, A. R. F., (55) 354, 360
POKROVSKAYA, S. V., (249) 391, 399; (266) 399; (249) 400; (266) 400, 410
POKROVSKII, Y. E., (36) 352
POLITYCHI, A., (112) 122
POMERANTZ, M. A., (26) 274
POMPLIANO. L. A., (106) 122
PORTNOY, W. M., (362) 430, 433, 450
POZHELA, Yu. K., (200) 377
PRATT, G. W., Jr., (66) 356
PRESNOV, V. A., (313) 415
PRESTON, J. S., (46) 286
PRICE, P. J., (6) 305
PRIMACHENKO, V. E., (219) 382; (233) 385

QUEISSER, H. J., (312) 415

READ, W. T., (18) 51; (16) 77; (26) 194; (81) 297
REDHEAD, P. A., (2) 90, 104; (62) 105
REDIKER, R. H., (126) 364
REINICKE, B., (339) 419
REYNOLDS, W. N., (97) 363, 364; (120) 364
RICHARDS, J. L., (115) 122
RICHARDSON, J. M., (164) 371
RICKEL, A. M., (216) 382, 391, 406
RITTNER, E. S., (2) 259, 260, 264, 266
ROBINS, H., (92) 120
ROBINSON, P. H., (13) 92, 93; (14) 93, 95; (13) 113, 114
ROLLIN, B. V., (139) 367
ROMANOV, O. V., (254) 395; (255) 396

Numbers in parentheses are reference numbers. References are listed at the end of each chapter, on the following pages: 13–14, 89, 124–127, 163–164, 208, 254–256, 301–303, 344–345, 459–470.

AUTHOR INDEX

RONALDI, M. J., (122) 124
ROSE, A., (27) 158, 162; (285) 405
ROSENBERG, A. J., (14) 93; (18) 93; (16) 93; (14) 95; (64) 105; (16) 113, 114
ROSI, F. D., (103) 363
ROSLYAKOV, V. V., (116) 364
ROSNER, O., (98) 121
ROSTOKER, N., (145) 368
RUBIN, L. G., (60) 104
RUBINSTEIN, R. N., (9) 215
RUDENOK, M. I., (286) 406
RUPPRECHT, G., (25) 192; (29) 229, 237; (80) 356; (240) 386, 387; (241) 387; (80) 391; (240) 391, 392; (241) 392
RZHANOV, A. V., (30) 196; (36) 204, 205; (116) 364; (129) 366; (211) 382; (212) 382; (223) 382; (226) 384; (227) 384; (228) 384, 388; (245) 388; (212) 391; (249) 391; (211) 398; (212) 398; (249) 398: (263) 398; (212) 399; (227) 399; (228) 399; (249) 399; (263) 299; (265) 399; (266) 399; (267) 399; (249) 400; (266) 400; (228) 401; (267) 401; (269) 401; (265) 402; (212) 407, 408; (227) 408; (266) 410; (212) 412

SAH, C.-T., (33) 201
SAMOGGIA, G., (246) 393; (247) 393
SANDOMIRSKII, V. B., (174) 372; (184) 373; (192) 375
SARBEI, O. G., (196) 377
SAUTTER, D., (88) 299; (133) 366; (134) 366; (133) 367
SAVAGE, S. D., (173) 372
SCHADE, H., (380) 441
SCHEER, J. J., (38) 279; (39) 279; (38) 281, 285; (39) 285; (397) 448; (398) 448; (399) 448; (398) 449; (397) 454; (409) 454
SCHEIBNER, E. J., (38) 97; (87) 119, (272) 402, 403, 410
SCHLIER, R. E., (29) 95; (30) 95; (37) 95; (29) 96; (30) 96, 97; (37) 97; (44) 99; (59) 104, 105; (30) 106; (41) 106; (30) 107, 109; (41) 109, 112, 113; (30) 117; (387) 443
SCHMIDT, P. F., (110) 122

SCHMIDT-TIEDEMANN, K. J., (47) 286
SCHOTTKY, W., (8) 4; (1) 128, 142
SCHRIEFFER, J. R., (7) 128; (6) 212, 244; (43) 244; (1) 304; (16) 329; (1) 338, 339; (41) 352; (47) 354; (41) 356, 357, 378, 390
SCHROEN, W., (312) 415
SCHULTZ, B. H., (118) 364
SCHUSTER, A., (2) 2
SCHWARTZ, B., (92) 120
SCHWUTTKE, G. W., (121) 124
SEILER, K., (134) 366
SEIWATZ, R., (17) 128, 145
SEKITA, Y., (189) 375
SELEZNEVA, M. A., (212) 382; (227) 384; (212) 391, 398, 399; (227) 399; (212) 407, 408; (227) 408; (212) 412
SERAPHIN, B., (112) 364
SERAPHIN, B. O., (61) 355
SERGEEV, S. I., (191) 375
SHARP, L. H., (76) 295
SHEFF, S., (111) 122
SHIBAIKE, T., (154) 368
SHILAL'NIKAS, V. I., (200) 377
SHOCKLEY, W., (15) 5; (17) 6; (19) 6; (23) 7; (24) 7; (15) 11; (1) 15; (16) 77; (4) 128; (21) 139; (3) 165, 169, 170; (23) 186; (26) 194; (33) 201; (3) 210; (35) 243; (1) 259, 262; (13) 271; (1) 272; (14) 272; (17) 272; (65) 355; (284) 405; (312) 415
SHOOTER, D., (31) 95, 96, 103
SIMON, R. E., (410) 454; (411) 454
SINYUKOV, M. P., (270) 401, 410
SIRRINE, R. C., (49) 354
SLOBODSKOY, A., (279) 403, 404; (282) 404
SMITH, R. A., (6) 15
SMITS, F. M., (292) 407
SNITKO, O. V., (219) 382; (220) 382; (237) 385; (238) 385, 388; (220) 391, 392; (238) 392; (301) 410, 414
SOCHANSKI, J., (194) 376; (195) 376
SOCHAVA, L. S., (161) 370
SOLOMON, R., (232) 385
SOMMER, A. H., (41) 281; (45) 285
SOMMER, H., (57) 104
SOMMERFELD, A., (4) 3; (10) 16

Numbers in parentheses are reference numbers. References are listed at the end of each chapter, on the following pages: 13–14, 89, 124–127, 163–164, 208, 254–256, 301–303, 344–345, 459–470.

SONDHEIMER, E. H., (3) 304, 322, 323
SOROKIN, O. V., (209) 382; (210) 382; (209) 383, 391
SOTINKOV, V. S., (345) 423
SPARKS, M., (23) 7; (109) 122
SPARNAAY, M. J., (65) 105; (404) 452; (405) 452; (406) 452; (407) 452
SPEAR, W. E., (244) 388
SPENKE, E., (4) 15; (1) 128; (5) 128; (1) 142
SPICER, W. E., (43) 284; (45) 285; (410) 454; (411) 454
SPRAGUE, J. L., (273) 402, 403; (279) 403, 404
STATZ, H., (9) 172; (46) 250; (50) 250–254; (45) 353; (46) 353; (48) 354; (45) 356; (46) 356; (48) 356; (45) 357; (46) 357; (48) 357; (46) 360; (48) 360; (86) 360; (89) 360; (91) 360; (48) 361; (89) 361; (91) 361; (46) 378; (48) 378; (86) 378; (46) 390; (48) 390; (86) 390; (46) 392, (86) 392, 399; (48) 407; (86) 407
STELLA, A., (16) 346
STELZER, I., (56) 354; (175) 372
STEVENSON, D. T., (28) 194; (108) 363, 364; (115) 364
STÖCKMANN, F., (180) 373
STONER, E. C., (24) 145
STRATTON, R., (49) 286; (53) 287–289; (49) 290
STRAUGHAM, W. B., (116) 123
STRIKHA, V. I., (119) 364
SUGIYAMA, K., (364) 434; (365) 434, 442
SUHL, H., (13) 271
SULLIVAN, M. V., (106) 122
SUN, R., (16) 329; (41) 352, 356, 357, 378, 390
SUNDBURG, W. J., (331) 418, 419
SURDUTS, A., (298) 410
SYNOROV, V. F., (288) 406

TAFT, E., (40) 280; (369) 436
TAKISHIMA, Y., (392) 444
TAMM, I. E., (14) 5, 11; (1) 165–170
TANNENBAUM, E., (87) 119; (272) 402, 403, 410
TEAL, G. K., (23) 7

TELLER, E., (17) 93
TEMPLETON, I. M., (139) 364
TERMAN, L. M., (53) 254; (278) 403, 404
THOMAS, D. G., (172) 372
THOMAS, H. A., (57) 104
THOMAS, J. E., Jr., (126) 364; (153) 368, 369
THOMSON, Sir W. (Lord Kelvin), (27) 274
TIKHONOV, V. I., (77) 356
TIPPLE, P. M., (120) 364
TOLANSKY, S., (90) 119
TOMÁŠEK, M., (13) 174; (14) 174; (13) 181; (14) 182
TOOTS, J., (3) 90, 99, 110
TRAPNELL, B. M. W., (63) 105, 114
TRESSLER, K. M., (21) 93
TURNER, D. R., (82) 118, (104) 121; (317) 415
TYLER, W. W., (19) 185

UNTERWALD, F., (40) 97, 98

VALDES, L. B., (21) 272
van der ZIEL, A., (86) 299
van LAAR, J., (38) 279; (39) 279; (38) 281, 285; (39) 285; (397) 448; (398) 448; (399) 448; (398) 449; (397) 454
van ROOSBROECK, W., (3) 260, 269–271; (22) 272
van RULER, J., (404) 452; (405) 452; (406) 452; (407) 452
van VLIET, K. M., (87) 299; (135) 366; (137) 367
VENNIK, J., (63) 355
VICKERS, A. E. J., (88) 119
VOL'KENSHTEIN, F. F., (179) 372, 373; (183) 373; (184) 373
von HIPPEL, A., (62) 290
VORONINA, G. F., (191) 375

WALLIS, G., (208) 381; (217) 382; (218) 382; (208) 390; (218) 396; (217) 398; (218) 399; (358) 430
WALLMARK, J. T., (186) 374; (277) 403–405; (186) 411
WALTERS, G. K., (83) 297; (84) 297; (83) 299

Numbers in parentheses are reference numbers. References are listed at the end of each chapter, on the following pages: 13–14, 89, 124–127, 163–164, 208, 254–256, 301–303, 344–345, 459–470.

WANG, S., (208) 381; (217) 382; (218) 382; (208) 390; (218) 396; (217) 398; (218) 399; (358) 430
WAREKOIS, E. P., (79) 117
WARFIELD, G., (280) 403, 404, 410
WATKINS, T. B., (80) 297; (28) 342; (15) 346; (72) 356; (155) 368, 370
WATTERS, R. L., (8) 268
WEBSTER, W. M., (16) 272
WEIMER, P. K., (12) 223
WEISZ, P. B., (170) 371
WERTHEIM, G. K., (20) 185; (10) 268
WICKERSHEIM, K. A., (9) 268
WIED, O. J., (273) 402, 403
WILBUR, J. M., (120) 124
WILLARDSON, R. K., (6) 92
WILLIAMS, R., (67) 291, 292; (325) 418
WILSON, A. H., (6) 3; (13) 18
WOLFF, G. A., (12) 92, 93, 108; (120) 124
WOLSKA, A., (128) 366
WOLSKY, S. P., (28) 94; (31) 95; (28) 96; (31) 96, 103; (60) 104; (61) 104; (71) 114
WOODBURY, H. H., (19) 185
WROTENBERY, P. T., (335) 419; (337) 419
WYSOCKI, C., (87) 360, 374, 410

YAGER, W. A., (81) 297
YARWOOD, J., (2) 90, 104
YOUNG, C. E., (16) 128, 154
YUNOVICH, A. E., (38) 244; (40) 244; (77) 356; (201) 378, 390

ZAININGER, K. H., (280) 403, 404, 410
ZDANUK, E. J., (31) 95, 96, 103; (60) 104; (61) 104; (71) 114
ZEMEL, J. N., (23) 139; (79) 296, 297; (9) 313, 315, 321–323; (13) 323; (26) 338; (25) 339; (26) 339; (30) 342
ZENER, C., (58) 289
ZHUZE, V. P., (209) 382; (210) 382; (209) 383, 391
ZWERDLING, S., (111) 122

Numbers in parentheses are reference numbers. References are listed at the end of each chapter, on the following pages: 13–14, 89, 124–127, 163–164, 208, 254–256, 301–303, 344–345, 459–470.

SUBJECT INDEX

Absorption coefficient, 41, 280
Absorption, optical, **41, 292**
Ac field effect – *see*
        High-frequency field effect
        Low-frequency field effect
Acceptor levels, 47, 57
Acceptor-like surface states, 132, 184
    fast states, 381, 384, 388, 390, 392, 423
    slow states, 352, 372, 375
Accumulation layer, 137 – *see also*
        Space-charge region
Activation and neutralization of fast surface states, 397, 401, 423, 455
Activation energy for
    charge transfer between slow states and interior, 360, 411
    formation of fast states by heating, 399
    gas adsorption, 374
    ion migration through oxide film, 411
    surface recombination velocity, 199, 381
    thermal emission from fast states, 386
Adsorption – *see also*
        Chemisorption
        Physical adsorption
    heat of, **105**
    measurements, 98, **103**
Adsorption, effect on electrical properties of clean surfaces, **449**
    dependence on coverage, 450, 453
    electron affinity and surface dipole, 99, 452, 459
    surface conductance, 429, 437, 441, 443, 450
    surface potential, 431, 441, 450
    surface-state density, 442, 451, 459
    work function, 100, 451
Adsorption, effect on electrical properties of real surfaces, **371**
    adsorbed molecules and slow states, 371, 410, 457

    adsorption kinetics, 373
    charge transfer theories for chemisorption, 371
    effect of water vapour, 360, 374, 398, 407, 410
    evidence for validity of charge transfer model, 373
    photoadsorptive effect, 373
    strong and weak chemisorption, 373
    work function and electron affinity, 375
Affinity – *see* Electron affinity
Ambients, effect on
    barrier height, 209, 349, 372, 374, 429, 437, 450
    electron affinity, 375, 436, 451, 454
    fast states, 398, 401
    photoelectric yield, 377, 454
    slow states, 360, 371, 374, 410
    surface conductance, 349, 374, 429, 437, 441, 443, 450
    surface recombination velocity, 363, 381, 385, 398, 430
    work function, 349, 375, 436, 451, 454
Ambipolar diffusivity, 84, 258
Ambipolar mobility, 84, 258
Anchoring of surface potential by surface states, 185
    slow states, 226, 253, 357, 361, 373
    states on clean surfaces, 425, 436, 445
Annealing, 50, 95
    effect on diffraction patterns, 109, 111
    effect on electronic structure, 440, 449
Anodic dissolution, 121
Anodic oxidation, 122
Antimonides, oxygen chemisorption on, 113
Arsenides, oxygen chemisorption on, 113
Asymmetric termination of potential at surface, 169
Auger effect in recombination, 77
Auger ejection, **101**

Boldface type indicates beginning of section or subsection in which subject is treated in detail.

Avalanche breakdown, 286, 370
Average mobility, **312**
  definition, 306
  in presence of magnetic field, 324
  partially specular scattering, **311**
  plotted against sample thickness, 314, 325
  simple considerations, **307**

Background pitting, 123
Band model
  effective mass, 33, 40
  motion of electrons and holes, **25, 33**
  one-dimensional lattice, **19, 166**
  one-electron models, **18**
  shape of energy bands, **34**
  three-dimensional lattice, **32, 171, 174, 178**
Band structure of Ge and Si, **37**
Barium oxide, effect on electron affinity, 377
Barrier height – *see also* Potential barrier
  anchoring by surface states, 185, 226, 253, 357, 361, 373, 425
  definition, 136
  effect of illumination, 157, 349, 420
  effect of surface states, **134**
  metal–semiconductor contact, **131**, 135, 361
  value at conductance minimum, 212
  variation by ambients, 209, 349, 372, 429, 437, 450
  variation by bulk doping, 428, 446
  variation by transverse fields, **129, 184**, 210, 352
Barrier-height determination
  channel conductance, 254, 354, 426
  contact potential and photoelectric threshold, 274, 428, 446
  electrode potential, 224, 417
  junction injection, 277
  pulsed field effect, 239, 332
  surface capacitance, 222, 419
  surface conductance, **211**, 417, 421, 426
  surface photovoltage, 277, 349
BET method, 93
Binding energy of chemisorbed gases, 105
Boltzmann transport equation, **67**
  surface scattering, 312, 323
Boron contamination of clean surfaces, 430, 440
Boundary conditions
  cyclic, 17, 25
  surface recombination, 259
  surface scattering, 312, 321, 323
  surface states, 168, 171, 176, 179
Bounded and unbounded carriers, 309, 315
Brattain–Bardeen gaseous cycle, 209, 349
Brillouin zone, 24, 26, 32, 37, 43
Bulk and surface effects, separation of, 209, 263
Bulk potential, definition, 136
Bulk recombination, 76
Bulk relaxation and collision times, 62, 307

Cadmium sulphide
  blocking contact, 223, 291
  effect of ambients on surface recombination, 364
  noise, 366
  surface capacitance, 418
Capacitance – *see also*
  Differential capacitance
  effective, 221, 233
  geometric, 221
  space-charge, 140, 220
  surface, **220**
  surface-state, 220
Capture cross section, 77, 190
  effective, 386, 394
  values for fast states, 390, 392, 423
Capture probability, 77, 190, 194, 201, 207
Carrier capture and energy dissipation, 204
Carrier densities
  excess surface densities, **149**
  in energy bands, 53, 57, 78, 80, 88
  in localized bulk states, 55, 58, 78, 159
  in surface states, 186, 188, 190, 194
  space-charge, 137, 156
Carrier transport, **61** – *see also* Mobility
Catalysis, 92, 372
Caustic soda etch for Si, 124
Cesium adsorption, effect on
  electron affinity, 281, 377
  photoelectric threshold, 454
  surface dipole moment, 454
  work function, 454
Channel conductance, **249** – *see also*
  Surface conductance
  bias dependence, 353
  effect of slow states, 353, 356
  energy-level diagram, 251
  measurement of, 250

SUBJECT INDEX

on clean surfaces, 437, 443
Charge relaxation time, **229**
   dependence on filament length, 232
   in semiconductor–electrolyte system, 233
   magnitude of, 189, 232
Chemical etching – *see* Etching, chemical
Chemical potential, 52
Chemical reactivity of surface, **90**
Chemisorption on clean surface, 91, **113**, 117
   carbon dioxide on GaSb, 113, 117
   effect on electrical properties, **449**
   hydrogen on Ge, 117
   iodine on Ge and Si, 113, 117
   measurements of, 98, 103
   oxygen on Ge and Si, 113
   oxygen on III–V compounds, 113
   phosphorus on Si, 113, 117
Clean and real surfaces, 90
Clean surfaces, electronic structure, **425**– *see also*
        Field-effect mobility
        Photoelectric emission and yield
        Surface conductance
        Surface states on clean surfaces
        Work-function
   bombardment–annealed Ge and Si, **429**, **440**
   cleaved Ge and Si, **437**, **443**
   comparison of cleaved and annealed surfaces, 440, 449
   dependence on crystal face, 430, 436, 442
   effect of adsorption, **449**
   surface transport, 337, 435
Clean surfaces, lattice structure and reactivity, **90**
   chemisorption, 91, **113**, 117
   cleanliness, 90, 98, 430, 440
   electron diffraction characteristics, 98, 106, 109, 112
   field emission patterns, 101
   lattice structure of diamond-type semiconductors, **105**
   physical adsorption, 91, 105, 450, 452
Clean surfaces, methods of measurement, **96**, **425**
   adsorption, 98, **103**
   contact potential and work function, **99**, **273**, 428

field-effect mobility, 426
field emission, **100**, 288, 443
field-ion microscopy, 101
heat of adsorption, **105**
low-energy electron diffraction, **96**
photoelectric yield and threshold, **99**, **279**, 428
surface conductance, 426
surface-state distribution, 429
Clean surfaces, preparation of, **92**
   cleavage, **92**
   crushing, **92**
   field desorption, 100
   heating in high vacuum, **94**
   heating in oxygen, 452
   hydrogen reduction, **93**
   ion bombardment and annealing, **95**
   regeneration, 94, 451
Cleavage, **92**
   evolution of gas from moving parts, 438
   plastic deformation, 437
Cleavage planes of crystals, 92
Cleaved, compared to annealed surface
   diffraction characteristics, 99, 112
   neutral level of surface states, 440, 449
   work function and photoelectric threshold, 440, 449
Cleaved surfaces – *see* Clean surfaces
Collector probe for effective-lifetime measurements, 272, 364
Collision time for surface scattering, 307, 309
Complete ionization, conditions for, 59
Complex s curves, 204, 382, 385, 394, 397, 400
Complex surface states
   occupation statistics, **185**
   surface recombination, **201**
Conchoidal fractures in polishing, 117
Conduction band – *see also* Band model
   definition, 29
   density of states, 36, 55
   effective density of states, 56
Conduction-band edge – *see* Energy-band edges
Conductivity, 64
Conductivity mobility – *see* Mobility
Constant-energy surfaces, 35, 39
Contact injection, 268
Contact potential, 99, 131, **273** – *see also* Work function

SUBJECT INDEX

change by ambients, 349, 374, 436, 452, 454
change by bulk doping, 428, 436, 439, 445
change with illumination, 275, 349, 409
measurement of, 274, 429
slow relaxation in, 356, 359, 376, 409
Continuity equation, 83
  solutions for bulk recombination, 83
  solutions for surface recombination, 258, 266
Continuous distribution of surface states, 378, 382, 399, 427, 433, 447, 457
Correlation mobility, 327
Coulomb integrals, 171, 175, 179
CP-4 and CP-4A etchants, 120
Cross section – see Capture cross section
Crushed powders
  adsorption on, 114, 452
  distribution of crystal planes, 93
Crushing, **92**
Crystal potential, 18, 32, 167, 171, 174
  deformation at surface, 167, 172, 181
Cyclic boundary conditions, 17, 25
Cyclotron frequency, 72, 324
Cyclotron resonance, 40
Cylindrical filament in field-effect experiment, 210, 441, 452

Dangling bonds
  contribution to surface dipole, 443
  effect on surface lattice structure, 111
  in models of surface states, 11, 173, 433, 458
  saturation by oxygen, 115, 459
Dash etchant, 121, 124
Dc field effect, **225** – see also Field effect
  derivation of surface-state distribution, 225
  effect of temperature drift, 226
  relaxation in, 356, 410
De Broglie wavelength, 16, 97
Debye length, 139
Debye length, effective, 139
  in presence of deep traps, 159, 162
  relation to width of space-charge region, 144, 327
Decoration techniques, 92
Deep traps, effect on potential barrier, **158**
  continuous distribution, **162**
  discrete set, **159**
Defects, 50

produced by bombardment, 352, 407
produced during polishing and sand-blasting, 117, 352
scattering by, 66, 344
Degenerate surface conditions, **145**
Dember potential, 276, 350
Density of states in energy bands, 36
Depletion layer, 137 – see also
  Space charge region
Detailed balance, principle of, 78
Diamond
  cleavage planes, 92
  surface-state calculations, 182
Diamond-type lattice
  diagram, 38, 48
  primitive cell, 37
Differential capacitance
  at semiconductor–electrolyte interface, 224, 418, 420
  confirmation of space-charge theory, 355, 418, 420
  effective, 222
  in MOS structure, 403
  measurement of, 221
  space-charge, 222
  surface, 222
  surface-state, 222
  vs. surface potential, 223, 421
Diffraction – *see* Electron diffraction
Diffuse and specular scattering, **305**, 343– see also
  Surface mobility
  Surface scattering
Diffusion constant, 80, 84
  ambipolar, 84, 258
  relation to mobility, 80
  units, 81
Diffusion length, 84, 259, 266
Diffusion of excess carriers, **79**, 258, 266
Diffusivity, ambipolar, 84, 258
Dimensionless potentials, definitions, 136
Dipole moment associated with absorbed atom, 375 – *see also* Electron affinity
  cesium, 454
  hydrogen, 453
  oxygen, 452
Direct and indirect optical transitions, 42, 281, 439, 446, 448
Dislocation lines
  density of, 52
  diagram, 51

revealed by etching, 120, 123
Dispersion in field-effect mobility, 247, 356
Displacement current in field-effect experiment, 228, 230, 234, 244, 249
  charge relaxation time, **229**
  semiconductor–electrolyte system, 232
Dissolution figures, 120, 122
Distribution function for free-carriers
  equilibrium, 52, 313
  in presence of electric and magnetic fields, 323
  in presence of electric fields, 68, 312
  surface scattering, 313, 321
Distribution function for localized bulk levels, 54, 58
Distribution function for surface states
  centres with excited levels, 188
  multiple-charge centres, 185
  single-charge centres, 182, 187
Donor levels, 47, 57
Donor-like surface states, 135, 184
  fast states, 384, 388, 390, 392
  slow states, 372, 374
Double-charge centres, **187, 201**
  occupation functions, 188
  surface recombination, 201
Drift length, 85
Drift mobility, 63
Drift of excess carriers, **79**, 85, 258, 266
Drift velocity, 63

Edge dislocations, 51
Effective charge distance
  definition, 140
  effect on surface scattering, 309, 434
  in presence of deep traps, 162
  plotted against barrier height, 146, 148
  relation to space-charge capacitance, 140, 231
Effective cross section, 386, 394
Effective Debye length – *see* Debye length, effective
Effective density of states, 56
Effective energy of localized levels
  double-charge levels, 187
  multiple-charge levels, 186
  single-charge levels, 54, 187
  with excited levels, 188, 205
Effective lifetime – *see* Sample effective lifetime
Effective mass
  along principal axes, 34
  density-of-states, 56
  in Ge and Si, 39, 41
  measurement of, 40
  tensor, 33
Effective mobility – *see* Surface mobility
Effective surface capacitance, 221, 233
Einstein relation, 80
Electrode potential, 417
Electrolyte–semiconductor interface – *see* Semiconductor–electrolyte interface
Electrolytic etching, 117, **121**
Electron affinity, 132, 274,
  changes due to adsorption, 281, 375, 436, 451, 453
  effect of annealing, 440, 449
  effect of extreme doping, 445
  measurement of, 274, 376, 428
  surface dipole moment, 375, 452
Electron density – *see* Carrier densities
Electron diffraction, low energy, **96**
  diffraction apparatus, 98
  diffraction patterns, 98, 109
Electron diffraction, medium energy
  etched Ge and Si surfaces, 118, 120
Electron escape depth, 280, 428, 446, 449
Electron microscopy techniques, 119, 397
Electron–phonon interactions, 64
Electronic structure of surface, **346**
Elovitch rate equation, 114
Emission of carriers from localized states, thermal, 78, 190, 194
Emission time constant, surface-state
  calculation of, 190, 245
  measurement of, **234**, 248
  temperature dependence, 385
Energy-band edges
  homogeneous semiconductor, 37
  in presence of electrostatic fields, 132, 190
  in presence of surface states, 134
  inhomogeneous semiconductor, 82
  surface channel, 251
Energy bands – *see also* Band model
  effect of lattice constant, 170
  of germanium oxide, 347, 409
Energy gap – *see* Forbidden gap
Escape depth, 280, 428, 446, 449
Etch pits, 120, **122**
Etchants for semiconductors, 120
Etching, chemical, **119**

SUBJECT INDEX

effect on surface mobility, 333
effect on surface-state structure, 395
formulae, 120
Etching, electrolytic, **121**
Excess reverse current in junction
    structures, 250, 354, 407
  anomalous effects introduced by water
    vapour, 360, 415
Excess surface-carrier densities, **149**
  effect of deep traps, 160
  graphical representation and approximate expressions, **150**
  in surface channels, 253
  measurement by infrared absorption, 293, 355
  measurement by pulsed field effect, 238, 241, 331, 335
  non-equilibrium conditions, 157
Exchange integrals, 171, 175, 179
Excited levels
  in recombination, 204, 384
  occupation statistics, 188
External field emission, **100, 286**
  emission patterns from clean Ge surface, 101
  experimental arrangement, 100
  shielding effect of surface states, 288, 443
Extrinsic semiconductor, 58

F functions, 139
  degernerate surface conditions, **145**
  effect of deep traps, 195
  extrinsic semiconductors, **141**
  intrinsic semiconductors, 139
  vs. barrier height, 146
Fast and slow surface states, 226, 348, **355**
  differences in capture times and density, 356
  in field effect, 356
Fast surface states, **377, 394** – *see also*
              Recombination centres
              Surface states
  acceptor- and donor-like, 381, 388, 423
  actual and effective energies, 379, 386, 393
  inner and outer sets, 394
  optical transitions, 389
  origin of, 397, 401, 423, 455
  parameters for aged Ge and Si surfaces, 390, 392
  parameters for semiconductor–electrolyte interface, 423
  physical location, 348, 357
  recombination and trapping centres, 381, 394, 397, 401, 423
Fast surface states, effect of surface treatment, **394**
  aging, 398
  bombardment, 407
  chemical etching, 395
  coating, 406
  electric fields, 407, 456
  electrolyte acidity, 424
  gaseous ambients, 398
  heat treatment, 399, 455
  metallic impurities, 396, 421, 455
  oxidation, 396, 402, 406
  temperature, 381, 408
Fermi–Dirac distribution function – *see*
              Distribution function
Fermi level
  definition, 52
  in extrinsic semiconductors, 59
  in intrinsic semiconductors, 56
Ferricyanide etchant, 121
Field desorption, 101
Field effect, **209** – *see also*
              Dc field effect
              High-frequency field effect
              Low-frequency field effect
              Pulsed field effect
  bridge circuits for measurement of, 228, 239
  experimental configuration, 210, 221, 224
  fast and slow relaxation, 356
Field-effect mobility, **244,** 426
  effect of oxygen adsorption, 442
  formulae, 244, 426
  frequency response, 247, 356
  in-phase component, 244, 356
  measurement of, 248
  on clean surfaces, 430, 432, 437, 442
  spurious effects introduced by stray capacitance, 438, 444
  temperature dependence, 433
Field-effect relaxation, fast – *see also*
              Pulsed field effect
              Relaxation processes
              Surface-state time constants
  decay characteristics, 356

frequency response, 356, 388
temperature dependence, 385
variation with barrier height, 388
Field-effect relaxation, slow – *see also* Slow surface states
  absence in high vacuum, 410
  absence on clean surfaces, 426, 451
  anomalous effects, 414
  decay characteristics, 356, 410
  dependence on oxide thickness, 360, 410
  distribution of time constants, 359
  effect of water vapour, 360, 410
  low-frequency response, 358
  non-linear effects, 358
  relation to 1/f noise, 369
  temperature dependence, 360, 411
Field-effect transistor – *see* MOS transistor
Field emission – *see*
  External field emission
  Internal field emission
Field ion microscopy, **100**
Flash filament technique, 103
Forbidden gap
  value for Ge and Si, 41, 57
  value for germanium oxide, 360, 409
Fractional-order diffraction maxima, 105
Free-carrier absorption, 42, 294
Free-electron approximation, 18, 172
Free energy – *see* Effective energy
Free orbitals – *see* Dangling bonds
Frenkel defects, 50
Fundamental absorption edge, 41
Fundamental-mode distribution of excess carriers, 262

G functions, 149
  degenerate surface conditions, 152
  extrinsic semiconductors, 152
  intrinsic semiconductors, 151
  vs. barrier height, 151, 154,
Gallium antimonide
  adsorption on, 113, 117
  electron-diffraction characteristics, 106, 109
  sticking coefficient for oxygen, 113
Gallium arsenide
  cleavage, 92
  dissolution figures, 122
Galvani potential, 416
Galvanomagnetic effects, **71, 295, 323**
  effective surface-mobility formalism, **326**
  experimental results, **339**
  measurement of, **295**
  solution of transport equation, **323**
Gaseous ambient cycle, 209, 349
Gauss's law, 129
Generation–recombination processes and noise, 299, 366
Germanium, bulk properties
  band structure, **37**, 41, 57
  carrier densities, 57
  mobilities, values for, 71
Germanium, clean surfaces
  adsorption, **113, 449**
  bombardment–annealed surfaces, **429**
  cleaved surfaces, **437**
  lattice structure, **105**
  photoelectric emission, 436, 438
  sticking coefficient for oxygen, 113
  surface mobility, 337, 435
  surface states, **429, 437**
  work function, 436, 439, 451
Germanium, real surfaces
  adsorption, **371**
  etching, **119, 121, 122**
  fast states, **355, 377, 394**
  germanium–electrolyte interface, **415**
  noise, **366**
  oxide film, 347, 360, 409
  slow states, **355, 408**
  structure and reactivity, **117**
  surface mobility, **329**
  surface recombination, **363, 377**
  work function and affinity, 375
Gouy layer, 416
Grain boundaries revealed by etching, 120
Group III–V compounds – *see*
  Gallium antimonide
  Gallium arsenide
  Indium antimonide

Half-order diffraction maxima, 105
Hall angle, 73, 324
Hall coefficient and Hall mobility, 74, 295, 324, **337**
  bulk and surface contributions, equivalent circuit for, 295
  bulk scattering, 74
  experimental data on Ge and Si surfaces, 337
  measurement of, 297
  surface scattering, 324, 326

SUBJECT INDEX

Hall effect – *see* Hall coefficient
Hall mobility – *see* Hall coefficient
Haynes–Shockley drift experiment, 86, 243, 272
Heat of adsorption, **105**, 114
Heavy and light holes, 39, 333, 435
Heitler–London approximation, 18, 171
Helmholtz layer, 416
High-field effects, **286**
   external field emission, 100, 287, 443
   impact ionization, 290
   internal field emission, 289, 412
   measuring techniques, 100, 290
   surface breakdown in p–n junctions, 361, 368
High-frequency field effect, **244** – *see also* Field-effect mobility
Hole density – *see* Carrier densities
Holes – *see* Positive holes
Hot carriers, 286, 377
Hydrogen adsorption, effect on
   surface conductance, 452
   surface dipole moment, 453, 459
   work function, 453
Hydrogen chemisorption, 117, 452
Hydrogen peroxide etchant, 121
Hydrogen reduction, **93**
Hydrophobic surfaces, 397, 411

Impact ionization, 280, 290, 368, 370
Impurities, 43, **46**
Indium antimonide
   cleavage of, 92
   electron diffraction characteristics, 106, 109
   sticking coefficient for oxygen, 113
Infrared absorption at surface, **292**
   adsorbed media, 294
   multiple reflection technique, 293
   space-charge layer, 294, 355, 362
   surface states, 294, 388
Infrared absorption, free carrier, 42, 273, 294, 355, 362
Interferometry techniques, 119
Internal field emission, 289
   band-to-band, 289
   from localized levels, 289
   from slow states, 409, 412
Interstitial atoms, 46, 50
Intrinsic Fermi level, 56
Intrinsic semiconductor, **55**

Inversion layer, 137 – *see also* Space-charge region
   communication with bulk, 216, 336
Ion-bombardment cleaning, **95**
Ion bombardment, effect on
   surface conductance, 352, 442
   surface recombination, 407
Ionization energy of localized centres, 47, 187
   excited states, 205
Ionized-impurity scattering, 66
Isotropic and anisotropic scattering, 62

Junction as collector probe, 272
Junctions – *see* P–n junctions

Kelvin method of contact-potential measurement, 274
Kronig–Penney model, **19**
   Tamm's modification of, **167**

Lattice defects, **50** – *see also* Defects
Lattice structure of clean surfaces, **105**
Lattice vibrations, **43**
   acoustical and optical branches, 45
   scattering by, 64
Lifetime of excess carriers, **75**
   effect of surface recombination, 259
   in bulk, 79
   measurement of, **263**
   sample effective lifetime, **258**
Light and heavy holes, 39, 333, 435
Line defects, 51
Localized levels, **57** – *see also* Surface states
   multiplicity, 54
   occupational statistics, 54, 182, 185, 188
Lorentz formula, 71
Low-energy electron diffraction, **96** – *see also* Electron diffraction, low energy
Low-frequency field effect, **226** – *see also* Field-effect mobility
   field-effect patterns for Ge, 352, 382
   measurement of, 227
   trapped carrier density *vs.* surface potential, 378, 380, 383, 400

Macropotential, 133
Magnetic measurements, **295**
   galvanomagnetic effects, **295**

paramagnetic resonance, **297**
Magnetoresistance, longitudinal, 74, 341
Magnetoresistance, transverse, 74
  bulk scattering, 74
  experimental data on Ge and Si surfaces, 338
  measurement of, 297
  surface scattering, 324, 327
Magneto-surface effects – *see* Surface mobility
Majority carriers, definition, 58
Maxwell–Boltzmann distribution law, 54, 80
Mean free path, 62
  unilateral mean free path, 308
Mean free path for hole–electron pair production, 280, 284
Mean free time between collisions, 62
Mean square carrier velocity, 65
  unilateral mean velocity, 308
Metal–semiconductor contact, 89, **131** – *see also* Rectification properties
  infrared absorption measurements, 362
Micropotential, 132, 375, 416
Midgap, 56
Minority carriers, definition, 58
Mobility, 63, 69 – *see also* Average mobility, Surface mobility
  ambipolar, 84, 258
  correlation mobility, 327
  drift mobility, 63
  Hall mobility, 74
  impurity scattering, 70
  isotropic and anisotropic, 70
  lattice scattering, 64, 69
  relation to diffusion constant, 81
  units, 64
MOS capacitor element, 403
MOS transistor
  as tool for studying surface states, 405
  current–voltage oscillograms, 405
  high-frequency performance, 234, 404
  schematic diagram, 405
Multiple-beam interferometry techniques, 119
Multiple-charge surface states, **185**

N-type semiconductor, definition, 58
Neutral level of surface states
  definition, 428
  position at clean Ge surface, 431, 435, 437, 439, 457
  position at clean Si surface, 441, 443, 447, 457
  temperature dependence, 448
Neutrality condition at surface, 135
Noise, **299**, **366**
  1/f noise, 300, 367
  G–R noise, 299, 366
  measurement of, 300
  spectrum, 367
Noise, 1/f, **299**, **366**
  effect of ambients and fields, 368, 370
  in accumulation and inversion layers, 368, 370
  in Hall effect, 368
  in p–n junctions, 368
  on clean silicon surface, 370
  role of slow states, 369
  temperature dependence, 368, 370
  theoretical models, 369
Non-degeneracy, conditions for, 54, 59
Non-equilibrium phenomena, **75** – *see also* Recombination, Relaxation processes, Surface recombination
  diffusion and drift of excess carriers, **79**
  lifetime of excess carriers, **75**
  quasi Fermi level, 82, 86, 156, 195, 252
  sample effective lifetime, **258**
  space-charge region, **156**
Non-spherical energy surfaces, 34, 70, 305
Nuclear magnetic I resonance, 298

Occupation function – *see* Distribution function
Occupation statistics, **52**
  surface states, **182, 185**
Ohm's law, 63
Omegatron, 104
One-electron models, **18**
Optical absorption, **41, 292**
Optical excitation, **41,** 292
Optical gap, 41
Optical measurements, **292**
Optical transitions involving surface states, 388, 408
Oxide film
  energy bands, 347, 408
  forbidden gap, 360, 409
  hydration of, 374, 411

SUBJECT INDEX

on clean surfaces, 451, 457
on etched surfaces, 118, 122
potential drop across, 348, 354, 419
thermally grown, 119, 402
thickness, 118, 122, 347, 354, 410
Oxygen adsorption
  measurement of, 104
  on III–V compounds, 114
  on Ge, 113, 115
  on Si, 114, 116
  sticking coefficients, 113
Oxygen adsorption, effect on
  electron affinity and surface dipole, 376, 451, 459
  electron-diffraction characteristics, 113
  photoelectric threshold, 452
  surface conductance, 352, 429, 441, 450
  surface potential, 352, 371, 431, 450
  surface-state density, 442, 451, 459
  work function, 349, 376, 436, 451
Ozone, effect on
  barrier height, 349, 352, 356
  fast surface states, 398, 401

P–n junctions, **86**, 128
  breakdown phenomena, 291, 361, 368
  excess reverse current, 250, 360, 407, 415
  revealed by etching, 119, 122
P–n–p junction structures, **249** – *see also* Channel conductance
P-type semiconductor, definition, 59
Pairing model for surface atoms, 107
Paramagnetic resonance, **297**
Partially diffuse scattering, **311**, **322**
Passivation of surfaces, 120, 396, 406
Phonon-aided field emission, 288, 360, 456
Phonon-aided optical transitions, 42, 281, 439, 448
Phonons, 46, 66, 70
Photoelectric emission, **279**
  bulk and surface scattering process, 282
  direct and indirect excitation, 281, 439, 448
  effect of band bending, 281, 439, 446, 449
  electron escape depth, 280, 428, 446, 449
  polarization effects, 439, 448
  surface-state emission, 446, 448
  valence-band and surface-state emission, 282, 429, 439, 448
Photoelectric threshold
  definition, 279

detection of surface cleanliness, 99
effect of annealing, 440, 449
effect of band bending, 280, 428, 446, 449
effect of cesium adsorption, 377, 454
values of, 436, 439, 445, 448
vs. bulk potential, 438, 445
Photoelectric yield
  definition, 280
  effect of bulk doping and band bending, 438, 445, 449
  measurement of, 285
  vs. photon energy, 438, 446, 448
Photoelectromagnetic (PEM) effect, 269
Physical adsorption, 91, 105, 371, 450
Plastic deformation, 52, 437
Point defects, 50
Poisson's equation, **138**
  approximate solutions, **141**
  boundary conditions, 138
  channel conductance, 254
  continuous trap distribution, **162**
  deep traps, **158**
  intrinsic semiconductor, 139
  non-equilibrium conditions, 157
  numerical solutions, **143**
  small disturbances, 138
Positive holes, **30**
  heavy and light, 40, 435
  motion of, **33**
Potential barrier, 133, **136**
  effect of traps, **158**
  quantization effects, 141, 434
  vs. distance from surface, 144
Potential drop across oxide film, 348, 354, 419
Preferential etching, 119, 122
Pressure measurements, 104
Primitive cell, 37
Principle of detailed balance, 78
Pulsed field effect, **229** – *see also* Field-effect relaxation, fast Relaxation processes
  charge relaxation time, **229**
  internal field emission, 291, 412
  measurement of, 228, 235
  surface mobility measurement, **238**, 243, 332

Quantization effects in steep barriers, 141, 434
Quasi Fermi level, 82, 156

definition, 82
in p-n junction, 86
in presence of surface recombination, 156, 195
in surface channel, 252

Radiative recombination, 77
Radioactive tracer analysis of surface, 397, 406, 423, 441
Rayleigh phonons, 344
Real surfaces, electronic structure, **377**, 390, 392, 423 – *see also*
    Fast surface states
    Semiconductor–electrolyte interface
    Slow surface states
basic model, **347**
effect of adsorption, **371**
fast and slow states, **355**
noise, **299**, **366**
surface recombination, **363**, 379
surface transport, **329**
Real surfaces, structure and reactivity, **117**
chemical etching, **119**
chemisorption, **371**
electrolytic etching, **121**
oxide film, 118, 122, 347, 360, 379, 402, 411
physical adsorption, 91, 371
preferential etching, **122**
preparation, **117**
smoothness, 118, 120
Recombination, **76** – *see also*
    Surface recombination
carrier lifetime, 77
direct and indirect, 76
Shockley–Read statistics, 77
Recombination centres – *see also*
    Fast surface states
    Surface recombination velocity
background distribution, 382, 397
discrete and continuous distribution, 382, 394, 399
parameters for Ge and Si, 390, 392, 423
statistical weight factor, 379, 393
Rectification properties of metal–semiconductor contact, 89
effect of slow states, 153, 356, 361
lack of dependence on work function, 361
Reduced wave vector, 24, 43

Reduction by hydrogen, **93**
Regeneration of initially clean surfaces, 94, 451
Relaxation processes involving surface states – *see also*
    Field-effect relaxation
    Pulsed field effect
interaction with both bands, **240**, 247
interaction with single band, **189**, **234**, 245
surface-state time constants, 192, 193, **234**, **245**, 248, 385
Relaxation time for carrier scattering, 62, 67, 312
bulk and surface, 307
lattice and impurity scattering, 66
weighted averages, 69, 73
Resonance – *see*
    Cyclotron resonance
    Paramagnetic resonance
Roughness factor of surface, 420
Rutherford scattering, 66

Sample effective lifetime, **258**
steady state, surface excitation, 260
steady state, uniform excitation, 259
transient excitation, 261
Sample effective lifetime, measurement of, **263**
collector probe, 272, 364
conductance methods, 264, 267
end effects, 266
infrared absorption, 273
PEM effect, 269
pulse-injection method, 268, 384
pulse-reverse method, 273
pulsed excitation, 268
Suhl effect, 271
surface photovoltage, 273
Scattering processes, **61**
diffuse and specular, 305
impurities and defects, 66
isotropic and anisotropic, 62
lattice vibrations, 64
randomizing action, 62, 305, 313, 439
relaxation time, 62, 66, 307, 312
Schottky barrier, 142, 291, 418
Schottky defects, 50
Schrödinger's equation, solutions for free electrons, **16**
Kronig–Penney model, **19**

SUBJECT INDEX

Shockley states, **178**
Tamm states, **166**, **174**
Screening of space-charge region by surface states, 185, 226, 253, 361, 373, 425
Screw dislocations, 51
Secondary emission and Auger effect, 102
Semiconductor–electrolyte interface, 223, **415**
  differential capacitance *vs.* barrier height, 418, 421
  effect of electrolyte acidity on surface states, 424
  effect of hydrogen and oxygen on surface states, 424
  equivalent circuit, 233
  parallel conductance through electrolyte, 225, 420
  potential distribution at interface, 416
  rectification properties, 416
  slow states, 419
  surface conductance *vs.* electrode potential, 417, 420
  surface dipole, 418
  surface photovoltage, 420
  surface recombination velocity *vs.* surface potential, 419, 421
  surface states introduced by metallic impurities, 421, 423
  trap-free surface, 420, 456
Shaping of sample by etchants, 119, 122
Shockley states, 170, 172, **178**
Silicon, bulk properties
  band structure, **37**, 41, 57
  carrier densities, 57
  mobilities, values for, 71
Silicon, clean surfaces
  adsorption, 113, **449**
  bombardment–annealed surfaces, **440**
  cleaved surfaces, **443**
  lattice structure, **105**
  photoelectric emission, 445, 448
  sticking coefficient for oxygen, 113
  surface states, **440**, **443**
  work function, 442, 445, 451, 453
Silicon, real surfaces
  adsorption, **371**
  etching, **119**, **121**, **122**
  fast states, **355**, **377**, **394**
  oxide film, 119, 402
  silicon–electrolyte interface, **415**
  slow states, **355**, **408**

structure and reactivity, **117**
surface mobility, **329**
surface recombination, **377**
Simple *s* curves, 198, 365, 380, 394, 399, 421
Single-charge surface states, 182, 187
Slow surface states, 134, 226, **355**, **408**
  absence on clean surfaces, 426, 451
  acceptor- and donor-like, 352, 371, 374
  activation energy for charge transfer, 360, 411
  anchoring by, 226, 253, 357, 361, 373
  capture times, 356
  energy position, 413
  in metal–semiconductor contact, 361
  lower-limit density, 358, 361
  optical transitions from, 408
  origin of, 348, 371, 410, 456
  physical location, 348, 358, 360, 410
  tunnelling from, 409, 412
Slow surface states, charge-transfer models
  electron diffusion, 411
  ionic migration, 411
  surface heterogeneity, 359, 369
  T–F emission, 360, 372, 456
  thermionic emission, 358, 370, 372
  tunnelling, 360, 372, 409, 412
Slow surface states, effect of
  ambients, 360, 371, 374, 410
  evacuation, 410
  heating in vacuum, 410
  imperfections and impurities in oxide, 410
  thick oxide films, 360, 403, 410
Space-charge capacitance, 140, 220, 222, 418, 420
Space-charge density, **136**
  degenerate conditions, **145**
  extrinsic semiconductors, **141**
  in presence of deep traps, 159
  in surface channels, 252
  intrinsic semiconductors, **139**
  non-equilibrium conditions, 157, 277
  vs. barrier height, 146, 278
Space-charge region, **128**
  charge exchange with surface states, **189**, **234**, **240**
  definitions, **136**
  degenerate conditions, **145**
  effective charge distance, 140, 146, 148
  effective Debye length, 139
  excess carrier densities ΔN and ΔP, **149**

experimental confirmation of theoretical characteristics, **349**, 418, 420
extrinsic semiconductors, **141,** 152
in presence of deep traps, **158**
infrared absorption in, 294, 355, 362
intrinsic semiconductors, 139, 151
non-equilibrium conditions, **156,** 277
origin of, **129, 134**
quantization effects, 141, 434
time constant for charging, 232
Specular and diffuse scattering, **305,** 343
Spherical energy surfaces, 36, 305, 433
Spin–lattice and spin–spin relaxation times, 299
Sputtering rates, 96
Statistical weight factor, 54, 187, 379, 393 – see also Effective energy
Steady-state carrier densities, 156
Sticking coefficient, 99, 104, 113, 117, 453
Storage effect, 412
Substitutional impurity, 46, 49
Suhl effect, 271
Superoxol etchant, 121, 124
Surface area, determination of, 93
Surface breakdown, 361, 368, 370
Surface capacitance – see
 Capacitance
 Differential capacitance
Surface conductance, **211** – see also
 Channel conductance
 Field effect
 Surface mobility
determination of barrier height, 215, 417, 421, 426
inversion layers, error in measurement, 219
measurement of, 214, 426
minimum in, 212
on clean surfaces, 430, 437, 441, 443
spurious effects from uncleaned faces, 426, 442, 453
variation by ambients, 349, 374, 429, 437, 441, 443, 450
vs. barrier height, 213, 417, 431
Surface damage, produced by
bombardment, 96, 352, 407
cleavage, 437, 448
polishing, 117, 352
Surface dipole, 99, 375 – see also
 Electron affinity
Surface heterogeneity, 359, 369

Surface mobility, **304**
definition, 306
effect on surface conductance, 213, 306, 354, 426, 431
experimental results, **329, 337**
galvanomagnetic effects, 323
measurement of, **238,** 243, **295,** 331, 337
partially diffuse scattering, 322
plotted against barrier height, 317, 319
quantization effects, 435
simple considerations, **307**
theoretical calculations for diffuse scattering, **312, 323**
Surface patchiness, 384
Surface photovoltage, 157, **273**
dependence on light intensity, 277, 351
dependence on surface potential, 350, 420
effect of fast states, 277, 350
effect of slow states, 349, 356, 410
measurement of, 275
on clean surfaces, 441
theory of, 157, 277
Surface potential – see Barrier height
Surface recombination, theory of, **194, 201**
double-charge centres, **201**
excited levels, **204**
large disturbances, 199
range of validity, 195, 201
small disturbances, 195
Surface recombination velocity – see also
 Recombination centres
complex *s* curves, 382, 385, 394, 397, 400
definition, 196, 259
experimental confirmation of theoretical model, **363**
measurement of, **257, 263**
simple *s* curves, 365, 380, 394, 399, 421
trapping effects, 385
vs. surface potential, 198, 365, 380, 383, 400, 421, 431
Surface recombination velocity, effect of
ambients, 363, 381, 385, 398, 430
bombardment, 407
bulk resistivity, 364
etching, 395
excitation level, 199, 366, 385, 419
fields, 364, 379, 382, 419, 421
heat treatment, 399
impurities, 397, 421, 424,
temperature, 364, 379
Surface scattering, **304** – *see also*

SUBJECT INDEX

Surface mobility
  collision time, 307
  diffuse and specular, **305**
  effect on distribution function, 312
  possible causes of, 344
Surface-state bands, 172, 178, 182, 433, 447
Surface-state capacitance – *see*
      Capacitance
      Differential capacitance
Surface-state parameters, methods of measurement
  channel conductance, **249**
  contact potential and photoelectric emission, **273, 279,** 428
  field effect, **225, 226, 229, 244**
  field-effect mobility, **244, 426**
  internal field emission, 289, 412
  optical effects, **292,** 388, 408
  surface capacitance, **220,** 421
  surface conductance, **211,** 426
  surface recombination, **194, 257**
Surface-state time constants
  calculation of, 190, 245
  measurement of, **234,** 248
  temperature dependence, 385
Surface states – *see*
      Fast surface states
      Slow surface states
      Surface states on clean surfaces
Surface states on clean surfaces
  conduction in, 434, 442
  dependence on crystal face, 430, 433, 458
  effect of annealing, 440, 449
  effect of oxygen adsorption, 442, 451, 459
  estimates of density, 432, 437, 442, 447, 457
  models for energy distribution, 427, 433, 447
  origin of, 458
  position of neutral level, 431, 437, 439, 441, 443, 447, 457
  recombination properties, 430
  trapped charge *vs.* surface potential, 445
Surface states, phenomenological description
  double-charge centres, **187, 201**
  effective energy, 182, 187
  excited levels, **188, 204**
  interaction with both bands, **194, 201, 240, 247**
  interaction with single band, **189,** 245

  multiple-charge centres, **185**
  single-charge centres, 187, **194**
Surface states, theory of
  conditions for existence, 177, 181
  density of, 172, 178, 181
  free-electron approximation, 172
  one-dimensional lattice, **166**
  Shockley states, 170, 172, **178**
  symmetrical and asymmetrical termination of potential, 169
  Tamm states, 167, 171, **174**
  three-dimensional lattice, **171**
  tight-binding approximation, **174**
Surface transport – *see* Surface mobility

T–F emission, 288, 360, 456
Tamm states, 167, 171, **174**
Thermal noise, 299
Thermal oxidation, 119, 402
Thermal velocity, 65
Thermionic emission, 288
Thermodynamic potential, 52
Thermoelectric power, 342, 352
Threshold energy for atom ejection, 96
Threshold energy for photoemission – *see* Photoelectric threshold
Tight-binding approximation, **174**
Transport equation, **67,** 312, 323
Transport processes at surface – *see* Surface mobility
Trap-free surface, 420, 456
Trapping effects, 76, 273, 385
Tunnel diode, 291
Tunnelling – *see* Internal field emission

Uniform surface-state distribution, 428, 433, 447
Unilateral mean free path, 308
Unilateral mean velocity, 308
Unipolar field-effect transistor – *see* MOS transistor
Unit cell, 37
  of diamond-type lattice, 38
Unsaturated bonds – *see* Dangling bonds

Vacancies, 50
Valence band – *see also* Band model
  definition, 29
  density of states, 36, 55
  effective density of states, 56

Valence-band edge – *see*
    Energy-band edges

Wannier functions, 181
Water-vapour adsorption, effect on
    barrier height, 349, 372, 374, 435
    fast states, 398, 401, 407
    ionic motion, 360, 415
    oxide film of Ge, 374, 411
    recombination centres, 398, 401
    reverse current in junction structures, 360, 407
    slow field-effect relaxation, 360, 410
    surface conductance, 352, 374, 435, 452
Wave number, 16, 19, 24
Wave vector of electron, 18, 32, 43

Wave vector of lattice vibrations, 43
White etchant, 120
Work function, 273 – *see also*
    Contact potential
    accuracy in measurement of, 429
    definition, 131
    dependence on crystal face, 436, 442
    detection of surface cleanliness, 100
    effect of adsorption, 436, 452, 454
    effect of annealing, 440, 449
    effect of bulk doping, 436, 439, 445
    vs. bulk potential, 439, 445

Zener effect, 289
Zinc oxide, surface capacitance measurements, 418